The Muscarinic Receptors

The Receptors

The Muscarinic Receptors, edited by ***Joan Heller Brown,*** 1989
The Serotonin Receptors, edited by ***Elaine Sanders-Bush,*** 1988
The alpha-2 Adrenergic Receptors, edited by ***Lee Limbird,*** 1988
The Opiate Receptors, edited by ***Gavril W. Pasternak,*** 1988
The alpha-1 Adrenergic Receptors, edited by ***Robert R. Ruffolo, Jr.,*** 1987
The GABA Receptors, edited by ***S. J. Enna,*** 1983

The Muscarinic Receptors

Edited by

Joan Heller Brown

University of California, San Diego
La Jolla, California

THE HUMANA PRESS • CLIFTON, NEW JERSEY

Library of Congress Cataloging in Publication Data

Main entry under title:

The Muscarinic receptors.

(The Receptors)
Includes bibliographies and indexes.
1. Muscarinic receptors. I. Brown, Joan Heller.
II. Series.
QP364.7.M87 1989
612.8'9—dc20
DNLM/DLC 89-15539
ISBN 0-89603-156-X

Crescent Manor
PO Box 2148
Clifton, NJ 07015

Printed in the United States of America

Preface

Research on muscarinic receptors is advancing at an extraordinary rate. Ten years ago, the existence of muscarinic receptor subtypes was a logical assumption with only scattered experimental support. The discovery that pirenzepine recognized apparent heterogeneity in muscarinic binding sites infused new life into the problem of subclassifying muscarinic receptors. Simultaneous advances in molecular biology created a frenzy to clone cell surface receptors. The muscarinic receptor succumbed surprisingly quickly, revealing its structure and that of at least four closely related gene products within a year. Our hope of obtaining clear evidence for muscarinic receptor subtypes was answered with a vengeance. Now a family of muscarinic receptors sits before us, asking to be understood. The bounty is as attractive to those who have not previously studied muscarinic receptors as to those who have dedicated their research careers to this subject.

The goal of this book is to ensure that the new generation of research will profit from the wisdom of the past. The tools of molecular biology are well suited to the tasks of characterizing the pharmacology, function, and regulation of the distinct muscarinic receptor subtypes. However, efficient and intelligent use of these tools is not possible, unless one understands the properties of the receptor, the molecular mechanisms by which it couples to effectors, and the ways that it is regulated. This volume contains extensive information on these subjects, presented and interpreted by distinguished authors who have all had an abiding interest in muscarinic receptor mechanisms. Some of the topics covered in individual chapters may be found in review articles, but much of the information presented here has not previously been compiled. Certainly, no other available volume analyzes the subject of muscarinic receptors in such depth.

The book begins with a scholarly introduction by Sir Arnold Burgen, who proposed the existence of muscarinic receptor subtypes more than 20 years ago. Section Two deals with the basic properties of muscarinic receptors as deduced from receptor purification, radioligand binding, analysis of functional responses, and molecular cloning. These data should be invaluable for those interested in designing or choosing specific receptor ligands. The third section describes coupling of the receptor to its biochemical effectors and details the functional responses that are regulated through the muscarinic receptor: adenylate cyclase inhibition, phosphoinositide hydrolysis, calcium mobilization, activation of cyclic GMP and eicosanoid production, and activation of various ion channels. Those interested in whether receptor subtypes define particular functional responses should profit from the wealth of information contained in these chapters. A final section describes the regulation of the muscarinic receptor by ligand occupancy, through association with other membrane proteins and in development. These phenomena provide an obvious point of departure for examining differences in the behavior of the muscarinic receptor subtypes. The book closes with a chapter that reflects the excitement anticipated to accompany future research into muscarinic receptor mechanisms.

Joan Heller Brown

Contents

Section 3: Biochemical Effectors Coupled to Muscarinic Receptors

Chapter 6

Muscarinic Cholinergic Receptor-Mediated Regulation of Cyclic AMP Metabolism

T. Kendall Harden

Chapter 7

Muscarinic Cholinergic Receptor Regulation of Inositol Phospholipid Metabolism and Calcium Mobilization

Joan Heller Brown and Patrick M. McDonough

Section 4: Regulation of Muscarinic Receptors

Section 5: Future Directions

Contributors

N. J. M. BIRDSALL • *Division of Physical Biochemistry, National Institute for Medical Research, Mill Hill, London, U. K.*
JOAN HELLER BROWN • *Department of Pharmacology, University of California, San Diego, La Jolla, California*
A. S. V. BURGEN • *Darwin College, Cambridge, U. K.*
DAVID B. BYLUND • *Department of Pharmacology, School of Medicine, University of Missouri, Columbia, Missouri*
T. KENDALL HARDEN • *Department of Pharmacology, School of Medicine, University of North Carolina at Chapel Hill, Chapel Hill, North Carolina*
YOAV I. HENIS • *Laboratory of Neurobiochemistry, Department of Biochemistry, The George S. Wise Faculty of Life Sciences, Tel Aviv University, Tel Aviv, Israel*
E. C. HULME • *Division of Physical Biochemistry, National Institute for Medical Research, Mill Hill, London, U. K.*
YOEL KLOOG • *Laboratory of Neurobiochemistry, Department of Biochemistry, The George S. Wise Faculty of Life Sciences, Tel Aviv University, Tel Aviv, Israel*
PATRICK M. McDONOUGH • *Department of Pharmacology, University of California, San Diego, La Jolla, California*
MICHAEL McKINNEY • *Abbott Laboratories, North Chicago, Illinois*
NEIL M. NATHANSON • *Department of Pharmacology, University of Washington, Seattle, Washington*
R. ALAN NORTH • *Vollum Institute, Oregon Health Sciences University, Portland, Oregon*
ELLIOTT RICHELSON • *Mayo Foundation, Rochester, Minnesota*
BJÖRN RINGDAHL • *Department of Pharmacology, School of Medicine, University of California, Los Angeles, California*

MICHAEL SCHIMERLIK • *Department of Biochemistry and Biophysics, Oregon State University, Corvallis, Oregon*
MORDECHAI SOKOLOVSKY • *Laboratory of Neurobiochemistry, Department of Biochemistry, The George S. Wise Faculty of Life Sciences, Tel Aviv University, Tel Aviv, Israel*
BARRY B. WOLFE • *University of Pennsylvania, School of Medicine, Department of Pharmacology, Philadelphia, Pennsylvania*

SECTION 1

HISTORICAL PERSPECTIVES

1

History and Basic Properties of the Muscarinic Cholinergic Receptor

A. S. V. Burgen

1. History

In the folk medicine of several cultures, mushrooms have been used in religious and magical ceremonies. One of these mushrooms is the fly agaric, Amanita muscaria (Wasson, 1973). The first scientific studies of this mushroom were made by Schmiedeberg and Koppe (1869), who showed that extracts of the mushroom could slow, and at higher concentrations arrest, the beat of the frog heart. Using this action as a guide, they purified the extract using the standard alkaloid purification methods then available and obtained a crystalline aurichloride salt. Since no microanalytic data were provided by them, it is unclear whether or not this was a pure compound of the alkaloid, which they named muscarine, but later evidence suggests that it was no more than 25% pure (Eugster, 1960). It was a reliable preparation that they then used for an extensive study of its pharmacology. They showed that, in addition to its action on the heart, it contracted the smooth muscle of the stomach and intestine, stimulated the secretion of tears, saliva and mucus, constricted the pupil, caused accommodation of the lens, and also produced dyspnea. Applied to the brain and spinal cord, it produced flaccidity and weakened peripheral reflexes. The action of the vagus on the heart was well known, having been described by the Webers in 1845, and Schmiedeberg and Koppe noted:

> *muscarine does not destroy the contractile force of the heart muscle, but only oppresses it and prevents its natural manifestation from becoming apparent. This can only be accomplished through an increase of natural resistances which have their origin in the vagus. The poison must produce such a marked stimulation of the vagus that the heart stops as it does during electrical stimulation of the vagus . . . section of both vagi in the neck has no effect on the cardiac actions of the poison . . . Tackling this question has become possible in view of the observations of von Bezold and Bloebaum which were later confirmed by Bidder and Keuchel. These authors showed that very low doses of atropine paralyse the vagal terminals . . . after the administration of a very small dose of atropine, muscarine is no longer able to arrest the heart (Holmstedt and Liljestrand, 1981).*

This interpretation thus rests on the idea that both atropine and muscarine acted on the vagal nerve endings (the idea of chemical transmission was not yet on the horizon), but nevertheless this book provides a remarkably comprehensive survey of the pharmacology of muscarine.

The chemistry of muscarine remained obscure, but in 1877, Schmiedeberg and Harnack found that, when choline was treated with fuming nitric acid, a substance was formed that appeared to have the same actions as muscarine. It was referred to as synthetic muscarine. Thereafter, this substance, which was readily prepared, was used as though it were the same as natural muscarine, and this led to some confusion in the pharmacological literature until Ewins (1914) showed that it was the nitrite ester of choline and Dale (1914) found that its pharmacology was different in that it had considerable nicotinic activity. There were further false leads in the chemistry (Wilkinson, 1961) that were only settled finally in 1957 when Kogl et al. established the tetrahydrofuran structure of muscarine with the correct stereochemistry by X-ray crystallography, and this was confirmed by synthesis by Hardegger and Lohse (1957). Eugster (1960) carried out an exhaustive study of the chemistry of the enantiomers and analogs of muscarine. This was accompanied by a detailed evaluation of the pharmacology by Waser (1962).

The original interest in Amanita muscaria lay in its psychopharmacological actions, but it is doubtful if muscarine contributes at all to this, which is probably because of other alkaloids in the mushroom such as ibotenic acid.

Choline was first isolated from lecithin in 1862 and when Hunt (1900) found a blood pressure lowering substance in extracts of the adrenal, he had seen the recent work of Halliburton and Mott (1899), in which choline was identified as the hypotensive agent in cerebrospinal fluid. He was therefore inclined to think that his material was also choline. Later, Hunt and Taveau (1906) realized that their material was clearly more active than authentic choline. Their material was also rather unstable, and this led them to suspect that they might be dealing with an ester of choline. They made acetylcholine (first synthesised by von Baeyer in 1867), and found that it was 10,000 times as active as choline in lowering blood pressure and that, like choline, its action was blocked by atropine. A brief description of the activity of propionyl, butyryl, valeryl, succinyl, and benzoylcholines was also given.

Curiously, it was not until 1929 that it was proved by Dale and Dudley that acetylcholine was a normal constituent of an animal tissue. In the course of extracting histamine from ox spleen, they noted the presence of a substance that lowered the blood pressure of the rabbit and that contracted the rabbit ileum. Its activity was too great to be accounted for by choline, and after a complicated series of purification steps, it was identified as acetylcholine through the isolation of its crystalline double platinichloride with choline. There was a surprisingly large amount present (around 10 mg/kg) and a similar amount was found in horse spleen. Its physiological role in this organ remains obscure.

The interest of Dale in acetylcholine came about through his systematic study of ergot (Dale, 1953) from which his colleague Ewins was able to isolate acetylcholine. In his paper of 1914, Dale studied a considerable number of esters and ethers of choline, and found that, in low doses, they mostly had well-marked depressor activity, which was prevented by previous treatment of the animal with atropine. However, larger doses of acetylcholine and some of the other compounds would then produce a rise of blood pressure. Nicotine was also known to have a pressor action that was not dependent on atropine and that was subject to marked tachyphylaxis, i.e., repeated doses produced smaller effects that were finally ineffective. When an animal had been rendered unresponsive to nicotine, acetylcholine would no longer have a pressor action in the presence of atropine. The relative activity of the substances as regards depressor and pressor actions varied; at the extremes, muscarine had only depressor activity and nicotine only pressor activity. The two activities of acetylcholine and the other substances hence were called "muscarine actions" and "nicotine actions" (later they were called muscarinic and nicotinic). Dale remarked, "The question of a possible physiological significance in the resemblance between the action of choline esters and the effects of certain divisions of the involuntary nervous system is of great interest, but one for the discussion of which little evidence is available. Acetylcholine is, of all the substances, the one whose action is most suggestive in this direction." Dale did not go further, although Elliott (1905) and Dixon (1907, 1909) had earlier suggested the possibility of a chemical mediation of autonomic effects, and indeed, Dixon had rather clear ideas of a postsynaptic receptive substance upon which atropine acted.

Experimental proof in favor of acetylcholine as a chemical transmitter at the vagal endings in the heart came from Loewi (1921). The experiment was very simple: two frog hearts were set up with Straub cannulas. The vagus nerve to one of the hearts was stimulated for a few minutes. The fluid was then removed from the cannula and used to replace the fluid in the cannula of the second (unstimulated) heart. It was found to reduce the amplitude of beat of the second heart. Transfer of fluid from a heart whose vagus had not been stimulated was without effect, nor was the effect seen if the second heart had been pretreated with atropine. In subsequent experiments, the "vagusstoff" was shown to be unstable in alkali and in serum. The substance was not choline, and by 1926, Loewi said "that in view of the extremely high activity of the vagal substance it might be thought that it is acetylcholine" (Loewi and Navaratil, 1926a). That same year, he showed that physostigmine sensitized the heart to the "vagusstoff" just as it did to vagus stimulation and to acetylcholine (Loewi and Navaratil, 1926b), but not to muscarine or choline. This was shown to be the result of reduced metabolism, as was the increased duration. The proof was growing and, over the next few years, largely because of the work of Dale and Feldberg, became totally convincing. The formation of acetylcholine in the heart in the presence of physostigmine was shown by Beznak (1934), and there was the beginning of a systematic study of acetylcholine synthesis in the brain by Quastel et al. (1936). More recent work has defined the synthetic enzyme cholineacetyl-transferase, which transfers acetyl groups from acetyl-CoA to choline. The acetylcholine is stored in vesicles in the nerve terminals, and althogh most of the detailed evidence about the behavior of the storage and release mechanisms has been obtained in endings related to nicotinic receptors at the neuromuscular junction and in ganglia, the behavior at muscarinic junctions is likely to be at least qualitatively similar (MacIntosh and Collier, 1976).

Plants that were later shown to contain antagonists for the muscarinic receptor have been known since antiquity for their poisonous properties. They include deadly nightshade (Atropa belladonna), thornapple (Datura stramonium), also known as Jimson weed, and devil's apple and henbane (Hyoscyamus niger), all members of the Solanaceae.

Nightshade was well known in the Middle Ages as an agent used by poisoners, and Linnaeus imaginatively gave it the name Atropa after the oldest of the three Fates who, in Greek mythology, cut the thread of life. The second part of the name came from an alleged practice of Italian ladies of instilling an infusion of the berries into the conjunctival sac causing pupillary dilation and thus making the eyes appear more luminous! Datura was used by thieves and prostitutes in the East to produce intoxication and immobilization. It may also have been used among the priestesses and acolytes in the Temple of Apollo in Delphi to produce a state of excitement. Certainly the picture of toxicity involving a rapid pulse, dry warm skin, dry mouth, dilated pupils, and central disturbances was well known, although the descriptions found in the old herbals rarely exhibited such brevity and coherence. Hyoscyamine was isolated from Datura and Atropine from Belladonna by Brandes in 1831 (Brandes, 1831a,b), and atropine was also isolated from Belladonna by Mein and by Geiger and Hesse in 1833; the structure of both were subsequently determined by Ladenburg (1883).

As mentioned earlier, the action of atropine in blocking the action of the vagus had already been described by von Bezold and Bloebaum before Schmiedeberg and Koppe's study of muscarine. The action on the eye was noted even earlier (Himly, 1802), and this was followed by several comprehensive treatises (Bayle, 1830; Schotten, 1842; Frazer, 1869).

2. The Range of Muscarinic Activity

Muscarinic actions are widely distributed in vertebrate organs (Table 1). Practically all smooth muscle is responsive, and both

Table 1
Sites of Muscarinic Action

Smooth muscle contraction
Stomach body
Small and large intestine
Bladder detrusor
Pupil constrictor
Ciliary muscle
Smooth muscle relaxation
Small blood vessels
Pyloric sphincter
Bladder sphincter
Cardiac effects
Decreased pacemaker frequency
Decreased force of auricular contraction
Inhibition of positive inotropic effects
Central nervous system
Excitory and inhibitory postsynaptic actions
Presynaptic inhibition
Glandular secretion
Stimulation, sweat, salivary, lachrymal, bronchial, pancreatic, and gastric glands

excitatory and inhibitory actions are found. These may be seen as a coordinated pair, as in the contraction of the detrusor and relaxation of the sphincter in the urinary bladder or the corresponding effects in the stomach. Blood vessels are nearly all relaxed by acetylcholine. The physiological meaning of this is obscure, since most vessels do not have a cholinergic innervation. Acetylcholine is also not a circulating hormone. On the other hand, vessels in the genitalia have an important cholinergic dilator innervation (nervi erigentes), and at least part of the dilatation that occurs during activity in secretory glands that supplies fluid and nutrients is the result of a muscarinic action. In the eye, the pupil constrictor and ciliary muscles have extensive cholinergic innervation. Among the secretory glands, the salivary, lacrimal, and sweat glands are strongly dependent on cholinergic innervation, and although gastric and pancreatic secretion are also stimulated through their cholinergic innervation, the dominant stimulus is hormonal. In the heart, the action on the pacemaker in the sinoatrial node is to cause a reduced rate of firing. In the atria, the force of contraction is reduced and the rate of repolarization of the action potential increased. In the

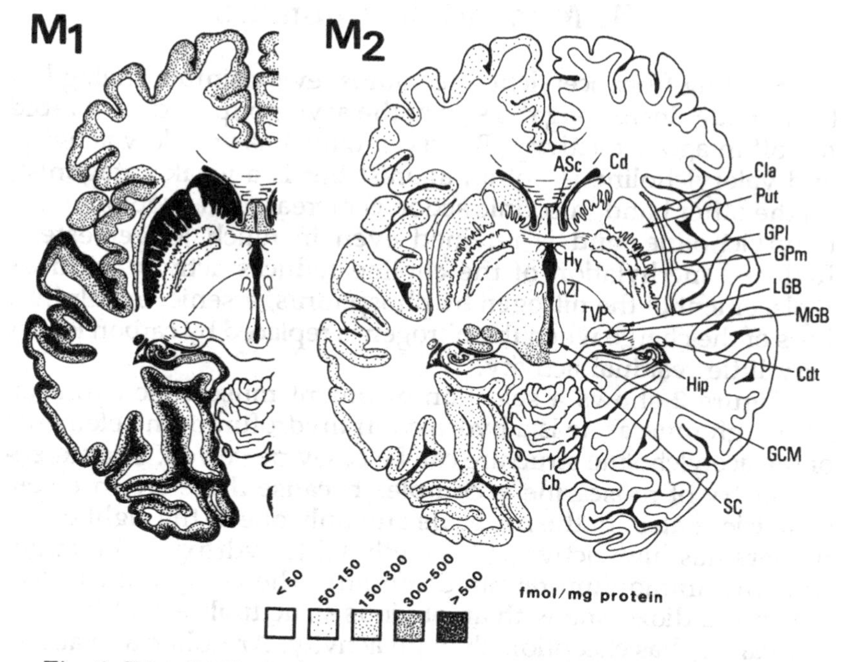

Fig. 1. Distribution of M_1 and M_2 muscarinic receptor subtypes in the human brain (Palacios et al., 1985).

ventricle of the frog and turtle, the force of contraction is also reduced, but in the mammalian ventricle, the action of acetylcholine is indirect. If the ventricle is stimulated by a positive inotropic agent such as adrenaline, this effect can be greatly reduced by acetylcholine.

Tracts containing acetylcholine and cells with muscarinic receptors and responsive to muscarinic agonists are widely distributed in the central nervous system (Rotter et al., 1979; Palacios et al., 1985; *see* Fig. 1); in fact, acetylcholine appears to be the most widely distributed central synaptic transmitter, and has been shown to be both excitatory and inhibitory (Brown, 1983; Shepherd, 1983).

Throughout the vertebrate phyla, muscarinic receptors seem to have broadly similar characteristics. In invertebrates, a distinct type of acetylcholine receptor has been found that does not fall clearly into the muscarinic/nicotinic classification and at which both atropine and tubocurarine are antagonists.

3. Muscarinic Agonists

Starting from acetylcholine, almost every simple analog has been made. Increasing the size of the acyl group leads to a rapid fall-off in agonist activity. Butyrylcholine has very low activity, and valerylcholine is not an agonist but is a weak antagonist; on the other hand, nicotinic action is increased in propionyl and butyrylcholine, and is retained even in much longer esters. Reduced methylation of the choline reduces activity, as does replacement of the nitrogen by phosphorus, arsenic, or sulphur. Loss of the charge when the nitrogen is replaced by carbon leaves very little residual activity.

Figure 2 shows a selection of potent muscarinic agonists. Although muscarine is a hydroxytetrahydrofuran, the elements of the acetylcholine structure can be easily perceived. The stereochemistry of muscarine is complex because there are three assymetric centers in the molecule and only one of the eight enantiomers has high activity (2-S methyl 3-R hydroxy 5-S methyltrimethylammonium-tetrahydrofuran). The compound F2268, which is a dioxolane with an obvious structural resemblance to muscarine, has exceptionally high activity. Arecoline and aceclidine also are obvious members of the family. The highly active oxotremorine is less obviously related, but is also very selective and has some affinity with pilocarpine. McN-A-343 is more remote although related to oxotremorine, but shows selectivity among muscarinic subtypes (*see below*). There is a huge amount of literature on acetylcholine analogs (Bovet and Bovet-Nitti, 1948; Barlow, 1964; Triggle and Triggle, 1976; Cannon, 1981). Muscarinic agonist properties are discussed in detail in Chapter 5 of this volume.

4. Muscarinic Antagonists

The characteristic features of acetylcholine antagonists lies in the introduction of bulky aromatic, heterocyclic, polycyclic, or even just large aliphatic groups on the acyl end of an acetylcholine-like base; alterations at the cationic end are of lesser significance. A selection from the very large number available are shown in Fig. 3. All the substances shown are very potent and discriminate selectively against nicotinic receptors by at least 1000-fold. Propylbenzilylcholine mustard (PrBCM) is an alkylat-

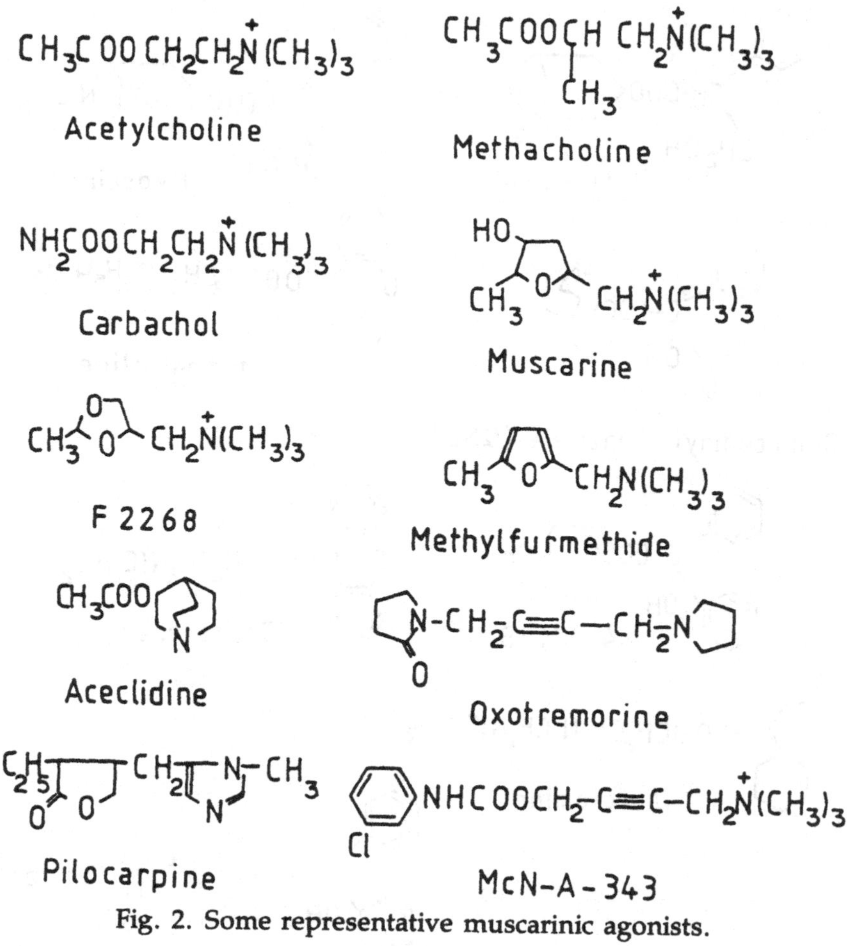

$CH_3COOCH_2CH_2\overset{+}{N}(CH_3)_3$

Acetylcholine

$CH_3COOCH(CH_3)CH_2\overset{+}{N}(CH_3)_3$

Methacholine

$NH_2COOCH_2CH_2\overset{+}{N}(CH_3)_3$

Carbachol

HO, O, CH_3, $CH_2\overset{+}{N}(CH_3)_3$

Muscarine

CH_3, O, O, $CH_2\overset{+}{N}(CH_3)_3$

F 2268

CH_3, O, $CH_2N(CH_3)_3$

Methylfurmethide

CH_3COO, N

Aceclidine

N-CH_2-C≡C-CH_2-N, O

Oxotremorine

C_2H_5, O, O, CH_2, N-CH_3, N

Pilocarpine

Cl, NHCOOCH$_2$-C≡C-CH$_2\overset{+}{N}(CH_3)_3$

McN-A-343

Fig. 2. Some representative muscarinic agonists.

ing compound that produces irreversible inactivation of the receptor. Comprehensive lists of antagonists are given in Bovet and Bovet-Nitti (1948), Barlow (1964), Triggle and Triggle (1976).

5. Receptor Binding Studies

The binding of ligands to muscarinic receptors has been studied extensively. There are basically five methods in use. First, a radiolabeled ligand may be brought into contact with an intact tissue, such as a strip of intestinal smooth muscle (Paton and Rang, 1965), and after wiping off the surface liquid, the

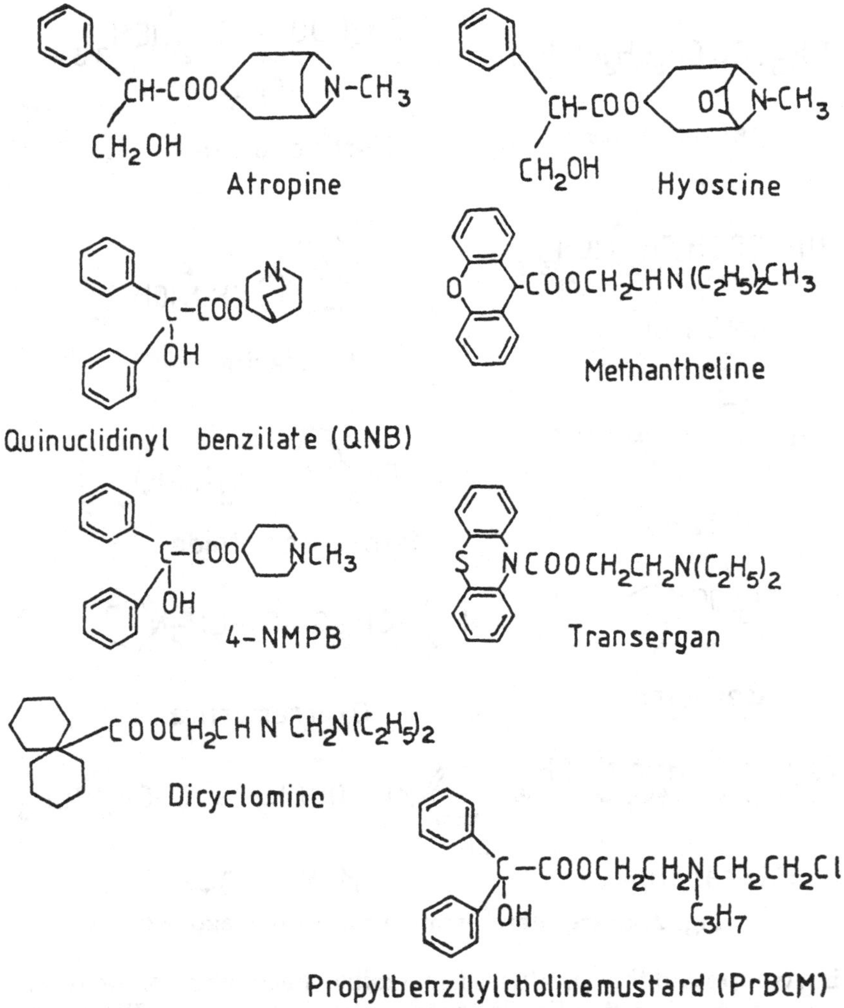

Atropine

Hyoscine

Quinuclidinyl benzilate (QNB)

Methantheline

4-NMPB

Transergan

Dicyclomine

Propylbenzilylcholinemustard (PrBCM)

Fig. 3. Some representative muscarinic antagonists.

radioactivity is measured. This can also be done with tissue slices, and in fact, most muscarinic antagonists are sufficiently metabolically stable that the same method can be applied to organ distribution after intravenous injection (Hammer et al., 1985; Hammer, 1986). From the dependence on ligand concentration, the affinity constant for the ligand and the total binding capacity can be derived. Allowance for nonspecific binding is made by addition of a swamping concentration of either the unlabeled ligand or of an unlabeled competitor. The same principle can be applied to subcellular membrane preparations, but this requires separation of the membranes from the excess fluid either by centrifugation or by filtration. Filtration has an advantage in reducing the level of nonspecific radioactivity from the interstitial water, but it is also liable to remove a ligand that is weakly bound to specific sites. Centrifugation does not yield low nonspecific levels, but accurately retains all the specific binding. The binding procedures are also readily applied to histological sections. Alkylating antagonists such as PrBCM may be used, but in this instance, it is the rate of uptake that is proportional to the concentration of the receptor and to its saturation in its intermediate reversible complex. Muscarinic receptors have been localized in vivo by the use of an antagonist labeled with a shortlived isotope (^{123}I) and single photon counting (Eckelman et al., 1984) or by positron emission tomography (Maziere et al., 1981).

Simple mass-action binding of a ligand in 1:1 complex is given by:

$$[D] + [R] \rightleftharpoons [DR]$$

where D is the concentration of free drug, R is the concentration of free receptor, and DR is the concentration of drug-receptor complex. Since $[R]_{TOT} = [DR] + [R]$, we have:

$$[DR] = \frac{[D] \cdot [R]_{TOT}}{K_D + [D]}$$

This gives the familiar sigmoid plot (using a logarithmic scale for D along the abscissa) as shown in Fig. 4a.

The experimental data can be applied to this equation to extract R_{TOT} and K_D. Alternatively, the equation can be transformed by cross multiplication to give:

$$\frac{[DR]}{[D]} = \frac{[R]_{TOT}}{K_D} - \frac{1}{K_D} \cdot [DR]$$

This indicates that, if we now plot $[DR]$ along the abscissa and $[DR]/[D]$ along the ordinate, we will have a straight line that intersects the abscissa at R_{TOT} (i.e., the total amount of receptor) and the ordinate at R_{TOT}/K_D (Fig. 4b). This is a useful transformation when computer fitting is not available. An excellent fit to this simple relationship is found with ordinary antagonists. A variant of this binding relationship can be used to assess binding of other antagonists by competition, where D is the labeled drug and A is the antagonist:

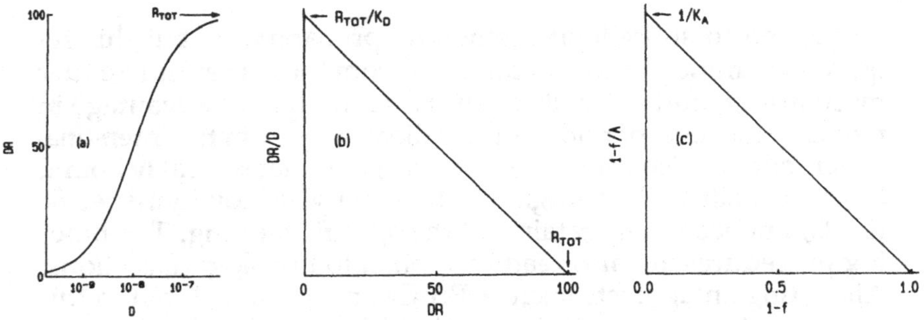

Fig. 4. Methods of plotting direct binding (a,b) and competitive ligands (c).

$$[D] + [A] + [R] \rightleftharpoons [DR] + [AR]$$

which breaks down into:

$$[D] + [R] \rightleftharpoons [DR]$$

$$[A] + [R] \rightleftharpoons [AR]$$

and this leads to:

$$[DR] = \frac{[D][R]_{TOT}}{[D] + K_D(1 + \frac{A}{K_A})}$$

If the concentration of D (the labeled antagonist) is reduced so that $[D] << K_D$, then the relative amount of the competing antagonist that is bound is given by:

$$f = \frac{1}{\frac{A}{K_A} + 1}$$

and the binding constant of the competitor is readily obtained. We can rearrange this equation to give a linear plot in the form:

$$(1 - f)/A = -(1 - f)\frac{1}{K_A} + \frac{1}{K_A}$$

and if we then plot $(1 - f)/A$ on the ordinate and $(1 - f)$ on the abscissa, the line intersects the ordinate at $1/K_A$ (Fig. 4c).

Since the amount of indicator ligand used is usually such that $D/K_D < 0.2$, the correction of the binding constant required is small and is usually ignored. The correlation between the affinity constants of antagonists determined in this way and that determined in vivo are excellent (Hulme et al., 1978; and *see below*).

Deviations from simple mass-action relationships were first noticed in competitive experiments between agonists and labeled antagonists (Burgen and Hiley, 1975). Here the inhibition curves were flatter than expected from mass action and sometimes showed very pronounced inflections (Birdsall et al., 1978, 1980; Fig. 5). These differences could have their origin in complexity of binding to a single receptor series, such as 2:1 complex, which occurs at the nicotinic receptor or because the muscarinic receptor exists in multiple states, each with its own affinity for agonists—states distinguished by agonists but not by antagonists. The first explanation is difficult to reconcile with inflections that do not correspond to integral fractions of receptor concentration. Secondly, an analysis based on the second approach has an internal consistency; for instance, the different properties of the receptor in the areas of the central nervous system fit well to differing proportions of three receptor substates of near identity in affinity. In the heart (and some other peripheral sites), guanine nucleotides like GppNHp shift the agonist binding curve to lower concentrations (Berrie et al., 1979); on analysis using the proposition of a mixture of states, it is found that only the proportions of the states are changed, the residual affinities of the states being unchanged (Burgen, 1987). The evidence so far available thus favors a heterogeneity of agonist binding sites thus:

$$\text{Agonist bound} = \sum_{1}^{n} \frac{[R]_{\mathrm{TOT,n}} \cdot [D]}{K_{\mathrm{D,n}} + [D]}$$

The test of how many components one requires to achieve an optimal fit to the experimental data requires statistical evaluation. In most sites, three components are clearly required, but present methods are inadequate to decide whether more than three states are really present. In the heart, the ratio of affinity between the three states seems to be rather constant and the ratio of the binding constant of the highest affinity binding state (superhigh) to the lowest affinity (low) is about 2500 (Burgen, 1987). The states have been mostly examined in competition experiments between agonists and antagonists. Since direct agonist binding is usually evaluated by the filtration method, the lower affinity binding may be missed because of rapid dissociation.

It has already been noted that GppNHp and GTP cause a shift in the population of receptor states in the heart to low affinity, and it is this low affinity state that is likely to correspond to the physiological situation because the endogenous concen-

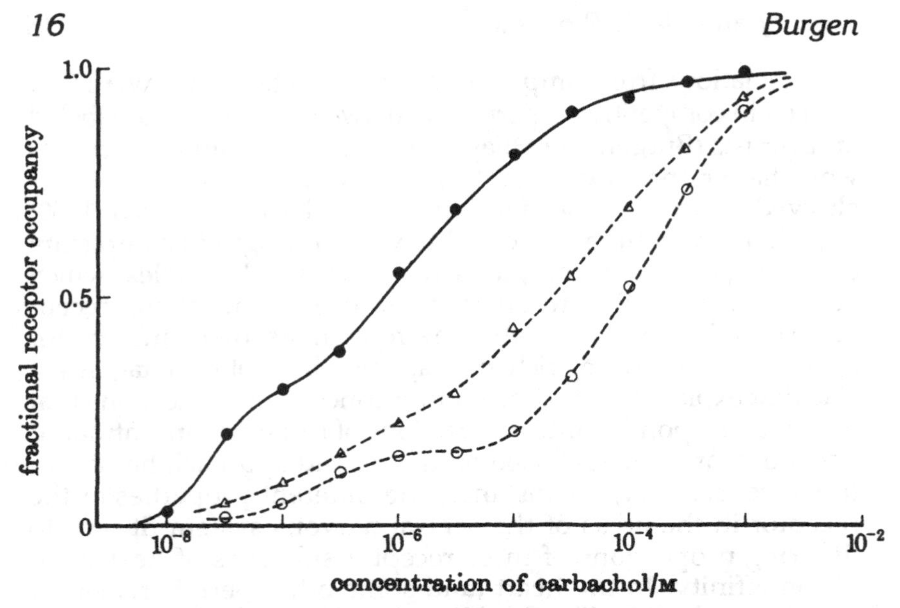

Fig. 5. The binding of carbachol to areas of the rat central nervous system in competition with labeled propylbenzilylcholine. ● cerebellar cortex; ▽ cerebral cortex; ○ hippocampus (Birdsall et al., 1980).

tration of GTP is adequate to ensure that this transformation is complete. In some tissues, such as the conducting tissue of the heart, GTP is without effect on the receptor population (Burgen et al., 1981; Hulme et al., 1981a), and in many other tissues, an intermediate effect may be seen, as in various areas of the central nervous system. In the usual ionic strengths, the binding of antagonists is unaffected, but in low ionic strength media, the binding of antagonists may no longer correspond to a simple mass-action relationship and GTP restores simple binding (Hulme et al., 1981b). The proportion of high affinity sites is dependent on the concentration of magnesium, and with increasing ionic strength, there is a progressive conversion of the superhigh and high states into the low state (Burgen, 1986).

6. Quantitation of Muscarinic Effects in Intact Tissues

Classical studies of muscarinic agents have been carried out on isolated smooth muscle preparations, such as the guinea-pig ileum. An agonist like acetylcholine produces a graded shortening of the muscle in response to increase in concentration, but the shape of the response curve depends also on the method of recording, i.e., the load on the tissue and how far one is recording tension or simple shortening. The dose–response curve may or may not come close to a pure mass-action form; however, some point on the curve, such as half maximum, can be used for assessment of agonist activity. For most agonists, the same level of maximum activity is found, and such agonists are called full agonists. There are usually some agonists, mostly the weaker ones, that are only capable of a smaller maximum response and these are referred to as partial agonists.

In the presence of reversible antagonists, the dose–response curve is shifted to a lower concentration range, and this usually yields a curve parallel to the control curve (when plotted on a log–dose scale) so there is a uniform dose ratio required to produce equivalent effects. Plotting *dose ratio* -1 against the concentration of antagonist then produces a linear plot passing through the origin whose slope gives the affinity of the antagonist. Strictly this relationship depends on the rapid attainment of equilibrium between agonist, antagonist, and receptor within the rather short time during which the response is being observed. However, it is usually the case that, in the control test, only a small proportion of the receptors need to be occupied in order to produce a maximum response, because the maximum response is determined by some function of the contractile process. We commonly refer to this situation as the presence of "spare receptors." When this is the situation, the kinetics of the attainment of equilibrium are rarely a problem, since a pseudoequilibrium gives essentially similar results to a true equilibrium. The presence of spare receptors can be demonstrated with irreversible antagonists such as PrBCM. Where there are no spare receptors, the attainment of equilibrium may offer difficulty.

In general, there is excellent agreement between the affinity for the receptor determined either by this method or direct binding in the case of antagonists. If spare receptors are present, the concentration of an agonist that produces a half-maximal contractile response is lower than that needed for half occupation of the receptor and is not, therefore, even an approximate guide to its affinity for the receptor. If the spare receptors are eliminated by appropriate treatment with an irreversible antagonist, the concentration of agonist producing a half maximal response is an approximate measure of the dissociation constant (i.e., the reciprocal of the affinity constant). In the case of intestinal smooth muscle, the values so obtained seem to correspond rather well with the affinity of the high affinity form of the receptor (Birdsall et al., 1978). The availability of different values of the spare receptor ratio in different receptor locations or in activating different effectors can be the basis of selectivity in agonist action. An interesting example of this has recently been reported with respect to two actions of muscarinic agonists in inhibiting adenylate cyclase and in stimulating phosphoinositide hydrolysis in the chick heart (Brown and Goldstein, 1986).

7. Physiological and Biochemical Actions of Muscarinic Agonists

7.1. Actions through Adenylate Cyclase

Activation of muscarinic receptors in the atrium causes inhibition of adenylate cyclase (Murad et al., 1962; Gilman, 1984; Katada et al., 1984). The receptor complex consists of the drug recognition unit, which is complexed with guanine nucleotide regulatory proteins (G) and the catalytic cyclase unit. The G proteins comprise a complex of an inhibitory alpha subunit α_i, and two other subunits, β and γ. In the unstimulated state, the α_i subunit binds GDP. On stimulation by muscarinic agonists, exchange of GTP for GDP occurs followed by dissociation of the complex into α_iGTP and $\beta\gamma$. Decrease in activity of the cyclase occurs either because of its association with α_iGPT or because of dissociation of the stimulatory combination α_sGPT.C into free C and recombined $\alpha_s\beta\gamma$. The details of this and other mechanisms by which muscarinic agonists decrease cyclic AMP are discussed in Chapter 6.

Cyclic AMP is an important regulator of phosphorylation of proteins in the cell through cyclic AMP-dependent protein kinases. In some systems such as the ventricular myocardium, when adenylate cyclase is stimulated by agents such as noradrenaline through a stimulatory G system, the activated G_i system is able to compete for the enzyme and so reduce subsequent activation.

7.2. Actions through Inositol Phosphates

A second mechanism of muscarinic receptor action is through activating a phospholipase C, which hydrolyses a membrane lipid phosphatidylinositol 4,5-biphosphate (PIP_2) to give two active substances, inositol 1,4,5-trisphosphate ($InsP_3$) and diacylglycerol (DAG) (Berridge, 1984). The activation of phospholipase C is also coupled to the mAChR through a GTP-binding protein. $InsP_3$ releases Ca^{2+} from binding sites in the endoplasmic reticulum; it is this increase in intracellular Ca that is responsible for contraction in smooth muscle. DAG activates protein kinase C. Muscarinic regulation of this effector system is discussed in detail in Chapter 7.

7.3. Actions on Potassium Permeability

A characteristic action on the heart is hyperpolarization of the membrane resulting from an increase in K permeability that shifts the resting potential closer to the K diffusion potential. The change in K^+ permeability is the result of activation of a channel identified as I_K (Loffelholz and Pappano, 1985). I_K appears to be under the direct influence of the muscarinic receptor and is present in the SA node and atrium, but in the mammalian ventricle is not controlled by the receptor. Opening of the K^+ channel is also regulated through a GTP-binding protein. There is conflicting evidence as to whether it is the α_i or free $\beta\gamma$ that opens these channels (Logothetis et al., 1987).

In the pacemaker region, the decrease in K permeability leads to a higher diastolic potential and a reduced rate of potential drift towards the firing threshold, and hence, to a slower rate of firing. An increase of K permeability also seems to underly muscarinic inhibition at central synapses. Activation by muscarinic agonists in cells in autonomic ganglia and central cells is mediated by inhibition of a K channel (M-current), which allows the resting potential to drift down slowly to the firing threshold (Brown, 1983).

7.4. Actions on Calcium Permeability

Muscarinic agonists increase the rate of repolarization of the action potential of the atrium and greatly decrease the duration of the action potential. The upstroke of the action potential is the result of sodium influx, but the late (plateau) phase is the result of Ca influx, and it is this that is reduced. The reduced Ca influx also means that the intracellular Ca is not raised so much, and this is responsible for the reduced contraction—the negative inotropic action (Loffelholz and Pappano, 1985; Trautwein et al., 1975). Both decreased and increased Ca fluxes have been reported in nerve cells. The regulation of various ion channels by the mAChR is the topic of Chapter 9.

8. Subtypes of the Muscarinic Receptor

The first indication that muscarinic receptors did not form a homogeneous class came from the observation that gallamine, a neuromuscular blocking agent to be regarded as a nicotinic drug, blocked the action of acetylcholine on the heart (Riker and Wescoe, 1951), but did not appreciably affect the action of acetylcholine elsewhere. Recent work has confirmed that gallamine has a selective action (Clark and Mitchelson, 1976) and has added the rabbit ear artery to the list of sensitive sites. Smooth muscles in the ileum and bladder are one or two orders of magnitude less sensitive. Related substances such as pancuronium, stercuronium, and anatruxonium behave in a similar way (Mitchelson, 1983). Receptor-binding studies have shown that gallamine and its congeners inhibit NMS from muscarinic binding sites, but do not behave in a simple competitive manner (Stockton et al., 1983). The differences are twofold: first, there is a maximum degree of inhibition of binding that can be attained, and secondly, the kinetics of NMS binding are considerably slowed. These results have been interpreted as evidence that gallamine interacts with a site on the receptor that is distinct from that binding NMS and is producing its effects by an allosteric conformational change. The affinity of gallamine for the receptor in different sites and the degree to which the receptor affinity can be reduced is in substantial agreement with the pharmacological effects found in intact tissues.

Discrepancies between the sensitivity of the ileum and the heart to apparently classical antagonists such as 4-DAMP were also reported (Barlow et al., 1976). However, the current interest in exploring subtypes of the receptor came principally from the introduction of pirenzepine, which was shown to inhibit gastric secretion in doses that were without effect on the heart rate or on the eye, although they did produce some diminution in saliva secretion (Leitold et al., 1977; Matsuo and Seki, 1979); this finding has been abundantly confirmed. In vitro binding studies showed differences between tissues and, indeed, evidence of more than one receptor type in several cases. This had not previously been encountered with any antagonist (Hammer et al., 1980). Analysis suggested that pirenzepine discriminated at least three receptor subtypes. To begin with, pirenzepine was used to divide receptors into M_1, which were sensitive, and M_2, which were insensitive; the actual discrimination is rather modest and no more than two orders of magnitude. In fact, this classification was first proposed by Goyal and Rattan in 1978 based on the behavior of the receptors in the oesophagus of the opossum. They noted that bethanechol had an opposite effect to vagal stimulation, although both were blocked by atropine, and also that McN-A 343 had a selective effect on what they called the M_1 receptor (Rattan and Goyal, 1983). McN-A 343 is indeed selective in other systems; it is an active agonist at the sympathetic ganglion and in contracting the guinea-pig taenia coli, but has little or no activity on the ileum or atrium (Eglen et al., 1987). In vitro binding showed a different mode of binding in the atrium and cerebral cortex (Birdsall et al., 1983). An analog of pirenzepine, AF-DX 116 (Fig. 6), has a higher affinity for receptors in the heart than in the exocrine glands, whereas receptors in the cerebral cortex were of intermediate affinity (Hammer et al., 1986). This agent being M_2 selective is thus able to discriminate between a cardiac type of "M_2" and a glandular type of "M_2." A new group of polymethylene tetramines (Cassinelli et al., 1986) are highly active, and discriminate between the receptors in the atrium and ileum. Methoctramine is selective for the atrium, whereas extending the interamine chain length from 8 carbons to 12 (Medotramine) reversed the selectivity.

It is evident that the pharmacological sensitivity of muscarinic receptors is sufficiently complex as to require the postulation of at least three subtypes, and it is by no means clear that we are at the end of this subdivision. There was always the possibility that the subtypes were not distinct proteins, but were the result of conformational changes in a single muscarinic receptor protein in response to the environmental effects of other proteins and membrane structures. It is now known that at least five functional muscarinic receptor subtypes exist with substantial differences in amino-acid sequence.

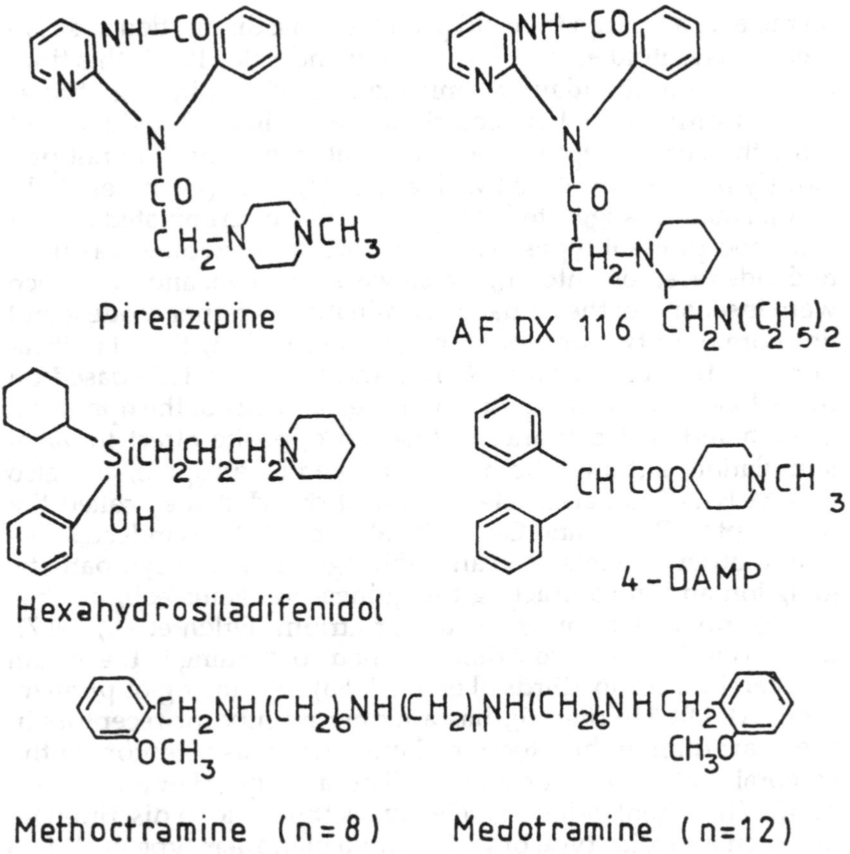

Pirenzipine

AF DX 116

Hexahydrosiladifenidol

4-DAMP

Methoctramine (n = 8) Medotramine (n=12)

Fig. 6. Selective antagonists of muscarinic receptors.

There is also evidence that these different subtypes may preferentially exert their effects through different second-messenger systems. Details of the molecular structures involved are described in Chapter 4.

References

Barlow, R. B. (1964) *Introduction to Chemical Pharmacology*, 2nd Ed., (Methuen, London).

Barlow, R. B., Berry, K. J., Glenton, P. A. M., Nickolaou, N. M., and Soh, K. S. (1976) A comparison of affinity constants for muscarinic sensitive acetylcholine receptors in guinea pig atrial pacemaker cells at 29 °C and in others at 29 °C and 37 °C. *Br. J. Pharmacol.* **58**, 613–620.

Bayle, A. J. (1830) On belladonna, in *Bibliotheque de Therapeutique*, p. 331–518 (Paris).

Berridge, M. J. (1984) Inositol triphosphate and diacylglycerol as second messengers. *Biochem. J.* **226**, 345–360.

Berrie, C. P., Birdsall, N. J. M., Burgen, A. S. V., and Hulme, E. C. (1979) Guanine nucleotides modulate muscarinic receptor binding in the heart. *Biochem. Biophys. Res. Commun.* **87**, 1000–1005.

Beznak, A. B. L. (1934) On the mechanism of autacoid function of parasympathetic nerve. *J. Physiol.* **82**, 129–153.

Birdsall, N. J. M., Burgen, A. S. V., and Hulme, E. C. (1978) The binding of agonists to brain muscarinic receptors. *Mol. Pharmacol.* **14**, 723–736.

Birdsall, N. J. M., Hulme, E. C., and Burgen, A. S. V. (1980) The character of muscarinic receptors in different regions of the brain. *Proc. Roy. Soc. London. B.* **207**, 1–12.

Birdsall, N. J. M., Burgen, A. S. V., Hulme, E. C., Stockton, J. M. and Zigmond, M. J. The effect of McN-A-343 on muscarinic receptors in the cerebral cortex and heart. *Br. J. Pharmacol.* **78**, 257–259.

Bovet, D. and Bovet-Nitti, F. (1948) *Medicaments du systeme nerveux vegatatif*. (Karger, Basel).

Brandes, R. (1831a) Ueber die giftige Substanz des Pilsenkrautes. *Annalen* **1**, 333–357.

Brandes, R. (1831b) Ueber des Atropin. *Annalen* **1**, 68–87.

Brown, D. A. (1983) Muscarinic excitation of sympathetic and central neurones. *Trends Pharmacol. Sci.* **Suppl**, 32–34.

Brown, J. H. and Goldstein, D. (1986) Differences in muscarinic receptor reserve for inhibition of adenylate cyclase and stimulation of phosphoinositide hydrolysis in chick heart cells. *Mol. Pharmacol.* **30**, 566–570.

Burgen, A. S. V. (1986) The effect of ionic strength on cardiac muscarinic receptors. *Br. J. Pharmacol.* **88**, 451–455.

Burgen, A. S. V. (1987) The effects of agonists on the components of the cardiac muscarinic receptor. *Brit. J. Pharmacol.* **92**, 327–332.

Burgen, A. S. V. and Hiley, C. R. (1975) The use of an alkylating antagonist in investigating the properties of muscarinic receptors, in *Cholinergic Mechanisms* (Waser, ed), Raven, N.Y.

Burgen, A. S. V., Hulme, E. C., Berrie, C. P., and Birdsall, N. J. M. (1981) The nature of the muscarinic receptors in the heart, in *Cell Membranes in Function and Dysfunction of Vascular Tissue* (Godfraind, T. and Meyer, P., eds.) Elsevier, North Holland.

Cannon, J. G. (1981) Cholinergics, in *Burger's Medicinal Chemistry* (Wolff, M. E., ed.), Wiley-Interscience, pp. 339–360.

Cassinelli, A., Melchiorre, C., and Quaglia, W. (1986) Selective inhibition of cardiac M_2 muscarinic receptors by polymethylene tetramines. *Proc. Int. Symp. Muscarinic Cholinergic Mechanisms*, Tel Aviv.

Clark, A. C. and Mitchelson, F. (1976) The inhibitory effect of gallamine

on muscarinic receptors. *Br. J. Pharmacol.* **58**, 323–373.
Dale, H. H. (1914) The action of certain esters and ethers of choline and their relation to muscarine. *J. Pharmacol. Exp. Ther.* **6**, 147–190.
Dale, H. H. (1953) *Adventures in Physiology*, (Pergamon, London).
Dale, H. H. and Dudley, F. H. (1929) The presence of histamine and acetylcholine in the spleen of the ox and the horse. *J. Physiol.* **68**, 97–123.
Dixon, W. E. (1907) On the mode of action of drugs. *Med. Magazine* **16**, 454–457.
Dixon, W. E. and Hamill, P. (1909) The mode of action of specific substances with special reference to secretion. *J. Physiol.* **38**, 314–336.
Eckelman, W. C., Reba, R. C., Rzeszotanski, W. J., Gibson, R. E., Hill, T., Holman, B. L., Budinger, T., Conklin, J. J., Eng. R., and Grisson, M. P. (1984) External imaging of cerebral muscarinic receptors. *Science* **223**, 291–293.
Eglen, R. M., Kenny, B. H., Michel, A. D., and Whiting, R. L. (1987) Muscarinic activity of McN-A-343 and its value in muscarinic receptor classification. *Br. J. Pharmacol.* **90**, 693–700.
Elliott, T. R. (1905) The innervation of the bladder and urethra. *J. Physiol.* **35**, 367–445.
Eugster, C. H. (1960) The chemistry of muscarine. *Adv. Organic Chem.* **2**, 427–455.
Ewins, A. J. (1914) Acetylcholine: a new active principal of ergot. *Biochem. J.* **8**, 44–49.
Frazer, T. R. (1869) An investigation into some previously undescribed tetanic symptoms produced by atropia in cold blooded animals with a comparison of the action of atropia in cold blooded animals and in mammals. *Trans. Roy. Soc. Edin.* **25**, 449–490.
Geiger, and Hesse, (1833) Fortgesetzge Versuche ueber Atropin. *Annalen* **6**, 44–65.
Gilman, A. G. (1984) G proteins and dual control of adenylate cyclase. *Cell* **36**, 577–579.
Goyal, R. K. and Rattan, S. (1978) Neurohumoral, hormonal and drug receptors for the lower oesophageal sphincter. *Gastroenterology* **74**, 598–619.
Halliburton, W. C. and Mott, F. W. (1899) Physiological action of choline and neurine. *Phil. Trans. Roy. Soc. B.* **191**, 211–267.
Hammer, R. (1986) In vivo Bindungstudien an muscarinisch Rezeptoren des Herzens der glatten Muskulatur und exocriner Drusen, Habilitationsschrift, Johan Wolfgang Goethe Universitat, Frankfurt.
Hammer, R. Berrie, C. P., Birdsall, N. J. M., Burgen, A. S. V., and Hulme, E. C. (1980) Pirenzepine distinguishes between different subclasses of muscarinic receptors. *Nature* **283**, 90–92.
Hammer, R., Giraldo, E., Schiavi, G. B., Monferini, E., and Ladinsky, H. (1986) Binding profile of a novel cardioselective muscarinic receptor antagonist AF-DX 116 to membranes of peripheral tissues and brain in the rat. *Life Sci.* **38**, 1653–1665.

Hammer, R. Ladinsky, H., and de Conti, L. (1985) In vivo labelling of peripheral muscarinic receptors. *Trends Pharmacol. Sci.* **Suppl**, 33–66.

Hardegger, E. and Lohse, F. (1957) Uber Muscarin, Synthese und absolut Konfigurations des Muscarines. *Helv. Chim. Acta.* **40**, 2383–2389.

Himly, F. K. (1802) Paralysis of the iris by an application of belladonna and its use in treating eye diseases, Paris.

Holmstedt, B. and Liljestrand, G. (1981) *Readings in Pharmacology*, Macmillan, New York, pp. 169–172.

Hulme, E. C., Berrie, C. P., Birdsall, N. J. M., and Burgen, A. S. V. (1981a) Interactions of muscarinic receptors with guanine nucleotides and adenylate cyclase, in *Drug Receptors and Their Effectors* (Birdsall, N. J. M., ed.), Macmillan, London.

Hulme, E. C., Berrie, C. P., Birdall, N. J. M., and Burgen, A. S. V. (1981b) Two populations of binding sites for muscarinic antagonists in the rat heart. *Eur. J. Pharmacol.* **73**, 137–142.

Hulme, E. C., Birdsall, N. J. M., Burgen, A. S. V., and Mehta, P. (1978) The binding of antagonists to brain muscarinic receptors. *Mol. Pharmacol.* **14**, 737–750.

Hunt, R. (1900) Note on a blood pressure lowering body in the suprarenal gland. *Am. J. Physiol.* **3**, xviii–xix.

Hunt, R. and Taveau, R. (1906) On the physiological action of certain choline derivatives and new methods for detecting choline. *Brit. Med. J.* **2**, 1788–1789.

Katada, T., Northup, J. K., Bokoch, G. M., Ui, M., and Gilman, A. G. (1984) The inhibitory guanine nucleotide-binding regulatory component of adenylate cyclase. *J. Biol. Chem.* **259**, 3578–3585.

Kogl, F., Salemink, C. A., Schouten, H., and Jellinek, F. (1957) Uber muscarin III. *Rec. Trav. Chim.* **76**, 109–127.

Landenburg, A. (1883) Die constitution des Atropins. *Annalen* **217**, 71–149.

Leitold, M., Engelhorn, R., Balhause, H., Kuhn, F. J., Ziegler, H., Schierok, H. J., Seidel, H., Guderlei, B., and Milotzke, W. (1977) General pharmacodynamic properties of secretion and ulcer inhibition by pirenzepine (LS 519). *Therapiewoche* **27**, 1551–1564.

Loewi, O. (1921) Ueber humorale Uebertragbarkeit der Herznervenwirkung (Mitteilung I). *Arch. Ges. Physiol.* **189**, 239–242.

Loewi, O. and Navaratil, E. (1926a) Ueber humorale Uebertragbarkeit der Herznervenwirkung (Mitteilung X). *Arch. Ges. Physiol.* **214**, 678–688.

Loewi, O. and Navaratil, E. (1926b) Ueber humorale Uebertragbarkeit der Herznervenwirkung (Mitteilung IX). *Arch. Ges. Physiol.* **214**, 689–696.

Loffelholz, K. and Pappano, A. J. (1985) The parasympathetic neuroeffector junction of the heart. *Pharmacol. Rev.* **37**, 1–24.

Logothetis, D. E., Kurachi, Y., Galper, J., Neer, E. J., and Clapham,

D. E. (1987) The $\beta\gamma$ subunits of GTP-binding proteins activate the muscarinic K^+ channel in heart. *Nature* **325**, 321–326.

MacIntosh, F. C. and Collier, B. (1976) Neurochemistry of cholinergic terminals. *Handb. Exper. Pharmacol.* **42**, 99–228.

Matsuo, Y. and Seki, A. (1979) Actions of pirenzepine dihydrochloride (LS 519) in gastric juice secretion, gastric motility and experimental gastric ulcer. *Arzneimittelforsch.* **29**, 1028–1035.

Maziere, M., Comar, D., Godot, J. M., Collard, P., and Cepeda, C. (1981) In vivo characteristics of myocardial muscarinic receptors by positron emission tomography. *Life Sci.* **29**, 2391–2397.

Mein (1833) Ueber die Darstellung des Atropin in weissen Krystallen. *Annalen* **6**, 67–72.

Mitchelson, F. (1983) Heterogeneity in muscarinic receptors: evidence from pharmacological studies with antagonists. *Trends Pharmacol. Sci.* **Suppl**, 12–16.

Murad, F., Chui, Y. M., Rall, R. W., and Sutherland, E. W. (1962) Adenylcyclase III. The effect of catecholamines and choline esters on the formation of adenosine 3′,5′-phosphate by preparations from cardiac muscle and liver. *J. Biol. Chem.* **237**, 1233–1238.

Palacios, J. M., Corteco, R., Probet, A., and Karobath, M. (1985) Mapping of subtypes of muscarinic receptors in the human brain with receptor autographic techniques. *Trends Pharmacol. Sci.* **Suppl**, 56–60.

Paton, W. D. M. and Rang, H. P. (1965) The uptake of atropine and related drugs by intestinal smooth muscle of the guinea pig in relation to acetylcholine receptors. *Proc. Roy. Soc. Lond. B* **163**, 1–44.

Quastel, J. H., Tennenbaum, M., and Wheatley, A. H. M. (1936) Choline ester formation in and choline esterase activities of tissues in vitro. *Biochem. J.* **30**, 1668–1681.

Rama Sastry, B. V. (1981) Anti-cholinergics, anti-spasmodics and anti-ulcer drugs. in *Burger's Medicinal Chemistry* (Wolff, M. E., ed.), Wiley-Interscience, pp. 361–412.

Rattan, S. and Goyal, R. K. (1983) Identification of M_1 and M_2 muscarinic receptor subtypes in the control of the lower oesophageal sphincter in the opossum. *Trends Pharmacol. Sci.* **Suppl**, 78–81.

Riker, W. F. and Wescoe, W. C. (1951) The pharmacology of flaxedil with observations on certain analogs. *Ann. N. Y. Acad. Sci.* **54**, 373–394.

Rotter, A., Birdsall, N. J. M., Burgen, A. S. V., Field, P. M., Hulme, E. C., and Raisman, G. (1979) Muscarinic receptors in the central nervous system of the rat. Technique for autoradiographic localization of the binding of 3H-propylbenzilylcholine and its distribution in the forebrain. *Brain Res. Rev.* **1**, 141–165.

Schmiedeberg, O. and Harnack, E. (1877) Uber die Synthese des Muscarins und uber muscarinartig wirkende Ammoniumbasen. *Arch. Exp. Path. Pharm.* **6**, 101–112.

Schmiedeberg, O. and Koppe, R. (1869) *Das Muscarin. Das giftige Alkaloid des Fliegenpilzes* (Vogel, Leipzig).

Schotten, L. F. T. (1842) *De effecte Atropin*, Marburg.

Shepherd, C. M. (1983) *Neurobiology*, Oxford, London.

Stockton, J. M., Birdsall, N. J. M., Burgen, A. S. V., and Hulme, E. C. (1983) Modification of the binding properties of muscarinic receptors by gallamine. *Mol. Pharmacol.* **23**, 551–557.

Trautwein, W., McDonald, T. P., and Tripathi, D. (1975) Calcium conductance and tension in mammalian ventricular muscle. *Eur. J. Physiol.* **354**, 55–74.

Triggle, D. J. and Triggle, C. R. (1976) *Chemical Pharmacology of the Synapse* (Academic Press, New York).

Waser, P. (1962) Chemistry and pharmacology of muscarines, muscarone and some related compounds. *Pharmacol. Rev.* **13**, 465–515.

Wasson, R. (1973) *Soma, Divine Mushroom of Immortality* (Harcourt, Brace, Jovanovich, New York).

Wilkinson, S. (1961) The history and chemistry of muscarine. *Quart. Rev. Chem. Soc.* **15**, 153–171.

[illegible] and Pepp[illegible] Das [illegible]
[illegible]
Schu[illegible]ten, [illegible]
[illegible]
[illegible] and [illegible]
[illegible]
[illegible]
Tu[illegible] [illegible] Ca[illegible]
[illegible] muscle [illegible]
[illegible] 37.
[illegible] Chemical [illegible] of the
[illegible] Academic Press [illegible]
W[illegible]ser, P. (1982) Chem[illegible] of [illegible]
[illegible] and [illegible] related [illegible] 13,
[illegible]
[illegible] (1971) [illegible] ultrastructure [illegible]
[illegible] New York.
[illegible] The [illegible] chemistry of connective tissue [illegible]
[illegible]

SECTION 2

Pharmacological Properties, Purification, and Cloning of Muscarinic Receptors

2

The Binding Properties of Muscarinic Receptors

N. J. M. Birdsall
and
E. C. Hulme

1. Introduction

Paton and Rang (1965) performed the first binding studies on muscarinic receptors when they labeled the receptors on whole smooth muscle strips using [^{3}H]atropine and [^{3}H]methylatropine as radioligands. The low specific activity of the ligands and the use of whole tissue limited the usefulness of their approach, and it took another nine years before muscarinic receptors on membrane preparations from brain and smooth muscle were labeled with the high specific activity reversible antagonists [^{3}H]dexetimide (Beld and Ariens, 1974) and [^{3}H]quinuclidinylbenzilate (Yamamura and Snyder, 1974a,b), and the irreversible antagonist [^{3}H]propylbenzilylcholine mustard (Burgen et al., 1974). The commercial availability of several radiolabeled muscarinic ligands and the ease of carrying out the binding studies have resulted in a plethora of papers describing the ever-increasing complexity of the muscarinic receptor system, which has been confirmed recently by the discovery of multiple molecular species of muscarinic receptors (Kubo et al., 1986a,b; Peralta et al., 1987; Bonner et al., 1987).

This chapter outlines the results of receptor binding studies of muscarinic receptor preparations, discusses anomalies, and where possible, provides unifying themes within the framework of existing knowledge. Reviews on the results of binding studies include Birdsall and Hulme, 1976, 1983, 1985; Sokolovsky, 1984; Hoss and Messer, 1986; Nathanson, 1987; and the three supplements on muscarinic receptor subtypes published in *Trends in Pharmacological Sciences* (Hirschowitz et al., 1984; Levine et al., 1986, 1988).

2. The Binding of Antagonists

2.1. Early Studies: A Simple Picture

The initial investigations of Yamamura and Snyder (1974b) demonstrated that there was a good correlation between the binding constants for several muscarinic antagonists measured in a binding assay on ileal microsomal membranes and the potency of these antagonists in blocking functional muscarinic responses in whole ileal strips. These results suggested that the authors were in fact observing the binding of the radioligand to muscarinic receptors and, furthermore, that these binding properties were unperturbed by tissue disruption and homogenization. More recent results, however, suggest that there may be major complications in the performance of binding assays on membrane preparations derived from smooth muscle. This will be discussed in a later section.

It also seemed from the early binding studies that muscarinic receptors were homogeneous. The binding properties of the brain receptors were similar to those deduced from the analysis of whole tissue responses in smooth muscle (Burgen et al., 1974; Yamamura and Snyder, 1974a). Indeed, we examined the binding of many antagonists to muscarinic receptors from rat cerebral cortex and found that they bound to an apparently uniform population of binding sites (Hulme et al., 1978). The Hill coefficients for binding were close to one, and the measured affinity constants were independent of the radioligand used to monitor the binding to muscarinic receptors. There was an excellent 1:1 agreement between the affinity constants of the antagonists measured in binding and functional assays. In that paper, we noted that the degree of agreement was particularly surprising in that the comparisons were made between tissues and species. We concluded that the results suggested the ligand binding subunits of central and peripheral muscarinic receptors were similar, if not identical, and that, on the basis of these studies and those in frog brain (Birdsall et al., 1980b), the properties of the receptor had been strongly conserved during the course of vertebrate evolution. An updated correlation diagram in which 60 antagonists have been examined is shown in Fig. 1. These antagonists cover a million-fold range in affinities, and the data confirm the identification of the binding sites as muscarinic receptors.

A correlation diagram of this type is quite instructive as a model in that two possible conclusions are often readily (but incorrectly) drawn from it. First, muscarinic receptors in cerebral cortex are identical to those in the ileum and, secondly, there is little if any species difference between muscarinic receptors (at least between the rat and guinea pig). Subsequent studies have shown the first conclusion to be incorrect; differences can be detected if appropriate selective drugs are tested in the two assay systems (*see* next section). Caution should therefore be applied to the interpretation of such correlations.

2.2. Recent Studies: A Very Complex Picture

As the precision of receptor binding studies increased and more and more tissues were examined, small regional variations in the affinity constants of the radiolabeled antagonists were noted (e.g., Kloog et al., 1979a,b). However, it was the novel antagonist pirenzepine that provided the clue to the existence of muscarinic receptor subtypes. The study of Hammer et al. (1980) showed that pirenzepine bound to muscarinic receptors in different tissues with quite different affinities, which were in agreement with its functional potency and selectivity in vivo and in vitro. The affinity of pirenzepine for the receptors in cardiac and smooth muscle tissue was relatively low ($K_D \sim 10^{-6}M$), whereas pirenzepine occupied 50% of the muscarinic binding sites in the cortex at a concentration of $5 \times 10^{-8}M$, which is fivefold lower than that found from functional studies on smooth muscle (e.g., Barlow et al., 1981). Furthermore, the binding curve for pirenzepine in the cortex exhibited a Hill coefficient of less than one, a finding that has been reproduced in numerous papers. The data could be fit to a two-site model, with approximately 60% of the sites having a high affinity for pirenzepine

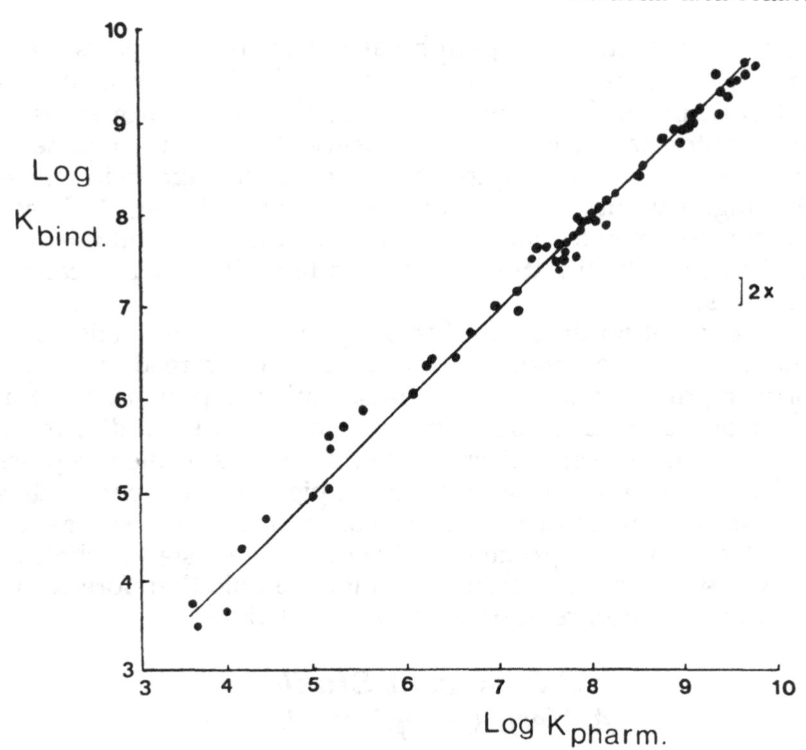

Fig. 1. Correlation between the binding affinity of 60 muscarinic antagonists for rat cerebral cortical muscarinic receptor binding sites and values of their potency (pA_2) in antagonizing muscarinic contraction of the longitudinal smooth muscle of the guinea pig ileum. The line drawn is the line of equivalence [reproduced from Birdsall et al. (1987b) with permission].

($K_D \sim 2 \times 10^{-8}M$) the remaining sites being of approximately 20-fold lower affinity. These results meant that the data point on the correlation diagram in Fig. 1 did not fall near the line of equivalence. However, many studies have shown that the considerable regional variation in the binding potency of pirenzepine and its binding profile is quite compatible with its pharmacological profile (e.g., Hammer et al., 1980; Hammer and Giachetti, 1982, 1984; Hirshowitz et al., 1984; Birdsall and Hulme, 1983, 1985; Levine et al., 1986, and references cited therein).

The existence of muscarinic receptor subclasses, which incidentally was either implicit in the data of previous workers (e.g., Riker and Wescoe, 1951; Roszkowski, 1961) or had been inferred on the basis of functional studies (e.g., Barlow et al., 1976;

Goyal and Rattan, 1978), had led to an extended search for novel muscarinic receptor subtypes and concomitantly to the discovery of new selective antagonists. As in many rapidly developing fields there is a problem of nomenclature: many different systems have been proposed (for a discussion, *see* Birdsall and Hulme, 1983, 1985). The most popular nomenclature, suggested by Hammer and Giachetti (1982) was to categorize those receptors with a high affinity for pirenzepine as M_1 and those with low affinity as M_2. However, this is clearly incorrect, because there are at least two subtypes of M_2 receptors ("cardiac" and "smooth muscle/glandular" types) (Barlow et al., 1976; Birdsall and Hulme, 1983, 1985; Mutschler and Lambrecht, 1984; Giachetti et al., 1986b; Hammer et al., 1986), and there may be subtypes of M_1 receptors (Lambrecht et al., 1987; Bonner et al., 1987).

The analysis of the results of binding studies to delineate the binding properties of individual muscarinic receptor subtypes has been and continues to be a somewhat convoluted process, because the number of receptor subtypes is not known and, in general, tissues do not contain a "pure" receptor subtype. The problems and analyses are discussed in the following sections.

2.3. Comparison of the Binding Properties of Receptors in Different Tissues

This is a simple method in which the affinity constants of ligands for muscarinic receptors in membrane preparations, derived from different tissues are compared. Such affinity constants are generally derived from competition studies using a nonselective antagonist as radioligand. As many of the radioligands in common use, e.g. [^{3}H]3-quinuclidinylbenzilate, [^{3}H]*N*-methylscopolamine, [^{3}H]*N*-methyl-4-piperidinylbenzilate, and [^{3}H]dexetimide have very high affinities for muscarinic receptors (K_D values in the range of $5 \times 10^{-10}M - 10^{-11}M$ depending on the ligand and the experimental conditions), it is not always possible to use concentrations of the radioligands that are considerably below the dissociation constant of the radioligand and below the concentration of receptors in the preparation. Hence, the IC_{50} value from the competition experiment may not be a good estimate of the dissociation constant. Correction factors have then to be applied to the IC_{50} value to allow for high occupancy of the receptor by the radioligand (Cheng-Prusoff correction—Cheng and Prusoff, 1973) and/or for depletion of the free concentration of the radioligand by its binding to the receptor (Cuatrecasas and Hollenberg, 1976). Both correction factors require an accurate knowledge of the affinity constant of the radioligand; any error in this value will be reflected in the calculated affinity constant. We have found that, in the central nervous system, where the receptor levels are high, a ligand of somewhat lower affinity ($K_D \sim 10^{-8}M$), [^{3}H]propylbenzilylcholine (Hulme et al., 1978; Hammer et al., 1980), is convenient. With low concentrations ($<10^{-9}M$) of this radioligand, the IC_{50} value of a competing ligand needs minimal correction to provide an estimate of the affinity constant, providing of course that the competing ligand itself does not have a very high affinity ($K_D < 10^{-9}M$).

The study of Hammer et al. (1980) showed that pirenzepine exhibited over 20-fold variation in its potency for competing for muscarinic receptors in different tissues. In the same paper, a small variation (fourfold) in [^{3}H]*N*-methylscopolamine affinity was detected, of a similar magnitude to that observed by Kloog et al. (1979b) for [^{3}H]*N*-methyl-4-piperidinylbenzilate. The antagonists trihexyphenidyl and dicyclomine (Nilvebrandt and Sparf, 1983, 1986; Potter et al., 1984; Tien and Wallace, 1985; Giachetti et al., 1986a) have a slightly lower (10-fold) variation in potency than is found for pirenzepine. The pattern of dissociation constants for these ligands is basically the same, cerebral cortex ~ hippocampus < glandular tissue < cerebellum ~ heart. Interestingly, both [^{3}H]quinuclidinylbenzilate (e.g., Watson et al., 1986) and secoverine, an antagonist postulated to exhibit selectivity in vivo (Zwagemakers and Claassen, 1980) and in vitro (Marchi and Raiteri, 1985), appear to be nonselective in binding assays (Nilvebrandt and Sparf, 1983; Choo and Mitchelson, 1985; Brunner et al., 1986).

A different selectivity pattern is found for antagonists that are cardioselective in functional tests. The first such drug examined was gallamine (Birdsall et al., 1981; Ellis and Hoss, 1982; Stockton et al., 1983), which exhibits a cardioselectivity in binding assays. However, its interaction with muscarinic receptors is complicated and controversial, and will be discussed later. A more recently developed antagonist, AF-DX 116, behaves as a competitive cardioselective antagonist in both functional (Giachetti et al., 1986b) and binding studies (Hammer et al., 1986a). Its pattern of dissociation constants mirrors that found for gallamine (heart ~ cerebellum < cerebral cortex < glandular tissue) and methoctramine (Melchiorre et al., 1987), a cardioselective antagonist of even greater selectivity than AF-DX 116 (Michel and Whiting, 1988).

2.4. Simplification of the Binding Pattern by the Use of Tritiated Selective Antagonists

In some tissues, several muscarinic receptor subtypes coexist. This is manifest by inhibition curves of certain selective antagonists having Hill coefficients less than 1. In some instances, it may be possible to dissect out the components of the binding curve by an appropriate nonlinear least squares fitting procedure. The affinities of the sites obtained by such an analysis can then be compared with those found in other tissues where the receptor population is apparently homogeneous. This approach depends on the choice of an appropriate model (2-site, 3-site, etc. or a nonmodel-dependent method of analysis, Tobler and Engel, 1983) and on the selectivity of the drug and the precision of the data. As an example, the data of Hammer et al. (1980) were analyzed to suggest that, in the cerebral cortex, ca. 60% of the binding sites had a high affinity for pirenzepine (K_D 1.7 × $10^{-8}M$), with the remaining sites having a lower affinity (K_D 2.5 × $10^{-7}M$) (Fig. 2).

The selectivity suggested by the analysis indicated that low concentrations of [^{3}H]pirenzepine ($10^{-8}M$) should occupy predominantly the high affinity sites and, hence, provide a good radioligand for examining just one subpopulation of binding sites. This was found to be the case (Hammer et al., 1980), with pirenzepine inhibition of [^{3}H]pirenzepine ($10^{-9}M$) binding giving a curve with a Hill coefficient of 1 and a dissociation constant of 1.7 × $10^{-8}M$, which was the value obtained by computer analysis of the inhibition curve using a nonselective antagonist as radioligand (Fig. 2).

This approach is a very good method for detecting and measuring the selectivity of ligands that do not discriminate well between the receptor subtypes. For such ligands, the inhibition curve against the nonselective radioligand cannot be distinguished experimentally from a simple mass action binding isotherm. However, the ratio of the IC_{50} values of the drug in competition with a selective antagonist, say [^{3}H]pirenzepine, to that found for a nonselective antagonist, say [^{3}H]propylbenzilylcholine, can be determined with high precision. Ratios of affinity

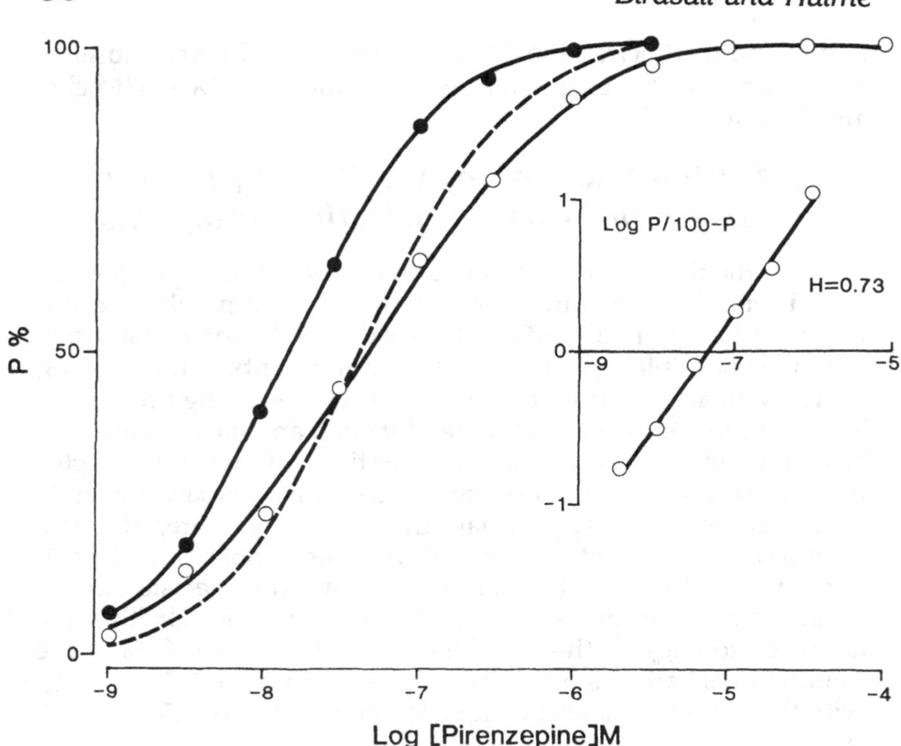

Fig. 2. The binding of pirenzepine to rat cerebral cortical receptors as determined by the ability of increasing concentrations of pirenzepine to inhibit the receptor-specific binding of the nonselective antagonist [^{3}H]propylbenzilylcholine ($10^{-9}M$, open symbols) and a low concentration of [^{3}H]pirenzepine ($3 \times 10^{-10}M$, closed symbols). Assay conditions are as described by Hammer et al. (1980). The dashed curve is a simple mass-action curve. The curve through the open symbols is a nonlinear, least squares fit to a two-site model with 60 ± 5% sites of high affinity ($K_D = 1.7 \pm 0.3 \times 10^{-8}M$) and the remaining sites having lower affinity ($K_D = 2.5 \pm 0.6 \times 10^{-7}M$). The line through the solid symbols is the nonlinear least squares fit to a one-site model with a K_D of $1.7 \times 10^{-8}M$. The inset shows a Hill plot of the [^{3}H]propylbenzilylcholine–pirenzepine competition data, the slope being 0.73 [reproduced from Birdsall et al. (1987b) with permission].

constants of >1.5 for an antagonist binding to different subtypes within a tissue can be estimated.

The dissociation constant for [^{3}H]pirenzepine is relatively low ($1\text{–}2 \times 10^{-8}M$, e.g., Watson et al., 1983; Luthinand Wolfe, 1984), although it can be decreased to ~ $2 \times 10^{-9}M$ by carrying out the binding assay in a low ionic strength 10 mM phosphate buffer (Watson et al., 1983, 1986). Very recently, [^{3}H]telenzepine has become available. This ligand has a selectivity comparable to that exhibited by pirenzepine, but it is about ten times more potent (Eltze et al., 1985; Schudt et al., 1988a,b). The radioligand has been shown to bind to the receptor binding sites that have a high affinity for pirenzepine with a dissociation constant in the range 5–12.5 $\times$ $10^{-10}M$, depending on the species (Schudt et al., 1988a,b). Using a similar approach, [^{3}H]AF-DX 116 has been used to label the cardiac receptor subtype previously suggested (Hammer et al., 1986, 1987a) to be present in cerebral cortical and cardiac membranes (Wang et al., 1987b; Watson, 1988).

2.5. Simplification and Analysis of the Binding Complexity by Taking Advantage of the Different Kinetics of [^{3}H]NMS Binding to the Receptor Subclasses

A rather clever method of simplifying the binding picture has been devised by Waelbroeck and her colleagues (1986, 1987). This method utilizes the differences in the kinetics of binding of radioligands to different muscarinic receptor subtypes and will be explained in some detail.

The kinetics of binding of antagonists are complex (Galper et al., 1977; Kloog and Sokolovsky, 1978; Laduron et al., 1979) and under some conditions are compatible with a two-step process involving a rapid binding followed by slow isomerization (Jarv et al., 1979; Kloog et al., 1979b). However, in many tissues there are mixtures of receptor subtypes, and this will complicate any kinetic analysis if each subtype has different kinetic properties. In fact, this seems to be the case, with the off-rates of [^{3}H]quinuclidinylbenzilate and [^{3}H]*N*-methyl-4-piperidinylbenzilate from muscarinic receptors in the forebrain and brain stem differing by factors of 2–4 (Potter et al., 1984; Kloog et al., 1979b). A much larger difference has been observed with [^{3}H]*N*-methylscopolamine ([^{3}H]NMS) (Hammer et al., 1986b). The half-time for dissociation of [^{3}H]NMS from muscarinic receptors in the lacrimal gland was reported to be 15 min at 30°C and ~ 1 min for cardiac receptors.

In their analysis of the dissociation kinetics of [^{3}H]NMS from muscarinic receptors in four regions of the rat brain and the rat heart, Waelbroeck et al. (1986) noticed that the amplitude of the slow component of [^{3}H]NMS dissociation in the cortex, striatum, and hippocampus matched the proportion of sites that had a low affinity for pirenzepine. In contrast, the [^{3}H]NMS off-rate for the receptors in the rat heart, which also have a low affinity for pirenzepine, was described by a single fast component.

A provisional nomenclature, previously suggested for muscarinic receptor subtypes discriminated by pirenzepine and other ligands (Birdsall et al., 1980a), was used by Waelbroeck et al. (1986, 1987). Receptors of the A type are present in the cortex, hippocampus, and striatum, and have a high affinity for pirenzepine and [^{3}H]NMS, and are equivalent to M_1 receptors. Receptors of the B type have a high affinity for NMS, intermediate affinity for pirenzepine, and are present in the same brain regions and in glandular tissue ("glandular" M_2 receptors). The third receptor subtype, C, has a low affinity for pirenzepine and NMS, and is the predominant species in the cerebellum and heart ("cardiac" M_2 receptor). The dissociation constants of [^{3}H]NMS and pirenzepine, together with the on- and off-rate constants estimated by Waelbroeck et al. (1986), are shown in Table 1.

The differences in equilibrium and kinetic properties of [^{3}H]NMS were exploited so that the binding and inhibition curves favored the A, B, or C sites, when they coexisted (Waelbroeck et al., 1986, 1987). For example, the binding of a ligand to B sites was monitored by initially performing a conventional competition experiment with the ligand using [^{3}H]NMS as a tracer. Instead of measuring the [^{3}H]NMS bound when the system had reached equilibrium, atropine was added to prevent the rebinding of [^{3}H]NMS to the more rapidly dissociating A and C sites. After an appropriate time (35 min), the bound [^{3}H]NMS was measured as a function of ligand concentration. The measured inhibition curve reflected predominantly the binding of the ligand to B sites, if there were comparable proportions of A, B, and C sites present in the membranes.

If pirenzepine was the competing ligand, the change in [^{3}H]NMS bound to A and B sites (estimated by two-site analysis of the competition curves) after 3–35 min was used to estimate the dissociation rates of [^{3}H]NMS from A and B sites (Table 1). By analyzing the amplitudes of these two rates from [^{3}H]NMS dissociation curves at different [^{3}H]NMS concentrations, the affinities of [^{3}H]NMS for the A and B sites were calculated and shown to be slightly different.

Table 1
Binding Properties of Three Postulated Muscarinic Receptor Subtypes Found in Rat Forebrain, Adapted from Waelbroeck et al. (1987)

	Receptor Types		
	A	B	C
[^{3}H]NMS dissociation constant (M)	1.4–2 × 10^{-10}	0.5–1 × 10^{-10}	5 × 10^{-10}
Pirenzepine dissociation constant (M)	0.5–1 × 10^{-8}	0.5–1 × 10^{-7}	3.3 × 10^{-7}
[^{3}H]NMS off-rate (min^{-1})	0.1	0.02	0.4
[^{3}H]NMS on-rate (M^{-1} min^{-1})	6 × 10^8	3 × 10^8	10^9

The monitoring and estimation of the binding of ligands to C sites when A and B sites were also present (e.g., in rat cerebral cortex) were found to be difficult. The proportion of the C sites tended to be underestimated because of lower affinity of [^{3}H]NMS for these sites, unless relatively high concentrations of [^{3}H]NMS (>5 n*M*) were used. In an attempt to circumvent this problem, the binding of the competing ligand to the receptors was allowed to come to steady-state, and then a relatively high concentration of [^{3}H]NMS (1 n*M*) was added for only a short time (1 min). It was assumed, from the estimated rate constants in Table 1, that the C sites would be overlabeled in relation to their proportion in the membranes. This indeed was observed. In fact, it seems that the use of a short incubation time with a *low* concentration of [^{3}H]NMS ($<10^{-10}M$) would have provided a greater selective labeling of the C sites, because under these conditions, the rate of approach to equilibrium would now be regulated by the off-rate, which exhibits a greater variability between the subtypes than the on-rate (Table 1). It should be noted that Waelbroeck and coworkers assumed *no* isomerization of the ligand-receptor complex and deduced that the apparent isomerization observed by other workers could have been explained by sequential binding of the tritiated ligands to these three receptors. The impending availability of single cloned receptor species stably expressed in cell lines, together with the techniques of measuring the kinetics of labeled or unlabeled ligands (Schreiber et al., 1985), will allow the determination of the rate constants for the binding and isomerization (if any) of ligands at individual receptor subtypes.

2.6. Effects of Incubation Conditions on Antagonist Binding

There are several factors that regulate the affinity of antagonists in binding assays. The lowering of the ionic strength will increase antagonist affinities (Birdsall et al., 1979a,b). This procedure is often exploited in attempts to increase the ratio of specific:nonspecific binding. In addition to a general effect of ionic strength on binding, tris ions (Murphy and Sastre, 1983) and other ions (Birdsall et al., 1979b) may have more specific effects.

The binding of antagonists is sensitive to temperature (Kloog and Sokolovsky, 1978; Barlow et al., 1979; Gies et al., 1986). In general, affinities increase on lowering the temperature, but the extent of the increase depends on the ligand and the experimental conditions. At the present time, it is not possible to predict the magnitude of the affinity changes.

Although the binding of antagonists in buffers of physiological ionic strength is not, in general, sensitive to guanine nucleotides, at low ionic strength the binding of antagonists to (particularly) the cardiac muscarinic receptor subtype is increased by GppNHp or GTP (Hulme et al., 1980, 1981a; Burgisser et al., 1982; Hosey, 1982). The precise mechanism is not understood at present, but it appears that, at low ionic strength, low concentrations of divalent cations such as Mg^{2+} can induce a state of the receptor that binds antagonists with low affinity (Hosey, 1982). Addition of guanine nucleotides converts the low affinity state of the receptor to one that can bind several different antagonists with approximately 20-fold higher affinity (Hulme et al., 1981a), giving values of affinity constants for antagonists comparable to those found for cortical receptors at low ionic strength and in the absence of GTP. It appears that under certain conditions the formation of a receptor-G protein complex (*see* section 3.3. for a more detailed description) is favored in the absence of agonist and presence of Mg^{2+}. This complex binds antagonists with low affinity. Addition of GppNHp presumably dissociates the receptor-G protein complex and results in high affinity antagonist binding to the uncoupled receptor. A consequence of this phenomenon is that antagonists under some conditions may have a positive action, that is, the uncoupling of a muscarinic receptor from a G-protein. In this connection, it is interesting that Soejima and Noma (1984) reported that, in "patches" from cells from the sino-atrial node, atropine will close spontaneously opening K^+ channels that are similar to those opened by acetylcholine via the intermediary action of a G-protein (Soejima and Noma, 1984; Pfaffinger et al., 1985; Breitwieser and Szabo, 1985). This would be a functional correlate of what is observed in the binding studies. A guanine nucleotide *decrease* in antagonist binding for muscarinic receptors in the rat adenohypophysis of male (but not female) rats has been reported (Avissar and Sokolovsky, 1981). This system is described in detail in Chapter 10.

Mattera et al. (1985) have also reported that guanine nucleotides enhance the binding of [^{3}H]QNB to purified heart sarcolemmal membranes. They observed apparent homotropic positive cooperativity of [^{3}H]QNB binding in the presence of GTP and no cooperativity in the absence of GTP. The data were interpreted in terms of a model in which the receptor is at least bivalent and exists in two states; one without and the other with cooperativity. The experiments of Mattera et al. were carried out under conditions of high depletion of [^{3}H]QNB. Attempts were made to account for depletion of the [^{3}H]QNB by estimating the free concentration of [^{3}H]QNB by subtraction of the bound [^{3}H]QNB in the pellet from the measured total radioactivity added. However, the bound [^{3}H]QNB and the nonspecific binding were only estimated after the sample had been diluted fivefold with 20% polyethylene glycol followed by 0.5 mg/mL bovine gamma-globulin, stood at 0–2°C for 10–20 mins, centrifuged at 2,400*g*, and the pellets resuspended and recentrifuged twice. These procedures will tend to underestimate the amount of specifically bound [^{3}H]QNB either by dissociation of the radioligand or by handling losses during this protocol. The true nonspecific binding would also have been underestimated. Both these effects will tend to lead to underestimation of the depletion of the [^{3}H]QNB, which, at low concentrations, they calculated to be up to 70%. If the maximum depletion were in fact somewhat higher (80–90%), the binding curve in the presence of GTP would be homogeneous, with no positive cooperativity, whereas that in the absence of GTP would exhibit apparent heterogeneity, as observed in the other studies. Boyer et al. (1986) have also suggested the presence of guanine-nucleotide-induced positive cooperativity of [^{3}H]QNB binding to heart membranes. In these experiments, there was relatively little depletion of free [^{3}H]QNB, although the observed cooperativity could have been explained if the binding of low concentrations of [^{3}H]QNB were not in equilibrium.

2.7. Drugs That Do Not Bind at the Same Site as Acetylcholine or Atropine

In 1951, Riker and Wescoe reported that the neuromuscular blocker gallamine exhibited a cardioselectivity in its side effects on muscarinic receptors. This action of gallamine was investigated in detail by Clark and Mitchelson (1976). In their whole-tissue functional studies, they concluded that gallamine and carbachol did not act competitively at atrial muscarinic receptors, but could not conclude as to whether it produced its effects by binding to the receptor or to an effector molecule. The binding of gallamine to membrane-bound muscarinic receptors was examined by Stockton et al. (1983). The cardioselective nature of the interaction was confirmed, and its binding to muscarinic receptors was shown not to be competitive with that of a series of agonists and antagonists; gallamine appeared to be capable of binding to another site that was allosterically coupled to the conventional site.

The manifestation of this allosteric (negative heterotropic cooperative) effect was that, when gallamine bound to muscarinic receptors, it changed the binding affinity of conventional muscarinic ligands from the value found in the absence of gallamine to another (generally lower) value. In an equilibrium "competition" assay, this interaction was manifest by the inability of high concentrations of gallamine to inhibit all of the specific binding of the ^{3}H-ligand. In fact, the inhibition curve reached a plateau of less than 100%. This plateau denotes the complete formation of the ternary complex of receptor, ^{3}H-ligand and gallamine. An analysis of a series of inhibition curves using different concentrations of the radioligand [^{3}H]NMS was quantitatively compatible with the allosteric model.

Besides affecting the equilibrium binding properties of muscarinic drugs, gallamine slows down the binding kinetics of the radioligand (Stockton et al., 1983; Dunlap and Brown, 1983). The extent of the slowing down depends on the concentration of gallamine. At high concentrations, the phenomenon is so pronounced that it can be very difficult to obtain equilibrium binding data. The change in kinetics is predicted from the allosteric model, and may be used as one criterion for deciding whether a drug interacts with the allosteric site or the conventional site.

A further interesting observation is that the binding of agents to the allosteric site is very sensitive to changes in ionic strength

(Birdsall et al., 1981; Ellis and Lenox, 1985; Nedoma et al., 1985). Up to 10^3-fold changes in gallamine affinity can be measured under conditions in which the binding of conventional antagonists is unaltered (Birdsall et al., 1987c).

A combination of these three criteria (inhibition curves not reaching 100%, alteration of dissociation rate of the radioligand, sensitivity of binding affinity to changes in ionic strength) has been used to delineate several agents that appear to bind to the allosteric site (summarized in Birdsall et al., 1987c). These include certain drugs, such as pancuronium (Dunlap and Brown, 1983) and clomiphen (Ben-Baruch et al., 1982), which appear to bind to both the conventional and the allosteric site, and the selective agonist of unusual structure, McN-A-343 (Birdsall et al., 1983a).

This is not an area of research that is free of controversy. For example, there have been reports that gallamine acts as a competitive antagonist (Ellis and Hoss, 1982; Ellis and Lenox, 1985; Burke, 1986). Part of the evidence presented was that gallamine did not change the maximal number of binding sites for [^{3}H]QNB. However, a drug cannot be assumed to be competitive solely because it does not change the number of binding sites for the radioligand. In addition, Ellis and Lenox (1985) could not demonstrate that gallamine at the concentrations used slows down the dissociation rate of [^{3}H]QNB. However, this slowing down was observed by Dunlap and Brown (1983).

The discrepancies appear to result in part from the fact that the allosteric interaction can only be detected when a *significant proportion of the receptor is present as the ternary complex,* that is, with both [^{3}H]QNB *and* gallamine bound to it. With a negative heterotropic interaction, much higher concentrations of gallamine are required to bind to the receptor-[^{3}H]QNB complex than to form the binary receptor-gallamine complex. Therefore, in order to demonstrate a negative cooperative allosteric interaction in kinetic experiments, higher concentrations of gallamine are required than might be thought intuitively.

Another problem can arise with ^{3}H-ligand-gallamine competition experiments. Because gallamine slows down the on-rate of the ^{3}H-ligand, it is sometimes possible for ^{3}H-ligand binding in the high concentration range of gallamine not to have reached equilibrium, whereas ^{3}H-ligand binding in the absence of gallamine or in the presence of low concentrations of gallamine is in equilibrium. This can result in the plateau of less than 100%

inhibition of binding not being observed if insufficient incubation time is allowed. Instead, inhibition reaches 100%, and the interaction appears competitive. For [^{3}H]NMS, a ligand with relatively fast kinetics, 3–5 h incubation at 30°C is often required. For [^{3}H]QNB, which has much slower kinetics, it may not be practicable to obtain a full equilibrium "competition" curve with gallamine.

The allosteric site does in fact provide the opportunity to design novel muscarinic drugs to bind to this second binding site and, thereby, modulate or tune muscarinic responses. There is a potential application for the enhancement of muscarinic responses in Alzheimer's disease (Birdsall et al., 1986b), possibly providing some therapeutic benefit.

2.8. Soluble Receptors

The detergent most commonly used to solubilize muscarinic receptors is digitonin. The first report was that of Beld and Ariens (1974), who used membranes with the receptors prelabeled with [^{3}H]dexetimide. Sometime later, Hurko (1978), Gorissen et al. (1978), and Aronstam et al. (1978) demonstrated that it was also possible to solubilize unliganded muscarinic receptors in digitonin. Batches of digitonin are notoriously variable in their solubility characteristics and in their ability to solubilize receptors. In part, this is the result of variable proportions of digitonin and gitonin in the sample (Repke and Matthies, 1980a,b). Methods are available for producing "soluble" digitonin (Repke and Matthies, 1980a; Janski et al., 1980). Some of the solubility problems may also be overcome by adding cholate to the digitonin (Cremo et al., 1981).

Other detergents such as cholate (in the presence of a muscarinic ligand, Haga et al., 1982), cholate/1*M* NaCl (Carson, 1982), lysophosphatidylcholine (Haga, 1980), Lubrol PX (Haga, 1982), and the zwitterionic detergent CHAPS (Gavish and Sokolovsky, 1982; Kuno et al., 1983) have been used successfully, but in general, the preparations are less stable than digitonin-solubilized receptors. The reported solubilization of muscarinic receptors by high salt (Alberts and Bartfai, 1976; Carson et al., 1977) has not been confirmed (Hurko, 1978; Aronstam et al., 1978).

The initial studies on the solubilized receptors indicated that the receptors had retained the general antagonist binding characteristics. However, with the emergence of the concept of muscarinic receptor subtypes, increased attention has been paid to the detailed binding properties of soluble muscarinic receptors from different tissues in order to determine whether the binding heterogeneity was retained.

In general, it has been reported that pirenzepine binding to muscarinic receptors in the cerebral cortex is relatively unaltered by solubilization in digitonin, especially if the binding is measured at low temperatures (0–4°C) (Laduron et al., 1981; Roeske and Venter, 1984; Berrie et al., 1985b, 1986; Luthin and Wolfe, 1985; Flynn and Potter, 1986; Baumgold et al., 1987; Mei et al., 1987; Wang et al., 1987a). However, qualitatively quite different conclusions have been drawn regarding the existence of pirenzepine binding heterogeneity within the solubilized preparation and the effects of temperature on the binding properties. Pirenzepine binding, assayed by competition with [^{3}H]QNB, has been reported to exhibit no heterogeneity, when either assayed at 4°C (Wang et al., 1987a) or at 32°C (Luthin and Wolfe, 1985). In both studies, however, [^{3}H]pirenzepine binding could also be observed, but this radioligand appeared to label a fewer number of sites than those labeled with [^{3}H]QNB, possibly indicating the existence of binding heterogeneity, which was not detected in the competition experiment. In contrast, Baumgold et al. (1987) found substantial binding heterogeneity in soluble bovine cerebral cortical preparations (n_H = 0.69, competition with [^{3}H]NMS, room temperature). Similarly, Berrie et al. (1985b) could detect high and low affinity pirenzepine binding sites in digitonin solubilized receptors from rat cerebral cortex (n_H = 0.75, competition with [^{3}H]NMS, 0°C), although pirenzepine selectivity had apparently been considerably attenuated from that found in membranes. The high affinity sites could be labeled by [^{3}H]pirenzepine.

It seems that the kinetics of [^{3}H]pirenzepine binding to solubilized cortical muscarinic receptors are very slow (Berrie et al., 1985b; Luthin and Wolfe, 1985; Mei et al., 1987), and these slow kinetics may cause problems in competition experiments if the binding of the high affinity ^{3}H-antagonist (usually QNB or NMS) approaches equilibrium more rapidly than the competing drug, pirenzepine. In many instances, the off-rate of the ^{3}H-ligand is very slow compared to the duration of the incubation, and thus, these competition experiments are often not in equilibrium over the whole concentration range. It is possible for the ^{3}H-ligand to bind more selectively than expected (in a nonequilibrium manner) to the receptors having a high affinity (and slow kinetics) for pirenzepine binding. In simple terms, the ^{3}H-ligand beats pirenzepine to the binding site and, once bound, acts as a pseudo-irreversible and not a reversible ligand. Higher concentrations of pirenzepine (increasing the k_{on}) are therefore required to compete on a kinetic basis than would be required in an equilibrium experiment. In some instances, this kinetic anomaly may be of a sufficient magnitude to obscure binding heterogeneity. One method of overcoming this anomaly is to allow pirenzepine binding to come to equilibrium before adding the ^{3}H-ligand to monitor the unoccupied receptors (Berrie et al., 1985b, 1986). It seems to us that, if binding is measured at low temperatures, the affinity of pirenzepine for the high affinity sites (M_1 sites) is essentially unaltered on solubilization: the affinity of the low affinity pirenzepine sites is somewhat increased and approaches that of the M_1 sites.

The stability of pirenzepine and the effect of increasing assay temperature on binding to digitonin solubilized cortical receptors have been described in several ways. It has been suggested that

1. M_1 sites are converted to low affinity pirenzepine sites (Roeske and Venter, 1984)
2. there is a nonselective loss of receptor binding (Luthin and Wolfe, 1985)
3. there is a selective loss of M_1 sites (Berrie et al., 1985b) and
4. the sites are stable and the affinity of pirenzepine decreases very considerably with increase in temperature, especially between 25° and 37°C (Mei et al., 1987).

The reasons for these differences are not clear: they may be because of differences in the solubilization and/or assay protocols. There seems a general tendency for pirenzepine binding to be weaker when assayed at higher temperatures, at least in the rat, but perhaps not in the rabbit (Flynn and Potter, 1986).

In contrast to the divergent results from the rat forebrain, there is general agreement that pirenzepine binding to rat myocardial receptors is increased on solubilization in digitonin when binding is assayed at low temperatures (Berrie et al., 1986; Birdsall et al., 1986a; Wang et al., 1987a; Mei et al., 1987). This phenomenon has also been described for muscarinic receptors solubilized from the porcine myocardium (Schimerlik et al., 1986) and bovine medulla-pons (Baumgold et al., 1987). The increase in pirenzepine affinity is such that it is possible to label the soluble myocardial receptors with [^{3}H]pirenzepine (Birdsall et al., 1986a; Wang et al., 1987a; Mei et al., 1987), whereas that was not possible for membrane-bound receptors. The soluble [^{3}H]PZ-receptor sites have been characterized by both binding and hydrodynamic studies (Birdsall et al., 1986a). The equilibrium binding properties of the soluble myocardial receptors, as far as pirenzepine is concerned, are very similar to those of the cortical receptors; that is, the differences found in membrane preparations and in functional studies have largely been abolished. However, the kinetics of [^{3}H]pirenzepine binding (and [^{3}H]NMS binding) to the soluble myocardial receptors are 4–10 times slower than that found for the soluble cortical receptors (Birdsall et al., 1986a). Smaller differences are reported by Wang et al. (1987a). In general, pirenzepine binding appears to be heterogeneous as adjudged by Hill coefficients being less than one in competition experiments with [^{3}H]QNB or [^{3}H]NMS, or the capacity estimates of [^{3}H]pirenzepine binding being less than that found for [^{3}H]NMS or [^{3}H]QNB. The complexity does not seem to be the result of receptor association with a G-protein (Birdsall et al., 1986a). It has been concluded that the membrane environment is one determinant that can affect the affinity of pirenzepine binding to the rat myocardial receptor. Removal of the constraint by solubilization allows the expression of high affinity pirenzepine binding (Birdsall et al., 1986a).

At this stage, a cautionary note should be introduced regarding the general conclusions about pirenzepine binding to soluble cortical and myocardial receptors. The temperature dependence studies of Mei et al. (1987) suggest that pirenzepine binding to the two soluble receptor preparations at 37 °C is low. If the temperature sensitivity of pirenzepine binding is *reversible*, one could conclude from their data that, at 37 °C, muscarinic receptors from rat heart are solubilized with no change in affinity, whereas those from rat forebrain are decreased in affinity on solubilization. These are the opposite conclusions to those drawn from the results at 4 °C! However, the extreme susceptibility of the soluble muscarinic receptor to proteolytic "nicking" (Berrie et al., 1985a) means that the results at higher temperatures should be treated with caution.

In agreement with the results on soluble membrane preparations, little difference in the pirenzepine binding properties of partially purified or purified receptor preparations from forebrain and heart has been found (Haga and Haga, 1985; Berrie et al., 1985a; Schimerlik et al., 1986; Wheatley et al., 1987; Baumgold et al., 1987; Haga et al., 1988). There seems to be a slight decrease in pirenzepine binding during purification, but whether this is the result of selective purification of a specific receptor subtype remains to be determined. It also seems that the redox state of the receptor may be important in determining pirenzepine affinity as DTT will strongly reduce pirenzepine binding (Wheatley et al., 1987; Haga et al., 1988).

So far, the discussion of the binding properties of muscarinic receptors has been limited in terms of both the tissues examined (mainly forebrain and heart) and the selective drug used (pirenzepine). On the basis of these studies alone, it might be inferred that all muscarinic receptor subtypes exhibit the same binding properties when solubilized in digitonin. In fact, pirenzepine binding to muscarinic receptors solubilized from the rat lacrimal gland is *decreased* about sixfold from that found in membranes to a value which is ~50-fold weaker than that found for soluble cortical and myocardial receptors (Berrie et al., 1986). In all three soluble receptor preparations, the affinity of AF-DX 116 is decreased and the cardioselectivity is retained, but in an attenuated form (Birdsall et al., 1987a). The selectivity of hexahydrosiladifenidol is abolished. The most robust selectivity retained on solubilization is that of gallamine, whose interaction with soluble receptors is not competitive, as found in membranes (Berrie et al., 1986; Pedder et al., unpublished results). The major conclusion is that it is necessary to avoid the use of one selective drug to characterize the effect of solubilization on the binding properties of muscarinic receptor subtypes. It appears that the structure-binding relationships of the subtypes have been altered on solubilization in digitonin. Analogous results have been observed for muscarinic receptor subtypes, solubilized with the zwitterionic detergent, CHAPSO (Poyner et al., 1987).

2.9. Differences in the Binding of Hydrophobic and Hydrophilic Ligands to Muscarinic Receptors

In the previous two sections, we have discussed topics in which there are elements of controversy. A third area has emerged in the last 2–3 years, and this has involved the finding that, in some but not all studies, higher numbers of muscarinic binding sites have been estimated when a hydrophobic ligand (e.g., [^{3}H]QNB) has been used as the radioligand instead of a hydrophilic ligand (e.g., [^{3}H]NMS). These findings have been interpreted in terms of existence of binding sites that are accessible to hydrophobic, but not hydrophilic ligands. In this context, tertiary amines are considered to be hydrophobic ligands, and quaternary ammonium ligands are considered to be hydrophilic.

There are several possible explanations that should be borne in mind when trying to analyze these results.

1. The specific activity of one (or both) of the radioligands may be incorrect. As capacity estimates are based solely on saturation curves, the values are totally dependent on the precise knowledge of specific activity. Some confusion in the literature may have arisen because one batch of [^{3}H]NMS (specific activity 84.8 Ci/mmol, and a daughter batch, 85 Ci/mmol) did not have the correct specific activity probably because of the presence of a high affinity, nonradioactive contaminant (Birdsall and Hulme, unpublished observations; R. Young, personal communication). Our sample of this batch appeared to have a specific activity of 35-40 Ci/mmol. Recent batches of [^{3}H]NMS have not had this problem.
2. In studies in whole cells, there may be some receptors on the outside of the cell and others internally located. Tertiary amines readily penetrate across the membrane, and some are actively taken up into cells (Gossuin et al., 1984). Quaternary ammonium ligands do not readily penetrate into cells and, therefore, only label cell surface receptors. This discriminatory behavior of polar ligands has also been found for β-adrenergic receptors (Stahelin and Simons, 1982).
3. In some membrane studies using [^{3}H]QNB, quaternary ligand competition curves have been found to be complex, indicating heterogeneity; using [^{3}H]NMS, quaternary ligand curves are mass action. In contrast, the competition curves for tertiary amines with both radioligands are simple. If [^{3}H]NMS labels fewer sites than [^{3}H]QNB, such a result points to the presence of sites, accessible to QNB and relatively inaccessible to NMS. A relatively trivial explanation for this finding is that there are receptors inside tightly sealed vesicles in the membrane preparation. It is possible to easily obtain such a preparation from the heart (Bers, 1979), a tissue on which many

of the studies have been carried out. The more interesting hypothesis is that there is a subset of receptors on the surface of membranes that do not bind quaternary ligands. Presumably these receptors will not bind acetylcholine and, thus, will be nonfunctional. As the acetylcholine/antagonist binding site is thought to be deeply buried within the membrane (Wheatley et al., 1988; Curtis et al., 1989), it is not inconceivable that the receptor could be modified so that access to the binding site is only possible via the phospholipid surrounding the receptor. Whether such a modification would be naturally occurring or an artifact consequent upon homogenization has not been examined.

4. Finally, there is another trivial explanation that, in some studies, there may be "nonspecific" [^{3}H]QNB binding sites that are inhibited with high affinity by tertiary amines, but not quaternary ammonium compounds.

The early study of Hulme et al. (1978) found that five tritiated antagonists, two tertiary amines, and three quaternary ligands bound to the same number of muscarinic receptor binding sites in membrane preparations from rat brain. Other studies have shown that the tertiary amines[^{3}H]scopolamine and [^{3}H]QNB, as expected, label the same number of sites (Burgermeister et al., 1978). However, Gibson et al. (1984) report a discrepancy between the binding of [^{3}H]QNB and an [^{125}I]derivative (also a tertiary amine). This latter result may reflect a problem in the estimation of the specific activity of the iodinated ligand.

In cell cultures, there have been several reports that, in the absence of a muscarinic agonist, the number of receptors, *measured on intact cells,* is independent of whether [^{3}H]QNB or [^{3}H]NMS is used as the radioligand. These include chick heart cells (Galper et al., 1982; Nathanson, 1983; Siegel and Fischbach, 1984) and 1321N1 cells (Evans et al., 1984). Feigenbaum and El-Fakahany (1985) have reported 50% *more* [^{3}H]NMS binding sites than [^{3}H]QNB sites on NIE115 cells, whereas Masters et al. (1985) and Brown and Goldstein (1986) found 50% *fewer* [^{3}H]NMS sites than [^{3}H]QNB on 1321N1 cells and chick heart cells, respectively. The latter two results disagree with the previously mentioned studies on the same systems.

In several systems, high levels of nonspecific binding of [^{3}H]QNB have been reported (Galper et al., 1982; Siegel and Fischbach, 1984; Evans et al., 1984; Gossuin et al., 1984; Harden et al., 1985). As a consequence, some studies have compared the binding of [^{3}H]NMS to whole cells with [^{3}H]QNB binding to lysed cells, where the nonspecific binding was found to be lower. For example, in intact 1321N1 cells, the levels of [^{3}H]NMS binding measured at 4° or 37°C were found to be same as the number of [^{3}H]QNB sites, estimated on membranes from lysed cells at 37°C (Evans et al., 1984; Harden et al., 1985). Somewhat surprisingly, Evans et al. (1984) reported that in lysed cells, [^{3}H]NMS detected only 30–50% of the sites detected by [^{3}H]QNB. This would imply a *loss* (or inaccessibility) of 50–70% of the [^{3}H]NMS sites on lysis. It is difficult to reconcile the results of the three studies on 1321N1 cells without further experimentation.

Brown and Goldstein (1986) have carried out an extensive analysis of the binding of [^{3}H]QNB and [^{3}H]NMS to intact dissociated embryonic chick heart cells in culture. They found that [^{3}H]NMS bound to a uniform population of sites with affinities agreeing with those observed in functional studies. Most of these sites could be alkylated by the irreversible antagonist, the (quaternary) aziridinium ion of propylbenzilylcholine mustard. In contrast, [^{3}H]QNB bound to approximately 80% more binding sites. The [^{3}H]QNB seemed to label the same sites labeled by [^{3}H]NMS, but also labeled additional sites. These had the same affinity for the tertiary ligands, atropine, scopolamine, dexetimide, and levetimide and oxotremorine, as the [^{3}H]NMS sites, but had a very low affinity (1000-fold weaker) for quaternary ligands such as NMS, *N*-methylatropine, and carbachol. The additional sites could not readily be labeled by the aziridinium ion of propylbenzilylcholine mustard. It appears therefore that, in this particular chick heart preparation, there are receptors that cannot readily be accessed by quaternary ligands. In our view, the most likely possibility is that these receptor binding sites are present on the inside of intracellular vesicles. In this connection, it is of interest that Siegel and Fischbach (1984) observed an apparently lower affinity "nonspecific" [^{3}H]QNB binding site in heart cells in culture. Binding to this site could be inhibited by atropine and not *N*-methylatropine. Furthermore, in 108 CC 15 cells, Gossuin et al. (1984) demonstrated that there was an uptake process for the tertiary ligands [^{3}H]QNB and [^{3}H]dexetimide, which was blocked by methylamine, chloroquine, high concentrations of atropine, and scopolamine, but not by NMS. It is not likely that the latter two phenomena could explain the results of Brown and Goldstein.

Agonist binding studies on whole cells, at temperatures at which internalization takes place, must inevitably be distorted by the fact that agonists will increase the rate of internalization. In "competition" studies with a [^{3}H]quaternary ligand such as NMS, the number of sites that [^{3}H]NMS can bind to will vary with the concentration of agonist, i.e., there will be competitive plus noncompetitive interactions. It may also be true that the [^{3}H]NMS-receptor complex does not "internalize" at the same rate as the unliganded receptor. Therefore, caution should be taken in the interpretation of the results of the binding of [^{3}H]tertiary and quaternary ligands to muscarinic receptors in whole cells.

Despite these caveats and the differences in the results of binding studies on membranes, [^{3}H]NMS has been used to monitor the receptors on cell surfaces. The fast muscarinic agonist induced desensitization can be monitored by the apparent loss of [^{3}H]NMS binding sites (Galper et al., 1982; Maloteaux et al., 1983; Harden et al., 1985; Feigenbaum and El-Fakahany, 1985) with very little, if any, loss of [^{3}H]QNB sites. On removal of the agonist, the [^{3}H]NMS sites can reappear relatively quickly; this can be slowed down by carrying out binding studies at lower temperatures (<20°C, Harden et al., 1985). In a recent paper, Fisher has demonstrated that the relative ability of tertiary and quaternary ligands to bind to muscarinic receptors on human SK-N-SH neuroblastoma cells is dependent on temperature, the presence of agonists, and cellular integrity (Fisher, 1988). The disappearance of cell surface receptors is associated with the specific transfer of muscarinic receptors from a plasma membrane fraction to a "light vesicle" fraction on sucrose density gradients (Harden et al., 1985). A slower nonreversible loss of [^{3}H]NMS and [^{3}H]QNB sites can be observed on longer exposure to agonists (Galper et al., 1982; Masters et al., 1985; Harden et al., 1985).

The type of behavior described for whole cells by Brown and Goldstein (1986) has been found by some authors for membrane preparations from rat brain (Lee and El-Fakahany, 1985a,b; Ellis and Lenox, 1985; El-Fakahany et al., 1986; Ensing and Zeeuw, 1986; Lee et al., 1986; Norman et al., 1986) and Torpedo electric organ (Dowdall et al., 1983). Some but not all of the quantitative differences between the number of sites labeled by the tritiated tertiary and quaternary antagonists could possibly be explained by problems regarding the precise specific activity of the ligands.

The heterogeneous [^{3}H]QNB/NMS competition curves certainly suggest the presence of receptor sites that are poorly accessed or inhibited by quaternary ligands. At present, the necessary experiments have not been carried out to determine whether the homogenization procedures generate tightly sealed vesicles. In this context, it is interesting that a preparation of coated vesicles from bovine brain has been shown to exhibit "anomalous" NMS binding (Silva et al., 1986).

2.10. Binding Studies on Smooth Muscle

In general, the results of binding studies on smooth muscle do not agree with those of functional studies. For example, the potency of gallamine in binding studies on smooth muscle has been found to be very similar to that found in the heart (Mitchelson, 1984; Choo et al., 1985a; Nedoma et al., 1985), although Stockton et al. (1983) found a value comparable to that observed in functional studies. Recently, Michel and Whiting (1988) have reported little if any difference in the affinities of a range of selective antagonists for cardiac receptors and receptors prepared from circular or longitudinal smooth muscle of the ileum. By analyzing [^{3}H]NMS/AF-DX 116 competition curves in ileal smooth muscle, Giraldo et al. (1987b) were able to delineate a minor population of sites with the characteristics of "smooth muscle/glandular" receptors, but the major population appeared to be the "cardiac" receptor. A similar finding has been reported for membrane preparations from bovine tracheal smooth muscle (Roffel et al., 1987). To our knowledge, no "cardiac" muscarinic receptor type has been described in functional studies. What are these "cardiac" receptor binding sites doing in smooth muscle?

In a study on the binding of [^{3}H]propylbenzilylcholine mustard to whole smooth muscle strips, Elliott et al. (1978) observed that distilled water treatment of the strips increased the number of binding sites, suggesting that cell lysis leads to the exposure of extra binding sites. Even in the original study of Paton and Rang (1965), there was an indication that the permeant antagonist [^{3}H]atropine labeled more sites than [^{3}H]*N*-methylatropine. In vivo studies of [^{3}H]NMS binding have also pointed to an anomalously low level of binding to smooth muscle, compared to that found in homogenates (Hammer et al., 1986b). The simplest interpretation of these data is that the "cardiac" receptor binding sites may be located intracellularly and not expresed on the cell surface. Another factor to be borne in mind is that the muscarinic receptors in the smooth muscle membrane preparations are probably proteolytically "nicked." In our studies, proteolysis of the receptor, assayed by SDS-PAGE of [^{3}H]propylbenzilylcholine mustard labeled receptors, could not be controlled (Birdsall et al., 1979a). It may be that the binding properties of the proteolysed smooth muscle receptor resembles those of the cardiac receptor. Other chemical or biochemical modifications of the receptors may lead to the real (or apparent) interconversion of the binding properties.

3. The Binding of Agonists

In retrospect, it is perhaps inevitable, in view of the diversity of muscarinic receptor subtypes and their antagonist binding properties, that the agonist binding properties of muscarinic receptors have not turned out to be simple. The picture has also been complicated by the fact that both receptor binding and function are mirrored in agonist binding.

3.1. Early Studies

The initial reports that agonist binding was complex resulted from studies on smooth muscle (Burgen and Hiley, 1974; Young, 1974; Ward and Young, 1978), although equivalent studies in the rat cerebral cortex suggested that the binding of agonists was simple (Burgen et al., 1974). The phenomenon was investigated in more detail by our group (Hulme et al., 1975; Birdsall and Hulme, 1976; Birdsall et al., 1978). In membranes from rat cerebral cortex, it was found that agonist binding curves to the muscarinic receptor deviated significantly from that predicted for a simple 1:1 interaction with a uniform population of binding sites. At that time, it appeared that antagonists were binding to a uniform population of receptor binding sites (Hulme et al., 1978). The apparent heterogeneity in agonist binding did not appear to be the result of homotropic negative cooperativity between agonist binding sites: alkylation of up to 90% of the binding sites with the irreversible antagonist propylbenzilylcholine mustard, in order to reduce site-site interactions, did not affect the agonist binding properties of the unalkylated sites. It appeared that the complex agonist binding curve resulted from the coexistence of more than one binding component of the receptor that had identical antagonist binding affinities but differing affinities for a given agonist. The competition curves could not be analyzed totally satisfactorily in terms of a two-site model, and three sites [''super high'' (SH), ''high'' (H), and ''low'' (L) affinity] were postulated (Birdsall et al., 1978, 1980c). The heterogeneity of a carbachol binding curve, estimated by competition with a ^{3}H-antagonist, is illustrated in Fig. 3. The binding curve extends over 5–6 orders of magnitude in concentration. Differentiation of the inhibition curve provides a visual picture of the three components, as a shoulder and two peaks (Fig. 3). One very interesting observation was that, when the inhibition curves of 21 different agonists were examined, the ability of agonists to discriminate between the SH, H, and L components appeared to be related to their efficacy (Hulme et al., 1975; Birdsall et al., 1978). For some agonists of high efficacy, the affinities of the SH, H, and L components can vary by up to 3000; for partial agonists, the ratio is much smaller (Birdsall et al., 1978, 1980c), and for antagonists such as atropine the ratio is 1. In other words, both binding and function were being detected in the binding curve, as stated in the introduction to this section.

Heterogeneity in agonist binding has been found in essentially all muscarinic binding studies on membranes. It has also been reported in some, but not all, whole tissue and whole cell studies (Young, 1974; Ward and Young, 1978; Galper et al., 1982; McKinney et al., 1985, but *see* for example, Siegel and Fischbach, 1984; Nathanson, 1983; Brown and Goldstein, 1986) and in autoradiographic localization studies on thin sections (Rotter et al., 1979; Wamsley et al., 1980). However, there is also a regional variation in the agonist binding properties, described originally by Aronstam et al. (1977, 1978), Kloog et al. (1979a), and Birdsall et al. (1980c), and found subsequently in many studies. In our original analysis (Birdsall et al., 1980c), we concluded that the regional variation in binding properties, found for several agonists in different brain regions, resulted primarily from variations in the proportions of SH, H, and L components, rather than from differences in the affinity of an agonist for a given state. There have also been many studies in which agonist binding data has been fitted to a two-site model, and regional variations in the estimated affinities of the high and low affinity sites reported. It is not clear whether some of the differences in affinity result from the fitting of the data to a two-site model, when a more

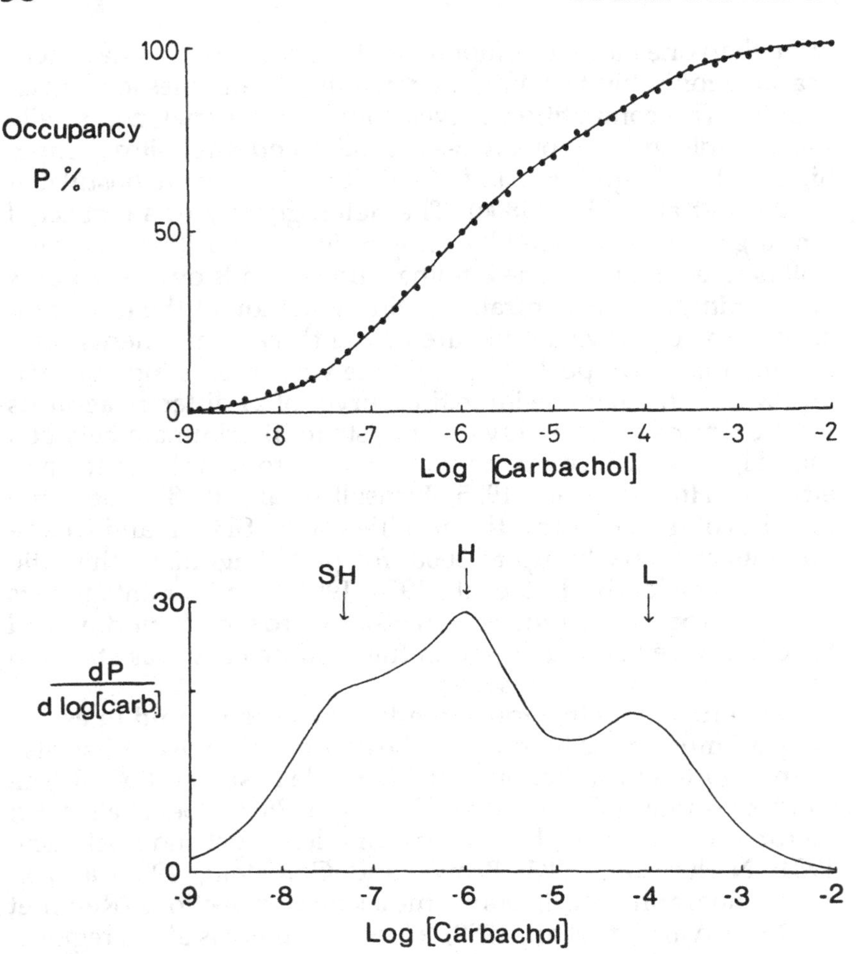

Fig. 3. Carbachol binding to muscarinic receptors from rat medulla-pons, assessed by the inhibition of the specific binding of the antagonist [^{3}H]propylbenzilylcholine (10^{-9}*M*) [Birdsall et al. (1980c)]. The lower part of the figure depicts the gradient of the curve through the experimental data points in the upper half of the figure. The affinities of the putative SH, H, and L sites are marked by arrows [reproduced from Birdsall and Hulme (1983) with permission].

complex model requiring more complete data would be more appropriate, or from genuine differences in affinity. The knowledge of the existence of muscarinic receptor subtypes and some understanding of the receptor mechanism now require a more detailed investigation of this problem.

The original studies by Birdsall et al. (1978, 1980c) suggested that the proportions of SH, H, and L components calculated from an agonist competition curve in the cerebral cortex did not depend on the agonist. In contrast, Burgen (1987) reported that the proportion of SH and L components detected in membranes from the rat heart did depend on the agonist; agonists of high efficacy induced a higher proportion of the SH component. The reasons for the differences are not clear. It may be the result of differences in the agonist binding properties of the different receptor species and systems in the two tissues, in particular, the apparently greater conformational flexibility of the cardiac receptor system.

3.2. Simplification of the Agonist Binding Properties

The irreversible blockade of subclasses of agonist binding components has provided one method by which the binding picture has been simplified. For example, membranes have been incubated with low concentrations of carbachol, which occupy predominantly SH and H sites and the unoccupied (L) receptor sites selectively alkylated by propylbenzilylcholine mustard (Birdsall et al., 1978). Examination of the residual unalkylated sites after washout of carbachol indicated that low affinity sites had been alkylated and the remaining sites exhibited a high affinity for carbachol. This finding argued against interconversion of the agonist binding site/states *under those experiment conditions*. Ehlert and Jenden (1985) have also performed analogous alkylations of the different receptor states using an alkylating derivative of the agonist oxotremorine. An important finding from this study and that of Christophe et al. (1986) is that the L state of the receptor, under certain conditions and in some tissues, is more rapidly alkylated than the higher affinity components, which would have been expected to be selectively blocked.

Another way of looking at the agonist binding heterogeneity is to examine the binding of the agonist directly using a tritiated agonist. If the agonist has a high efficacy, it is possible to selectively label the highest affinity component of the receptor system, and monitor the binding of ligands to that component. Ligands that have been used include [^{3}H]oxotremorine-*M* (Hulme et al., 1975; Birdsall et al., 1978, 1980c; Waelbroeck et al., 1982; Harden et al., 1983; Gillard et al., 1987), [^{3}H]cis-methyldioxolane (Ehlert et al., 1980a; Nukada et al., 1983a; Vickroy et al., 1984; Closse et al., 1987), and [^{3}H]acetylcholine (Hulme et al., 1975; Birdsall and Hulme, 1976; Birdsall et al., 1978; Gurwitz et al., 1983, 1984b, 1985; Kellar et al., 1985). In most instances, the tritiated agonist seems to label the highest affinity state of the receptor, and the binding properties of the site are essentially the same as those inferred from a multi-site fit of the agonist/^{3}H-antagonist competition curve (e.g., Birdsall et al., 1978, 1980c; Waelbroeck et al., 1982). The predominant site labeled is probably the SH state of the receptor. (A certain amount of confusion may arise in reading some papers in which two-site fits have been used to analyze data, and the high and low affinity sites have been termed H and L; in some instances, these may correspond to the terms SH and H used in other papers.)

Under certain conditions, however, it may be possible to label both SH and H receptor states by using higher concentrations of ^{3}H-agonist (Birdsall et al., 1978). A new assay method involving both centrifugation and filtration of the [^{3}H]cis-methyldioxolane-receptor complex (Galper et al., 1987) seems very promising.

In order to determine whether the agonist binding properties of muscarinic receptors are different, it is preferable to obtain direct estimates of the structure-binding relationships of agonists for a specific state of a single receptor subtype, rather than carrying out model-dependent computer fits to complex agonist binding data. This approach will be made easier in the near future with the expression of the cDNA for cloned receptor subtypes in various cell lines (Bonner et al., 1987; Peralta et al., 1987). At the present time, relatively few studies have been carried out on membranes and receptors derived from cloning experiments. Under specified conditions, the sites in cerebral cortical membranes, labeled with [^{3}H]pirenzepine, appear to correspond to a relatively uniform population of the L state of the M_1 receptor. The affinity of most agonists for this receptor subtype seems to be much lower than for the sites/states of the receptors in the cerebral cortex, which have a lower affinity for pirenzepine (Birdsall and Hulme, 1983; Birdsall et al., 1984). The exception is the selective agonist McN-A-343 (Roszkowski, 1961), which appears to have a higher affinity for the L sites of the M_1 receptors than for the agonist binding sites (mainly SH and H) of the "non-M_1" receptors. Whether this selectivity in the binding assay is a manifestation of the pharmacological selectivity of McN-A-343 remains to be determined. In accordance with these data is the finding that certain actions of McN-A-343 are antagonized by low concentrations of pirenzepine (Hammer and Giachetti, 1982; Choo et al., 1985b). The study of Evans et al. (1984) indicated, again under rigidly specified conditions, that agonist binding to the L state of muscarinic receptors in membranes of two cell lines, 1321N1 and NG108-15, could be compared. These cell lines appear to express different receptor subclasses and have different effector mechanisms. Differences in the rank order of affinities of five agonists were found. In particular, the ratio of affinities of carbachol and methacholine for binding to membranes from NG108-15 and 132N1 cells was 1.2 and 0.3, respectively. Somewhat surprisingly (although the ionic conditions were quite different), no selectivity was found when agonist binding to whole cells was measured.

Vickroy et al. (1984) used [^{3}H]cis-methyldioxolane to label SH receptor sites in membranes from rat heart and cerebral cortex. The radioligand did not seem to label a uniform population of sites assessed either by competition with agonists or pirenzepine. Nevertheless, a cortical selectivity was seen as regards the affinity of pilocarpine and McN-A-343 (putative selective M_1 agonists), but not acetylcholine, carbachol, and oxotemorine, for the high affinity [^{3}H]cis-methyldioxolane binding sites. There were suggestions in this study that M_1 (A) sites may have been labeled. A similar study by Gillard et al. (1987) using [^{3}H]oxotremorine-*M* as the radioligand and somewhat different incubation conditions (50 m*M* phosphate buffer, 2 m*M* Mg^{2+}, 25 °C, 10 min cf. 10 m*M* phosphate, no Mg^{2+}, 25 °C, 120 min for Vickroy et al. 1984) found that C sites and B + C sites (*see* Table 1 for terminology) were labeled in the heart and cortex, respectively. As in the study of Vickroy et al. (1984), no difference in carbachol and oxotremorine affinity was detected, but neither was any pilocarpine selectivity detected. Closse et al. (1987) used yet another different set of conditions (2 m*M* Mg^{2+}, 2 m*M* Ca^{2+}, 50 m*M* tricine buffer, 22 °C, 30 min) and could also label M_1 sites as well as sites with a low affinity for pirenzepine. A modest cardioselectivity of oxotremorine and arecoline could be demonstrated. We have also examined the binding of agonists to the sites in the cerebral cortex and myocardium that are labeled by [^{3}H]oxotremorine-*M* (Birdsall and Hulme, 1986). The conditions (EDTA washed membranes, 20 m*M* HEPES, 1 m*M* Mg^{2+}, 30 °C, 15 min) were chosen to maximize [^{3}H]oxotremorine-*M* binding. In this study, using ten agonists and five antagonists, a cardioselectivity of oxotremorine and a cortical selectivity of methylfurmethide were demonstrated. In this connection, it is of interest that Newberry and coworkers (1985, 1987) have shown that methylfurmethide has a selective depolarizing action in the rat superior cervical ganglion (blocked with high affinity by pirenzepine; Brown et al., 1980) with little if any of the accompanying hyperpolarizing action observed for most other agonists (a response with a "cardiac" receptor pharmacology).

It is clear that agonist binding is very sensitive to the ions present (Birdsall et al., 1979b; Hulme et al., 1983; Burgen, 1986, and the above references) and to the precise incubation conditions. To date, no consensus has emerged regarding appropriate methods for detecting selectivity in agonist binding studies.

3.3. Modulation of the Agonist Binding Properties by Divalent Cations and Nucleotides

The reader may have noted some variability in our terminology regarding the nature of the species found in agonist binding studies. They have been called "components," "sites," and "states." This reflects a deficiency in our understanding of precisely how agonist binding heterogeneity arises and the complexity of the system(s).

An advance in our understanding came from the discovery that muscarinic agonist binding in the myocardium was decreased by GTP and related guanine nucleotides (Berrie et al., 1979a; Rosenberger et al., 1979; Wei and Sulakhe, 1979). This finding has been confirmed by many other groups, and guanine nucleotide regulation of muscarinic agonist binding has been found in the majority of the tissues where muscarinic binding sites have been detected. Some exceptions are the conduction tissue of the bovine heart (Burgen et al., 1981), a population of binding sites in the cerebral cortex that have a low affinity for pirenzepine (Birdsall et al., 1984), ciliary muscle (Barany et al., 1982), and the male adenohypophysis (Avissar and Sokolovsky, 1981). There is also one report that GTP can increase agonist binding (Avissar and Sokolovsky, 1981).

At about the same time or slightly before the GTP effects were described, it was found that divalent ions could selectively enhance agonist and not antagonist binding (Aronstam et al.,

1978; Birdsall et al., 1979b; Gurwitz and Sokolovsky, 1980; Wei and Sulakhe, 1980a; Hulme et al., 1981b). In many instances, the presence of divalent ions such as Mg^{2+} or Mn^{2+} was found to be necessary for a GTP effect on agonist binding to be observed.

The dual regulation of agonist binding by divalent ions and guanine nucleotides is a characteristic, originally reported for the glucagon receptor (Rodbell et al., 1971), of a large superfamily of receptors. These receptors can interact with G-proteins, a family of heterotrimeric GTP binding proteins that mediate receptor–response coupling (for a brief review, *see* Gilman 1985). Divalent ions are thought to promote receptor-G-protein coupling in the absence of GTP. When GTP is present, this coupling is disrupted and the G-protein activated. The binding of an agonist to the receptor facilitates the binding of GTP to the G-protein in a catalytic manner by lowering the activation energy for GTP binding.

The effect of Mg^{2+} and GTP on carbachol binding in the myocardium is illustrated in Fig. 4. The agonist binding curves have been analyzed in terms of a three-site model (SH, H, and L states of the receptor). In the absence of divalent cations, H states are predominant. The effect of Mg^{2+} (1 m*M*) is to decrease the IC_{50} by increasing the proportion of the SH state. Subsequent addition of GTP produces a 40-fold increase in the IC_{50} of carbachol, and most of the binding sites are of the L type. The extent of these modulations is very variable, and depends not only on the tissue from which the membranes are prepared, but also on the method of preparation of the membranes. For example, larger effects are observed if the membranes are treated with EDTA (Hulme et al., 1983) or, in some instances, by the presence of a sulphydryl reducing agent (Halvorsen and Nathanson, 1984).

The regulation of agonist binding by divalent cations is a complex process. At low concentrations, agonist binding is enhanced, but at higher concentrations it is decreased (Hulme et al., 1983). The shape of the dose–effect curves depends on the metal ion examined. For example, Mg^{2+} has a comparable or greater enhancing effects vis-a-vis equivalent concentrations of Ca^{2+}, whereas Ca^{2+} has a more pronounced inhibitory effect at higher concentrations (>1 m*M*) (Hulme et al., 1983). Furthermore, the effect of guanine nucleotides is strongly potentiated by the presence of low concentrations of free Mg^{2+}, but less so by Ca^{2+}

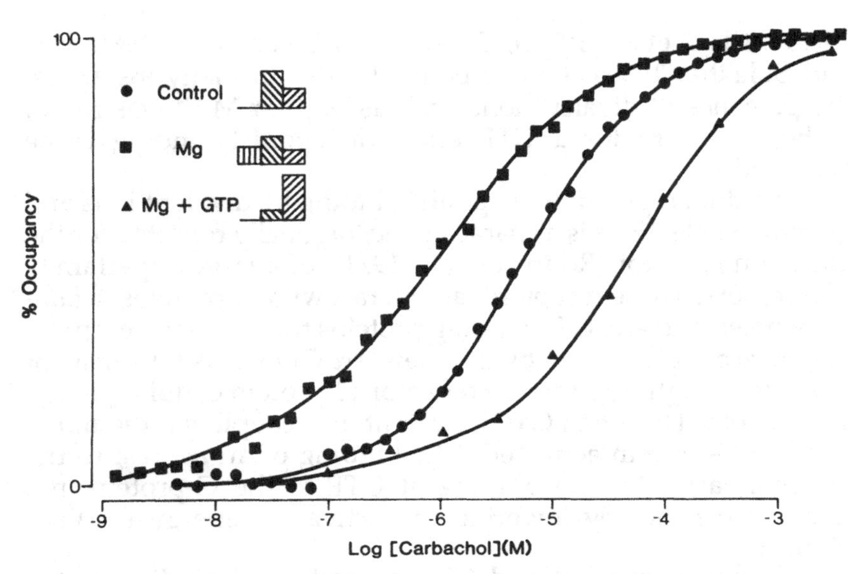

Fig. 4. Modulation of the agonist binding properties of myocardial muscarinic receptors by Mg^{2+} and GTP. Occupancy concentration curves for the binding of carbachol were generated from [^{3}H]NMS–carbachol competition experiments at 30°C and are corrected for the receptor occupancy by the ^{3}H-ligand. The incubation medium was 100 m*M* NaCl/20 m*M* HEPES pH 7.4 (●) with added Mg^{2+} (1 m*M*) (■) or Mg^{2+} (1 m*M*) + GTP (10^{-4}*M*) (▲). The inset shows the estimated proportions of SH, H, and L sites obtained by a nonlinear, least squares fitting procedure applied to the three sets of data [reproduced from Birdsall et al. (1984) with permission].

(Hulme et al., 1983). This effect of Mg^{2+} may be mimicked by Mn^{2+}, Co^{2+}, and Ni^{2+} (Gurwitz and Sokolovsky, 1980). In examining the detailed characteristics of the modulatory effects of ions and nucleotides, it should be noted that potential artifacts are, first, the retention of substantial amounts of Mg^{2+} and Ca^{2+} bound to atrial membrane prepared in the absence of high concentrations of EDTA (Wei and Sulakhe, 1980b), and secondly, the high-affinity agonist binding states of the receptor are not always stable under the assay conditions (Waelbroeck et al., 1982; Aronstam and Greenbaum, 1984).

It appears that micromolar concentrations of GppNHp (a hydrolysis-resistant analog of GTP) will not inhibit muscarinic agonist binding unless (1) there are SH and H sites for the GppNHp to act on and (2) a sufficiently high concentration of free Mg^{2+} is available to allow the nucleotide to bind with high affinity (Hulme et al., 1983). In some tissues such as the heart, it appears that, especially at low temperatures, SH and H states may be formed in the absence of *added* divalent ions, whereas in other tissues exogenous divalent ions are required. In many instances, it has been convenient to monitor these effects on agonist binding by using ^{3}H-agonists to label selectively the highest affinity state of the receptor (e.g., Waelbroeck et al., 1982; Hulme et al., 1983; Harden et al., 1983; Vickroy et al., 1984; Nukada et al., 1983a; Gurwitz et al., 1984b, 1985; Kellar et al., 1985; Gillard et al., 1987; Galper et al., 1987). One surprising feature to emerge from these studies is that, whereas ^{3}H-agonist binding to myocardial receptors is almost totally abolished by guanine nucleotides, only 25–60% of the binding of ^{3}H-agonists to the SH state of the cortical and hippocampal receptors can be inhibited (Hulme et al., 1983; Nukada et al., 1983a; Vickroy et al., 1983; Gurwitz et al., 1984b, 1985; Kellar et al., 1985; Gillard et al., 1987). There seems to be a population of binding sites in these tissues that are "locked" into a high affinity binding state for agonists, and do not respond to Mg^{2+} and GTP. Our studies have suggested that, in the cerebral cortex, these sites have a low affinity for pirenzepine (Birdsall et al., 1984). The explanation of this locking process is not known, but it may involve sulphydryl groups on the receptor (Gurwitz et al., 1984a; Haga et al., 1988). It should be noted that agonist binding heterogeneity has been found in purified porcine cortical receptor preparations where no G-protein is present (Haga et al., 1988). It therefore appears that muscarinic receptors are intrinsically capable of existing in different conformational states, even in the absence of G-proteins.

3.4. Models for Agonist Binding Heterogeneity

The original studies found that a multi-site fit to the agonist binding data (generally a three-site fit, with precise data at many concentrations) provided a self-consistent description of the binding data (Hulme et al., 1978; Birdsall et al., 1980c), certainly as far as the SH and L sites were concerned. The H site has always been somewhat enigmatic, since there has been no way of measuring binding to this site, independent of the other sites; it is a product of computer derived curve fitting. Gurwitz et al. (1985) and Sokolovsky et al. (1986) have suggested that a two-site model provides an adequate description of the binding of [^{3}H]-ACh, whereas oxotremorine-*M* and cis-methyldioxolane can distinguish between SH and H sites. Other workers have found that a three-site model is required to fit ACh-[^{3}H]-antagonist competition curves adequately (Birdsall et al., 1980c; McMahon and Hosey, 1985; Burgen, 1987).

Most of the detailed modeling/analysis has been carried out on myocardial receptors because of the apparent homogeneity of the receptors and the large effects of GTP. Burgen (1987) analyzed the binding of 11 agonists to rat cardiac muscarinic receptors and found that three components (SH, H, and L) were required to describe the binding. The proportion of the H component was found to be a constant (~40%), independent of the agonist examined. In contrast, the proportions of SH and L sites varied in a complementary manner depending on the agonist. Under these incubation conditions, the maximum proportion of the SH component was 36%, and the minimum value 2%. The effect of GppNHp was to reduce the H population by 63%, and about 85% of the total receptors were of the L type. The affinity constants of the H and L receptor populations were not affected by GppNHp. Analogous results have been reported by McMahon and Hosey (1985). The mathematics of the analysis suggests that the apparent conversions are from SH → L and H → L, rather than being explained by SH → H′ and H → L, where H′ and H are not identical species. This latter mechanism has been proposed by Mizushima et al. (1987) on the basis of the effects of GppNHp on carbachol binding after the L population of sites had been selectively alkylated by propylbenzilylcholine mustard, the SH and H sites being protected by carbachol. However, the carbachol affinity constants reported for the SH and H sites are very much lower (20-fold) than those found by other workers (e.g., Burgen, 1987).

All these multi-site analyses suffer somewhat in the sense that there is no mechanistic interpretation of the data. Clearly, the system is complicated involving, *as a minimum*, agonist, receptor, G-protein, Mg^{2+}, and Mg–GTP. It is also not known how many of the receptor states interconvert under specific experimental conditions. Ehlert (1985) was able to explain the receptor binding properties of a series of muscarinic agonists and their inhibitory effects on adenylate cyclase in rabbit myocardial membranes, using a ternary complex model (Jacobs and Cuatrecasas, 1976; Wreggett and De Lean, 1984). This model (Scheme I) assumes the receptor, R, and the G-protein, G, are free to move

SCHEME I

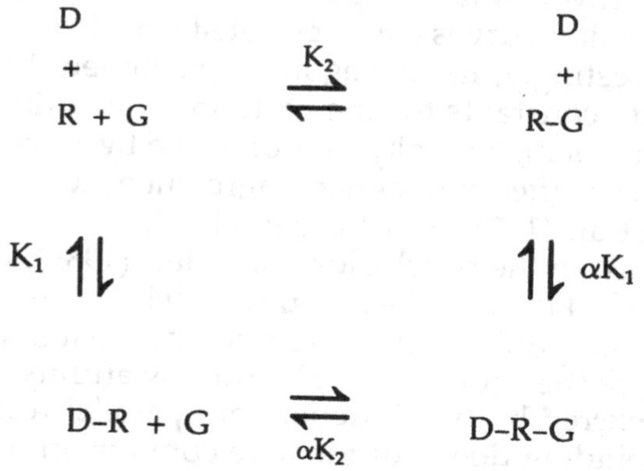

$$
\begin{array}{ccc}
\begin{array}{c} D \\ + \\ R + G \end{array} & \underset{}{\overset{K_2}{\rightleftharpoons}} & \begin{array}{c} D \\ + \\ R\text{–}G \end{array} \\
K_1 \updownarrow & & \updownarrow \alpha K_1 \\
D\text{–}R + G & \underset{\alpha K_2}{\rightleftharpoons} & D\text{–}R\text{–}G
\end{array}
$$

Ternary complex model for the interaction of a muscarinic drug, D, with a receptor, R, and G-protein, G.

laterally in the plane of the membrane and interact to form a complex, R–G. A muscarinic agonist, D, has an allosteric effect on equilibrium involving R, G, and R–G, indicated by a cooperativity factor, α. The low affinity state of the receptor, found in the presence of GTP, may be equated with the D–R complex formed by dissociation of the D–R–G–GTP complex and activation of the G-protein (as G*–GTP). The high affinity state of the receptor is a manifestation of the formation of the D–R–G complex, but its microscopic affinity constant, αK_1, does not necessarily equal the high affinity constant obtained by a multi-site analysis of the data. Since this scheme specifies interconversion of all complexes, a heterogeneous agonist binding curve is only predicted if [G] < [R]. Ehlert (1985) estimated that the ratio $[G]_{total}/[R]_{total}$ was 0.8. The "local" concentration of G-proteins in the membrane implied by this analysis is 60–240-fold less than the macroscopic density of G-proteins found in bovine heart (Sternweis and Robishaw, 1984). If this scheme is correct, it suggests that only a small pool of G-proteins is able to interact with muscarinic receptors. In order to fit the binding and functional data, it was necessary to assume that very little of the receptor was precoupled to the G-protein in the absence of agonist ($K_2 < 10^{-2}$). However, under different conditions, the receptor may be precoupled to G. If $\alpha << 1$, heterogeneous antagonist binding and GTP effects on antagonist binding are predicted, and these are found experimentally (Hulme et al., 1980, 1981a; Burgisser et al.,

1982; Hosey, 1982). It should be noted that the rabbit myocardial receptor system examined by Ehlert (1985) was apparently simpler than myocardial receptors from other species, since the agonist inhibition curves could be satisfactorily fit to a two-site model. Interestingly, for the agonists examined, both the ratio of the affinity constants for the high and low affinity sites and the proportion of high affinity sites obtained by a two-site fit were related to their efficacy in general agreement with the findings of Birdsall et al. (1978) and Burgen (1987).

In contrast to the conclusions of Ehlert (1985), Galper et al. (1987) presented data on the kinetics of [^{3}H]cis-methyldioxolane binding to embryonic chick heart receptors that suggested that the binding of the agonist to high and low affinity forms of the receptor occurred by two independent parallel reactions. That is, agonist binding does not mediate conversion of the low affinity receptor state to the high affinity state: in other words, the high affinity (coupled) form of the receptor preexists. They also suggested that the conversion by GTP of the high affinity form to low affinity form involved an intermediate agonist–receptor–G-protein–GTP complex, and the order of binding of agonist and GTP during formation of this intermediate was random.

The ternary complex model has been examined in great detail in two papers (Lee et al., 1986 and Wong et al., 1986). They concluded that there were very considerable constraints on the parameters in the scheme that would give rise to heterogeneous agonist binding curves and, furthermore, that only certain combinations of parameters obtained by a two-site fit to the binding curve would be compatible with a ternary model. They concluded that the available data on β-adrenergic receptors (Kent et al., 1980) fit the model, but not the data on D_2 dopamine receptors (Sibley and Creese, 1983) or their data on hamster myocardial muscarinic receptors (Wong et al., 1986). Caution should therefore be applied to the precise interpretation of parametric values obtained from a multisite model or a ternary complex model.

3.5. Modulation of the Agonist Binding Properties by Other Agents

In the previous sections, the allosteric effects of divalent cations and guanine nucleotides on agonist binding have been considered in terms of changes in receptor–G protein coupling. There is also evidence from a large number of studies that various agents, particularly protein modification reagents, will also affect agonist binding and/or coupling to G-proteins. Only a selected number of these agents will be discussed.

The most extensively studied reagent is *N*-ethylmaleimide (NEM). This sulphydryl reagent was first shown by Aronstam et al. (1977) to increase the binding of carbachol to forebrain muscarinic receptors selectively without affecting antagonist binding. This result was confirmed by Carson (1980), Ehlert et al. (1980b), and in several more recent studies. The effect has been interpreted in terms of an interconversion of L to H states (Aronstam et al., 1977) or by a selective increase in affinity of the L sites without interconversion of sites/states. Antagonists do not inhibit the rate of alkylation, but interestingly, agonists enhance the rate of alkylation by NEM and can decrease by tenfold the concentration at which NEM is effective in increasing agonist binding (Aronstam et al., 1977; Vauquelin et al., 1982). This suggests that agonists induce a conformational change in the receptor system and thereby alter the reactivity of a sulphydryl group. When the SH affinity state of the receptor was labeled by [^{3}H]cis-methyldioxolane, Vickroy et al. (1983) observed an increase in radioligand binding to cerebral cortical receptors at high concentrations of NEM (1 m*M*). However, Nukada et al. (1983b) demonstrated that the picture was more complicated. They found that, in the porcine caudate nucleus, [^{3}H]cismethyldioxolane binding was *decreased* at low concentrations of NEM (10^{-5}–$10^{-4}M$) and *increased* at higher concentrations (10^{-4}–$10^{-3}M$). The guanine nucleotide (and divalent cation) sensitive binding of [^{3}H]cis-methyldioxolane was abolished at low concentrations of NEM, whereas the guanine nucleotide insensitive binding component was enhanced at high concentrations.

At cardiac receptors (or "cardiac-type" receptors present in the hindbrain), NEM abolishes the effects of Mg^{2+} and GTP on agonist binding (Wei and Sulakhe, 1980b; Harden et al., 1982; McMahon and Hosey, 1983, 1985). The effects of GTP are abolished by 1 m*M* NEM, and result in "steeper" and less potent inhibition curves for agonists such as acetylcholine or carbachol. Higher concentrations of NEM (~10 m*M*) increase agonist binding and abolish the effects of Mg^{2+} on agonist binding. Harden et al. (1983) observed that [^{3}H]oxotremorine-*M* binding to rat myocardial receptors was inhibited by 20–50 μM NEM. The effect was to decrease the apparent number of SH sites and not the affinity of the sites.

It seems that there are at least two effects of NEM on brain and cardiac receptors. At low concentrations, receptor–G-protein interactions are disrupted, quite possibly by the alkylation of the G-protein, whereas at higher concentrations NEM can increase the affinity of several agonists.

DTNB [5,5′-dithiobis (2-nitrobenzoic acid)] increases agonist binding in the heart (Uchida et al., 1984; Mizushima et al., 1987) and smooth muscle (Uchida et al., 1982). The effect has been reported to be similar to that of NEM in that the GTP effect is abolished, at least in smooth muscle, and the number of SH binding sites are increased. Analogous effects are also found for diamide (Gurwitz et al., 1984a), Cu^{2+} ions (Gurwitz et al., 1984a) and possibly Cd^{2+} (Hedlund and Bartfai, 1979).

The action of another sulphydryl reagent, *p*-chloromercuribenzoate (PCMB), has also been investigated in some detail (Aronstam et al., 1978; Birdsall et al., 1983a,b). Associated with muscarinic receptor systems in rat cerebral cortex are at least three classes of sulphydryl groups that are modified by this reagent (Birdsall et al., 1983b,c). At low concentrations, PCMB produces a selective change in the binding of agonists. This is qualitatively similar to the effect of NEM in that the effect cannot be inhibited by agonists or antagonists, but for some agonists, decreases in affinity are observed. The structure binding relationships of both the H and L sites, observed under "noncoupling conditions" (i.e., no added Mg^{2+}) are changed (Birdsall et al., 1983c). At higher concentrations of PCMB, antagonist and agonist binding is affected. The structure–binding relationships of antagonists have been altered: the affinity of some antagonists is increased, whereas that of other antagonists is decreased over 100-fold. The binding of agonists is greatly decreased and binding heterogeneity eliminated. In fact, the binding differences between the cortical and cardiac receptors seem to have been eliminated. These two effects of PCMB can be inhibited by agonists and antagonists and reversed by dithiothreitol, which in itself has little if any effect on the binding properties of muscarinic receptors in the cerebral cortex. It was also found that PCMB could inactivate the binding sites (Aronstam et al., 1978; Birdsall et al., 1983b), an effect that could be inhibited by antagonists and partially reversed by dithiothreitol. The binding sites with a high affinity for pirenzepine (and a low affinity for agonists under these conditions) were lost selectively. It appears that the interaction of PCMB with these sulphydryl groups, which are probably not present within the binding domain of the receptor, produce different conformational modifications of the ligand binding site.

Tetranitromethane, a reagent that nitrates tyrosine residues, has also been shown to selectively increase high affinity acetylcholine binding to forebrain muscarinic receptors (Gurwitz and Sokolovsky, 1985). This effect was reported to be blocked by antagonists, but not by agonists. Interestingly, neither the binding of oxotremorine nor of oxotremorine-*M* was affected. This reagent is worthy of further investigation, particularly now that the sequences of muscarinic receptor subtypes are becoming available.

The interaction of monovalent ions, particularly Na^+, with muscarinic receptors has been investigated. The impetus for these studies was the finding that the inhibition of adenylate cyclase, mediated by muscarinic (Lichtshtein et al., 1979; Jakobs et al., 1979) and other receptors (Limbird, 1981), was dependent on the presence of monovalent ions. For α_2-adrenergic receptors, Na^+ has been shown to selectively decrease agonist binding, and this effect persists in soluble receptor preparations (Limbird et al., 1982). An equivalent selective effect of Na^+ has been reported by Rosenberger et al. (1980b) for rat heart muscarinic receptors. Other studies in rat brain (Birdsall et al., 1979) and heart (Wei and Sulakhe, 1980; Hulme et al., 1981b; Harden et al., 1983) have not detected this phenomenon. However, a selective effect of NH_4^+ on agonist binding to chick cardiac receptors has been reported (Hosey, 1983; McMahon and Hosey, 1983, 1985), but no functional correlates of this phenomenon have been reported. In conduction tissue of the bovine heart, Burgen et al. (1981) observed that agonist binding was almost unaffected by GTP, but was selectively modulated by K^+ ions. This finding is enigmatic in that the effector system in the conduction tissues of the heart is a K^+ channel (Soejima and Noma, 1984), but the coupling to the K^+ channel is thought to be mediated by a G-protein (Pfaffinger et al., 1985; Breitwieser and Szabo, 1985; Yatani et al., 1987; Codina et al., 1987), and yet little if any effect of GppNHp on agonist binding in this tissue was observed (Burgen et al., 1981).

Higher concentrations of monovalent ions (0.3–2*M*) have been reported to decrease agonist and antagonist binding and eliminate agonist binding heterogeneity in the cerebral cortex and heart (Berrie et al., 1979b; Hosey, 1983; Burgen, 1986). This approach probably involves the uncoupling of receptor–effector complexes to form the L state, maybe even of those sites with high affinity for agonists that are not uncoupled by GTP. A small selective inhibitory effect of fluoride ions (?AlF_4^-) on agonist binding to mouse cardiac muscarinic receptors has also been reported (Barritt et al., 1982). The significance of this phenomenon in terms of receptor–G-protein coupling has not been investigated.

There have been several reports that ligands that bind to other receptors or effector systems may modulate muscarinic agonist binding. These include ligands that act at β-adrenergic receptors (Rosenberger et al., 1980a), VIP receptors (Lundberg et al., 1982), and the voltage-dependent Na^+ channel (Cohen-Armon et al., 1985a,b, 1988). There are other reports in which antagonist binding is affected by drugs that act at dopamine receptors (Ehlert et al., 1981), oestrogen receptors (Sokolovsky et al., 1981; Ben-Baruch et al., 1982), calcium channels (Sokolovsky et al., 1986), and K^+ channels (Lai et al., 1985). The mechanistic implications for the effects on agonist binding in terms of agonist–receptor complexes competing for limited pools of G-proteins or subcomponents (i.e., βγ subunits) have been discussed elsewhere (Birdsall, 1982). It is much more difficult to explain indirect interactions affecting antagonist binding unless the drugs are binding to the receptor (maybe to the "gallamine" site).

3.6. Soluble Receptors

The majority of the earlier studies on the agonist binding properties of soluble muscarinic receptors indicated that the binding of potent agonists was of low affinity when the detergent used was digitonin (Hurko, 1978; Gorissen et al., 1978), digitonin/cholate (Herron et al., 1982; McMahon and Hosey, 1985), cholate (Carson, 1982; Haga et al., 1982), or CHAPS (Gavish and Sokolovsky, 1982). Other studies have shown that, in some preparations from bovine, rabbit, and rat brain, agonist binding is heterogeneous (Wenger et al., 1985) and sensitive (Kuno et al., 1983; Flynn and Potter, 1986) or insensitive to guanine nucleotides (Luthin and Wolfe, 1985). Regional variations in the agonist binding properties of soluble muscarinic receptors have also been reported (Wenger et al., 1985; Flynn and Potter, 1986).

The finding that the formation of the nucleotide-sensitive high affinity agonist binding state of myocardial muscarinic receptors was favored by low temperature, low ionic strength, and the presence of divalent cations such as Mg^{2+} and Mn^{2+} (Hulme et al., 1983; Potter et al., 1984) suggested that these were the conditions that might favor solubilization of a receptor–G-protein complex. Under these conditions, it has been possible to solubilize myocardial and cerebral cortical muscarinic receptors prelabeled with [^{3}H]oxotremorine-*M* using digitonin-cholate (Harden et al., 1983), digitonin (Berrie et al., 1984), and CHAPSO (Poyner et al., 1987). In some instances, it has also been possible to solubilize a high affinity agonist binding state of muscarinic receptors in the absence of an agonist and to label this state of the receptor with [^{3}H]oxotremorine-*M after* solubilization (Berrie et al., 1984, 1986; Poyner et al., 1987). In other words, agonist binding to the receptor may not be a prerequisite for formation of a receptor–G-protein complex. The binding of [^{3}H]oxotremorine-*M* is of high affinity ($K_D \sim 10^{-9}M$) and is sensitive to guanine nucleotides: up to 60% of the soluble myocardial receptors can be in this state. The effect of guanine nucleotides is to dramatically decrease the affinity of oxotremorine-*M*; most of the sites have an affinity 30,000-fold lower than that found for the [^{3}H]oxotremorine-*M*-labeled sites in the absence of guanine nucleotide. In contrast, the affinity of the [^{3}H]oxotremorine-*M*-labeled sites in CHAPSO is only decreased 100-fold by GppNHp (Poyner et al., 1987). The site labeled with [^{3}H]oxotremorine-*M* has a higher sedimentation coefficient than the sites labeled by a ^{3}H-antagonist in the presence of GppNHp, and probably represents a complex between the ligand binding subunit of the receptor and a G-protein (Berrie et al., 1984).

Purified receptor preparations from porcine heart, forebrain, and rat forebrain have also been reported to have heterogeneous agonist binding properties (Peterson et al., 1984; Schimerlik et al., 1986; Haga and Haga, 1985; Haga et al., 1988; Berrie et al., 1986), despite the fact that the initial solubilized species did not show any heterogeneity (Cremo et al., 1981; Berrie et al., 1985b). Even in the absence of a G-protein, muscarinic receptors seem to be intrinsically capable of existing in different conformational states.

4. General Conclusions

Binding studies have made major contributions to our current understanding of muscarinic receptor subclasses and their functions. It is clear that the binding properties of these receptors are very sensitive to the incubation conditions, e.g., temperature, ionic composition of the incubation medium, and membrane environment. Hence, great care needs to be taken in the design of the experiments and interpretation of the data, especially with regard to the linkage between binding and function.

References

Alberts, P. and Bartfai, T. (1976) Muscarinic acetylcholine receptor from rat brain: partial purification and characterization. *J. Biol. Chem.* **251**, 1543–1547.

Aronstam, R. S. and Greenbaum, L. M. (1984) Guanine nucleotide sensitivity of muscarinic acetylcholine receptors from rat brainstem is eliminated by endogenous proteolytic activity. *Neurosci. Lett.* **47**, 131–137.

Aronstam, R. S., Abood, L. G., and Hoss, W. (1978) Influence of sulfhydryl reagents and heavy metals on the functional state of the muscarinic acetylcholine receptor in rat brain. *Mol. Pharmacol.* **14**, 575–586.

Aronstam, R. S., Hoss, W., and Abood, L. G. (1977) Conversion between configurational states of the muscarinic receptor in rat brain. *Eur. J. Pharmacol.* **46**, 279–282.

Aronstam, R. S., Schuessler, D. C., Jr., and Eldefrawi, M. E. (1978) Solubilization of muscarinic acetylcholine receptors of bovine brain. *Life Sci.* **23**, 1377–1382.

Avissar, S. and Sokolovsky, M. (1981) Guanine nucleotides preferentially inhibit binding of antagonists (male) and agonists (female) to muscarinic receptors of rat adenohypophysis. *Biochem. Biophys. Res. Comm.* **102**, 753–760.

Barany, E., Berrie, C. P., Birdsall, N. J. M., Burgen, A. S. V., and Hulme, E. C. (1982) The binding properties of the muscarinic receptors of the cynomolgous monkey ciliary body and the response to the induction of agonist subsensitivity. *Brit. J. Pharmacol.* **77**, 731–739.

Barlow, R. B., Berry, K. J., Glenton, P. A. M., Nikolau, N. M., and Soh, K. S. (1976) A comparison of affinity constants for muscarine-sensitive acetylcholine receptors in guinea-pig atrial pacemaker cells at 29°C and in ileum at 29°C and 37°C. *Brit. J. Pharmacol.* **58**, 613–620.

Barlow, R. B., Birdsall, N. J. M., and Hulme, E. C. (1979) Temperature coefficients of affinity constants for the binding of antagonists to muscarinic receptors in the rat cerebral cortex. *Brit. J. Pharmacol.* **66**, 587–590.

Barlow, R. B., Caulfield, M. P., Kitchen, R., Roberts, P. M., and Stubley, J. K. (1981) The affinities of pirenzepine and atropine for functional muscarinic receptors in guinea-pig atria and ileum. *Brit. J. Pharmacol.* **73**, 183–184.

Barritt, D., Yamamura, H. I., and Roeske, W. R. (1982) Muscarinic receptor binding in mouse heart: the selective modulatory effect of fluoride ion on agonist binding. *Life Sci.* **30**, 875–877.

Baumgold, J., Merril, C., and Gershon, E. S. (1987) Loss of pirenzepine regional selectivity following solubilization and partial purification of the putative M_1 and M_2 muscarinic receptor subtypes. *Mol. Brain Res.* **2**, 7–14.

Beld, A. J. and Ariens, E. J. (1974) Stereospecific binding as a tool in attempts to localize and isolate muscarinic receptors. 2. Binding of (±)-benzetimide, (−)-benzetimide and atropine to a fraction from bovine tracheal smooth-muscle and to bovine caudate-nucleus. *Eur. J. Pharmacol.* **25**, 203–209.

Ben-Baruch, G., Schreiber, G., and Sokolovsky, M. (1982) Cooperativity pattern in the interaction of the antiestrogen drug clomiphene with the muscarinic receptors. *Mol. Pharmacol.* **21**, 287–293.

Berrie, C. P., Birdsall, N. J. M., Burgen, A. S. V., and Hulme, E. C. (1979a) Guanine nucleotides modulate muscarinic receptor binding in the heart. *Biochem. Biophys. Res. Comm.* **87**, 1000–1004.

Berrie, C. P., Birdsall, N. J. M., Burgen, A. S. V., and Hulme, E. C. (1979b) Ionic perturbation of agonist binding to brain muscarinic receptors. *Brit. J. Pharmacol.* **66**, 470–471.

Berrie, C. P., Birdsall, N. J. M., Dadi, H. K., Hulme, E. C., Morris, R. J., Stockton, J. M., and Wheatley, M. (1985a) Purification of the muscarinic acetylcholine receptor from rat forebrain. *Trans. Biochem. Soc.* **13**, 1101–1103.

Berrie, C. P., Birdsall, N. J M., Hulme, E. C., Keen, M., and Stockton, J. M. (1985b) Solubilization and characterization of high and low affinity pirenzepine binding sites from rat cerebral cortex. *Brit. J. Pharmacol.* **85**, 697–703.

Berrie, C. P., Birdsall, N. J. M., Hulme, E. C., Keen, M., and Stockton, J. M. (1984) Solubilization and characterization of guanine nucleotide sensitive muscarinic agonist binding sites from rat myocardium. *Brit. J. Pharmacol.* **82**, 853–861.

Berrie, C. P., Birdsall, N. J. M., Hulme, E. C., Keen, M., Stockton, J. M., and Wheatley, M. (1986) Muscarinic receptor subclasses: the binding properties of the soluble receptor sites. *Trends Pharmacol. Sci.* **Suppl. II**, 8–13.

Bers, D. M. (1979) Isolation and characterization of cardiac sarcolemma. *Biochim. Biophys. Acta* **555**, 131–146.

Birdsall, N. J. M. (1982) Can different receptors interact directly with each other? *Trends Neurosci.* **5**, 137.

Birdsall, N. J. M. and Hulme, E. C. (1976) Biochemical studies of muscarinic acetylcholine receptors. *J. Neurochem.* **27**, 7–14.
Birdsall, N. J. M. and Hulme, E. C. (1983) Muscarinic receptor subclasses. *Trends Pharmacol. Sci.* **4**, 459–463.
Birdsall, N. J. M. and Hulme, E. C. (1985) Multiple muscarinic receptors: further problems in receptor classification. *Trends Autonom. Pharmacol.* **3**, 17–34.
Birdsall, N. J. M. and Hulme, E. C. (1986) Differences in the agonist binding properties of muscarinic receptor subpopulations in the rat cerebral cortex and myocardium, in *Basic and Therapeutic Strategies in Alzheimer's Disease and Other Age-related Neuropsychiatric Disorders* (Fisher, A., ed.) Plenum, N.Y. 565–574.
Birdsall, N. J. M., Burgen, A. S. V., and Hulme, E. C. (1978) The binding of agonists to brain muscarinic receptors. *Molec. Pharmacol.* **14**, 723–736.
Birdsall, N. J. M., Burgen, A. S. V., and Hulme, E. C. (1979a) A study of the muscarinic receptor by gel electrophoresis. *Brit. J. Pharmacol.* **66**, 337–342.
Birdsall, N. J. M., Burgen, A. S. V., Hulme, E. C., and Wells, J. W. (1979b) The effect of ions on the binding of agonists and antagonists to muscarinic receptors. *Brit. J. Pharmacol.* **67**, 371–377.
Birdsall, N. J. M., Burgen, A. S. V., Hammer, R., Hulme, E. C., and Stockton, J. M. (1980a) Pirenzepine—a ligand with original binding properties to muscarinic receptors. *Scand. J. Gastroenterol.* **Suppl. 66**, 1–4.
Birdsall, N. J. M., Burgen, A. S. V., and Hulme, E. C. (1980b) The binding properties of muscarinic receptors in the brain of the frog (*R. Temporaria*). *Brain Res.* **184**, 385–393.
Birdsall, N. J. M., Hulme, E. C., and Burgen, A. S. V. (1980c) The character of the muscarinic receptors in different regions of the rat brain. *Proc. R. Soc. Lond. B.* **207**, 1–12.
Birdsall, N. J. M., Burgen, A. S. V., Hulme, E. C., and Stockton, J. M. (1981) Gallamine regulates muscarinic receptors in the heart and cerebral cortex. *Brit. J. Pharmacol.* **74**, 798P.
Birdsall, N. J. M., Burgen, A. S. V., Hulme, E. C., Stockton, J. M., and Zigmond, M. J. (1983a) The effect of McN-A-343 on muscarinic receptors in the cerebral cortex and heart. *Brit. J. Pharmacol.* **78**, 257–259.
Birdsall, N. J. M., Burgen, A. S. V., Hulme, E. C., and Wong, E. H. F. (1983b) The effects of *p*-chloromercuribenzoate on muscarinic receptors in the cerebral cortex. *Brit. J. Pharmacol.* **80**, 187–196.
Birdsall, N. J. M., Burgen, A. S. V., Hulme, E. C., and Wong, E. H. F. (1983c) The effect of *p*-chloromercuribenzoate on the structure-binding relationships of muscarinic receptors in the rat cerebral cortex. *Brit. J. Pharmacol.* **80**, 197–204.
Birdsall, N. J. M., Hulme, E. C., and Keen, M. (1986a) The binding

of pirenzepine to digitonin-solubilized muscarinic acetylcholine receptors from the rat myocardium. *Brit. J. Pharmacol.* **87**, 307–316.

Birdsall, N. J. M., Hulme, E. C., Kromer, W., Peck, B. S., Stockton, J. M., and Zigmond, M. J. (1986b) Two drug binding sites on muscarinic receptors, in *New Concepts in Alzheimer's Disease* (Briley, M., Kato, A., and Weber, M., eds.) MacMillan, 103–121.

Birdsall, N. J. M., Hulme, E. C., Keen, M., Pedder, E. K., Poyner, D., Stockton, J. M., and Wheatley, M. (1987a) Soluble and membrane-bound muscarinic acetylcholine receptors. *Biochem. Soc. Symp.* **52**, 23–32.

Birdsall, N. J. M., Hulme, E. C., Keen, M., Stockton, J. M., Pedder, E. K., and Wheatley, M. (1987b) Can complex binding phenomena be resolved to provide a safe basis for receptor classification? in *Perspectives on Receptor Classification* (Black, J. W., Jenkinson, D. H., and Gerskowitch, V. P., eds.) Liss, N.Y., pp. 61–69.

Birdsall, N. J. M., Hulme, E. C., Kromer, W., and Stockton, J. M. (1987c) A second drug binding site on muscarinic receptors. *Fed. Proc.* **46**, 2525–2527.

Birdsall, N. J. M., Hulme, E. C., and Stockton, J. M. (1984) Muscarinic receptor heterogeneity. *Trends Pharmacol. Sci.* **Suppl. I**, 4–8.

Bonner, T. I., Buckley, N. J., Young, A. C., and Brann, M. R. (1987) Identification of a family of muscarinic acetylcholine receptor genes. *Science* **237**, 527–532.

Boyer, J. L., Martinez-Carcamo, M., Monroy-Sanchez, J. A., Posadas, C., and Garcia-Sainz, J. A. (1986) Guanine nucleotide-induced positive cooperativity in muscarinic cholinergic antagonist binding. *Biochem. Biophys. Res. Commun.* **134**, 172–177.

Breitwieser, G. E. and Szabo, G. (1985) Uncoupling of cardiac muscarinic and β-adrenergic receptors from ion channels by a guanine nucleotide analogue. *Nature* **317**, 538–540.

Brown, D. A., Forward, A., and Marsh, S. (1980) Antagonist discrimination between ganglionic and ileal muscarinic receptors. *Brit. J. Pharmacol.* **71**, 362–364.

Brown, J. H. and Goldstein, D. (1986) Analysis of cardiac muscarinic receptors recognized selectively by nonquaternary but not by quaternary ligands. *J. Pharmacol. Exp. Ther.* **238**, 580–586.

Brunner, F., Waelbroeck, M., and Christophe, J. (1986) Secoverine is a nonselective muscarinic antagonist on rat heart and brain receptors. *Eur. J. Pharmacol.* **127**, 17–25.

Burgen, A. S. V. (1986) The effect of ionic strength on cardiac muscarinic receptors. *Brit. J. Pharmacol.* **88**, 451–455.

Burgen, A. S. V. (1987) The effects of agonists on the components of the cardiac muscarinic receptor. *Brit. J. Pharmacol.* **92**, 327–332.

Burgen, A. S. V. and Hiley, C. R. (1974) Two populations of acetylcholine receptors in guinea-pig ileum. *Brit. J. Pharmacol.* **51**, 127P.

Burgen, A. S. V., Hiley, C. R., and Young, J. M. (1974) The properties of muscarinic receptors in mammalian cerebral cortex. *Brit. J. Pharmacol.* **51**, 279–285.

Burgen, A. S. V., Hulme, E. C., Berrie, C. P., and Birdsall, N. J. M. (1981) The nature of the muscarinic receptors in the heart, in *Cell Membrane in Function and Dysfunction of Vascular Tissue* (Godfraind, T. and Meyer, P., eds.) Elsevier/North Holland Biomedical Press, Holland, pp. 15–25.

Burgermeister, W., Klein, W. L., Nirenberg, M., and Witkop, B. (1978) Comparative binding studies with cholinergic ligands and histrionicotoxin at muscarinic receptors of neural cell lines. *Mol. Pharmacol.* **14**, 751–767.

Burgisser, E., De Lean, A., and Lefkowitz, R. J. (1982) Reciprocal modulation of agonist and antagonist binding to muscarinic cholinergic receptors by guanine nucleotides. *Proc. Natl. Acad. Sci. USA* **79**, 1732–1736.

Burke, R. E. (1986) Gallamine binding to muscarinic M_1 and M_2 receptors, studied by inhibition of [^{3}H]pirenzepine and [^{3}H]quinuclidinylbenzilate binding to rat brain membranes. *Mol. Pharmacol.* **30**, 58–68.

Carson, S. (1980) Differential effect of *N*-ethylmaleimide on muscarinic agonist binding in rat and bovine brain membranes. *FEBS Lett.* **109**, 81–84.

Carson, S. (1982) Cholate-salt solubilization of bovine brain muscarinic receptors. *Biochem. Pharmacol.* **31**, 1806–1809.

Carson, S., Goodwin, S., Massoulie, J., and Kato, G. (1977) Solubilisation of atropine-binding material from brain. *Nature* **266**, 176–178.

Cheng, Y. C. and Prusoff, W. H. (1973) Relationship between the inhibition constant (K_i) and the concentration of inhibitor which causes 50 percent inhibition (I_{50}) of an enzymatic reaction. *Biochem. Pharmacol.* **22**, 3099–3108.

Choo, L. K. and Mitchelson, F. J. (1985) Comparison of the affinity constant of some muscarinic receptor antagonists with their displacement of [^{3}H]quinuclidinyl benzilate binding in atrial and ileal longitudinal muscle of the guinea-pig. *J. Pharmac. Pharmacol.* **37**, 656–658.

Choo, L. K., Leung, E., and Mitchelson, F. (1985a) Failure of gallamine and pancuronium to inhibit selectively (−)-[^{3}H] quinuclidinyl benzilate binding in guinea pig atria. *Can. J. Physiol. Pharmacol.* **63**, 200–208.

Choo, L. K., Mitchelson, F., and Vong, Y. M. (1985b) The interaction of McN-A-343 with pirenzepine and other selective muscarinic receptor antagonists at a prejunctional muscarinic receptor. *Naunyn-Schmiedeberg's Arch. Pharmacol.* **328**, 430–438.

Christophe, J., de Neef, P., Robberecht, P., and Waelbroeck, M. (1986) Propylbenzilylcholine mustard is selective for rat heart muscarinic receptors having a low affinity for agonists. *Brit. J. Pharmacol.* **88**, 63–70.

Clark, A. L. and Mitchelson, F. (1976) The inhibitory effect of gallamine on muscarinic receptors. *Brit. J. Pharmacol.* **58**, 323–331.

Closse, A., Bittiger, H., Langenegger, D., and Wanner, A. (1987) Binding studies with [^{3}H]-cis-methyldioxolane in different tissues. *Naunyn-Schmiedeberg's Arch. Pharmacol.* **335**, 372–377.

Codina, J., Yatani, A., Grenet, D., Brown, A. M., and Birnbaumer, L. (1987) The α-subunit of the GTP binding protein G_K opens atrial potassium channels. *Science* **236**, 442–445.

Cohen-Armon, M., Garty, H., and Sokolovsky, M. (1988) G-protein mediates voltage regulation of agonist binding to muscarinic receptors: effects on receptor–Na^+–channel interaction. *Biochemistry* **27**, 368–374.

Cohen-Armon, M., Henis, Y. I., Kloog, Y., and Sokolovsky, M. (1985a) Interactions of quinidine and lidocaine with rat brain and heart muscarinic receptors. *Biochem. Biophys. Res. Commun.* **127**, 326–332.

Cohen-Armon, M., Kloog, Y., Henis, Y. I., and Sokolovsky, M. (1985b) Batrachotoxin changes the properties of the muscarinic receptor from rat brain and heart: possible interaction(s) between muscarinic receptors and sodium channels. *Proc. Natl. Acad. Sci. USA* **82**, 3524–3527.

Cremo, C. R., Herron, G. S., and Schimerlik, M. I. (1981) Solubilisation of the atrial muscarinic acetylcholine receptor: a new detergent system and rapid assays. *Anal. Biochem.* **115**, 331–338.

Cuatrecasas, P. and Hollenberg, M. D. (1976) Membrane receptors and hormone action. *Adv. Protein Chem.* **30**, 251–451.

Curtis, C. A. M., Wheatley, M., Bansal, S., Birdsall, N. J. M., Eveleigh, P., Pedder, E. K., Poyner, D., and Hulme, E. C. (1989) Propylbenzilylcholine mustard labels an acidic residue in transmembrane helix 3 of the muscarinic receptor. *J. Biol. Chem.* **264**, 489–495.

Dowdall, M. J., Strange, P. G., and Golds, P. R. (1983) Muscarinic acetylcholine receptors in *Torpedo* electric organ: effect of guanine nucleotides. *J. Neurochem.* **41**, 556–561.

Dunlap, J. and Brown, J. H. (1983) Heterogeneity of binding sites on cardiac muscarinic receptors induced by the neuromuscular blocking agents gallamine and pancuronium. *Molec. Pharmacol.* **24**, 15–22.

Ehlert, F. (1985) The relationship between muscarinic receptor occupancy and adenylate cyclase inhibition in the rabbit myocardium. *Mol. Pharmacol.* **28**, 410–421.

Ehlert, F. J. and Jenden, D. J. (1985) The binding of a 2-choroethylamine derivative of oxotremorine (BM 123) to muscarinic receptors in the rat cerebral cortex. *Mol. Pharmacol.* **28**, 107–119.

Ehlert, F. J., Dumont, Y., Roeske, W. R., and Yamamura, H. I. (1980a) Muscarinic receptor binding in rat brain using the agonist [^{3}H]-cis-methyldioxolane. *Life Sci.* **26**, 961–967.

Ehlert, F. J., Roeske, W. R., and Yamamura, H. I. (1980b) Regulation of muscarinic receptor binding by guanine nucleotides and N-ethylmaleimide. *J. Supramolec. Struct.* **14**, 149–162.

Ehlert, F. J., Roeske, W. R., and Yamamura, H. I. (1980) Striatal muscarinic receptors: regulation by dopaminergic agonists. *Life Sci.* **28**, 2441–2448.

El-Fakahany, E. E., Ramkumar, V., and Lai, W. S. (1986) Multiple binding affinities of *N*-methylscopolamine to brain muscarinic acetylcholine receptors: differentiation from M_1 and M_2 receptor subtypes. *J. Pharmacol. Exp. Ther.* **238**, 554–563.

Elliot, J. M., Taylor, P. J., and Young, J. M. (1978) Changes in muscarinic ligand binding to intestinal muscle strips produced by pre-exposure to hypotonic conditions. *Brit. J. Pharmacol.* **30**, 27–35.

Ellis, J. and Hoss, W. (1982) Competitive interaction of gallamine with multiple muscarinic receptors. *Biochem. Pharmacol.* **31**, 873–876.

Ellis, J. and Lenox, R. H. (1985) Characterisation of the interactions of gallamine with muscarinic receptors from brain. *Biochem. Pharmacol.* **34**, 2214–2217.

Eltze, M., Gonne, S., Riedel, R., Schlotke, B., Schudt, C., and Simon, W. A. (1985) Pharmacological evidence for selective inhibition of gastric acid secretion by telenzepine, a new antimuscarinic drug. *Eur. J. Pharmacol.* **112**, 211–224.

Ensing, K. and Zeeuw, R. A. (1986) Different behavior toward muscarinic receptor binding between quaternary anticholinergics and their tertiary analogues. *Pharmac. Res.* **3**, 327–332.

Evans, T., Smith, M. M., Tanner, L. I., and Harden, T. K. (1984) Muscarinic cholinergic receptors of two cell lines that regulate cyclic AMP metabolism by different molecular mechanisms. *Mol. Pharmacol.* **26**, 395–404.

Feigenbaum, P. and El-Fakahany, E. E. (1985) Regulation of muscarinic cholinergic receptor density in neuroblastoma cells by brief exposure to agonist: Possible involvement in desensitization of receptor function. *J. Pharmacol. Exp. Therapeut.* **233**, 134–140.

Fisher, S. K. (1988) Recognition of muscarinic cholinergic receptors in human SK-N-SH neuroblastoma cells by quaternary and tertiary ligands is dependent upon temperature, cell integrity and the presence of agonists. *Molec. Pharmacol.* **33**, 414–422.

Flynn, D. D. and Potter, L. T. (1986) Effect of solubilization on the distinct binding properties of muscarinic receptors from rabbit hippocampus and brain stem. *Mol. Pharmacol.* **30**, 193–199.

Galper, J. B., Klein, W., and Catterall, W. A. (1977) Muscarinic acetylcholine receptors in developing chick heart. *J. Biol. Chem.* **252**, 8692–8699.

Galper, J. B., Aziekan, L. C., O'Hara, D. S., and Smith, T. W. (1982) The biphasic response of muscarinic cholinergic receptors in cultured heart cells to agonists: Effects on receptor number and affinity in intact cells and homogenates. *J. Biol. Chem.* **257**, 10344–10356.

Galper, J. B., Haigh, L. S., Hart, A. C., O'Hara, D. S., and Livingston, D. J. (1987) Muscarinic cholinergic receptors in the embryonic chick heart: Interaction of agonist receptor and guanine nucleotides studied by an improved assay for direct binding of the muscarinic agonist [^{3}H]cismethyldioxolane. *Mol. Pharmacol.* **32**, 230–240.

Gavish, M. and Sokolovsky, M. (1982) Solubilisation of muscarinic acetylcholine receptor by zwitterionic detergent from rat brain cortex. *Biochem. Biophys. Res. Comm.* **109**, 819–824.

Giachetti, A., Giraldo, E., Ladinsky, H., and Montagna, E. (1986a) Binding and functional properties of the selective M_1 muscarinic receptor antagonists trihexyphenidyl and dicyclomine. *Brit. J. Pharmacol.* **89**, 83–90.

Giachetti, A., Micheletti, R., and Montagna, E. (1986b) Cardioselective profile of AF-DX 116, a muscarine M_2 receptor antagonist. *Life Sci.* **38**, 1663–1672.

Gibson, R. E., Rzeszotarski, W. J., Jagoda, E. M., Francis, B. E., Reba, R. C., and Eckelman, W. C. (1984) [^{125}I] 3-Quinuclidinyl 4-iodobenzilate: a high affinity high specific activity radioligand for the M_1 and M_2 acetylcholine receptors. *Life Sci.* **34**, 2287–2296.

Gies, J.-P., Ilien, B., and Laduron, P. (1986) Muscarinic acetylcholine receptor: thermodynamic analysis of the interaction of agonists and antagonists. *Biochim. Biophys. Acta* **889**, 103–115.

Gillard, M., Waelbroeck, M., and Christophe, J. (1987) Muscarinic receptor heterogeneity in rat central nervous system II. Brain receptors labelled by [^{3}H]oxotremorine-*M* correspond to heterogeneous M_2 receptors with very high affinity for agonists. *Mol. Pharmacol.* **32**, 100–108.

Gilman, A. G. (1985) G proteins and dual control of adenylate cyclase. *Cell* **36**, 577–579.

Giraldo, E., Hammer, R., and Ladinsky, H. (1987a) Distribution of muscarinic receptor subtypes in rat brain as determined in binding studies with AF-DX 116 and pirenzepine. *Life Sci.* **40**, 833–840.

Giraldo, E., Monferini, E., Ladinsky, H., and Hammer, R. (1987b) Muscarinic receptor heterogeneity in guinea pig intestinal smooth muscle: binding studies with AF-DX 116. *Eur. J. Pharmacol.* **141**, 475–477.

Gorissen, H., Aerts, G., and Laduron, P. (1978) Characterisation of digitonin-solubilised muscarinic receptor from rat brain. *FEBS Lett.* **96**, 64–68.

Gossuin, A., Maloteaux, J.-M., Trout, A., and Laduron, P. (1984) Differentiation between ligand trapping into intact cells and binding on muscarinic receptors. *Biochim. et Biophys. Acta.* **804**, 100–106.

Goyal, R. K. and Rattan, S. (1978) Neurohumoral, hormonal and drug receptors for the lower esophageal sphincter. *Progress in Gastroenterology* **75**, 598–619.

Gurwitz, D. and Sokolovsky, M. (1980) Agonist-specific reverse regulation of muscarinic receptors by transition metal ions and guanine nucleotides. *Biochem. Biophys. Res. Comm.* **96**, 1296–1305.

Gurwitz, D. and Sokolovsky, M. (1985) Increased agonist affinity is induced by tetranitromethane-modified muscarinic receptors. *Biochemistry* **24**, 8086–8093.

Gurwitz, D., Baron, B., and Sokolovsky, M. (1984a) Copper ions and diamide induce a high affinity guanine nucleotide-insensitive state for muscarinic receptors. *Biochem. Biophys. Res. Commun.* **120**, 271–277.

Gurwitz, D., Kloog, Y., and Sokolovsky, M. (1984b) Recognition of the muscarinic receptor by its endogenous neurotransmitter: binding of [^{3}H]acetylcholine and its modulation by transition metal ions and guanine nucleotides. *Proc. Natl. Acad. Sci. USA* **81**, 3650–3654.

Gurwitz, D., Kloog, Y., and Sokolovsky, M. (1983) Saturable [^{3}H]acetylcholine binding to rat cerebral cortex muscarinic receptors: increased binding induced by transition metals is reversed by Gpp(NH)p. *J. Neurochem.* **41**, S140.

Gurwitz, D., Kloog, Y., and Sokolovsky, M. (1985) High affinity binding of [^{3}H]acetylcholine to muscarinic receptors: regional distribution and modulation by guanine nucleotides. *Mol. Pharmacol.* **28**, 297–305.

Haga, K. and Haga, T. (1985) Purification of the muscarinic acetylcholine receptor from porcine brain. *J. Biol. Chem.* **260**, 7927–7935.

Haga, T. (1980) Solubilisation of muscarinic acetylcholine receptors by *L*-α-lysophosphatidylcholine. *Biomed. Res.* **1**, 265–268.

Haga, T. (1982) Characterization of muscarinic acetylcholine receptors solubilized by *L*-α-lysophosphatidylcholine and Lubrol PX, in *Pharmacologic and Biochemical Aspects of Neurotransmitter Receptors* (Yoshida, H. and Yamamura, H., eds.) John Wiley and Sons, Chichester, 43–58.

Haga, T., Nukada, T., and Haga, K. (1982) Solubilization of the muscarinic acetylcholine receptor by sodium cholate: stabilization of the receptor by muscarinic ligands. *Biomed. Res.* **3**, 695–698.

Haga, T., Haga, K., Berstein, G., Nishiyama, T., Uchiyama, H., and Ichiyama, A. (1988) Molecular properties of muscarinic receptors. *Trends Pharmacol. Sci.* **Suppl. III**, 12–18.

Halvorsen, S. W. and Nathanson, N. M. (1984) Ontogenesis of physiological responsiveness and guanine nucleotide sensitivity of cardiac muscarinic receptors during chick embryonic development. *Biochem.* **23**, 5813–5821.

Hammer, R. and Giachetti, A. (1982) Muscarinic receptor subtypes: M_1 and M_2 biochemical and functional characterization. *Life Sci.* **31**, 2991–2998.

Hammer, R. and Giachetti, A. (1984) Selective muscarinic receptor antagonists. *Trends Pharmacol. Sci.* **5**, 18–20.

Hammer, R., Berrie, C. P., Birdsall, N. J. M., Burgen, A. S. V., and Hulme, E. C. (1980) Pirenzepine distinguishes between subclasses of muscarinic receptor. *Nature (Lond.)* **283**, 90–92.

Hammer, R., Giraldo, E., Schiavi, G. B., Monferini, E., and Ladinsky, H. (1986a) Binding profile of a novel cardioselective muscarinic receptor antagonist, AF-DX 116, to membranes of peripheral tissues and brain in the rat. *Life Sci.* **38**, 1653–1662.

Hammer, R., Ladinsky, H., and De Conti, L. (1986b) *In vivo* labelling of peripheral muscarinic receptors. *Trends Pharmacol. Sci.* **Suppl. II**, 33–38.

Harden, T. K., Meeker, R. B., and Martin, M. W. (1983) Interaction of a radiolabelled agonist with cardiac muscarinic cholinergic receptors. *J. Pharm. Exp. Ther.* **227**, 570–577.

Harden, T. K., Scheer, A. G., and Smith, M. M. (1982) Differential modification of the interaction of cardiac muscarinic cholinergic and beta-adrenergic receptors with a guanine nucleotide binding component(s). *Mol. Pharmacol.* **21**, 570–580.

Harden, T. K., Petch, L. A., Travnelis, S. F., and Waldo, G. L. (1985) Agonist-induced alteration in the membrane form of muscarinic cholinergic receptors. *J. Biol. Chem.* **260**, 13060–13066.

Hedlund, B. and Bartfai, T. (1979) The importance of thiol- and disulphide groups in agonist and antagonist binding to the muscarinic receptor. *Mol. Pharmacol.* **15**, 531–544.

Herron, G. S., Miller, S., Manley, W.-L., and Schimerlik, M. I. (1982) Ligand interactions with the solubilized porcine atrial muscarinic receptor. *Biochem.* **21**, 515–520.

Hirschowitz, B. I., Hammer, R., Giachetti, A., Kierns, J. J., and Levine, R. R. (1984) Subtypes of muscarinic receptors. *Trends Pharmacol. Sci.* **Suppl. I**, 1–103.

Hosey, M. M. (1982) Regulation of antagonist binding to cardiac muscarinic receptors. *Biochem. Biophys. Res. Com.* **107**, 314–321.

Hosey, M. M. (1983) Regulation of ligand binding to cardiac muscarinic receptors by ammonium ion and guanine nucleotides. *Biochim. Biophys. Acta.* **757**, 119–127.

Hoss, W. and Messer, W. (1986) Multiple muscarinic receptors in the CNS: significance and prospects for future research. *Biochem. Pharmacol.* **35**, 3895–3901.

Hulme, E. C., Burgen, A. S. V., and Birdsall, N. J. M. (1975) Interactions of agonists and antagonists with the muscarinic receptor. *INSERM* **50**, 49–50.

Hulme, E. C., Berrie, C. P., Birdsall, N. J. M., and Burgen, A. S. V. (1981a) Two populations of binding sites for muscarinic antagonists in the rat heart. *Eur. J. Pharmacol.* **73**, 137–142.

Hulme, E. C., Berrie, C. P., Birdsall, N. J. M., and Burgen, A. S. V. (1981b) Interactions of muscarinic receptors with guanine nucleotides and adenylate cyclase, in *Drug Receptors and Their Effectors* (Birdsall,

N. J. M., ed.) Macmillan, London, 23–24.

Hulme, E. C., Birdsall, N. J. M., Berrie, C. P., and Burgen, A. S. V. (1980) Muscarinic receptor binding: modulation by nucleotides and ions, in *Neurotransmitters and Their Receptors* (Littauer, U. Z., Dudai, Y., Silman, I., Teichberg, V. I., and Vogel, Z., eds.) John Wiley and Sons, Chichester, pp. 241–250.

Hulme, E. C., Birdsall, N. J. M., Burgen, A. S. V., and Mehta, P. (1978) The binding of antagonists to brain muscarinic receptors. *Molec. Pharmacol.* **14**, 737–750.

Hulme, E. C., Berrie, C. P., Birdsall, N. J. M., Jameson, M., and Stockton, J. M. (1983) Regulation of muscarinic agonist binding by cations and guanine nucleotides. *Eur. J. Pharmacol.* **94**, 59–72.

Hurko, O. (1978) Specific [^{3}H]-quinuclidinyl benzilate binding activity in digitonin-solubilised preparations from bovine brain. *Arch. Biochem. Biophys.* **190**, 434–445.

Jacobs, S. and Cuatrecasas, P. (1976) The mobile receptor hypothesis and "cooperativity" of hormone binding, application to insulin. *Biochem. Biophys. Acta.* **443**, 482–495.

Jakobs, K.-H., Aktories, K., and Schultz, G. (1979) GTP-dependent inhibition of cardiac adenylate cyclase by muscarinic cholinergic agonists. *Naunyn-Schmiedebergs Arch. Pharmac.* **310**, 113–119.

Janski, A. M., Cornell, N. W., and Yeh, H. J. P. (1980) Preparation of water-soluble digitonin that retains its ability to complex with cholesterol. *Fedn. Proc.* **39**, 1721.

Jarv, J., Hedlund, B., and Bartfai, T. (1979) Isomerization of the muscarinic receptor-antagonist complex. *J. Biol. Chem.* **254**, 5595–5598.

Kellar, K. J., Martino, A. M., Hall, D. P., Schwartz, R. D., and Taylor, R. L. (1985) High affinity binding of [^{3}H]acetylcholine to muscarinic cholinergic receptors. *J. Neurosci.* **5**, 1577–1582.

Kent, R. S., De Lean, A., and Lefkowitz, R. J. (1980) A quantitative analysis of *Beta*-adrenergic receptor interactions: resolution of high and low affinity states of the receptor by computer modelling of ligand binding data. *Mol. Pharmacol.* **17**, 14–23.

Kloog, Y. and Sokolovsky, M. (1978) Studies on muscarinic acetylcholine receptors from mouse brain: characterization of the interaction with antagonists. *Brain Res.* **144**, 31–48.

Kloog, Y., Egozi, Y., and Sokolovsky, M. (1979a) Regional heterogeneity of muscarinic receptors of mouse brain. *FEBS Lett.* **97**, 265–268.

Kloog, Y., Egozi, Y., and Sokolovsky, M. (1979b) Characterisation of muscarinic acetylcholine receptors from mouse brain: evidence for regional heterogeneity and isomerization. *Mol. Pharmacol.* **15**, 545–558.

Kubo, T., Fukuda, K., Mikami, A., Maeda, A., Takahashi, H., Mishina, M., Haga, T., Haga, K., Ichiyama, A., Kangawa, K., Kojima, M., Matsuo, H., and Numa, S. (1986a) Cloning, sequencing and expression of the complementary DNA encoding the muscarinic acetylcholine receptor. *Nature* **323**, 411–416.

Kubo, T., Maeda, A., Sugimoto, K., Akiba, I., Mikami, A., Takahashi, H., Haga, T., Haga, K., Ichiyama, A., Kangawa, K., Matsuo, H., Hirose, T., and Numa, S. (1986b) Primary structure of porcine cardiac muscarinic acetylcholine receptor deduced from the cDNA sequence. *FEBS Lett.* **209**, 367–372.

Kuno, T., Shirakawa, O., and Tanaka, C. (1983) Regulation of the solubilized bovine cerebral cortex muscarinic receptor by GTP and Na^+. *Biochem. Biophys. Res. Comm.* **112**, 948–953.

Laduron, P. M., Leysen, J. E., and Gorissen, H. (1981) Muscarinic receptor: multiple sites or unitary concept? *Arch. Int. Pharmacodyn.* **249**, 319–321.

Laduron, P. M., Verwimp, M., and Leysen, J. E. (1979) Stereospecific *in vitro* binding of [^{3}H]-dexetimide to brain muscarinic receptors. *J. Neurochem.* **32**, 421–427.

Lai, W. S., Ramkumar, V., and El-Fakahany, E. E. (1985) Possible allosteric interaction of 4-aminopyridine with rat brain muscarinic acetylcholine receptors. *J. Neurochem.* **44**, 1936–1942.

Lambrecht, G., Mutschler, E., Moser, U., Riotte, J., Wagner, M., and Wess, J. (1987) Heterogeneity in muscarinic receptors: evidence from pharmacological and electrophysiological studies with selective antagonists, in *Muscarinic Cholinergic Mechanisms* (Cohen, S. and Sokolovsky, M., eds.) Freund, Tel Aviv/London, pp. 245–253.

Lee, J.-H. and El-Fakahany, E. E. (1985a) Anomalous binding of [^{3}H]-*N*-methylscopolamine to rat brain muscarinic receptors. *Eur. J. Pharmacol.* **110**, 263–266.

Lee, J.-H. and El-Fakahany, E. E. (1985b) Heterogeneity of binding of muscarinic receptor antagonists in rat brain homogenates. *J. Pharmacol. Exp. Ther.* **233**, 707–714.

Lee, T. W. T., Sole, M. J., and Wells, J. W. (1986) Assessment of a ternary model for the binding of agonists to neurohumoral receptors. *Biochemistry* **25**, 7009–7020.

Levine, R. R., Birdsall, N. J. M., Giachetti, A., Hammer, R., Iversen, L. L., Jenden, D. J., and North, R. A. (eds.) (1986) Subtypes of muscarinic receptors. *Trends Pharmacol. Sci.* **Suppl. II**, 1–97.

Levine, R. R., Birdsall, N. J. M., North, R. A., Holman, M., Watanabe, A., and Iversen, L. L. (eds.) (1988) Subtypes of muscarinic receptors. *Trends Pharmacol. Sci.* **Suppl. III**, 1–92.

Lichtshtein, D., Boone, G., and Blume, A. (1979) Muscarinic receptor regulation of NG-108-15 adenylate cyclase: requirement for Na^+ and GTP. *J. Cyclic Nucleotide Res.* **5**, 367–375.

Limbird, L. E. (1981) Activation and attenuation of adenylate cyclase. *Biochem. J.* **195**, 1–13.

Limbird, L. E., Speck, J. L., and Smith, S. K. (1982) Sodium ion modulates agonist and antagonist interactions with the human platelet alpha$_2$-adrenergic receptor in membrane and solubilized preparations. *Mol. Pharmacol.* **21**, 609–617.

Lundberg, J. M., Hedlund, B., and Bartfai, T. (1982) Vasoactive intestinal polypeptide enhances muscarinic ligand binding in cat submandibular salivary gland. *Nature* **295**, 147–149.

Luthin, G. R. and Wolfe, B. B. (1984) Comparison of ^{3}H-pirenzepine and ^{3}H-quinuclidinylbenzilate binding to muscarinic cholinergic receptors in rat brain. *J. Pharmacol. Exp. Ther.* **228**, 648–655.

Luthin, G. R. and Wolfe, B. B. (1985) Characterization of [^{3}H]pirenzepine binding to muscarinic cholinergic receptors solubilized from rat brain. *J. Pharmacol. Exp. Ther.* **234**, 39–44.

McKinney, M., Stengstrom, S., and Richelson, E. (1985) Muscarinic responses and binding in a murine neuroblastoma clone. *Mol. Pharmacol.* **27**, 223–235.

McMahon, K. M. and Hosey, M. M. (1983) Potentiation of monovalent cation effects on ligand binding to cardiac muscarinic receptors in *N*-ethylmaleimide treated membranes. *Biochem. Biophys. Res. Comm.* **111**, 41–46.

McMahon, K. M. and Hosey, M. M. (1985) Agonist interactions with cardiac muscarinic receptors: effects of Mg^{2+}, guanine nucleotides and monovalent cations. *Mol. Pharmacol.* **28**, 400–409.

Maloteaux, J.-M. Gossuin, A., Pauwels, P. J., and Laduron, P. M. (1983) Short-term disappearance of muscarinic cell surface receptors in carbachol-induced desensitization. *FEBS Lett.* **156**, 103–107.

Marchi, M. and Raiteri, M. (1985) Differential antagonism by dicyclomine, pirenzepine and secoverine at muscarinic receptor subtypes in the rat frontal cortex. *Eur. J. Pharmacol.* **107**, 287–288.

Masters, S. B., Quinn, M. T., and Brown, J. H. (1985) Agonist induced densensitization of muscarinic receptor-mediated ion efflux without concomitant desensitization of phosphoinositide hydrolysis. *Mol. Pharmacol.* **27**, 325–332.

Mattera, R., Pitts, B. J. R., Entman, M. L., and Birnbaumer, L. (1985) Guanine nucleotide regulation of a mammalian myocardial muscarinic receptor system. *J. Biol. Chem.* **260**, 7410–7421.

Mei, L., Wang, J.-X., Roeske, W. R., and Yamamura, H. I. (1987) Thermodynamic analyses of pirenzepine binding to membrane-bound and solubilized muscarinic receptors from rat forebrain and heart. *J. Pharmacol. Exp. Ther.* **242**, 991–1000.

Melchiorre, C., Cassinelli, A., and Quaglia, W. (1987) Differential blockade of muscarinic receptor subtypes by polymethylene tetramines. Novel class of selective antagonists of cardiac M-2 muscarinic receptors. *J. Med. Chem.* **30**, 201–204.

Michel, A. D. and Whiting, R. L. (1988) Direct binding studies on ileal and cardiac muscarinic receptors. *Brit. J. Pharmacol.* **92**, 755–767.

Mitchelson, F. (1984) Heterogeneity in muscarinic receptors: evidence from pharmacologic studies with antagonists. *Trends Pharmacol. Sci.* **Suppl. I**, 12–16.

Mizushima, A., Uchida, S., Zhou, X.-M., Kagiya, T., and Yoshida, H. (1987) Cardiac M_2 receptors consist of two different types, both regulated by GTP. *Eur. J. Pharmacol.* **135**, 403–409.

Murphy, K. M. M. and Sastre, A. (1983) Obligatory role of a tris/choline allosteric site in guanine nucleotide regulation of [^{3}H]-*L*-QNB binding to muscarinic acetylcholine receptors. *Biochem. Biophys. Res. Comm.* **113**, 280–285.

Mutschler, E. and Lambrecht, G. (1984) Selective muscarinic agonists and antagonists in functional tests. *Trends Pharmacol. Sci.* **Suppl. I**, 39–44.

Nathanson, N. M. (1983) Binding of agonists and antagonists to muscarinic acetylcholine receptors on intact cultured heart cells. *J. Neurochem.* **41**, 1545–1549.

Nathanson, N. M. (1987) Molecular properties of the muscarinic acetylcholine receptor. *Ann. Rev. Neurosci.* **10**, 195–236.

Nedoma, J., Dorofeeva, N. A., Tucek, S., Shelkovnikov, S. A., and Danilov, A. F. (1985) Interaction of the neuromuscular blocking drugs alcuronium, decamethonium, gallamine, pancuronium, ritebronium, tercuronium and *d*-tubocurarine with muscarinic acetylcholine receptors in the heart and ileum. *Naunyn-Schmiedeberg's Arch. Pharmacol.* **3239**, 176–181.

Newberry, N. R. and Priestley, T. (1987) Pharmacological differences between two muscarinic responses of the rat superior cervical ganglion in vitro. *Brit. J. Pharmacol.* **92**, 817–826.

Newberry, N. R., Priestley, T., and Woodruff, G. N. (1985) Pharmacological distinction between muscarinic responses on the isolated superior cervical ganglion of the rat. *Eur. J.Pharmacol.* **116**, 191–192.

Nilvebrandt, L. and Sparf, B. (1983) Differences between binding affinities of some antimuscarinic drugs in the parotid gland and those in the urinary bladder and ileum. *Acta Pharmacol. Toxicol.* **52**, 30–38.

Nilvebrandt, L. and Sparf, B. (1986) Dicyclomine, benzhexol and oxybutynin distinguish between subclasses of muscarinic binding sites. *Eur. J. Pharmacol.* **123**, 133–143.

Norman, A. B., Blaker, S. N., Thal, L., and Creese, I. (1986) Effects of aging and cholinergic deafferentation on putative muscarinic cholinergic receptor subtypes in rat cerebral cortex. *Neurosci. Lett.* **70**, 289–294.

Nukada, T., Haga, T., and Ichiyama, A. (1983a) Muscarinic receptors in porcine caudate nucleus. I. Enhancement by nickel and other cations of [^{3}H]-cis-methyldioxolane binding to guanyl nucleotide-sensitive sites. *Mol. Pharmacol.* **24**, 366–373.

Nukada, T., Haga, T., and Ichiyama, A. (1983b) Muscarinic receptors in porcine caudate nucleus. II. Different effects of *N*-ethylmaleimide on [^{3}H]cis-methyldioxolane binding to heat labile (guanyl nucleotide-sensitive) sites and heat-stable (guanyl nucleotide-insensitive sites). *Mol. Pharmacol.* **24**, 374–379.

Paton, W. D. M. and Rang, H. P. (1965) The uptake of atropine and related drugs by intestinal smooth muscle of the guinea-pig in relation to acetylcholine receptors. *Proc. R. Soc. London* **163B**, 1–44.

Peralta, E. G., Winslow, J. W., Peterson, G. L., Smith, D. H., Ashkenazi, A., Ramachandran, J., Schimerlik, M. I., and Capon, D. J. (1987) Primary structure and biochemical properties of an M_2 muscarinic receptor. *Science* **236**, 600–605.

Peterson, G. L., Herron, G. S., Yamaki, M., Fullerton, D. S., and Schimerlik, M. J. (1984) Purification of the muscarinic acetylcholine receptor from porcine atria. *Proc. Natl. Acad. Sci. USA* **81**, 4993–4997.

Pfaffinger, P. J., Martin, J. M., Hunter, D. D., Nathanson, N. M., and Hille, B. (1985) GTP-binding proteins couple cardiac muscarinic receptors to a K^+ channel. *Nature* **317**, 536–538.

Potter, L. T., Flynn, D. D., Hanchett, H. E., Kalinowski, D. L., Luber-Narod, J., and Mash, D. C. (1984) Independent M_1 and M_2 receptors: ligands, autoradiography and functions. *Trends Pharmacol. Sci.* **Suppl. I**, 22–31.

Poyner, D., Pedder, E. K., Eveleigh, P., Birdsall, N. J. M., Hulme, E. C., Stockton, J. M., Curtis, C., and Wheatley, M. (1987) Solubilisation of the muscarinic receptor in CHAPSO: a comparison with the receptor in membranes and in digitonin solution. *Protides of the Biological Fluids* **35**, 181–184.

Repke, H. and Matthies, H. (1980a) Biochemical characterization of solubilized muscarinic acetylcholine receptors. *Brain Res. Bull.* **5**, 703–709.

Repke, H. and Matthies, H. (1980b) Preparative separation of digitonin and gitonin and characterization of its detergent properties. *Pharmazie* **35**, 233–234.

Riker, W. F. and Wescoe, W. C. (1951) The pharmacology of flaxedil with observations on certain analogs. *Ann. N.Y. Acad. Sci.* **54**, 373–394.

Rodbell, M., Krans, H. M. J., Pohl, S. L., and Birnbaumer, L. (1971) The glucagon-sensitive adenyl cyclase system in plasma membranes of rat liver. IV. Effects of guanyl nucleotides on binding of ^{125}I-glucagon. *J. Biol. Chem.* **246**, 1872–1876.

Roeske, W. R. and Venter, J. C. (1984) The differential loss of ^{3}H-pirenzepine vs ^{3}H-quinuclidinylbenzilate binding to soluble rat brain muscarinic receptors indicates that pirenzepine binds to an allosteric state of the muscarinic receptor. *Biochem. Biophys. Res. Commun.* **48**, 950–957.

Roffel, A. F., in't Hout, W. G., de Zeeuw, R. A., and Zaagsma, J. (1987) The M2 selective antagonist AF-DX 116 shows high affinity for muscarine receptors in bovine tracheal membranes. *Naunyn-Schmiedeberg's Arch. Pharmacol.* **335**, 593–595.

Rosenberger, L. B., Roeske, W. R., and Yamamura, H. I. (1979) The regulation of muscarinic cholinergic receptors by guanine nucleotides in cardiac tissue. *Eur. J. Pharmacol.* **56**, 179–180.

Rosenberger, L. B., Yamamura, H. I., and Roeske, W. R. (1980a) The regulation of cardiac muscarinic cholinergic receptors by isoproterenol. *Eur. J. Pharmacol.* **65**, 129–130.

Rosenberger, L. B., Yamamura, H. I., and Roeske, W. R. (1980b) Cardiac muscarinic cholinergic receptor binding is regulated by Na^+ and guanyl nucleotides. *J. Biol. Chem.* **255**, 570–577.

Roszkowski, A. P. (1961) An unusual type of sympathetic ganglionic stimulant. *J. Pharmacol. Exp. Ther.* **132**, 156–170.

Rotter, A., Birdsall, N. J. M., Burgen, A. S. V., Field, P. M., Hulme, E. C., and Raisman, G. (1979) Muscarinic receptors in the central nervous system of the rat. I. Technique for autoradiographic localisation of the binding of [^{3}H]propylbenzilylcholine mustard and its distribution in the forebrain. *Brain Res. Bull.* **1**, 141–165.

Schimerlik, M. I., Miller, S., Peterson, G. L., Rosenbaum, L. C., and Tota, M. R. (1986) Biochemical studies on muscarinic receptors in porcine atrium. *Trends Pharmacol. Sci.* **Suppl. II**, 2–7.

Schreiber, G., Henis, Y. I., and Sokolovsky, M. (1985) Analysis of ligand binding to receptors by competition kinetics. *J. Biol. Chem.* **260**, 8789–8794.

Schudt, C., Auriga, C., Birdsall, N. J. M., and Boer, R. (1988a) Are M1 selective antimuscarinics capable of prolonged protection of the receptor from released acetylcholine? *Pharmacology,* **37**, 32–39.

Schudt, C., Auriga, C., Kinder, B., and Birdsall, N. J. M. (1988b) The binding of [^{3}H]-telenzepine to muscarinic acetylcholine receptors in calf brain. *Eur. J. Pharmacol.*, **145**, 87–90.

Sibley, D. R. and Creese, I. (1983) Interactions of ergot alkaloids with anterior pituitary D-2 dopamine receptors. *Mol. Pharmacol.* **23**, 585–593.

Siegel, R. E. and Fischbach, G. D. (1984) Muscarinic receptors and responses in intact embryonic chick atrial and ventricular heart cells. *Dev. Biol.* **101**, 346–356.

Silva, W. I., Andres, A., Schook, W., and Puszkin, S. (1986) Evidence for the presence of muscarinic acetylcholine receptors in bovine brain coated vesicles. *J. Biol. Chem.* **261**, 14788–14796.

Soejima, M. and Noma, A. (1984) Mode of regulation of the ACh-sensitive K-channel by the muscarinic receptor in rabbit atrial cells. *Pflugers Arch.* **400**, 424–431.

Sokolovsky, M. (1984) Muscarinic receptors in the central nervous system. *Int. Rev. Neurobiol.* **25**, 139–183.

Sokolovsky, M., Egozi, Y., and Avissar, S. (1981) Molecular regulation of receptors: interaction of β-estradiol and progesterone with the muscarinic system. *Proc. Natl. Acad. Sci. USA* **89**, 5554–5558.

Sokolovsky, M., Cohen-Armon, M., Egozi, Y., Gurwitz, D., Henis, Y. I., Kloog, Y., Moscona-Amir, E., and Schreiber, G. (1986) Modulation of muscarinic receptors and their interactions. *Trends Pharmacol. Sci.* **Suppl. II**, 39–43.

Stahelin, M. and Simons, P. (1982) Rapid and reversible disappearance of β-adrenergic cell surface receptors. *EMBO J.* **1**, 187–190.
Sternweis, P. C. and Robishaw, J. D. (1984) Isolation of two proteins with high affinity for guanine nucleotides from membranes of bovine brain. *J. Biol. Chem.* **259**, 13806–13813.
Stockton, J., Birdsall, N. J. M., Burgen, A. S. V., and Hulme, E. C. (1983) Modification of the binding properties of muscarinic receptors by gallamine. *Mol. Pharmacol.* **23**, 551–557.
Tien, X.-Y. and Wallace, L. J. (1985) Trihexyphenidyl—further evidence for muscarinic receptor subclassification. *Biochem. Pharmacol.* **34**, 588–590.
Tobler, H. J. and Engel, G. (1983) Affinity spectra: a novel way for the evaluation of equilibrium binding experiments. *Naunyn-Schmiederberg's Arch. Pharmacol.* **322**, 183–192.
Uchida, S., Matsumoto, K., Mizushima, A., Osugi, T., Higuchi, H., and Yoshida, H. (1984) Effects of guanine nucleotide and sulphydryl reagent on subpopulations of muscarinic acetylcholine receptors in mammalian hearts: possible evidence for interconversion of super-high and low affinity agonist binding sites. *Eur. J. Pharmacol.* **100**, 291–298.
Uchida, S., Matsumoto, K., Takeyasu, K., Higuchi, H., and Yoshida, H. (1982) Molecular mechanism of the effects of guanine nucleotide and sulphydryl reagent on muscarinic receptors in smooth muscles studied by radiation inactivation. *Life Sci.* **31**, 201–209.
Vauquelin, G., Andre, C., DeBacker, J.-P., Laduron, P., and Strosberg, A. D. (1982) Agonist-mediated conformational changes of muscarinic receptors in rat brain. *Eur. J. Biochem.* **125**, 117–124.
Vickroy, T. W., Roeske, W. R., and Yamamura, H. I. (1984) Pharmacological differences between the high-affinity muscarinic agonist binding states of the rat heart and cerebral cortex labelled with (+)-[^{3}H]cis-methyldioxolane. *J. Pharmacol. Exp. Therap.* **229**, 747–755.
Vickroy, T. W., Yamamura, H. I., and Roeske, W. R. (1983) Differential regulation of high-affinity agonist binding to muscarinic sites in the rat heart, cerebellum and cerebral cortex. *Biochem. Biophys. Res. Comm.* **116**, 284–290.
Waelbroeck, M., Gillard, M., Robberecht, P., and Christophe, J. (1986) Kinetic studies of [^{3}H]-*N*-methylscopolamine binding to muscarinic receptors in the rat central nervous system: evidence for the existence of three classes of binding sites. *Mol. Pharmacol.* **30**, 305–314.
Waelbroeck, M. M., Gillard, P., Robberecht, P., and Christophe, J. (1987) Muscarinic receptor heterogeneity in rat central nervous system. I. Binding of four selective antagonists to three muscarinic receptor subclasses: a comparison with M2 cardiac muscarinic receptors of the C type. *Mol. Pharmacol.* **32**, 91–99.
Waelbroeck, M., Robberecht, P., Chatelain, P., and Christophe, J.

(1982) Rat cardiac muscarinic receptors. 1. Effects of guanine nucleotides on high- and low-affinity binding sites. *Mol. Pharmacol.* **21**, 581–588.

Wamsley, J. K., Zarbin, M. A., Birdsall, N. J. M., and Kuhar, M. J. (1980) Muscarinic cholinergic receptors: autoradiographic localization of high and low affinity agonist binding sites. *Brain Res.* **200**, 1–12.

Wang, J.-X., Mei, L., Yamamura, H. I., and Roeske, W. R. (1987a) Solubilization with digitonin alters the kinetics of pirenzepine binding to muscarinic receptors from rat forebrain and heart. *J. Pharmacol. Exp. Ther.* **242**, 981–990.

Wang, J.-X., Roeske, W. R., Gulya, K., Wang, W., and Yamamura, H. I. (1987b) [^{3}H]-AF-DX 116 labels subsets of muscarinic cholinergic receptors in rat brain and heart. *Life Sci.* **4**, 1751–1760.

Ward, D. M. and Young, J. M. (1978) Ligand binding to muscarinic receptors in intact longitudinal muscle strips from guinea-pig intestine. *Brit. J. Pharmacol.* **61**, 189–197.

Watson, M. (1988) [^{3}H] AF-DX 116 selectively identifies a subpopulation of muscarinic cholinergic receptors in rat cerebral cortex. *Trends Pharmacol. Sci.* **Suppl. III**, 90–91.

Watson, M., Yamamura, H. I., and Roeske, W. R. (1983) A unique regulatory profile and regional distribution of [^{3}H]pirenzepine binding in the rat provide evidence for distinct M_1 and M_2 muscarinic receptor subtypes. *Life Sci.* **32**, 3001–3011.

Watson, M., Roeske, W. R., and Yamamura, H. I. (1986) [^{3}H]Pirenzepine and (–)-[^{3}H]quinuclidinyl benzilate binding to rat cerebral cortical and cardiac muscarinic cholinergic sites. II. Characterization and regulation of antagonist binding to putative muscarinic subtypes. *J. Pharm. Exp. Therap.* **237**, 419–427.

Wei, J.-W. and Sulakhe, P. V. (1979) Agonist-antagonist interactions with rat atrial muscarinic cholinergic receptor sites: differential regulation by guanine nucleotides. *Eur. J. Pharmacol.* **58**, 91–92.

Wei, J.-W. and Sulakhe, P. V. (1980a) Cardiac muscarinic cholinergic receptor sites: opposing regulation by divalent cations and guanine nucleotides of receptor-agonist interaction. *Eur. J. Pharmacol.* **62**, 345–347.

Wei, J.-W. and Sulakhe, P. V. (1980b) Requirement for sulfhydryl groups in the differential effects of magnesium ion and GTP on agonist binding of muscarinic cholinergic receptor sites in rat atrial membrane fractions. *Naunyn-Schmiederberg's Arch. Pharmcol.* **314**, 51–59.

Wenger, D. A., Parthasarthy, N., and Aronstam, R. S. (1985) Regional heterogeneity of muscarinic acetylcholine receptors from rat brain is retained after detergent solubilization. *Neurosci. Lett.* **54**, 65–70.

Wheatley, M., Birdsall, N. J. M., Curtis, C., Eveleigh, P., Pedder, E. K.,

Poyner, D., Stockton, J. M. and Hulme, E. C. (1987) The structure and properties of the purified muscarinic acetylcholine receptor from rat forebrain. *Biochem. Soc. Trans.* **15**, 113–116.

Wheatley, M., Hulme, E. C., Birdsall, N. J. M., Curtis, C. A. M., Eveleigh, P., Pedder, E. K., and Poyner, D. (1988) Peptide mapping studies on muscarinic receptors: receptor structure and location of the ligand binding site. *Trends Pharmacol. Sci.* **Suppl. III**, 19–24.

Wong, H.-M. S., Sole, M. J., and Wells, J. W. (1986) Assessment of mechanistic proposals for the binding of agonists to cardiac muscarinic receptors. *Biochemistry* **25**, 6995–7008.

Wreggett, K. A. and De Lean, A. (1984) The ternary complex model: its properties and application to ligand interactions with the D_2-dopamine receptor of the anterior pituitary gland. *Mol. Pharmacol.* **26**, 214–227.

Yamamura, H. I. and Snyder, S. H. (1974a) Muscarinic cholinergic binding in rat brain. *Proc. Natl. Acad. Sci., USA* **71**, 1725–1729.

Yamamura, H. I. and Snyder, S. H. (1974b) Muscarinic cholinergic receptor binding in the longitudinal muscle of the guinea pig ileum with [^{3}H]quinuclidinyl benzilate. *Mol. Pharmacol.* **10**, 861–867.

Yatani, A., Codina, J., Brown, A. M., and Birnbaumer, L. (1987) Direct activation of mammalian atrial muscarinic receptors by GTP regulatory protein, G_K. *Science* **235**, 207–211.

Young, J. M. (1974) Desensitization and agonist binding to cholinergic receptors in intestinal smooth muscle. *FEBS Lett.* **46**, 354–356.

Zwagemakers, J. M. A. and Claasen, V. (1980) Pharmacology of secoverine, a new spasmolytic agent with specific antimuscarinic properties. Part 1: Antimuscarinic and spasmolytic effects. *Arzneini.-Forsch.* **30**, 1517–1526.

3

Muscarinic Receptor Purification and Properties

Michael Schimerlik

1. Solubilization and Purification of Muscarinic Receptors

1.1. Solubilization of Muscarinic Receptors (mAChR)

The solubilization and purification of mAChR has proven to be a rather difficult problem because of a low density of receptor sites, poor yields of solubilized protein upon detergent extraction, and receptor instability in many commonly used, inexpensive detergents. Initial studies indicated that the detergent digitonin was capable of solubilizing muscarinic binding sites from the brain (Beld and Ariens, 1974; Hurko, 1978; Aronstam et al., 1978; Gorissen et al., 1978). Since most commercial preparations of digitonin are heterogeneous, containing 70–80% digitonin, 10–20% gitonin and tigonin, plus 5–15% minor saponins, the composition of the solubilizing agent is poorly defined. The variability in efficiency of receptor solubilization and the solubility of the detergent itself in aqueous solution are most probably related to differing digitonin compositions that, in turn, depend on the lot number and supplier. In studies on receptor solubilization from the rat brain (Repke and Matthies, 1980), optimal results were obtained using a defined mixture of 3 digitonin/2 gitonin (w/w).

Reports of solubilization of brain mAChR using other detergent systems have also been published. The zwitterionic detergent CHAPS (3-[(3-cholamidopropyl)-dimethylammonio]-1-propane sulfonate) was effective in solubilizing about 15% of the muscarinic binding sites in rat (Gavish and Sokolovsky, 1982) and bovine brain (Kuno et al., 1983) preparations, whereas slightly higher yields were obtained from porcine caudate nucleus using L-α-lysophosphatidylcholine (Haga, 1983). In a study of the effects of various detergent systems on receptor solubilization from rat brain, Hulme et al. (1983) found that the alkylpolyoxyethylene detergents C12E10 or C12E9 could solubilize about 20% of the muscarinic binding sites, and that supplementation of CHAPS or cholate with 1*M* NaCl, or cholate with NaCl and egg lecithin dramatically increased the efficiency of extraction and the stability of the solubilized protein.

mAChR have also been solubilized from porcine atrial tissue (Cremo et al., 1981) using a detergent system composed of digitonin and cholate in a 5:1 (w/w) ratio. Recoveries of muscarinic antagonist binding sites were approximately 90% compared to 40% for saturated digitonin solutions used under the same experimental conditions. Additional experiments (Peterson and Schimerlik, 1984) indicated that the efficiency of receptor extraction was strongly dependent on the ratio of protein and/or membrane lipid to detergent. Thus, it was possible to remove selectively contaminating proteins prior to solubilization of the muscarinic receptor in a second extraction step to give an overall yield of about 70% and a threefold purification. Similar results have been found for porcine atrial mAChR (50% recovery, twofold purification by double extraction) using the detergent dodecyl-β-D-maltoside (Peterson et al., 1988). Solubilization of rat myocardial membranes by digitonin (precipitated from dimethylsulfoxide by ether/chloroform) at 0°C in the presence of 1 m*M* Mg^{2+}, using a detergent to protein ratio of 1:1 to 2:1 gave a recovery of antagonist binding sites of 80–90% (Berrie et al., 1984).

The properties of the solubilized preparation appear to be strongly dependent on both the composition of the detergent system, the method used for solubilization, and the tissue source of mAChR. Ligand binding studies of digitonin-solubilized preparations from calf cerebral cortex and caudate (Hurko, 1978) and rat forebrain (Gorissen et al., 1981) showed a single class of high affinity antagonist binding sites that bound agonists with low affinity. The agonist binding properties of the solubilized preparation were different than the membrane-bound receptor, which exists in low, high, and possibly superhigh agonist affinity states. The high affinity agonist sites are thought to be the result of receptor association with one or more type of guanine nucleotide binding protein (Berrie et al., 1979; Rosenberger et al., 1980; Sokolovsky et al., 1980). Similarly, homogeneous, low affinity agonist binding was observed for preparations from porcine caudate nucleus solubilized by lysophosphatidylcholine (Haga, 1983) and porcine atria solubilized in digitonin plus cholate (Herron et al., 1982).

In contrast to the above results, mAChR solubilized from rat brain using CHAPS or digitonin (Gavish and Sokolovsky, 1982; Baron et al., 1985), or bovine cerebral cortex by CHAPS (Kuno et al., 1983) showed heterogeneous binding of agonists and, in the latter case, a selective decrease in agonist affinity in the presence of Na^+ plus GTP was found. Rat myocardial mAChR solubilized as described above (Berrie et al., 1984) showed a selective decrease in affinity for the muscarinic agonist [^{3}H]oxotremorine-*M* at high affinity binding sites upon addition of GppNHp.

1.2. Purification of Muscarinic Receptors

Highly purified preparations of active mAChR have been reported from porcine atria (Peterson et al., 1984), porcine cerebellum (Haga and Haga, 1985), rat forebrain (Berrie et al., 1985), and chick heart (Kwatra and Hosey, 1986). Specific purification procedures varied somewhat depending on the tissue of interest; however, chromatography on anion exchange resins such as DEAE-Biogel or DEAE-Sephacel, hydroxylapatite, and gel permeation HPLC using a combination of TSK 4000 SW and TSK 3000 SW have proven effective as has affinity chromatography using wheat-germ agglutinin-agarose and a muscarinic affinity column containing the ligand 3-(2′-aminobenzhydryloxy) tropane coupled to sepharose 6B via a 1,4 butanediol diglycidyl ether spacer arm (Haga and Haga, 1983). Purification of inactive mAChR from calf forebrain has been reported utilizing a dexetimide affinity resin (Andre et al., 1983). Specific activities estimated for the highly purified protein range from about 11–13 μmol of [^{3}H]L-QNB or [^{3}H]*N*-methylscopolamine bound/g of protein, corresponding to one ligand binding site/77–91 kdaltons. Analysis of the purified mAChR by SDS polyacrylamide gel electrophoresis indicated a single polypeptide (apparent mol wt 70 kdaltons for the preparations from calf brain (Andre et al., 1983) and porcine brain (Haga and Haga, 1985), whereas two peptides (apparent mol wt, about 80 and 14 kdaltons) were found in the atrial preparation (Peterson et al., 1984). For the porcine brain and atrial preparations, it was shown that the high mol wt polypeptide was covalently labeled by [^{3}H]PrBCM indicating that it contained the muscarinic ligand binding site, whereas more recent studies (Peterson et al., 1986) indicate that the low mol wt component found in the atrial preparation is a contaminant.

2. Ligand Binding Properties of Purified Muscarinic Receptors

Ligand binding studies of the purified mAChR from porcine cerebrum (Haga and Haga, 1985) and atria (Schimerlik et al., 1986) have indicated that the dissociation constants for most agonists and antagonists do not vary significantly between the purified and initial detergent solubilized states (summarized in Table 1). In general, both the heart and brain preparations show only low affinity binding for agonists; however, in the atrial preparation, high affinity agonist binding (K_D for carbamylcholine, 1–5 μM) has been observed for up to 60% of the total L-QNB binding sites (Peterson et al., 1984). The reason for this is not known, although the consistently Gaussian distribution of mAChR and the absence of detectable lower mol wt polypeptides on SDS-polyacrylamide gels suggest that proteolysis is not the cause. The possibility of true heterogeneity in the mAChR preparation or that exposure to high agonist concentrations during elution from the affinity resin (or aging during storage) causes formation of high affinity dimers or a conformation change to a high affinity form cannot be discounted at this time. Kinetic studies of [^{3}H]L-QNB binding to the purified atrial mAChR (Schimerlik, unpublished) indicated that the reaction still obeyed the same two-step mechanism found for the solubilized porcine atrial mAChR (Herron et al., 1982); however, the preequilibrium dissociation constant for the first step increased by about fourfold, and the reverse rate constant (k_{-1}) for the isomerization step decreased by a factor of about 50. It is of interest that, in going from the membrane-bound to detergent-solubilized and highly purified state (Table 2), the overall affinity for [^{3}H]L-QNB remains relatively constant, whereas the kinetic constants for the binding of that ligand show considerable variation. Whether

Table 1
Equilibrium Dissociation Constants for Muscarinic Ligands Binding to Purified mAChR from Porcine Cerebrum and Atria

Ligand	Dissociation constant (*M*)	
	Cerebrum[a]	Atria[b]
Agonists		
Acetylcholine	5.2×10^{-5}	8.0×10^{-6}[c]
Carbamylcholine	1.8×10^{-4}	7.1×10^{-5}
Oxotremorine	7.0×10^{-6}	2.2×10^{-6}
Arecoline	2.7×10^{-5}	N.D.[d]
Pilocarpine	9.4×10^{-6}	1.0×10^{-5}
cis-methyldioxolane	5.4×10^{-5}	N.D.[d]
Acetyl-β-methylcholine	N.D.[d]	2.3×10^{-5}[c]
Antagonists		
L-QNB	N.D.[d]	6.1×10^{-11}
N-methylscopolamine	1.2×10^{-9}	N.D.[d]
L-hyoscyamine	N.D.[d]	4.8×10^{-10}
Pirenzepine	2.9×10^{-7}	2.1×10^{-7}
Dexetimide	2.0×10^{-9}	6.0×10^{-10}
Levetimide	N.D.[d]	3.1×10^{-6}
Atropine	7.8×10^{-9}	N.D.[d]
Isopropamide	7.0×10^{-9}	N.D.[d]
Scopolamine	N.D.[d]	8.5×10^{-9}

[a]0.1% w/v digitonin, 20 m*M* potassium phosphate pH 7.0 (Haga and Haga, 1985). Determined by displacement of [^{3}H]NMS.

[b]0.4% w/v digitonin, 0.08% w/v cholate, 10 m*M* sodium phosphate, 1 m*M* EDTA, 0.1 m*M* PMSF, pH 7.4 (Schimerlik et al., 1986). Determined by displacement of [^{3}H]L-QNB.

[c]same as [b], pH 6.9.

[d]N.D. means not done.

these effects are the result of receptor delipidation, changes in association with other detergent solubilized or membrane-bound proteins, or alterations in mAChR conformation remains open to speculation.

3. Physical Properties and Structural Studies of Muscarinic Receptors

3.1. Glycoprotein Properties

Affinity chromatography on various lectin resins and digestion by glycosidases have been useful in establishing the glyco-

Table 2
$[^3H]$L-QNB Binding to Porcine Atrial mAChR

	(nM)	(min⁻¹)		(nM)
Preparation*	K	k_1	k_{-1}	K_{ov}
Membrane bound	0.66	0.33	0.02	0.043
Detergent solubilized	5.7	0.24	0.01	0.250
Purified	22.0	0.12	1.7×10^{-4}	0.061

*All experiments in 10 m*M* Na Phosphate, 1 m*M* EDTA 0.1 m*M* PMSF, pH 7.4. The detergent solubilized and purified preparations also contained 0.4% w/v digitonin, 0.08% w/v cholate. Data were analyzed according to the mechanism

$R + Q \overset{K}{\rightleftharpoons} RQ \underset{k_{-1}}{\overset{k_1}{\rightleftharpoons}} (RQ)'$ where R is the mAChR and Q $[^3H]$L-QNB.

protein properties of the mAChR. Solubilized atrial mAChR bound almost completely to wheat-germ agglutinin affinity columns and about 50–60% of the bound mAChR was recovered by elution with *β-N*-acetylglucosamine (Herron and Schimerlik, 1983). Treatment with neuraminidase reduced binding to the resin by about 50% (in digitonin plus cholate buffers), indicating either heterogeneity of the attached carbohydrate residues or the accessibility of sialic acid groups to the enzyme. Neuraminidase digestion also enhanced specific binding to *Ricinus communis* agglutinin I and II affinity resins, suggesting that cleavage by the glycohydralase had exposed additional carbohydrate residues. Studies of mAChR solubilized from bovine cerebral cortex by CHAPS (Shirakawa et al., 1983) found specific binding to wheat-germ agglutinin and *Ricinus communis* lectin columns, but also detected specific binding to Con A and *Lens culinaris* agglutinin resins. Whether these results indicate a difference in glycosylation between the atrial and cerebral cortex mAChR or are due to different detergent systems and/or lectin columns is not known. Digestion of bovine cerebral cortex mAChR by endoglycosidases D and H (Rauh et al., 1986) reduced the apparent mol wt of PrBCM-labeled receptor from 73 to about 69 kdaltons, whereas neuraminidase treatment increased the isoelectric point from pH 4.3 to about pH 4.5. A reduction in apparent mol wt by about 6 kdaltons was also observed for both high (86 kdaltons) and low (72 kdaltons) mol wt receptors from chick retina (Large et al., 1985b). In summary, it seems clear that the mAChR are glycoproteins; however, the actual extent of glyosylation, the degree of carbohydrate heterogeneity, and the explicit role of carbohydrate groups in receptor structure, processing, and expression are not yet resolved.

3.2. Physical Properties

Much useful information regarding mAChR structure and properties has been obtained utilizing physical methods such as sucrose gradient centrifugation in H_2O/D_2O mixtures, gel filtration chromatography, isoelectric focusing, radiation inactivation analysis, and SDS polyacrylamide gel electrophoresis. mAChR solubilized from rat brain synaptosomes in digitonin/gitonin showed an isoelectric point of about pH 4.6 (Repke and Matthies, 1980). mAChR solubilized from a variety of tissues and species in 2% sodium dodecyl sulfate, followed by isoelectric focusing in Triton X-100 gave identical values of pH 5.9 for the isoelectric point, which decreased to pH 4.2 upon solubilization in digitonin (Venter et al., 1984). Whether this variability is the result of detergent binding, differential association of lipids, and or other proteins in the mAChR-mixed micelle or because of changes in receptor environment, resulting in alterations of the pK_a values for charged residues is unknown. Developmental studies of avian retinal mAChR (Large et al., 1985a) showed that the isoelectric point of the digitonin-solubilized protein changed from predominantly pH 4.25 to pH 4.50 in going from 13-d embryos to newly hatched chicks, indicating that measurements of this type are sensitive indicators of alterations in mAChR structure.

Several studies have been reported in which the properties of covalently labeled unpurified mAChR were investigated by SDS-polyacrylamide gel electrophoresis. After covalent incorporation of [^{3}H]PrBCM, mAChR from rat, guinea-pig, and frog brain were shown by analysis of electrophoretic data to have an apparent mol wt of about 80 kdaltons (Birdsall et al., 1979). Smooth muscle preparations were more susceptible to proteolysis, mainly generating a 60-kdalton species, but also showed a peak at 80 kdaltons. An electrophoretic analysis of [^{3}H]PrBCM-labeled mAChR from a variety of tissues and species (Venter, 1983) showed an apparent mol wt of 80 kdaltons, in agreement with the results of radiation-inactivation studies. Similar peptide maps were also found upon digestion with trypsin or papain suggesting that the various receptor species were similar in structure.

Recently, in tissues other than those examined above, some evidence for structural heterogeneity or structural alterations during development has been presented. Electrophoresis of [^{3}H]-PrBCM-labeled mAChR from rat pancreas, lacrimal, and parotid glands (Hootman et al., 1985) showed major peaks of apparent mol wt 118, 87, and 105 kdaltons, respectively, compared to IM-9 lymphocytes (114 kdaltons) and NG108 cells (72 kdaltons). Incorporation of label into lower mol wt polypeptides was enhanced when PrBCM labeling was done in enzymatically dissociated cells. Large et al. (1985b) found that mAChR in the avian central nervous system existed mainly in an 86-kdalton form prior to synaptogenesis, shifting to a lower apparent mol wt (72 kdaltons) after synaptogenesis.

Photoaffinity labeling techniques have also been utilized in structural studies of mAChR. Labeling of detergent-solubilized porcine atrial muscarinic receptors by [^{3}H]*p*-azidoatropine methyl iodide (Cremo and Schimerlik, 1984) gave a single specifically labeled band of apparent mol wt of 75 kdaltons. Studies utilizing [^{3}H]*N*-methyl-4-piperidyl *p*-azidobenzilate (Amitai et al., 1982) showed specific labeling of an 86-kdalton species in membranes from rat cerebral cortex. Additional studies with the latter probe (Avissar et al., 1982; Avissar et al., 1983) indicated the presence of a higher mol wt species having an apparent mol wt of 160 kdaltons, and that it might be possible to chemically cleave the 86-kdalton polypeptide into two 40-kdalton species. At this time, these results have not been completely explained in terms of mAChR structure-function. In general, electrophoretic studies showing heterogeneity of labeled mAChR should be viewed conservatively, since they may be subject to artifacts from possible endogenous proteolysis, chemical cleavage of peptide bonds or associated carbohydrates, intra- or interchain disulfide formation (Dadi and Morris, 1984), or protein aggregation.

A detailed electrophoretic analysis of the atrial mAChR (Peterson et al., 1986) indicated that the polypeptide of apparent mol wt of 70–80 kdaltons showed anomolous migration in SDS-polyacrylamide gels compared to mol wt standards. The deviation was attributed to both an abnormally high charge density resulting from excess binding of SDS compared to standard proteins and a large deviation of the retardation coefficient from that expected for a rod-shaped molecule. Analysis of the electrophoretic results by a semi-empirical method indicated a mol wt for the protein portion of the molecule of 50–60 kdaltons.

Hydrodynamic measurements such as gel filtration chromatography and sedimentation through linear 5–20% w/v sucrose gradients in H_2O and D_2O have been used to estimate physical parameters such as sedimentation coefficient, Stoke's radius, frictional ratio, and mol wt for the detergent-solubilized mAChR. Data for rat forebrain mAChR prelabeled with PrBCM prior to extraction in a variety of detergent systems (Berrie et al., 1984) and purified porcine atrial mAChR exchanged into Triton X-405 (Peterson et al., 1986) are summarized in Table 3. These data suggest that large variations in the frictional ratio, Stokes radius, and indeed, the calculated mol wt can be found depending on the detergent system employed. It is also of interest that the mAChR appears capable of binding large amounts of detergent—up to about 1 g of detergent/g of protein for Triton X-405 and lysophosphatidylcholine. When the calculated mol wt of the atrial mAChR was corrected for an estimated 26.8% contribution of carbohydrate (Peterson et al., 1986), the value of 50–52 kdaltons agreed quite well with that determined by electrophoresis.

By altering solubilization conditions to include Mg^{2+} and by lowering the temperature of extraction, it was possible to solubilize a species having a higher sedimentation coefficient (13.4 *s* in digitonin) from rat heart than that found for muscarinic binding sites prelabeled with antagonists in the presence of GppNHp (11.6 *s*). The binding sites solubilized in this manner showed high affinity binding of the muscarinic agonist oxotremorine-*M* that was sensitive to GppNHp, indicating that the higher mol wt species was a complex between the mAChR and a guanine nucleotide binding protein (Berrie et al., 1984). Solubilization of rat brain mAChR in CHAPS (Baron et al., 1985) also showed two molecular forms having sedimentation coefficients of 9.9 *s* and 14.9 *s*, respectively. Treatment of the membranes with GppNHp prior to solubilization eliminated the higher mol wt species, suggesting that the 14.9 *s* species was a complex between the mAChR and a guanine nucleotide binding protein. Sedimentation of the preparation from rat cerebral cortex solubilized in either CHAPS or digitonin in the presence of Cu^{2+} resulted in an enhancement of the concentration of higher mol wt receptor form (Baron et al., 1985). Solubilization of the porcine atrial mAChR in the presence of agonists showed a higher mol wt species in sucrose gradient (apparent mol wt 140 kdaltons) that was lost when detergent extraction was done in the presence of GppNHp (Schimerlik et al., 1986) or when agonist was removed by dialysis.

Table 3
Hydrodynamic Properties of mAChR

Parameter	Atrial mAChR[a], Triton X-405, 0.35% w/v	Rat forebrain[b] Lubrol PX, 0.4% w/v	LPC, 0.4% w/v[c]	Digitonin, 2% w/v	0.3% w/v cholate/ 1M NaCl/0.25% w/v PC[d]
Partial specific volume of protein plus detergent (cm^3/g)	0.813	0.786	0.840	0.710	0.808
S_{20}, w(s)	5.30	3.88	4.96	10.76	3.78
Stokes radius (nm)	4.29	7.01	8.29	6.83	7.05
Frictional ratio	1.21	1.97	1.81	1.58	1.91
Bound detergent (g/g)	1.01	0.30	1.16	—[e]	0.42
Mol wt of mAChR alone ($\times 10^{-3}$)	68.2–70.9	100–125	119–154	—[e]	75–89

[a]From Peterson et al., 1986.
[b]For PrBcM prelabeled mAChR (Berrie et al., 1984).
[c]Lysophosphatidylcholine.
[d]Phosphatidylcholine.
[e]Cannot be calculated.

Treatment of purified atrial mAChR with carbachol also resulted in a higher mol wt species that corresponded to a receptor dimer; reduction with dithiothreitol regenerated the monomeric series (Schimerlik et al., 1986). It is interesting to note that neither the CHAPS nor digitonin solubilized higher mol wt species (Baron et al., 1985) or the porcine atrial high mol wt mAChR form in Triton X-405 (Schimerlik et al., 1986) were sensitive to guanine nucleotides once solubilized, indicating that under these conditions dissociation of mAChR and guanine nucleotide binding protein does not occur. A difference in mol wt has also been observed in radiation inactivation experiments using bovine cerebral cortex membranes, when the target size for [^{3}H]L-QNB binding (91 kdaltons) was compared to that for [^{3}H]pirenzepine binding (157 kdaltons), leading those authors (Shirakawa and Tanaka, 1985) to conclude that the high affinity pirenzepine binding was probably coupled to other components in the membrane.

4. Cloning of Muscarinic Receptors

4.1. Amino-Acid Sequence and Structure Deduced from cDNA Clones

The complete amino-acid sequence of mAChR from porcine brain (Kubo et al., 1986a) and porcine atria (Kubo et al., 1986b; Peralta et al., 1987a), as deduced from cDNA clones, is shown in Fig. 1. The mol wts of the protein portion of the porcine atrial and brain mAChRs determined from the sequence data are 51,670 and 51,416 daltons, respectively. Analysis of 5′ sequence data upstream of the porcine atrial mAChR coding region in several cDNA clones (not shown) indicated three classes of leader exons (Peralta et al., 1987a), suggesting differential RNA processing. Both proteins lack an amino terminal signal sequence (Blobel and Doberstein, 1975). Analysis of the hydropathic character of the atrial and brain mAChR sequence by the methods of Kyte and Doolittle (1982), Engelman et al. (1986), and Lawson et al. (1984), shown in Fig. 2a–2c, respectively, suggests that both proteins traverse the membrane seven times, with transmembrane region seven having the lowest hydrophobicity. This appears to be a characteristic of membrane receptors of known structure that interact with guanine-nucleotide binding proteins, such as bovine and Drosophila rhodopsin, bacteriorhodopsin (Henderson and

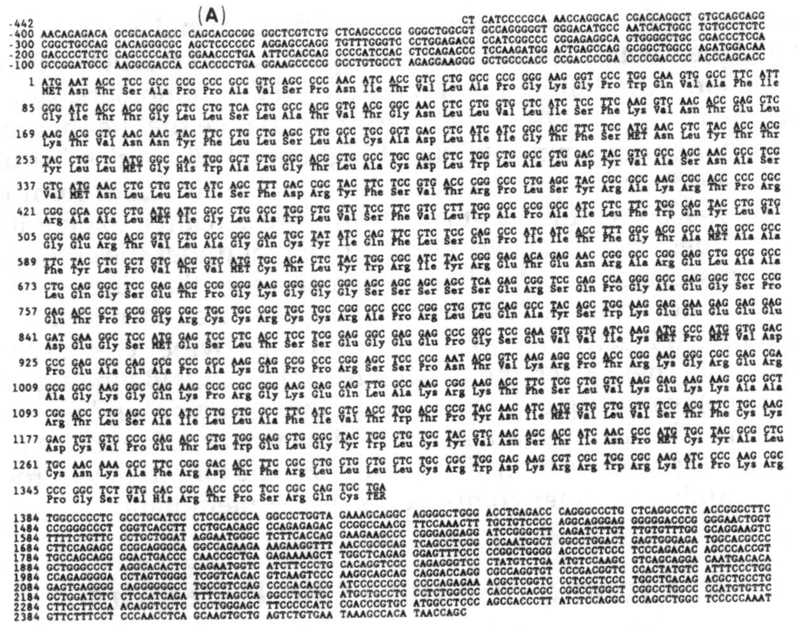

Fig. 1. Nucleic acid and corresponding amino-acid sequences of porcine brain (A, Kubo et al., 1986a) and atrial (B, Peralta et al., 1987a) muscarinic receptors. Where available, additional untranslated nucleic acid sequence is included.

Unwin, 1975; Ovchinnikov, 1982; Dratz and Hargrave, 1983; Zuker et al., 1985), mammalian visual rhodopsins (Nathans et al., 1986), and the β-adrenergic receptor (Dixon et al., 1986; Yarden et al., 1986). The relatively extensive sequence identity between these proteins has led to a suggestion that the genes encoding the mAChR, β-adrenergic receptor(s), and rhodopsin arose from a common ancestor (Kubo et al., 1986a). The membrane orientation of the mAChR, by analogy with the structurally well-characterized rhodopsin (Ovchinnikov, 1982; Dratz and Hargrave, 1983), has the amino terminus oriented on the extracellular side of the membrane and the carboxyl terminus on the cytoplasmic side (Fig. 3a,b). Alignment of the heart and brain mAChR sequences by the procedure of Wilbur and Lipman (1983) using a minimal gap penalty to obtain maximal alignment indicated about 50% identity of the two amino-acid sequences (Fig. 4). In general, the greatest conservation of amino-acid identity was found in membrane spanning regions; however, a high degree of sequence identity was also found in cytoplasmic loops

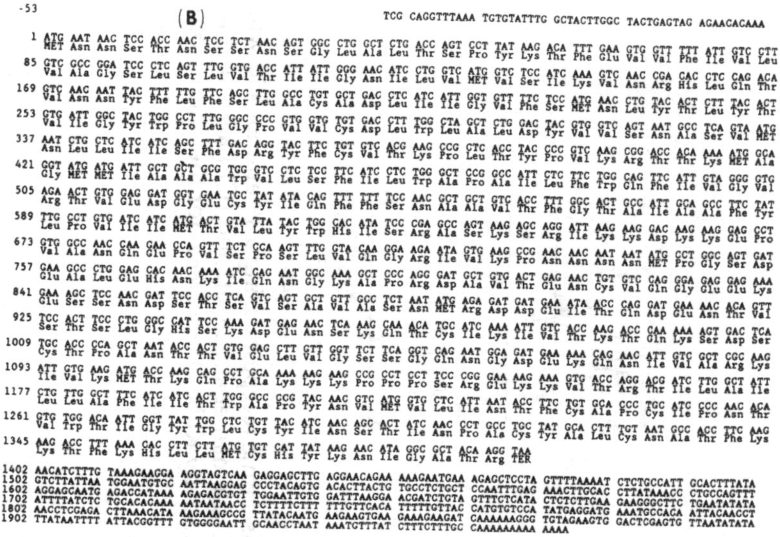

(B)

-53 TCG CAGGTTTAAA TGTGTATTTG GCTACTTGGC TACTGAGTAG AGAACACAAA

1 ATG AAT AAC TCC ACC AAC TCC TCT AAC AGT GGC CTG GCT CTG ACC AGT CCT TAT AAG ACA TTT GAA GTG GTT TTT ATT GTC CTT
MET Asn Asn Ser Thr Asn Ser Ser Asn Ser Gly Leu Ala Leu Thr Ser Pro Tyr Lys Thr Phe Glu Val Val Phe Ile Val Leu

85 GTC GCC GGA TCC CTC AGT TTG GTG ACC ATT ATT GGG AAC ATC CTG GTC ATG GTC TCC ATC AAA GTC AAC CGA CAC CTC CAG ACA
Val Ala Gly Ser Leu Ser Leu Val Thr Ile Ile Gly Asn Ile Leu Val MET Val Ser Ile Lys Val Asn Arg His Leu Gln Thr

169 GTC AAC AAT TAC TTT TTG TTC AGC TTG GCC TGT GCT GAC CTC ATC ATT GGT GTT TTC TCC ATG AAC CTG TAC ACT CTT TAC ACT
Val Asn Asn Tyr Phe Leu Phe Ser Leu Ala Cys Ala Asp Leu Ile Ile Gly Val Phe Ser MET Asn Leu Tyr Thr Leu Tyr Thr

253 GTG ATT GGC TAC TGG CCT TTG GGC CCC GTG GTG TGT GAC CTT TGG CTA GCT CTG GAC TAC GTG GTC AGT AAT GCC TCA GTA ATG
Val Ile Gly Tyr Trp Pro Leu Gly Pro Val Val Cys Asp Leu Trp Leu Ala Leu Asp Tyr Val Val Ser Asn Ala Ser Val MET

337 AAT CTG CTC ATC ATC AGC TTT GAC AGG TAC TTC TGT GTC ACG AAG CCG CTC ACC TAC CCC GTC AAG CGG ACC ACA AAA ATG GCA
Asn Leu Leu Ile Ile Ser Phe Asp Arg Tyr Phe Cys Val Thr Lys Pro Leu Thr Tyr Pro Val Lys Arg Thr Thr Lys MET Ala

421 GGT ATG ATG ATT GCT GCT GCG TGG GTC CTC TCC TTC ATC CTC TGG GCT CCG GCC ATT CTC TTC TGG CAG TTC ATT GTA GGG GTG
Gly MET MET Ile Ala Ala Ala Trp Val Leu Ser Phe Ile Leu Trp Ala Pro Ala Ile Leu Phe Trp Gln Phe Ile Val Gly Val

505 AGA ACT GTG GAG GAT GGT GAA TGC TAT ATA CAG TTT TTT TCC AAC GCT GCT GTC ACC TTT GGC ACT GCC ATT GCA GCC TTC TAT
Arg Thr Val Glu Asp Gly Glu Cys Tyr Ile Gln Phe Phe Ser Asn Ala Ala Val Thr Phe Gly Thr Ala Ile Ala Ala Phe Tyr

589 TTG CCT GTG ATC ATC ATG ACT GTA TTA TAC TGG CAC ATA TCC CGA GCC AGT AAG AGC AGG ATT AAG AAG GAC AAG AAG GAG CCT
Leu Pro Val Ile Ile MET Thr Val Leu Tyr Trp His Ile Ser Arg Ala Ser Lys Ser Arg Ile Lys Lys Asp Lys Lys Glu Pro

673 GTG GCC AAC CAA GAA CCA GTT TCT CCA AGT TTG GTA CAA GGA AGA ATA GTG AAG CCG AAC AAC AAT AAT ATG CCT GGC AGT GAT
Val Ala Asn Gln Glu Pro Val Ser Pro Ser Leu Val Gln Gly Arg Ile Val Lys Pro Asn Asn Asn Asn MET Pro Gly Ser Asp

757 GAA GCC CTG GAG CAC AAC AAA ATC CAG AAT GGC AAA GCT CCC AGG GAT GCT GTG ACT GAG AAC TGT GTC CAG GGA GAG GAG AAA
Glu Ala Leu Glu His Asn Lys Ile Gln Asn Gly Lys Ala Pro Arg Asp Ala Val Thr Glu Asn Cys Val Gln Gly Glu Glu Lys

841 GAA AGC TCC AAC GAT TCC ACC TCA GTC AGT GCT GTT GCC TCT AAT ATG AGA GAT GAT GAA ATA ACC CAG GAT GAA AAC ACA GTT
Glu Ser Ser Asn Asp Ser Thr Ser Val Ser Ala Val Ala Ser Asn MET Arg Asp Asp Glu Ile Thr Gln Asp Glu Asn Thr Val

925 TCC ACT TCC CTG GGC CAT TCC AAA GAT GAG AAC TCA AAG CAA ACA TGC ATC AAA ATT GTC ACC AAG ACC CAA AAA AGT GAC TCA
Ser Thr Ser Leu Gly His Ser Lys Asp Glu Asn Ser Lys Gln Thr Cys Ile Lys Ile Val Thr Lys Thr Gln Lys Ser Asp Ser

1009 TGC ACC CCA GCT AAT ACC ACT GTG GAG CTT GTT GGT TCT TCA GGT CAG AAT GGA GAT GAA AAA CAG AAC ATT GTC GCT CGC AAG
Cys Thr Pro Ala Asn Thr Thr Val Glu Leu Val Gly Ser Ser Gly Gln Asn Gly Asp Glu Lys Gln Asn Ile Val Ala Arg Lys

1093 ATT GTG AAG ATG ACC AAG CAG CCT GCA AAA AAG AAG CCG CCT CCT TCC CGG GAA AAG AAA GTG ACC AGG ACG ATC TTG GCT ATT
Ile Val Lys MET Thr Lys Gln Pro Ala Lys Lys Lys Pro Pro Pro Ser Arg Glu Lys Lys Val Thr Arg Thr Ile Leu Ala Ile

1177 CTG TTG GCT TTC ATC ATC ACT TGG GCC CCG TAC AAC GTC ATG GTG CTC ATT AAT ACC TTC TGT GCA CCC TGC ATC CCC AAC ACA
Leu Leu Ala Phe Ile Ile Thr Trp Ala Pro Tyr Asn Val MET Val Leu Ile Asn Thr Phe Cys Ala Pro Cys Ile Pro Asn Thr

1261 GTG TGG ACA ATT GGT TAT TGG CTC TGT TAC ATC AAC AGC ACT ATC AAC CCT GCC TGC TAT GCA CTT TGT AAT GCC ACC TTC AAG
Val Trp Thr Ile Gly Tyr Trp Leu Cys Tyr Ile Asn Ser Thr Ile Asn Pro Ala Cys Tyr Ala Leu Cys Asn Ala Thr Phe Lys

1345 AAG ACC TTT AAA CAC CTT CTT ATG TGT CAT TAT AAG AAC ATA GGC GCT ACA AGG TAA
Lys Thr Phe Lys His Leu Leu MET Cys His Tyr Lys Asn Ile Gly Ala Thr Arg TER

1402 AACATCTTTG TAAAGAAGGA AGGTAGTCAA GAGGAGCTTG AGGAACAGAA AAAGAATGAA AGAGCTCCTA GTTTTAAAAT CTCTGCCATT GCACTTTATA
1502 GTCTTATTAA TGGAATGTGC AATTAAGGAG CCCTACAGTG ACACTTACTG TGCCTCTGCT CCAATTTGAG AAACTTGCAC CTTATAAACC CTGCCAGTTT
1602 AGGAGCAATG AGACCATAAA AGAGACGTGT TGGAATTGTG GATTTAAGGA ACGATCTGTA GTTTCTCATA CTCTCTTGAA GAAGGGCTTC TGAATATATA
1702 ATTTTATCTC TGCACACAAA AATAATAACC TCTTTTCTTT TTTGTTCACA TTTTTGTTAC CATGTGTCCA TATGAGGATG AAATGCCACA ATTACAACCT
1802 AACCTCGAGA CTTAAACATA AAGAAAGCCG TTATACAATG AAGAAGTGAA GAAAGAAGAT CAAAAAAGGG TGTAGAAGTG GACTCGAGTG TTAATATATA
1902 TTATAATTTT ATTACGGTTT GTGGGGAATT GCAACCTAAT AAATGTTTAT CTTTCTTTGC CAAAAAAAAA AAAA

1-2 and 3-4. These regions are also very similar in sequence to the corresponding regions in the β-adrenergic receptor (Dixon et al., 1986; Yarden et al., 1986) and rhodopsin (Ovchinnikov, 1982; Dratz and Hargrave, 1983), indicating that they may be involved in recognition of guanine-nucleotide binding proteins. The second and third transmembrane regions each contain a single anionic amino acid in both brain (aspartate 71 and 105) and heart (aspartate 69 and 103) that corresponds to aspartic acid residues found in transmembrane regions two and three for the β-adrenergic receptor (Dixon et al., 1986; Yarden et al., 1986) and that found in transmembrane region two of rhodopsin (Ovchinnikov, 1982; Dratz and Hargrave, 1983). Based on sequence data of peptides labeled with PrBCM, it has been proposed (Hulme and Birdsall, 1986) that these residues are located at or near the muscarinic ligand binding site. Results from oligonucleotide-directed mutagenesis experiments on the gene of the hamster β-adrenergic receptor (Strader et al., 1987) suggest that the aspartate residue near the top of transmembrane helix two (corresponding to aspartate residues 105 and 103 of the brain and heart mAChR, respectively) forms a salt bridge with the quarternary ammonium group of agonists and antagonists, although this has not yet been proven for the mAChR. The lack of sequence identity in the large cytoplasmic loop 5-6 (composed of 156 and 180 amino acids for the brain and atrial mAChR, respectively) be-

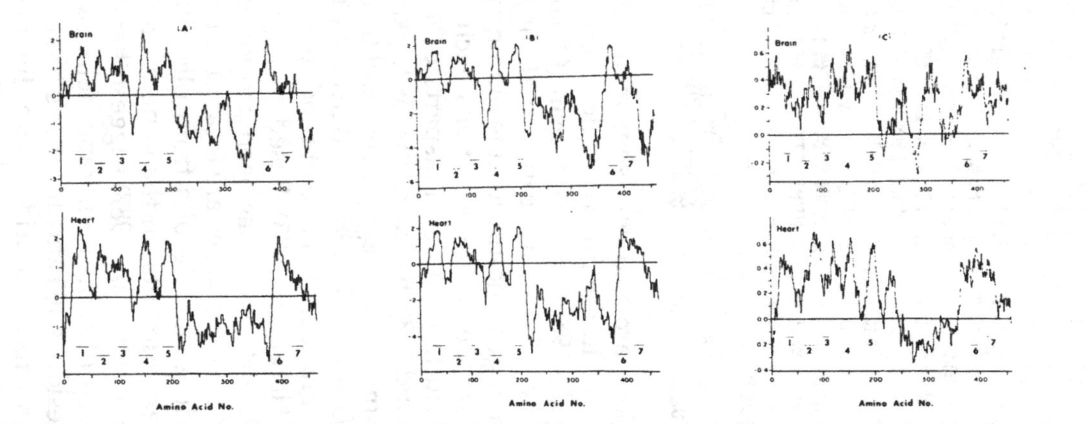

Fig. 2. Hydropathy plot analysis of porcine brain and heart muscarinic receptors using a 19 amino-acid window and the polarity scales of (A) Kyte and Doolittle (1982), (B) Engleman et al. (1986), and (C) Lawson et al. (1984). Transmembrane regions, designated by a bar, were those given by Kubo et al. (1986a) for the brain receptor and Peralta et al. (1987a) for the atrial receptor.

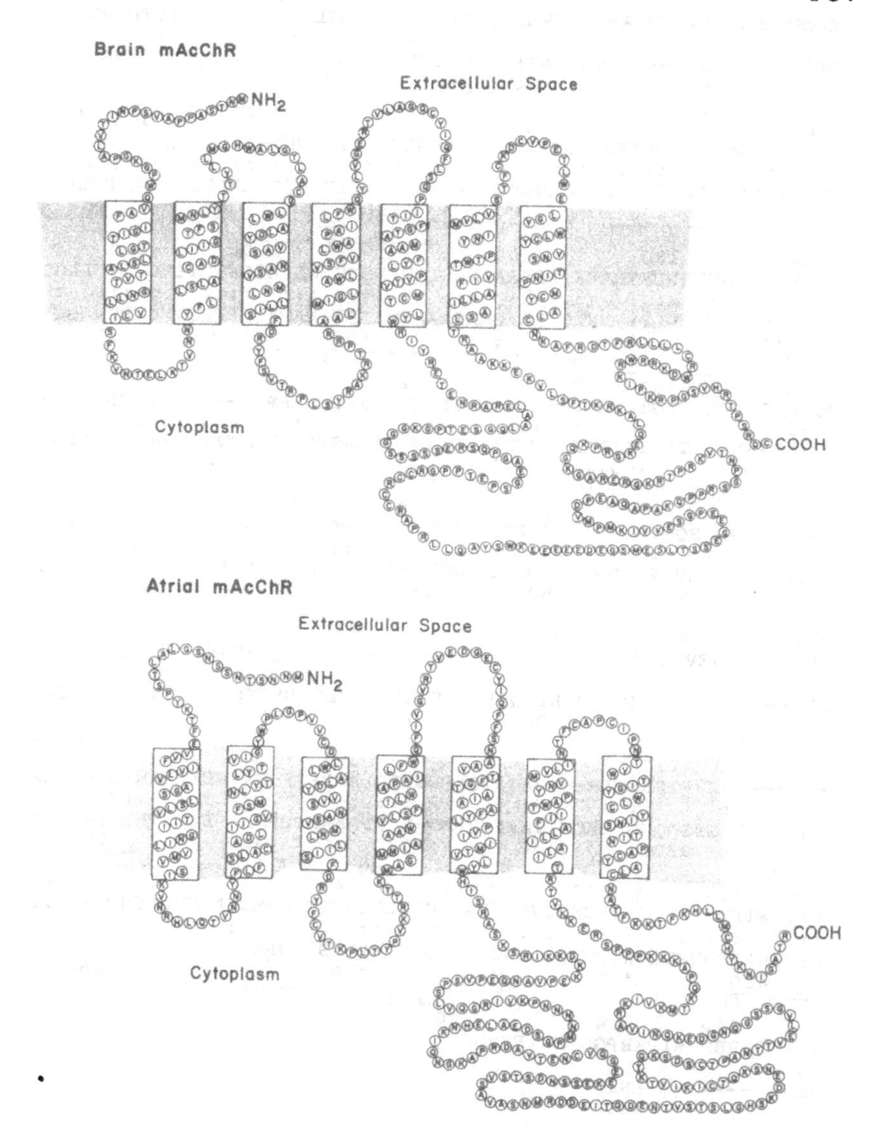

Fig. 3. Schematic drawing of the brain and heart muscarinic receptor orientation within the membrane. Transmembrane regions were assigned as in Fig. 2.

tween the two proteins and also in comparison to the β-adrenergic receptor (Dixon et al., 1986; Yarden et al., 1986) may indicate that this region plays a role in the coupling of these proteins to different effector systems.

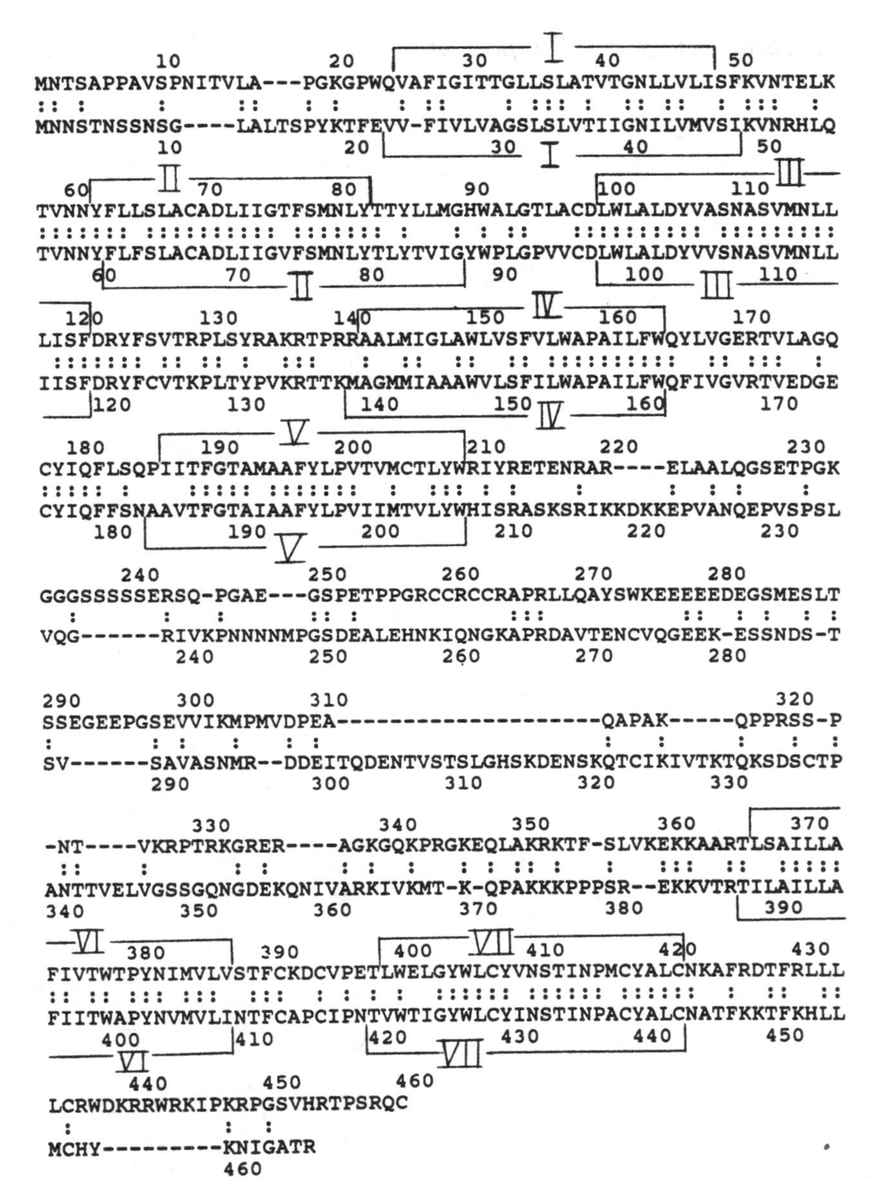

Fig. 4. Alignment of porcine brain (upper) and atrial (lower) mAChR amino-acid sequences according to the procedure of Wilbur and Lipman (1983) with a K-tuple size of one, window size = 20 and gap penalty = 1. Transmembrane sequences, assigned as in Fig. 2, are indicated by a line and Roman numeral, and matched amino acids are shown by two dots.

More recently, the genes for three additional mAChR subtypes composed of 478, 589 (Bonner et al., 1987; Peralta et al., 1987b), and 532 (Bonner et al., 1988) amino acids have been cloned and sequenced. Current data suggest that the expression of the different subtypes of mAChR are tissue specific (Peralta et al., 1987b). Overall, the regions of highest sequence identity for all five subtypes are located from the first transmembrane segment through the fifth transmembrane segment and in transmembrane segments six and seven. The amino-acid sequence shows a poorer correspondence at the carboxyl and amino termini, and is quite small in cytosolic loop 5-6. Further discussion of the structure and properties of mAChR subtypes can be found in Chapter 4 of this volume.

Possible extracellular sites for glycosylation (Hubbard and Ivatt, 1981) can be found near the amino terminal of both the brain mAChR (asparagine 2 and 12) and the atrial mAChR (asparagine 2, 3 and 6). Cytoplasmic domains containing potential phosphorylation sites for cyclic AMP (cAMP)-dependent protein kinase (Krebs and Beavo, 1979) can also be found in both brain (threonine 330 and 354, serine 356 and 451) and atrial (threonine 137 and 386) sequences. By analogy with rhodopsin (Wilden and Kuhn, 1982; Kelleher and Johnson, 1986) and the beta-adrenergic receptor (Benovic et al., 1985, 1986a, 1986b), the carboxyl termini of brain (threonine 428, 455; serine 451 and 457) and atrial (threonine 446, 450, 465 and tyrosine 459) mAChR contain several hydroxylamino acids that may be phosphorylated by protein kinases in vivo.

With the exception of arginine, asparagine, and histidine, the amino-acid compositions of the heart and brain mAChR are quite similar (Table 4). The partial specific volume for both proteins based on amino-acid composition alone is about 0.74 cc/g. Assuming 26% w/w carbohydrate (Peterson et al., 1986) reduces the partial specific volume to 0.725 and 0.732 for the heart and brain mAChR, respectively. The calculated values for the Z function (Barrantes, 1975) agree quite well with that predicted for an integral membrane protein ($Z = 0.52 \pm 0.11$).

4.2. Receptor Expression

mRNA coding for the brain mAChR has been transiently expressed in *Xenopus* oocytes (Kubo et al., 1986a). Injected oocytes showed 3–7 fmol of muscarinic ligand binding sites per oocyte compared to controls (< 0.1 fmol/oocyte). Muscarinic agonists

Table 4
Amino-Acid Composition of Protein Portion of mAChR

Amino Acid	Mol%	
	Porcine cerebrum	Porcine atria
Ala	8.5	7.5
Arg	7.2	2.8
Asp	2.0	3.2
Asn	3.3	7.3
Cys	3.3	2.8
Gly	6.3	4.3
Glu	5.7	3.6
Gln	2.8	2.8
His	0.4	1.3
Ile	4.1	7.9
Leu	11.5	7.7
Lys	5.0	6.9
Met	2.6	2.8
Phe	3.5	3.9
Pro	6.7	4.7
Ser	7.6	7.9
Thr	7.4	7.9
Trp	2.8	1.9
Tyr	3.5	3.4
Val	5.9	9.2
Partial specific vol[a]	0.740	0.743
Hydrophobicity[b]		
$H\Phi_{ave}$(kcal/mol)	1.244	1.241
Z	0.454	0.469

[a]Based on amino-acid composition alone.
[b]Calculated according to Barrantes (1975).

caused an increased current, after a delay of about 1 to 10 s, that appeared to be caused largely by an increase in chloride ion conductance (peak inward current at 1 μM ACh, −70 mV membrane potential = 1.6 μA). Muscarinic antagonists blocked the conductance increase in a reversible manner, although nicotinic acetylcholine receptor antagonists had no effect. Ligand binding studies using oocyte extracts gave dissociation constants for L-QNB (64 p*M*), atropine (0.38 n*M*), and pirenzepine (12 n*M*) in agreement with values typically found for brain mAChR (Hulme et al., 1978; Hammer et al., 1980). The high affinity binding of pirenzepine agrees quite well with the expected properties of a mAChR of the M_1 subtype (Chapter 4).

Injection of the mRNA from the atrial mAChR into oocytes (Fukuda et al., 1987) resulted in the expression of mAChR having a lower affinity for pirenzepine (500 n*M*), but a higher affinity for AF-DX 116 (0.4 μ*M*) than the brain mAChR, which is also consistent with the pharmacological distinction between the two subtypes. Addition of acetylcholine to oocytes in which the atrial mAChR was expressed appeared to activate an initial oscillatory chloride current similar to the brain mAChR. In addition, however, a second current, carried mainly by sodium and potassium ions, was found. These results indicate that the different mAChR were capable of coupling to differing physiological responses in *Xenopus* oocytes.

Transfection of Chinese hamster ovary cells with a plasmid containing the coding region of the porcine atrial mAChR and the mouse dihydrofolate reductase gene permitted coamplification of mAChR and dihydrofolate reductase in the presence of methotrexate (Ashkenazi et al., 1987). Homogenates of transfected cells showed specific activities of 10 nmol of $[^3H]$L-QNB bound/g protein compared to nontransfected cells, which did not bind L-QNB (< 5 pmol/g protein). Ligand binding studies permitted calculation of the dissociation constants for L-QNB (63 p*M*), atropine (4 n*M*), and pirenzepine (1 μ*M*). These values agree with those found for the membrane-bound porcine atrial mAChR (Searles and Schimerlik, 1980; Schimerlik et al., 1986), and the low affinity for the selective antagonist pirenzepine indicated that the mAChR expressed in the transformed cell is of the M_2 subtype (Chapter 4).

The expressed M_2 receptor in Chinese hamster ovary cells was able to couple to both inhibition of adenylate cyclase activity as well as the stimulation of phosphoinositide hydrolysis (Ashkenazi et al., 1987). Inhibition of adenylate cyclase activity appeared to be more efficiently coupled to the receptor than stimulation of phosphoinositide hydrolysis, since the latter response showed a much stronger dependence on receptor concentration in vivo. Both responses were mediated by pertussis-toxin-sensitive guanine-nucleotide binding regulatory proteins. However, the uncoupling of the mAChR from adenylate cyclase inhibition appeared to be much more sensitive to pertussis-toxin-catalyzed ADP-ribosylation than uncoupling from the phosphoinositide response. These results suggest that different guanine-nucleotide binding proteins may be involved as signal transducers for the two responses. Furthermore, the interactions of the mAChR with the guanine-nucleotide binding protein coupled to phosphoinositide hydrolysis were weaker than those observed for coupling to adenylate cyclase inhibition.

Expression of M_1–M_4 mAChR subtypes in embryonic kidney cells (Peralta et al., 1988) and NG108-15 neuroblastoma-glioma cells (Fukada et al., 1988) has shown that specific subtypes appear to be selective either for mediating the activation of phosphoinositide hydrolysis or the inhibition of adenylate cyclase. The M_5 subtype, expressed in Chinese hamster ovary cells couples most efficiently to the stimulation of phosphoinositide metabolism (Bonner et al., 1988). In the NG108-15 cell system, those subtypes that coupled to phosphoinositide metabolism also caused activation of a calcium-dependent potassium current and inhibition of the M-current suggesting a molecular mechanism for those muscarinic responses. Activation of M_1 mAChR transfected into A9 L cells resulted in a hyperpolarization of the membrane potential by activation of a calcium-dependent potassium channel and, to a lesser extent, activation of a chloride conductance (Brann et al., 1987; Jones et al., 1988).

5. Reconstitution of Muscarinic Receptors with Guanine Nucleotide Binding Proteins

Purified bovine brain guanine-nucleotide binding proteins G_o and G_i have been reconstituted into lipids with detergent-solubilized mAChR prepared from the same tissue (Florio and Sternweis, 1985). Prior to reconstitution, mAChR solubilized in deoxycholate were resolved from over 95% of the guanine-nucleotide binding activity by chromatography on DEAE sephacel in the presence of cholate, glycerol, and either atropine or oxotremorine in order to stabilize the solubilized protein. Fractions eluted into 10:1 w/w egg phosphatidylcholine:cholesterol were treated with XAD-2 beads, and muscarinic ligands were removed by gel filtration prior to addition of G proteins (in Lubrol 12A9). Of the methods tried for reconstitution (gel filtration, sucrose gradients, and dilution), and dilution method yielded the highest recovery. Addition of a large molar excess (about 1000-fold) of either G_i or G_o resulted in the conversion of the mAChR from the low affinity state for agonists, found for reconstituted resolved mAChR alone, to that of high affinity agonist binding. The IC_{50} for various agonists increased by five- to 100-fold upon addition of GTP, whereas antagonist binding was not affected by guanine nucleotides. Addition of purified $G_{o\alpha}$ in 100-fold molar excess over mAChR also resulted in a GTP-sensitive increase in agonist affinity as shown by a decrease of threefold in the IC_{50} value for oxotremorine, whereas addition of a 100-fold excess of purified $G_{o\beta}$ subunit had no effect. Addition of $G_{o\alpha}$ plus $G_{o\beta}$ caused a ninefold decrease in the IC_{50} value for oxotremorine, indicating the $G_{o\beta}$ subunit synergized the effects of the $G_{o\alpha}$ subunit and that the G_o holoprotein may interact more strongly with the mAChR than the α subunit alone. The order of potency of guanine nucleotides in inducing the interconversion of high to low agonist affinity was GTPγS >> GppNHp ~ GTP > GDP >> ITP.

Reconstitution of purified porcine brain mAChR (in a mixture of digitonin and dimyristoyl phosphatidylcholine) and purifed rat brain G_i (60-fold molar excess) into crude brain phospholipids by gel filtration chromatography resulted in a preparation that showed a twofold stimulation of GTPase activity by the muscarinic agonist carbachol, which was inhibited by the antagonist atropine and attenuated by pertussis-toxin-catalyzed ADP-ribosylation of G_i (Haga et al., 1985). Studies of purified porcine brain mAChR and purified rat brain G_i (Kurose et al., 1986) reconstituted by gel filtration at a final molar ratio of 2:15 of G_i to mAChR showed a twofold increase in the maximal apparent first order rate constant for GTPγS binding to G_i in the presence of the agonist carbachol, with no significant difference in half maximal GTPγS concentration for the effect. Analysis of the steady-state GTPase activity in the reconstituted preparation showed an increase in V_{max} in the presence of carbachol by about 20% at 50 μM $MgCl_2$ or 50% at 20 mM $MgCl_2$. A slight increase (30–50%) in the value(s) of the K_m for GTP in the presence of carbachol was also found at both magnesium concentrations. When the mAChR was reconstituted with ADP-ribosylated G_i, agonist-stimulated binding of GTPγS was totally inhibited, as was the coupling of the G_i and mAChR to form the high affinity agonist binding complex. These results agree with the notion that ADP-ribosylation functionally uncouples the mAChR from G_i.

Reconstitution of purified porcine brain mAChR with G_i and G_o purified from porcine brain (Haga et al., 1986) indicated a heterogeneous population of agonist binding sites. Saturation with either G_i or G_o at a 20-fold molar excess over mAChR converted about 50–70% of the agonist binding sites to the high affinity form. The effects of G_i and G_o were not additive, indicating that the same mAChR was capable of interacting with both guanine-nucleotide binding proteins. About 10–20% of the mAChR remained in the high affinity form at saturating levels of GppNHp. GDP was also capable of inducing the shift from high to low agonist affinity, although the half-maximal values were about twofold higher than for GTP.

Purified porcine atrial mAChR have been reconstituted with purified G_i from the same tissue into a mixture of phosphatidylserine:phosphatidylcholine:cholesterol (1:1:0.1 w/w) at a final receptor to G_i ratio of 1:5 (Tota et al., 1987). This molar ratio was found to be saturating and, under these conditions, about 50% of the agonist binding sites were in the high affinity state. Addition of GppNHp resulted in the conversion of the mAChR from the high affinity ($K_D \sim 1\ \mu M$) to low affinity ($K_D \sim 100\ \mu M$) state for carbachol.

The dissociation rate for [$\alpha - {}^{32}P$]GDP in the presence of muscarinic agonists showed a rapid kinetic phase corresponding to about 30% of the total G_i binding sites (Tota et al., 1987). The fast phase was observed only in the presence of carbachol, and the estimated increase in dissociation rate of GDP was about 50-fold faster than the single kinetic phase found for L-hyoscyamine or in the absence of ligands. GDP affinity, but not the affinity of GTPγS, was reduced by a factor of 40–50 in the presence of carbachol for approximately 30% of the total binding sites. These results suggest that about 30% of the reconstituted G_i proteins are coupled to mAChR in the reconstituted preparation and that muscarinic agonists selectively increase the dissociation rate of GDP from G_i. The order of potency for guanine nucleotides in inducing the shift from high to low affinity for agonists was GTPγS > GTP > GDP.

Analysis of the steady-state GTPase activity of the reconstituted preparation showed a fourfold agonist induced increase in the apparent maximal velocity with a corresponding threefold increase in apparent K_m (Tota et al., 1987). GDP release was the rate-limiting step in the absence of muscarinic agonists since $k_{cat} \approx k_{off}$ for GDP. Analysis of the kinetic data indicated that the increase in the steady-state V_{max} in the presence of carbachol was largely the result of the increase in the dissociation rate of GDP; however, possible contributions from other steps in the mechanism cannot be ruled out at this time. Finally, if one assumes that about 30% of the total G_i incorporated into lipids are interacting with about 50% of the reconstituted mAChR, and that the velocity observed in the presence of carbachol is actually the sum of the stimulated plus background rates, it is possible to estimate a turnover number of 2 min^{-1} for the agonist-stimulated GTPase activity of G_i (compared to about 0.2 min^{-1} in the presence of antagonists) and that one mAChR mol is capable of interacting with 3–5 mol of G_i.

Since the possibility of mAChR phosphorylation has been suggested as a possible in vivo regulatory mechanism (Burgoyne, 1980, 1981; Ho and Wang, 1985; Ho et al., 1986), and chick heart mAChR have been shown to be phosphorylated in vivo (Kwatra and Hosey, 1986), it was of interest to examine the effects of several protein kinases in vitro to determine their ability to phosphorylate purified and reconstituted mAChR. Studies utilizing purified porcine atrial mAChR in detergent solution indicated that the catalytic subunit of cAMP-dependent protein kinase, but not protein kinase C, cGMP-dependent protein kinase or myosin light-chain kinase could phosphorylate the mAChR up to about 0.6 mol P_i/mol protein (Rosenbaum et al., 1987). About 50% of the L-QNB binding sites were lost under these conditions; however, dephosphorylation by calcineurin was capable of completely reversing the loss of binding activity in agreement with in vivo results of Ho and Wang (1985). mAChR phosphorylation under these conditions was not ligand-dependent, and phosphorylation of the protein had no effect on the dissociation constant for L-QNB or the binding of the agonist carbachol.

When purified porcine atrial mAChR alone was reconstituted into lipids, the kinase specificity was the same as in detergent; however, the extent of apparent phosphorylation increased dramatically to about 10 mol P_i/mol [^{3}H]L-QNB sites, a value indicating that the protein was a much better substrate for cAMP-dependent protein kinase under these conditions (Rosenbaum et al., 1987). In this case, there was no loss of L-QNB binding sites, and as found in detergent solution, there was no effect of phosphorylation on the ligand-binding properties of the mAChR. Phosphorylation in the presence of muscarinic ligands did not alter the stoichiometry of incorporated label.

When the mAChR were reconstituted with purified G_i (1:5 ratio), the apparent stoichiometry of phosphorylation was the same as that for mAChR alone, in the presence of L-hyoscyamine or in the absence of ligands. However, phosphorylation in the presence of carbachol increased the apparent stoichiometry of ^{32}P incorporation to 20 mol/mol [^{3}H]L-QNB sites (Rosenbaum et al., 1987), indicating that the agonist-occupied receptor adopts a conformation either in a complex with G_i or upon transiently associating with G_i that is more susceptible to phosphorylation by the catalytic subunit of cAMP-dependent protein kinase. A ten- to 12-fold increase in ^{32}P incorporation into chick heart mAChR in vivo has also been found (Kwatra and Hosey, 1986) in the presence of agonists.

These studies suggest that the conformation of the mAChR is stabilized by reconstitution into lipids, since no binding sites are lost upon phosphorylation and, furthermore, that association or contact of the agonist-occupied mAChR with G_i induces an mAChR conformation in which additional phosphorylation sites are exposed. The observation that phosphorylation of the reconstituted mAChR did not cause a loss of binding sites implies that, if true, in vivo phosphorylation represents a signal for internalization or for modulating receptor–effector interactions as opposed to the inhibition of ligand binding. It is important to note, however, that studies such as these may be subject to several possible artifacts because of the high concentrations of kinases (0.38 μM) used, the lipid composition of the reconstituted system compared to native membranes, and the absence of other possible components found in an in vivo signaling system. Furthermore, although these studies showed complete specificity for the catalytic subunit of cAMP-dependent protein kinase, the possibility that there is a specific kinase such as observed for the beta-adrenergic receptor (Benovic et al., 1986a) remains open.

Acknowledgments

The assistance of Jack Kramer and the Oregon State University Central Service Laboratory of the Center for Gene Research and Biotechnology with computer programming of hydropathy plots and sequence comparisons, helpful discussions with Dr. Gary Peterson, and expert typing of Barbara Hanson, and the support of grant number HL23632 from the National Heart, Lung and Blood Institute are gratefully acknowledged.

References

Amitai, G., Avissar, S., Balderman, D., and Sokolovsky, M. (1982) Affinity labeling of muscarinic receptors in rat cerebral cortex with a photolabile antagonist. *Proc. Nat. Acad. Sci. USA* **79**, 243–247.

Andre, C., De Backer, J. P., Guillet, J. C., Vanderheyden, P., Vanquelin, G., and Strosberg, A. D. (1983) Purification of muscarinic acetylcholine receptors by affinity chromatography. *EMBO Journ.* **2**, 499–504.

Aronstam, R. S., Schuessler, D. C., Jr., and Eldefrawi, M. E. (1978) Solubilization of muscarinic acetylcholine receptors of bovine brain. *Life Sci.* **23**, 1377–1382.

Ashkenazi, A., Peralta, E. G., Winslow, J. W., Ramachandran, J., and Capon, D. J. (1989) Functionally distinct G proteins selectively couple different receptors to PI hydrolysis in the same cell. *Cell* **56**, 487–493.

Ashkenazi, A., Winslow, J. W., Peralta, E. G., Peterson, G. L., Schimerlik, M. I., Capon, D. J., and Ramachandran, J. (1987) An M_2 muscarinic receptor subtype coupled to both adenylyl cyclase and phosphoinositide turnover. *Science* **238**, 672–674.

Avissar, S. Amitai, G., and Sokolovski, M. (1983) Oligomeric structure of muscarinic receptors is shown by photoaffinity labeling: subunit assembly may explain high- and low-affinity agonist states. *Proc. Nat. Acad. Sci. USA* **80**, 156–159.

Avissar, S., Moscona-Amir, E., and Sokolovsky, M. (1982) Photoaffinity labeling reveals two muscarinic receptor macromolecules associated with the presence of calcium in rat adenohypophysis. *Febs. Lett.* **150**, 343–346.

Baron, B., Gavish, M., and Sokolovski, M. (1985) Heterogeneity of solubilized muscarinic cholinergic receptors: Binding and hydrodynamic properties. *Arch. Biochem. Biophys.* **240**, 281–296.

Barrantes, F. J. (1975) The nicotinic cholinergic receptor: Different compositions evidenced by statistical analysis. *Biochem. Biophys. Res. Comm.* **62**, 407–414.

Beld, A. J. and Ariens, E. J. (1974) Stereospecific binding as a tool in attempts to localize and isolate muscarinic receptors. *Eur. J. Pharmacol.* **25**, 203–209.

Benovic, J. L., Pike, L. J., Cerione, R. A., Staniszewski, C., Yoshima, T., Codina, J., Caron, M. G., and Lefkowitz, R. J. (1985) Phosphorylation of the mammalian β-adrenergic receptor by cyclic AMP-dependent protein kinase. *J. Biol. Chem.* **260**, 7094–7101.

Benovic, J. L., Strasser, R. H., Caron, M. G., and Lefkowitz, R. J. (1986a) β-Adrenergic receptor kinase: Identification of a novel protein kinase that phosphorylates the agonist-occupied form of the receptor. *Proc. Nat. Acad. Sci. USA* **83**, 2797–2801.

Benovic, J. L., Mayor, Jr., F., Somers, R. L., Caron, M. G., and Lefkowitz, R. J. (1986b) Light-dependent phosphorylation of rhodopsin by β-adrenergic receptor kinase. *Nature* **321**, 869–872.

Burgoyne, R. D. (1980) A possible role of synaptic-membrane protein phosphorylation in the regulation of muscarinic acetylcholine receptors. *Febs. Lett.* **122**, 288–292.

Burgoyne, R. D. (1981) The loss of muscarinic acetylcholine receptors in synaptic membranes under phosphorylating conditions is dependent on calmodulin. *Febs. Lett.* **127**, 144–148.

Berrie, C. P., Birdsall, N. J. M., Burgen, A. S. V., and Hulme, E. C. (1979) Guanine nucleotides modulate muscarinic receptor binding in the heart. *Biochem. Biophys. Res. Comm.* **87**, 1000–1005.

Berrie, C. P., Birdsall, N. J. M., Hulme, E. C., Keen, M., and Stockton, M. (1984) Solubilization and characterization of guanine nucleotide-sensitive muscarinic agonist binding sites from rat myocardium. *Br. J. Pharmacol.* **82**, 853–861.

Berrie, C. P., Birdsall, N. J. M., Dadi, H. K., Hulme, E. C., Morris, R. J., Stockton, J. M., and Wheatley, M. (1985) Purification of the muscarinic acetylcholine receptor from rat forebrain. *Biochem. Soc. Trans.* **13**, 1101–1103.

Birdsall, N. J. M., Burgen, A. S. V., and Hulme, E. C. (1979) A study of the muscarinic receptor by gel electrophoresis. *Br. J. Pharmacol.* **66**, 337–342.

Blobel, G. and Dobberstein, B. (1975) Transfer of proteins across membranes. II. Reconstitution of functional rough microsomes from heterologous components. *J. Cell Biol.* **67**, 852–862.

Bonner, T. I., Buckley, N. J., Young, A. C., and Brann, M. R. (1987) Identification of a family of muscarinic acetylcholine receptor genes. *Science* **237**, 527–532 (*see also Science* **237**, 1628 for complete DNA sequences).

Bonner, T. I., Young, A. C., Brann, M. R., and Buckley, N. J. (1988) Cloning and expression of the human and rat m5 muscarinic acetylcholine receptor genes. *Neuron* **1**, 403–410.

Brann, M. R., Buckley, N. J., Jones, P. S. V., and Bonner, T. I. (1987) Expression of a cloned muscarinic receptor in A9 L cells. *Molec. Pharmacol.* **32**, 450–455.

Cremo, C. R., Herron, G. S., and Schimerlik, M. I. (1981) Solubilization of the atrial muscarinic receptor: A new detergent system and rapid assays. *Anal. Biochem.* **115**, 331–338.

Cremo, C. and Schimerlik, M. I. (1984) Photoaffinity labeling of the solubilized, partially purified muscarinic acetylcholine receptor from porcine atria by *p*-azidoatropine methyl iodide. *Biochemistry* **23**, 3494–3501.

Dadi, H. K. and Morris, R. J. (1984) Muscarinic cholinergic receptor of rat brain. Factors influencing migration in electrophoresis and gel filtration in sodium dodecyl sulfate. *Eur. J. Biochem.* **144**, 617–628.

Dixon, R. A. F., Kobilka, B. K., Strader, D. J., Benovic, J. L., Dohlman, H. G., Frielle, T., Bolanowski, M. A., Bennett, C. D., Rands, E., Diehl, R. E., Mumford, R. A., Slater, E. E., Sigal, I. S., Caron, M. G., Lefkowitz, R. J., and Strader, C. D. (1986) Cloning of the gene and cDNA for mammalian β-adrenergic receptor and homology with rhodopsin. *Nature* **321**, 75–79.

Dratz, E. A. and Hargrave, P. A. (1983) The structure of rhodopsin and the rod outer segment disc membrane. *Tr. Biochem. Sci.* **8**, 128–131.

Engelman, D. M., Steitz, T. A., and Goldman, A. (1986) Identifying nonpolar transbilayer helices in amino acid sequences of membrane proteins. *Ann. Rev. Biophys. Biophys. Chem.* **15**, 321–353.

Florio, V. A. and Sternweis, P. C. (1985) Reconstitution of resolved muscarinic cholinergic receptors with purified GTP-binding proteins. *J. Biol. Chem.* **260**, 3477–3483.

Fukuda, K., Kubo, T., Akiba, I., Maeda, A., Mishina, M., and Numa, S. (1987) Molecular distinction between muscarinic acetylcholine receptor subtypes. *Nature* **327**, 623–625.

Fukuda, K., Higashida, H., Kubo, T., Maeda, A., Akiba, I., Bujo, H., Mishina, M., and Numa, S. (1988) Selective coupling with K^+ currents of muscarinic acetylcholine receptor subtypes in NG108-15 cells. *Nature* **335**, 355–358.

Gardner, A. L., Choo, L. K., and Mitchelson, F. (1988) Comparison of the effects of some muscarinic agonists on smooth muscle function and phosphatidylinositol turnover in the guinea-pig taenia caeci. *Br. J. Pharmacol.* **94**, 199–211.

Gavish, M. and Sokolovsky, M. (1982) Solubilization of muscarinic acetylcholine receptor by zwitterionic detergent from rat brain cortex. *Biochem. Biophys. Res. Comm.* **009**, 819–824.

Gorissen, H., Aerts, G., and Laduron, P. (1978) Characterization of digitonin-solubilized muscarinic receptor from rat brain. *Febs. Lett.* **96**, 64–68.

Gorissen, H., Aerts, G., Ilien, B., and Laduron, P. (1981) Solubilization of muscarinic acetylcholine receptors from mammalian brain: An analytical approach. *Anal. Biochem.* **14**, 33–41.

Haga, T. (1983) Characterization of muscarinic acetylcholine receptors solubilized by L-α-lysophosphatidylcholine and lubrol PX, in *Pharmacologic and Biochemical Aspects of Neurotransmitter Receptors* (Yoshida, H. and Yamamura, H. I., eds.) John Wiley, N.Y., 43–58.

Haga, K. and Haga, T. (1983) Affinity chromatography of the muscarinic acetylcholine receptor. *J. Biol. Chem.* **258**, 13575–13579.

Haga, K. and Haga, T. (1985) Purification of the muscarinic acetylcholine receptor from porcine brain. *J. Biol. Chem.* **260**, 7927–7935.

Haga, K., Haga, T., and Ichiyama, A. (1986) Reconstitution of the muscarinic acetylcholine receptor: guanine nucleotide-sensitive high affinity binding of agonists to purified muscarinic receptors reconstituted with GTP-binding proteins (G_i and G_o). *J. Biol. Chem.* **261**, 10133–10140.

Haga, K., Haga, T., Ichiyama, A., Katada, T., Kurose, H., and Ui, M. (1985) Functional reconstitution of purified muscarinic receptors and inhibitory guanine nucleotide regulatory protein. *Nature* **316**, 731–733.

Hammer, R., Berrie, C. P., Birdsall, N. J. M., Burgen, A. S. V., and Hulme, E. C. (1980) Pirenzepine distinguishes between different subclasses of muscarinic receptors. *Nature* **283**, 90–92.

He, X., Wu, X., and Baum, B. J. (1988) Protein kinase C differentially inhibits muscarinic receptor operated Ca^{2+} release and entry in human salivary cells. *Biochem. Biophys. Res. Commun.* **152**, 1062–1069.

Henderson, R. and Unwin, P. T. (1975) Three-dimensional model of purple membrane obtained by electron microscopy. *Nature* **257**, 28–32.

Herron, G. S. and Schimerlik, M. I. (1983) Glycoprotein properties of the solubilized atrial muscarinic acetylcholine receptor. *J. Neurochem.* **41**, 1414–1420.

Herron, G. S., Miller, S., Manley, W-L., and Schimerlik, M. I. (1982) Ligand interactions with the solubilized porcine atrial muscarinic receptor. *Biochemistry* **21**, 515–520.

Ho, A. K. S. and Wang, J. H. (1985) Calmodulin regulation of cholinergic muscarinic receptor: Effects of calcium and phosphorylating states. *Biochem. Biophys. Res. Comm.* **133**, 1193–1200.

Ho, A. K. S., Shang, K., and Duffield, R. (1986) Calmodulin regulation of the cholinergic receptor in the rat heart during ontogeny and senescence. *Mech. Ageing Develop.* **36**, 143–154.

Hootman, S. R., Picado-Leonard, T. M., and Burnham, D. B. (1985) Muscarinic acetylcholine receptor structure in acinar cells of mammalian exocrine glands. *J. Biol. Chem.* **260**, 4186–4194.

Hubbard, S. C. and Ivatt, R. J. (1981) Synthesis and processing of asparagine-linked oligosaccharides. *Ann. Rev. Biochem.* **50**, 555–583.

Hulme, E. C. and Birdsall, N. J. M. (1986) Distinctions in acetylcholine receptor activity. *Nature* **323**, 396–397.

Hulme, E. C., Birdsall, N. J. M., Burgen, A. S. V., and Mehta, P. (1978) The binding of antagonists to brain muscarinic receptors. *Mol. Pharmacol.* **14**, 737–750.

Hulme, E. C., Berrie, C. P., Haga, T., Birdsall, N. J. M., Burgen, A. S. V., and Stockton, J. (1983) Solubilization and molecular characterization of muscarinic acetylcholine receptors. *J. Receptor Res.* **3**, 301–311.

Hurko, O. (1978) Specific [^{3}H]quinuclidinyl benzilate binding activity in digitonin-solubilized preparations from bovine brain. *Arch. Biochem. Biophys.* **190**, 434–445.

Jones, P. S. V., Barker, J. L., Bonner, T. I., Buckley, N. J., and Brann,

M. R. (1988) Electrophysiological characterization of cloned m1 muscarinic receptors expressed in A9 L cells. *Proc. Natl. Acad. Sci. USA* **85**, 4056–4060.

Kelleher, D. J. and Johnson, G. L. (1986) Phosphorylation of rhodopsin by protein kinase C in vitro. *J. Biol. Chem.* **261**, 4749–4757.

Krebs, E. G. and Beavo, J. A. (1979) Phosphorylation-dephosphorylation of enzymes. *Ann. Rev. Biochem.* **48**, 923–959.

Kubo, T., Fukuda, K., Mikami, A., Maeda, A., Takahashi, H., Mishina, M., Haga, T., Haga, K., Ichiyama, A., Kangawa, K., Kojima, M., Matsuo, H., Hirose, T., and Numa, S. (1986a) Cloning, sequencing and expression of complementary DNA encoding the muscarinic acetylcholine receptor. *Nature* **323**, 411–416.

Kubo, T., Maeda, A., Sugimoto, K., Akiba, I., Mikami, A., Takahashi, H., Haga, T., Haga, K., Ichiyama, A., Kangawa, K., Matsuo, H., Hirose, T., and Numa, S. (1986b) Primary structure of porcine cardiac muscarinic acetylcholine receptor deduced from the cDNA sequence. *Febs. Lett.* **209**, 367–372.

Kuno, T., Shirakawa, O., and Tanaka, C. (1983) Regulation of the solubilized bovine cerebral cortex muscarinic receptor by GTP and Na^+. *Biochem. Biophys. Res. Comm.* **112**, 948–953.

Kurose, H., Katada, T., Haga, T., Haga, K., Ichiyama, A., and Ui, M. (1986) Functional interaction of purified muscarinic receptors with purified guanine nucleotide regulatory proteins reconstituted in phospholipid vesicles. *J. Biol. Chem.* **261**, 6423–6428.

Kwatra, M. M. and Hosey, M. M. (1986) Phosphorylation of the cardiac muscarinic receptor in intact chick heart and its regulation by a muscarinic agonist. *J. Biol. Chem.* **261**, 12429–12432.

Kyte, J. and Doolittle, R. F. (1982) A simple method for displaying the hydrophobic character of a protein. *J. Mol. Biol.* **157**, 105–132.

Large, T. H., Cho, N. J., De Mello, F. G., and Klein, W. L. (1985a) Molecular alteration of a muscarinic acetylcholine receptor system during synaptogenesis. *J. Biol. Chem.* **260**, 8873–8881.

Large, T. H., Rauh, J. J., De Mellow, F. G., and Klein, W. L. (1985b) Two molecular weight forms of muscarinic acetylcholine receptors in the avian central nervous system: Switch in predominant form during differentiation of synapses. *Proc. Nat. Acad. Sci. USA*, **82**, 8785–8789.

Lawson, E. Q., Sadler, A. J., Harmatz, D., Brandau, D. T., Micanovic, R., MacElroy, R. D., and Middaugh, C. R. (1984) A simple experimental model for hydrophobic interactions in proteins. *J. Biol. Chem.* **259**, 2910–2912.

Lindmar, R., Loffelholz, K., and Sandmann, J. (1988) The mechanism of muscarinic hydrolysis of choline phospholipids in the heart. *Biochem. Pharmacol.* **37**, 4689–4695.

Monsma, F. J., Abood, L. G., and Hoss, W. (1988) Inhibition of phosphoinositide turnover by selective muscarinic antagonists in the

rat striatum. *Biochem. Pharmacol.* **37**, 2437–2443.
Nathans, J., Thomas, D., and Hogness, D. S. (1986) Molecular genetics of human color vision: The genes encoding blue, green, and red pigments. *Science* **232**, 193–202.
Ovchinnikov, Y. A. (1982) Rhodopsin and bacterio-rhodopsin: structure-function relationships. *FEBS Lett.* **148**, 179–191.
Peralta, E. G., Winslow, J. W., Peterson, G. L., Smith, D. H., Askenazi, A. Ramachandran, J., Schimerlik, M. I., and Capon, D. J. (1987a) Primary structure and biochemical properties of an M_2 muscarinic receptor. *Science* **236**, 600–605.
Peralta, E. G., Ashkenazi, A., Winslow, J. W., Smith, D. H., Ramachandran, J., and Capon, D. J. (1987b) Distinct primary structures, ligand binding properties and tissue specific expression of four human muscarinic acetylcholine receptors. *EMBO J.* **6**, 3923–3929.
Peralta, E. G., Ashkenazi, A., Winslow, J. W., Ramachandran, J., and Capon, D. J. (1988) Differential regulation of PI hydrolysis and adenylyl cyclase by muscarinic receptor subtypes. *Nature* **334**, 434–437.
Peterson, G. L. and Schimerlik, M. I. (1984) Large scale preparation and characterization of membrane-bound and detergent-solubilized muscarinic acetylcholine receptor from pig atria. *Prep. Biochem.* **14**, 33–74.
Peterson, G. L., Herron, G. S., Yamaki, M., Fullerton, D. S., and Schimerlik, M. I. (1984) Purification of the muscarinic acetylcholine receptor from porcine atria. *Proc. Natl. Acad. Sci. USA* **81**, 4993–4997.
Peterson, G. L., Rosenbaum, L. C., and Schimerlik, M. I. (1988) Solubilization and hydrodynamic properties of porcine atrial muscarinic acetylcholine receptor in dodecyl-β-D-maltoside. *Biochem. Journ.* **255**, 553–560.
Peterson, G. L., Rosenbaum, L. C., Broderick, D. J., and Schimerlik, M. I. (1986) Physical properties of the purified cardiac muscarinic acetylcholine receptor. *Biochemistry* **25**, 3189–3202.
Rauh, J. J., Lambert, M. P., Cho, N. J., Chin, H., and Klein, W. L. (1986) Glycoprotein properties of muscarinic acetylcholine receptors from bovine cerebral cortex. *J. Neurochem.* **46**, 23–32.
Repke, H. and Matthies, H. (1980) Biochemical characterization of solubilized muscarinic acetylcholine receptors. *Brain Res. Bull.* **5**, 703–709.
Rosenberger, L. B., Yamamura, H. I., and Roeske, W. R. (1980) Cardiac muscarinic cholinergic receptor binding is regulated by Na^+ and guanine nucleotides. *J. Biol. Chem.* **255**, 820–823.
Rosenbaum, L. C., Malencik, D. A., Tota, M. R., Anderson, S. R., and Schimerlik, M. I. (1987) Phosphorylation of the porcine atrial muscarinic receptor by cAMP-dependent protein kinase. *Biochemistry* **26**, 8183–8188.
Schimerlik, M. I., Miller, S., Peterson, G. L., Rosenbaum, L. C., and

Tota, M. R. (1986) Biochemical studies on muscarinic receptors in porcine atrium. *Tr. Pharmacol. Sci. February Supple.*, 1–7.

Searles, R. P. and Schimerlik, M. I. (1980) Ligand interactions with membrane-bound porcine atrial muscarinic receptor(s). *Biochemistry* **19**, 3407–3413.

Sher, E., Gotti, C., Pandiella, A., Madeddu, L., and Clementi, F. (1988) Intracellular calcium homeostasis in a human neuroblastoma cell line: Modulation by depolarization, cholinergic receptors, and α-latrotoxin. *J. Neurochem.* **50**, 1708–1713.

Shirakawa, O. and Tanaka, C. (1985) Molecular characterization of muscarinic receptor subtypes in bovine cerebral cortex by radiation inactivation and molecular exclusion h.p.l.c. *Br. J. Pharmac.* **86**, 375–383.

Shirakawa, O., Kuno, T., and Tanaka, G. (1983) The glycoprotein nature of solubilized muscarinic acetylcholine receptors from bovine cerebral cortex. *Biochem. Biophys. Res. Comm.* **115**, 814–819.

Sokolovsky, M., Gurwitz, D., and Galron, R. (1980) Muscarinic receptor binding in mouse brain: Regulation by guanine nucleotides. *Biochem. Biophys. Res. Comm.* **94**, 487–492.

Strader, C. D., Sigal, I. S., Register, R. B., Candelore, M. R., Rands, E., and Dixon, R. A. F. (1987) Identification of residues required for ligand binding to the β-adrenergic receptor. *Proc. Nat. Acad. Sci. USA* **84**, 4384–4388.

Tota, M. R., Kahler, K. R., and Schimerlik, M. I. (1987) Reconstitution of the purified porcine atrial muscarinic acetylcholine receptors with purified porcine atrial inhibitory guanine nucleotide binding protein. *Biochemistry* **26**, 8175–8182.

Venter, J. C. (1983) Muscarinic cholinergic receptor structure. Receptor size, membrane orientation and absence of major phylogenetic structural diversity. *J. Biol. Chem.* **258**, 4842–4848.

Venter, J. C., Eddy, B., Hall, L. M., and Fraser, C. M. (1984) Monoclonal antibodies detect the conservation of muscarinic cholinergic receptor structure from Drosophila to human brain and detect possible structural homology with α_1-adrenergic receptors. *Proc. Nat. Acad. Sci. USA* **81**, 272–276.

White, H. L. (1988) Effects of acetylcholine and other agents on ^{32}P-prelabeled phosphoinositides and phosphatidate in crude synaptosomal preparations. *J. Neurosci. Res.* **20**, 122–128.

Wilbur, W. J. and Lipman, D. J. (1983) Rapid similarity searches of nucleic acid and protein data banks. *Proc. Nat. Acad. Sci. USA* **80**, 726–730.

Wilden, U. and Kuhn, H. (1982) Light-dependent phosphorylation of rhodopsin: Number of phosphorylation sites. *Biochemistry* **21**, 3014–3022.

Yarden, Y., Rodriguez, H., Wong, S.K.-F., Brandt, D. R., May, D. C., Burnier, J., Harkins, R. N., Chen, E. Y., Ramachandrin, J., Ullrich,

A., and Ross, E. M. (1986) The avian β-adrenergic receptor: Primary structure and membrane topology. *Proc. Nat. Acad. Sci. USA* **83**, 6795–6799.

Zuker, C. S., Cowman, A. F., and Rubin, G. M. (1985) Isolation and structure of a rhodopsin gene from D. melanogaster. *Cell* **40**, 851–858.

4

Subtypes of Muscarinic Cholinergic Receptors:

Ligand Binding, Functional Studies, and Cloning

Barry B. Wolfe

1. Introduction

Acetylcholine released from neurons has effects on many biological processes. These effects are mediated through two major classes of receptors termed nicotinic and muscarinic cholinergic receptors. There is currently strong evidence supporting the concept that both of these major classes of cholinergic receptors are themselves comprised of distinct subtypes of receptors. For example, the nicotinic receptors on autonomic ganglia have very different properties from those on skeletal muscle, suggesting that there are at least two subtypes of the nicotinic cholinergic receptor (Taylor, 1985). The genes encoding the subunits of nicotinic receptors have been isolated and sequenced from both muscle and nerve, and indeed these molecules have distinct primary structures (Boulter et al., 1985, 1986). Similarly, the existence of subtypes of mAChR has long been postulated on the basis of distinct pharmacologic profiles of some muscarinic receptors, and recently, several genes have been isolated and sequenced that appear to code for distinct subtypes of muscarinic receptors (Kubo et al., 1986a,b; Peralta et al., 1987a,b; Bonner et al., 1987, 1988; Braun et al., 1987; Akiba et al., 1988). Thus, it appears that there are at least five distinct genes that code for proteins that have the properties of muscarinic receptors in both the human and rat genome (Bonner et al., 1987, 1988). This chapter will attempt to put into historical perspective the various reports that have led us to our current thinking with regard to subtypes of muscarinic receptors.

2. Physiological Studies

As early as the mid-1960s, suggestions had been made regarding the possibility that subtypes of muscarinic receptors exist. Thus, Burgen and Spero (1968) examined the ratio of potencies of a variety of agonists at muscarinic receptors to stimulate two responses, contraction of guinea-pig ileum and stimulation of potassium efflux from the same tissue. The ratio of agonist potencies ranged from 1.5 to 1000, and these authors suggested that the properties of the receptors subserving these responses may be different. On the other hand, in the same study, they demonstrated that each of the several muscarinic receptor antagonists examined was essentially equipotent at blocking these two responses. Their inability to find ''selective'' antagonists weakened the hypothesis that subtypes of muscarinic receptors existed, since ''selective'' effects of agonists could be explained by, for example, different ratios of ''spare receptors'' existing for the two responses measured.

The earliest reports of selective antagonists at muscarinic receptors came from Fisher et al. (1976a,b). These authors synthesized several antagonists, some of which showed a degree of specificity for certain muscarinic functions both in vivo and in vitro. These compounds, fused quinuclidine-cyclohexane derivatives, were not very potent antagonists of muscarinic receptors compared to, for example, atropine. However, unlike atropine, these compounds displayed differences in potency in antagonizing oxotremorine-induced ileal contractions, salivary gland secretion, and CNS-mediated tremors.

Similarly, Barlow et al. (1976) measured the affinity of several antagonists for blocking muscarinic receptors in guinea-pig ileum and atrium. They described a compound, 4-diphenylacetoxy-*N*-methyl piperidine methiodide (4-DAMP), that showed 20-fold selectivity in blocking muscarinic receptor-mediated responses in the ileum compared to the heart. Structure–activity relationships indicated that drugs containing quaternary amino groups were more selective than drugs containing tertiary amino groups.

In other experiments, Gardier et al. (1978) presented physiological evidence for two distinct types of muscarinic receptor in the superior cervical ganglion of the cat. In these experiments, the pharmacological properties of two responses mediated by muscarinic receptors in the superior cervical ganglion were compared. Gallamine and pancuronium appeared to be inhibitors at the muscarinic receptors on dopaminergic interneurons in the ganglion that control contraction of the nictitating membrane. Other muscarinic receptors on postganglionic neurons are responsible for a receptor-mediated slow depolarization of the ganglion. Gallamine and pancuronium had no effect on this latter response. Gardier et al. (1978) concluded that muscarinic receptors on dopaminergic interneurons have different properties than do those muscarinic receptors on the postganglionic neurons in the cat superior cervical ganglion, and that drugs such as gallamine and pancuronium can distinguish between these receptors. Clark and Mitchelson (1976) also compared these types of drugs for inhibiting muscarinic receptor-mediated responses in guinea-pig atrium and ileum. They found, for example, that gallamine is 10–100-fold more potent on atrial receptors than ileal receptors and concluded that the receptors may be distinct. It appears, on the basis of several reports, that gallamine may interact allosterically with muscarinic receptors (Stockton et al., 1983; Dunlap and Brown, 1983). Allosteric effects are dealt with in detail in Chapters 2 and 10.

Goyal and Rattan (1978) measured the potencies of McN-A343 and bethanechol at muscarinic receptors on interneurons in the lower esophageal sphincter of the opossum and compared them to the potencies measured for stimulation of muscarinic receptors in the smooth muscle of the esophageal sphincter itself. These authors found that McN-A343 was markedly more potent at the neuronal receptor as compared to the smooth muscle receptor, whereas bethanechol had the opposite selectivity. They suggested that this selectivity was consistent with the hypothesis that subtypes of muscarinic receptors exist. They proposed that ganglionic receptors be termed M_1 and smooth muscle receptors be termed M_2. This nomenclature has become standardized in the field. It is, however, important to note here as well as below that, since the results of the molecular biological experiments have been published (Bonner et al., 1987), it is clear that M_1 and M_2 "receptors" (pharmacologically defined receptors) are likely to consist of more than one molecular component (i.e., receptor). The nomenclature for the receptors whose structures are known from cloned genes has been suggested by Bonner et al. (1987) to be m1 through m4, and so on, noting the use of lower case "m" to distinguish them from receptors based on the pharmacology of pirenzepine.

A major breakthrough in the characterization of muscarinic receptor subtypes resulted from the introduction of the compound pirenzepine. This drug has turned out to be the most selective compound yet described for subtypes (15–30-fold). Hammer and Giachetti (1982) have examined the relative affinities of pirenzepine and atropine for blocking muscarinic receptors in rat autonomic ganglia (that mediate an increase in blood pressure) and muscarinic receptors in rat atria (that mediate a decrease in heart rate). Although atropine was equipotent in blocking these responses, pirenzepine showed approximately 20-fold selectivity for the ganglionic receptors vs the cardiac receptors. Radioligand binding experiments performed in the same study confirmed the differential affinity of pirenzepine for muscarinic receptors in the two tissues. These types of experiments provided the impetus to define M_1 and M_2 receptors as having high or low affinity for pirenzepine, respectively. It should be noted that, although the sequences for the m1–4 receptors reported by Peralta et al. (1987b) appear to be essentially identical to those reported by Bonner et al. (1987), the nomenclature for the m3 and m4 receptors has been reversed by the two groups. Thus, Bonner's m3 is Peralta's m4 and vice versa. This review will utilize the nomenclature proposed by Bonner et al. (1987), since their m3 receptor has been shown to have high affinity for the compound hexahydrosiladifenidiol (Akiba et al., 1988). This compound has been used to define pharmacologically the M3 receptor reported to be in glandular tissue and ileum (Mutschler and Lamrecht, 1984).

Several studies have investigated the possibility that M_1 and/or M_2 receptors are selectively associated with either a pre- or postsynaptic anatomical localization. For example, Fuder (1982) measured the affinity of pirenzepine and other antagonists for blocking muscarinic receptors in the isolated rat and rabbit heart. Fuder (1982) characterized receptors located on cardiac myocytes controlling heart rate and, in the same preparation, the responses of receptors located presynaptically on noradrenergic nerve terminals. These latter receptors regulate release of norepinephrine in the heart. Fuder (1982) reported that the affinity of pirenzepine for the muscarinic receptors controlling both responses was identical, and both pre- and postsynaptic receptors appeared to be of the M_2 subtype. On the other hand, Raiteri et al. (1984) examined presynaptic receptors in three preparations from rat brain and found differences in the affinity of pirenzepine for affecting transmitter release. They concluded that some prejunctional receptors were of the M_1 subtype and some were of the M_2 subtype. Thus, Raiteri et al. (1984) measured the affinity of pirenzepine and other antagonists such as atropine for receptors in synaptosomal preparations from

1. Cerebral cortex, where autoreceptors control release of acetylcholine
2. Hippocampus, where autoreceptors also control release of acetylcholine and
3. Striatum, where presynaptic heteroreceptors control release of dopamine.

Pirenzepine was found to be a weak antagonist at both autoreceptor preparations, whereas, in contrast, it was 100 times more potent in the heteroreceptor preparation. Thus, it appears that functional receptors of both high and low affinity for pirenzepine (M_1 and M_2) can exist presynaptically.

Mutschler and Lamrecht (1984) have presented data on several agonist and antagonists that appear to affect selectively some muscarinic responses vs others. Thus, they have shown that muscarinic receptors in rat sympathetic ganglia that control release of norepinephrine in vasculature are of the M_1 subtype in that they have high affinity for pirenzepine. Conversely, they showed that muscarinic receptors in the atria and ileum have low affinity for pirenzepine and are thus classified as M_2. On the other hand, several of the antagonists that they tested appeared to distinguish between the receptors in the atria and ileum. The degree of selectivity of most of the drugs was not large (2–10-fold), but one compound in particular, hexahydro siladifenidol, was nearly 30-fold more potent at inhibiting muscarinic receptor-mediated responses in rat ileum than in rat atria. These authors suggested that M_2 receptors may need to be divided into subgroups.

Similarly, the compound AF-DX 116 appears to not only distinguish between M_1 and M_2 receptors, but also recognizes cardiac M_2 receptors with approximately 30-fold higher affinity than, for example, M_2 receptors located on smooth muscles or secretory glands (Giachetti et al., 1986). These authors measured the affinity of this antagonist for rat ganglionic and cardiac muscarinic receptors using the same paradigms as Hammer and Giachetti (1982) described above. AF-DX 116 was 30–50-fold selective for cardiac receptors. In the same study, these authors examined, using isolated, perfused organ preparations, the potency of AF-DX 116 for inhibiting muscarinic receptor-mediated inhibition of beating rate in isolated guinea-pig atrium, muscarinic receptor-mediated contractions of guinea-pig ileum and rat bladder, and muscarinic receptor-mediated stimulation of acid secretion in rat stomach. AF-DX 116 was approximately 30-fold selective for cardiac responses vs each of the others. Again, these data are consistent with the idea that more than two types of muscarinic receptors exist and that the receptors called M_2 should be subclassified.

Although these types of data (i.e., most of the above) are suggestive regarding the existence of subtypes of muscarinic receptors, they need to be interpreted with caution. Selectivity of action of a drug in a whole animal or in in vitro organ preparations can be the result of factors other than the existence of receptor subtypes such as tissue-specific differences in degradation, uptake, sequestration, of diffusion barriers. For example, O'Donnell and Wanstall (1976) compared the effects of a variety of drugs selective for β-2 adrenergic receptors on relaxation of bronchial smooth muscle vs vascular smooth muscle and found marked differences in potencies of antagonists, even though these responses are both thought to be primarily the result of interactions with β-2 adrenergic receptors. When inhibitors of extraneuronal uptake were administered, these differences in pharmacological specificity disappeared, implying that the observed differences were the result of tissue-specific differences in extraneuronal uptake.

3. Radioligand Binding Studies

One approach that minimizes some of the problems associated with whole animal or whole organ experiments involves the use of radioligand binding to membranes prepared from tissues. Using these techniques, several early studies demonstrated that several muscarinic receptor antagonists have the same affinity for muscarinic receptors in membrane preparations from many organs (Yamamura et al., 1974; Snyder et al., 1975; Hulme et al., 1976). Similarly, Beld et al. (1975) measured the affinity of atropine and the stereoisomers of *N*-isopropylatropine in tracheal muscle, parotid gland, and caudate nucleus using radioligand binding techniques, and found each of the antagonists tested to have essentially identical affinities in each of the tissues. These authors concluded that the muscarinic receptors were equivalent in these tissues and suggested that subtypes may not exist.

In addition to radioligand binding studies examining the affinity of antagonists, several studies have investigated the properties of agonist binding, and concluded that the binding is complex and that this complexity may (or may not) be related to subtypes of receptors. Birdsall et al. (1978, 1980) examined the binding of agonists both directly using the labeled agonist [^{3}H]oxotremorine-*M* ([^{3}H]OXO-*M*), and indirectly using agonist inhibition of binding of the labeled antagonist [^{3}H]propylbenzilylcholine. These authors found that [^{3}H]OXO-*M* binds to membranes from rat cerebral cortex with at least two distinct affinities. Thus, the Rosenthal plot was curvilinear, and dissociation kinetics were not monophasic (Birdsall et al., 1980). In agreement with this, Birdsall et al. (1978) showed that inhibition of radiolabeled antagonist binding by the agonist carbachol did not follow the law of mass action and the data could best be fit by the existence of three sites for carbachol that were termed low (L), high (H), and super high (SH) sites to designate their relative affinities for carbachol. Whether these sites with apparently different affinities for agonists are in some way related to subtypes of muscarinic receptors defined either by the affinity of pirenzepine or by the molecular biology experiments is still questionable. It is, however, clear that at least part of the complexity of agonist binding to muscarinic receptors is related to the fact that these receptors interact with guanine nucleotide binding regulatory proteins (G proteins). Thus, Berrie et al. (1979) and Rosenberger et al. (1980) showed that guanine nucleotides modulate the apparent affinity of agonists at muscarinic receptors. This modulation of agonist affinity is common among receptors coupled to adenylate cyclase or phosphoinositide breakdown, and is a reflection of the fact that agonist binding induces the formation of a ternary complex consisting of agonist-receptor-G protein. When this complex is formed to an appreciable extent, as in a membrane binding assay performed in the absence of guanine nucleotides, one always observes the appearance of multiple affinity states for agonists, some of which are modulated by the presence of guanine nucleotides (Boeynaems and Dumont, 1977; Jacobs and Cuatrecasas, 1976). This observation is true even in systems that contain only a single receptor subtype (Minneman and Molinoff, 1980). Thus, as is the case in physiological experiments, it is also true in radioligand binding experiments that one should avoid the use of agonists to define the existence of subtypes of receptors. On the other hand, it is clear from several studies (*see below*) that there is *some* (undefined at the molecular level) relationship between low affinity agonist binding and high affinity binding of pirenzepine to muscarinic receptors.

As with physiological studies of muscarinic receptors, the introduction of pirenzepine to radioligand binding studies opened a new area of investigation. Thus, Hammer et al. (1980) showed that pirenzepine inhibits radiolabeled antagonist binding in a biphasic manner in some tissues. For example, in rat cerebral cortex one can analyze the shallow inhibition curve produced by pirenzepine, assuming that the complexity of the binding is the result of the existence of two noninteracting sites that have different affinities for pirenzepine, and find that the high affinity sites (10 n*M*) comprise most (60–80%) of the sites and that the low affinity sites (300 n*M*) comprise the rest. Several investigators have shown that this ratio varies from tissue to tissue, for example, rat heart and cerebellum contain essentially pure populations of low affinity (M_2) sites (Hammer et al., 1980; Watson et al., 1982; Luthin and Wolfe, 1984a). Not long after the introduction of pirenzepine to radioligand binding experiments, the drug itself was radiolabeled and [^{3}H]pirenzepine ([^{3}H]PZ) was compared to standard radioligands such as [^{3}H]-quinuclidinylbenzilate ([^{3}H]QNB) (Watson et al. 1982, 1983; Luthin and Wolfe, 1984a). These studies showed that the binding of [^{3}H]PZ was to a single class of sites of high affinity (10 n*M*) and no evidence of curvilinearity was present in Rosenthal plots. The density of binding sites for [^{3}H]PZ was less than the density of binding sites for [^{3}H]QNB in all tissues tested. The pharmacological profile of drugs inhibiting [^{3}H]PZ binding was identical to that seen for [^{3}H]QNB with the exception that pirenzepine inhibited [^{3}H]PZ binding with a single high affinity (10 n*M*). These data were interpreted to mean that the assay for [^{3}H]PZ binding only allows measurement of a subset of [^{3}H]QNB binding sites, those with high affinity for pirenzepine. The fact that the low affinity sites are not detected in this assay is assumed to be the result of rapid dissociation from these sites during separation of bound and free ligand. The fact that the percentage of high affinity sites for pirenzepine, as determined by computer-assisted analysis of pirenzepine inhibition of [^{3}H]QNB binding, yielded estimates of the density of sites that matched those determined by using [^{3}H]PZ directly indicated that this interpretation was correct (Luthin and Wolfe, 1984a). Thus, the measurement of the density of [^{3}H]QNB sites yields an estimate of the sum of M_1 + M_2 sites and, by definition, the density of [^{3}H]PZ sites yields an estimate of the M_1 sites. The differences between the densities of sites determined with the two ligands yields the density of M_2 sites. It is important to note again that M_1 sites are not necessarily equivalent to m1 receptors as defined by the molecular cloning experiments.

Some studies examined alternative (to subtypes) possibilities to explain the observations discussed above. Since it had been shown previously that the binding of antagonists to muscarinic receptors involves a ligand-induced isomerization of the receptor (Jarv et al., 1979), Luthin and Wolfe (1984a) measured the kinetic rate constants of association (k_1), dissociation (k_{-1}), and isomerization (k_2,k_{-2}) for pirenzepine and QNB interacting with muscarinic receptors assuming the following model is correct:

$$L + R \underset{k_{-1}}{\overset{k_1}{\rightleftharpoons}} LR \underset{k_{-2}}{\overset{k_2}{\rightleftharpoons}} LR'$$

The hypothesis being tested was that, if the species measured in the ligand binding assay was LR′ and not LR, as had been suggested by several previous workers (Jarv et al., 1979), the ratio of k_{-2}/k_2 (i.e., the equilibrium constant for the isomerization) might be such that only a finite ($<<100$) percentage of the [^{3}H]PZ bound is to the isomerized state and that the ligand bound to the nonisomerized state would be considered free since its rate of dissociation is rapid. Although the data supported this model for antagonist binding to muscarinic receptors, the values obtained for the isomerization step indicated that more than 90% of the ligand bound is bound to the isomerized form of the receptor and that this model cannot explain the lower receptor density obtained with [^{3}H]PZ compared to [^{3}H]QNB.

Another possibility for the differences between the densities of binding sites for [^{3}H]QNB and [^{3}H]PZ is that [^{3}H]PZ is sensitive to a particular conformation of a single muscarinic receptor protein (Roeske and Venter, 1984). This study reported that, using solubilized receptor preparations, [^{3}H]PZ binding sites were very thermolabile and were lost rapidly even at low temperatures, whereas in the same preparation, [^{3}H]QNB binding sites were stable (Roeske and Venter, 1984). This result is consistent with the idea that [^{3}H]PZ binding is to a specific conformation of muscarinic receptor and [^{3}H]QNB binding is insensitive to this hypothetical conformational change. Alternatively, it could be postulated that the binding of [^{3}H]PZ is to a subtype of muscarinic receptors that is more thermolabile than other subtypes. If so, it would be expected that [^{3}H]QNB binding would be lost to an equal extent as that seen for [^{3}H]PZ. This was not the result reported by Roeske and Venter (1984). The findings in this (Roeske and Venter, 1984) report, however, were not confirmed in two other publications. Thus, Berrie et al. (1985) and Luthin and Wolfe (1985) reported, in contrast to the earlier study of Roeske and Venter (1984), the [^{3}H]QNB binding was, in fact, equally thermolabile as [^{3}H]PZ binding. The data from these latter studies are consistent with the idea that [^{3}H]PZ binding is related to subtypes of muscarinic receptors.

On the other hand, investigators have reported that, upon solubilization of cardiac plasma membranes that initially contain essentially no measurable high affinity sites for pirenzepine, the solubilized preparation contains significant percentages (50–60%) of high affinity binding sites for pirenzepine measured either directly or indirectly (Berrie et al., 1986). When these receptors are purified, the high affinity component is selectively lost during purification. These observations could have at least two distinct explanations: (1) High affinity binding of pirenzepine involves ligand-induced association of the receptor(s) with another (unidentified) protein. This association is somehow prohibited in cardiac membranes, and solubilization eliminates this prohibition. Purification eliminates the other (unidentified) protein. (2) The affinity of various muscarinic receptor subtypes for pirenzepine is altered differently upon solubilization. The heart contains a mixture of subtypes, and some subtypes increase their affinity, whereas others decrease their affinity. One subtype is selectively purified.

Data obtained from solubilization of muscarinic receptors from rat lacrimal gland only seem to confuse this issue. Thus, Berrie et al. (1986) reported that, for rat lacrimal gland, a tissue that normally displays only low affinity pirenzepine sites, there is, if anything, a small *decrease* in the affinity of pirenzepine upon solubilization. Thus, when inhibition curves for pirenzepine are examined in *membranes* from heart and lacrimal gland the IC_{50} values are approximately 3 μM and 1 μM, respectively. In *solubilized* preparations of these two tissues, the IC_{50} values are approximately 0.1 μM and 3μM, respectively. A possible explanation for all of these results is the existence of several (four or more) subtypes of muscarinic receptors, each with its own distinct affinity for pirenzepine and each having its affinity for pirenzepine modulated by solubilization in a distinct manner.

In support of this hypothesis, recent experiments (Luthin et al., 1988) utilizing antibodies selective for a unique region (C-terminus) of the (cloned) m1 receptor have shown that [^{3}H]PZ binding sites solubilized from rat cortex can be immunoprecipitated well (40%), whereas none (0%) of the [^{3}H]PZ binding sites solubilized from rat heart will precipitate with this antibody. Thus, it seems clear that [^{3}H]PZ binding sites, at least in solubilized preparations, contain more than just m1 muscarinic receptors, and that solubilization is not unmasking "cryptic" m1 receptors that normally exist in the heart, but are somehow not able to be assayed in membrane preparations. Additionally, this result, together with those discussed above, strongly suggests that the affinity of pirenzepine for muscarinic receptors is a function of membrane environment.

When one examines the distribution of high and low affinity binding sites for pirenzepine, the percentage of, for example, M_1 sites in membranes from various tissues varies markedly. Thus, in the rat cerebral cortex, striatum, and hippocampus the percentage is high (60–90%), whereas in the cerebellum, heart, ileum, and pons-medulla, the percentage is low (0–20%) (Watson et al. 1982, 1983; Luthin and Wolfe, 1984a). Although the percentage of M_1 sites is usually low in peripheral tissues, it has been shown that human stellate ganglia contains approximately 50% M_1 sites (Watson et al., 1984). This observation is in agreement with the several reports studying functional responses, noted above, demonstrating the some of the receptors in autonomic ganglia have high affinity for pirenzepine. Additionally, it has been reported that the majority (78%) of the muscarinic receptors in rabbit lung have a high (5 nM) affinity for [^{3}H]PZ (Bloom et al., 1987). Whether these M_1 receptors are on ganglia or smooth muscle remains to be determined.

Studies of the distribution of M_1 and M_2 binding sites have also been reported using autoradiographic techniques. Thus, Yamamura et al. (1983) reported that muscarinic receptors ([^{3}H]QNB binding sites) exist in the substantia gelatinosa and the rest of the dorsal horn of the spinal cord of the rat. On the other hand, [^{3}H]PZ binding sites appear to be mainly localized in the substantia gelatinosa and not in the dorsal horn. Wamsley et al. (1984) confirmed this observation and extended these studies to regions of the rat brain. These authors utilized a semiquantitative autoradiographic approach, and showed that areas such as the superior colliculus, facial nucleus, nucleus tractus solitarius and the hypoglossal nucleus have markedly lower amounts of [^{3}H]PZ binding relative to [^{3}H]QNB binding. Cortes and Palacios (1986) have quantitatively measured the density of muscarinic sites for [^{3}H]*N*-methylscopolamine ([^{3}H]NMS). This radioligand appears not to have markedly different affinities for M_1 and M_2 sites, but may label fewer sites because of the fact that it is a quaternary amine and does not cross cell membranes to label internalized receptors (Lee and El-Fakahany, 1985; Brown and Goldstein, 1986). These investigators examined inhibition curves generated with carbachol and pirenzepine in more than 20 discrete regions of the rat brain. These inhibition curves were computer modeled to generate estimates of the ratios of high and low affinity binding sites. High affinity binding sites for pirenzepine represented a clear majority (>60%) of the sites in the caudate-putamen, nucleus accumbens, hippocampus CA1, dentate gyrus, subiculum, nucleus amygdaloideus basalis, frontal cortex layers II-III and V-VI, and primary olfactory cortex. Low affinity binding sites for pirenzepine represented a clear majority (>60%) of the sites in the nucleus of the diagonal band, lateral hypothalamic area, ventrolateral thalamic nucleus, zona incerta, anterior pretectal area, superior and inferior colliculus, pontine nuclei, motor trigeminal nucleus, and facial nucleus. These same types of calculations were performed using inhibition curves for carbachol, and the relationship between high affinity sites for pirenzepine and high affinity sites for carbachol was examined. There appeared to be a strong (r^2 = 0.87) inverse correlation between these sites. Thus, areas with a high percentage of high affinity sites for pirenzepine (M_1 sites) have, in general, a low percentage of high affinity sites for carbachol. This relationship has prompted several investigators to redefine the meaning of the term "M_2 binding sites." Thus, Potter et al. (1984) define M_2 sites as those muscarinic receptors that have high affinity for agonists (either carbachol or oxotremorine-*M*) and retain the more conventional definition of M_1 sites. These authors have extensively studied various perturbants of receptor–effector coupling and have suggested that tissue pretreated with *N*-ethylmaleimide in an EDTA-containing buffer contains only receptors that are uncoupled from guanine nucleotide binding proteins. They suggest that this preparation is useful for pharmacologically distinguishing M_1 from M_2 sites by labeling receptors with [^{3}H]QNB in the presence of a specific concentration of either pirenzepine to block M_1 sites or carbachol to block M_2 sites. Alternatively, they suggest labeling M_2 sites for autoradiographic analysis with [^{3}H]OXO-*M* without pretreatment with NEM/EDTA. This assay would have some degree of specificity for receptors coupled to G proteins. Although this latter statement is true, it is also true that [^{3}H]OXO-*M* labeling of muscarinic receptors in membranes from rat cerebral cortex is only partially (50%) sensitive to the addition of exogenous guanine nucleotides (Martin et al., 1984). This is in contrast to [^{3}H]OXO-*M* labeling of membranes from rat heart, which is completely sensitive to guanine nucleotides (Harden et al., 1983). The meaning of these observations is still unclear. Nonetheless, Potter et al. (1984) report that binding sites in rat brain labeled by [^{3}H]OXO-*M* and thus by their definition M_2 coincide with the distribution of cholinergic cell bodies and fibers. For example, the motor nuclei of cranial nerves, medial septum, diagonal band of Broca, dorsal fornix, fimbria, and magnocellular neurons of the ventral globus pallidus are most densely labeled with this ligand and are also areas that contain high levels of choline acetyl transferase (Fibiger, 1982).

Using the most widely accepted current definitions of subtypes of binding sites (i.e., high and low affinity for pirenzepine), investigators have studied the possibility that the binding sites are regulated differentially. Thus, Evans et al. (1985) demonstrated that M_2 sites in the murine heart develop with a different time course than do M_1 sites in the cortex. Lee and Wolfe (1985) examined this same paradigm except that the data were all obtained in rat cerebral cortex. These authors have also reported distinct developmental profiles for the M_1 and M_2 binding sites. Both studies found that M_2 sites are nearly at adult levels at birth.

This was true whether M_2 sites were in murine heart or rat cortex. In contrast, both studies found that M_1 sites in cortex developed more slowly and did not reach adult levels until approximately 3 wk after birth. Another study involving regulation of these binding sites reported that chronic administration of atropine to rats resulted in a selective increase (70%) in the density of M_2 sites in rat cerebral cortex with no change in the density of M_1 sites in the same tissue (Lee and Wolfe, 1986). These types of results are consistent with the ideas that the binding sites represent distinct proteins under separate regulatory control.

Recently, reports have appeared that examine the pharmacological properties of a novel compound, AF-DX 116, using radioligand binding techniques (Hammer et al., 1986). This compound appears to be a simple competitive inhibitor of muscarinic receptors, and shows selectivity for the receptors in heart as compared to the CNS, smooth muscle, or exocrine glands. Thus, this drug has an approximate K_D value of 100 n*M* in heart and cerebellum, 800 n*M* in cerebral cortex, and 3000 n*M* in exocrine glands such as the submandibular and lacrimal glands (Hammer et al., 1986). Since pirenzepine has low affinity for muscarinic receptors in the heart, cerebellum, and exocrine glands, the receptors in these tissues have been classified as M_2. However, the data with AF-DX-116 are more consistent with the hypothesis that M_2 binding sites represent more than one receptor. Thus, Hammer et al. (1986) suggest subdividing the M_2 binding sites into classes termed B and C sites to indicate low and high affinity sites for AF-DX 116, respectively, M_1 sites would be termed A sites in this new classification scheme. Waelbroeck et al. (1987), using sophisticated kinetic arguments, have also suggested subdividing the M_1/M_2 classification scheme into A, B, and C sites. Doods et al. (1987) using methodology similar to Hammer et al. (1986) suggest a reclassification in which classical M_2 sites are subclassified into M_2 sites, which have high affinity for AF-DX 116 and M_3 sites, which have low affinity for AF-DX 116. Taken together, it seems that the pharmacological evidence for the existence of at least three subtypes of muscarinic receptors is extensive and, fortunately, molecular cloning experiments confirmed this hypothesis and extended it. Additionally, as discussed above, these data provide for a more rational nomenclature for subtypes of muscarinic receptors (Bonner et al., 1987).

4. Molecular Cloning of Genes for Muscarinic Receptors

To date (Winter, 1988), there have been five distinct gene sequences published that appear to code for muscarinic receptors (Kubo et al., 1986a,b; Peralta et al., 1987a,b; Bonner et al., 1987; Braun et al., 1987; Bonner et al., 1988). Genomic blots from human and rat DNA indicate that it is likely that more than five distinct subtypes of muscarinic receptors exist in these species (Bonner et al., 1987). The first reports of cloned genes (Kubo et al., 1986a,b; Peralta et al., 1987a) came from efforts using highly purified preparations of receptors from either pig cerebral cortex (Haga and Haga, 1983) or pig atria (Peterson et al., 1984). In each case, the purified receptors were enzymatically digested and resulting peptide fragments purified and partially sequenced. Using this sequence information, Kubo et al. (1986a) and Peralta et al. (1987a) synthesized oligonucleotides that were used to probe cDNA libraries to obtain clones coding for muscarinic receptors. Kubo et al. (1986a) showed that a clone obtained from a pig brain cDNA library did, indeed, code for a protein with the properties of a muscarinic receptor. Frog oocytes when injected with mRNA, made by transcribing the isolated cDNA, expressed receptors on their surface that, when stimulated by acetylcholine, gated ion channels. This acetylcholine-stimulated alteration in ion flow was sensitive to atropine. Binding of [^{3}H]QNB was blocked by low concentrations of pirenzepine (K_D = 10 nM). Thus, these authors suggested that they had isolated the gene coding for a receptor with the properties of the M_1 receptor or binding site. It has since been suggested (Bonner et al., 1987) that this receptor be termed the m1 receptor (a muscarinic receptor protein of known sequence) to distinguish it from the M_1 receptor (a muscarinic receptor protein or proteins having high affinity for pirenzepine). Another gene was isolated and sequenced by starting from a purified preparation of receptors, this time from pig atria. This gene was shown to code for a muscarinic receptor by transfecting Chinese hamster ovary cells with a plasmid containing the cloned DNA and determining the properties of the newly expressed receptors via radioligand binding (Peralta et al., 1987a). These receptors bound [^{3}H]QNB with high affinity, and this binding was inhibited by atropine with high affinity and by pirenzepine with low affinity (1.1 μM). Thus, it was suggested that this gene codes for a muscarinic receptor protein with the properties of the M_2 receptor. Again, it was later suggested that this receptor of known primary sequence be termed the m2 receptor for reasons discussed above (Bonner et al., 1987). At the same time that this gene was being isolated and sequenced from information derived from receptors purified from pig atria, Numa's group had noted that several polypeptide fragments that had been isolated and sequenced from the muscarinic receptor protein purified from pig cerebral cortex were not coded for by the gene that they had isolated (m1, Kubo et al., 1986a). Thus, these workers reasoned that their purified protein from cortex must contain a mixture of subtypes with the predominant one being the m1 receptor. Based on these (two) peptide sequences that did not appear in the m1 receptor, they synthesized oligonucleotide probes that were used to screen the several clones that they had previously isolated by using the labeled m1 clone as a probe (Kubo et al., 1986b). Clones that hybridized to these oligonucleotide probes were then used as probes with which to screen a cDNA library derived from pig heart. Full-length clones isolated from this library were radiolabeled and used as probes of Northern blots. These experiments indicated that mRNA coding for this receptor was to be found in heart and medulla-pons, but not in cerebral cortex or striatum and the authors suggested that this cDNA codes for the m2 muscarinic receptor.

Using the information published in the reports described above, Bonner et al. (1987) synthesized oligonucleotide probes and obtained several full-length clones from human and rat libraries. Of the clones they isolated and sequenced, they found four unique sequences. Although there were minor differences most likely resulting from species differences, one clone matched the m1 sequence from pig brain and one clone matched the m2 sequence from pig heart. Two other unique clones were found and termed m3 and m4. Each of these clones was expressed in COS-7 cells, and the properties of the receptors expressed were examined using radioligand binding. Each of the receptors had high affinity for [^{3}H]QNB and atropine, whereas the m1 and m4 receptors had high affinity for pirenzepine, the m2 receptor had low affinity for pirenzepine, and the m3 receptor had intermediate affinity for pirenzepine. Thus, it seems likely that m1 and m4 receptors fit the pharmacological definition of M_1 receptors, whereas m2 receptors fit the pharmacological definition of M_2 receptors. Where m3 muscarinic receptors fit into the M_1/M_2 classification scheme is not yet clear. Recently, Bonner et al. (1988) reported the cloning and sequencing of the gene for the rat and human m5 muscarinic receptor. This receptor was shown to have intermediate to low affinity for pirenzepine and AF-DX116, respectively. Detectable quantities of mRNA coding for this receptor were not found in rat heart, salivary glands, ganglia, or in several areas of the rat brain. Thus, currently, it is unclear where, or if, this receptor is expressed. On the other hand, Brann et al. (1988a) and Buckley et al. (1988) have demonstrated that mRNA coding for m1, m2, m3, and m4 receptors is found in a number of brain structures. Using the technique of *in situ* hybridization, these authors demonstrated that mRNA for the m1 receptor is high in hippocampus, dentate gyrus, olfactory bulb, amygdala, olfactory tubercule, and piriform cortex. mRNA for m2 receptors, in general, was in low abundance in the brain with the highest levels being found in the medial septum and diagonal band, areas of high concentrations of cholinergic cell bodies. mRNA for m3 receptors was found to be the highest in the hippocampus and the superficial layers of the cerebral cortex. In contrast, mRNA for m4 receptors was found in highest abundance in the caudate putamen and the plexiform layer of the olfactory bulb. This distinct localization of mRNA suggests that these receptors may subserve distinct physiological and behavioral effects.

Fukada et al. (1987) have injected mRNA coding for either the pig m1 and the pig m2 receptor into frog oocytes and examined the electrophysiological response to stimulation by acetylcholine. Oocytes expressing m1 receptors showed a rapid depolarization that was completely blocked by intracellular injection of EGTA, implying that the response depends on a rise in intracelluar calcium. On the other hand, oocytes expressing m2 receptors depolarized more slowly in response to acetylcholine, and this response was not dependent upon intracellular calcium since EGTA applied intracellularly had minimal effects. In another report examining the ability of a specific subtype of muscarinic receptor to couple to effector systems, Ashkenazi et al. (1987) demonstrated that pig m2 receptors expressed in Chinese hamster ovary cells were capable of modulating both the breakdown of phosphoinositides and the inhibition of adenylate cyclase. These responses, however, were not coupled to the m2 receptor with the same efficiency. Thus, it took at least 100-fold higher levels of m2 receptors to obtain stimulation of phosphoinositide breakdown than to obtain inhibition of adenylate cyclase. Several reports have appeared recently describing the second messengers affected by muscarinic receptor stimulation in cells containing a single, defined subtype of receptor (Bonner et al., 1988; Brann et al., 1988b; Fukuda et al., 1988; Li and Wolfe, 1988; Mei et al., 1989; Peralta et al., 1988; Shapiro et al., 1988). From these studies, it appears that the m1, m3, and m5 receptors are coupled to the stimulation of phosphoinositide breakdown, whereas the m2 and m4 receptors appear to be coupled preferentially to the inhibition of adenylate cyclase activity. A recent report (Stein et al., 1988) suggests, however, that this relationship may be more complex. They showed that cells transfected with a plasmid carrying the gene encoding the m1 receptor not only displayed carbachol-stimulated phosphoinositide breakdown, but also carbachol-mediated inhibition of cyclic AMP accumulation. Additionally, the effect on cyclic AMP was eliminated by prior treatment of the cells with pertussis toxin, whereas the effect on phosphoinositide metabolism was unaffected by pertusis toxin, implying that these two responses may be mediated via separate G proteins. In this regard, Ashkenazi et al. (1989) have demonstrated in an elegant study that muscarinic (and other) receptors can activate phosphoinositide breakdown via two distinct G proteins, one of which is sensitive to pertussis toxin and one of which is not. Thus, they demonstrated that m2 receptors can couple to this response, albeit weakly (cf Ashkenazi et al., 1987), in a manner that is completely pertussis toxin sensitive, whereas in the same cells, m1 receptors couple strongly to this second messenger system and only a small portion of the signal from the m1 receptor is toxin sensitive. A more complete discussion of this area is to be found in other chapters in this volume.

A figure illustrating the primary amino-acid sequence of the porcine m1 and m2 receptors is included in the chapter by Schimerlik in this volume. This figure illustrates the current hypothesis regarding the secondary structure of these molecules. Once the primary sequence of the first (m1) receptor was deduced from the DNA sequence, it was noted that the m1 receptor had a great deal of homology with the recently sequenced β-2 adrenergic receptor (Dixon et al., 1986, Kubo et al., 1986a). It had already been noted that the β-2 adrenergic receptor had a high degree of homology with rhodopsin (Dixon et al., 1986). Not only was there a high degree of homology of primary structure between each of these molecules, but the hydropathicity profile of each molecule was similar. Thus, each molecule has seven stretches of approximately 20 to 25 amino acids of high hydrophobicity, and it has been suggested that these seven hydrophobic regions are likely to be membrane-spanning regions. The seven membrane-spanning regions appear to be the most highly conserved portions of the molecules, with these regions being conserved not only across the various subtypes of muscarinic receptor, but also with regard to the β2 adrenergic receptor and rhodopsin (Dixon et al., 1986). Assuming that the *N*-terminus is on the outside of the cell as has been shown for a number of membrane bound proteins, this suggests that there are four regions on the outside of the cell, termed o1–o4 by Bonner et al. (1987), with the o1 region representing the *N*-terminus. There are two or three, depending on the subtype, highly conserved asparagine residues in the o1 region. These have been postulated to be sites of glycosylation. It has been known for some time that muscarinic receptors are highly glycosylated *in situ* (Herron and Schimerlik, 1983). The rest of the o1 region is relatively unique from subtype to subtype. The other regions postulated to be on the outside of the cell (o2–o4) are moderately well conserved between the various subtypes. On the inside of the cell are postulated to be four regions, termed i1 through i4 by Bonner et al. (1987). The i1 and i2 loops are moderately well conserved between subtypes, whereas the i3 loop appears completely unique for each subtype. The i4 loop (the C-terminus) is very similar when comparing the m2 and m4 receptors or when comparing the m1 and m3 receptors. The i3 loop, in addition to being structurally unique, is also much larger than the other regions comprising approxmately one-fourth of the entire sequence of each receptor (Bonner et al., 1987). It is of interest to note that the first 18 amino acids of the i3 loop of the m2 and m4 receptors share a high degree of homology with each other, whereas these same regions in the m1, m3, and m5 receptors are nearly identical for those three proteins (*see* Bonner et al., 1988). Since this grouping is identical to the grouping with respect to second-messenger modulation (*see above*), it is tempting to speculate that this portion of the protein interacts specifically with guanine-nucleotide-binding regulatory proteins and, thus, determines the second-messenger system to be modulated.

5. Conclusions

Although it has been at least 20 years since the initial speculation that subtypes of muscarinic receptors may exist, it is only very recently that real progress has been made to demonstrate definitively that this hypothesis is correct. The introduction of the selective antagonists pirenzepine and, more recently, AF-DX 116 produced data that convincingly indicated that more than one, and probably more than two, muscarinic receptors exist. The molecular cloning experiments have shown that this number is even larger, and where it will stop is unclear. Now that we are convinced that subtypes of muscarinic receptors are a reality, it has become apparent that our current pharmacological tools are inadequate, in the presence of a mixture of subtypes, to measure the density or properties of a single subtype. Although genetic experiments have shown that m3 and m4 receptors can be made by cells injected with the appropriate mRNA, there are as yet no demonstration that these receptors are expressed naturally. Until we develop new tools, either pharmacological or immunological, with which to assay for each of these proteins independently of the others, we will be unable to determine if, where, and how much of each of these receptors is expressed.

Acknowledgment

This work was supported by a grant from the NIH (GM31155). BBW is an Established Investigator of the American Heart Association (83-202).

References

Akiba, I., Kubo, T., Maeda, A., Bujo, H., Nakai, J., Mishina, M., and Numa, S. (1988) Primary structure of porcine muscarinic acetylcholine receptor III and antagonist binding studies. *FEBS Lett.* **235**, 257–261.

Ashkenazi, A., Winslow, J. W., Peralta, E. G., Peterson, G. L., Schimerlik, M. I., Capon, D. J., and Ramachandran, J. (1987) An M_2 muscarinic receptor subtype coupled to both adenylyl cyclase and phosphoinositide turnover. *Science* **238**, 672–675.

Ashkenazi, A., Peralta, E. G., Winslow, J. W., Ramachandran, J., and Capon, D. J. (1989) Functionally distinct G proteins selectively couple different receptors to PI hydrolysis in the same cell. *Cell* **56**, 487–493.

Barlow, R. B., Berry, K. J., Glenton, P. A. M., Nikolaou, N. M., and Soh, K. S. (1976) A comparison of affinity constants for muscarine-sensitive acetylcholine receptors in guinea pig atrial pacemaker cells at 29°C and in ileum at 29°C and 37°C. *Br. J. Pharmacol.* **58**, 616–620.

Beld, A. J., Van Den Hoven, S., Wouterse, A. C., and Zegers, M. A. P. (1975) Are muscarinic receptors in the central and peripheral nervous system different? *Eur. J. Pharmacol.* **30**, 360–363.

Berrie, C. P., Birdsall, N. J. M., Burgen, A. S. V., and Hulme, E. C. (1979) Guanine nucleotides modulate muscarinic receptor binding in the heart. *Biochem. Biophys. Res. Comm.* **87**, 1000–1005.

Berrie, C. P., Birdsall, N. J. M., Hulme, E. C., Keen, M., and Stockton, J. M. (1985) Solubilization and characterization of high and low affinity pirenzepine binding sites from rat cerebral cortex. *Br. J. Pharmacol.* **85**, 697–703.

Berrie, C. P., Birdsall, N. J. M., Hulme, E. C., Keen, M., Stockton, J. M., and Wheatley, M. (1986) Muscarinic receptor subclasses: The binding properties of the soluble receptor binding sites, in Subtypes of Muscarinic Receptors. *Trends Pharmacol. Sci. Supplement*, pp 8–13.

Birdsall, N. J. M., Burgen, A. S. V., and Hulme, E. C. (1978) The binding of agonists to brain muscarinic receptors. *Mol. Pharmacol.* **14**, 723–736.

Birdsall, N. J. M., Hulme, E. C., and Burgen, A. S. V. (1980) The character of the muscarinic receptors in different regions of the rat brain. *Proc. R. Soc. Lond. B* **207**, 1–12.

Bloom, J. W., Halonen, M., Lawrence, L. J., Rould, E., Seaver, N. A., and Yamamura, H. I. Characterization of high affinity (^{3}H)-pirenzepine and (^{3}H)-quinuclidinyl benzilate binding to muscarinic cholinergic receptors in rabbit peripheral lung. *J. Pharmacol. Exp. Ther.* **240**, 51–58.

Boeynaems, J. M. and Dumont, J. E. (1977) The two-step model of ligand-receptor interaction. *Mol. Cell. Endocrinol.* **7**, 33–47.

Bonner, T. I., Buckley, N. J., Young, A. C., and Brann, M. R. (1987) Identification of a family of muscarinic acetylcholine receptor genes. *Science* **237**, 527–532.

Bonner, T. I., Young, A. C., Brann, M. R., and Buckley, N. J. (1988) Cloning and expression of the human and rat m5 muscarinic acetylcholine receptor genes. *Neuron* **1**, 403–410.

Boulter, J., Evans, K., Goldman, D., Martin, G., Treco, D., Heinemann, S., and Patrick, J. (1986) Isolation of a cDNA clone coding for a possible neural nicotinic acetylcholine receptor alpha-subunit. *Nature* **319**, 368–373.

Boulter, J., Luyten, W., Evans, K., Mason, P., Ballivet, M., Goldman, D., Stengelin, S., Martin, G., Heinemann, S., and Patrick, J. (1985) Isolation of a clone coding for the alpha-subunit of a mouse acetylcholine receptor. *J. Neurosci.* **5**, 2545–2552.

Brann, M. R., Buckley, N. J., and Bonner, T. I. (1988a) The striatum and cerebral cortex express different muscarinic receptor mRNA's. *FEBS Lett.* **230**, 90–94.

Brann, M. R., Conklin, B. R., Penelope-Jones, S. V., Dean, N. M., Collins, R. M., Bonner, T. I., and Buckley, N. J. (1988b) Cloned muscarinic receptors couple to different G-proteins and second messengers. *Soc. Neurosci. Abstr.* **14**, 600.

Braun, T., Schofield, P. R., Shivers, B. D., Pritchett, D. B., and Seeburg, P. H. (1987) A novel subtype of muscarinic receptor identified by homology screening. *Biochem. Biophys. Res. Comm.* **149**, 125–132.

Brown, J. H. and Goldstein, D. (1986) Analysis of cardiac muscarinic receptors recognized selectively by nonquaternary but not by quaternary ligands. *J. Pharmacol. Exp. Ther.* **238**, 580–586.

Buckley, N. J., Bonner, T. I., and Brann, M. R. (1988) Localization of a family of muscarinic receptors mRNAs in rat brain. *J. Neurosci.* **8**, 4646–4652.

Burgen, A. S. V. and Spero, L. (1968) The action of acetylcholine and other drugs on the efflux of potassium and rubidium from smooth muscle of guinea pig intestine. *Br. J. Pharmacol.* **34**, 99–115.

Cortes, R. and Palacios, J. M. (1986) Muscarinic cholinergic receptor subtypes in the rat brain. I. Quantitative autoradiographic studies. *Brain Res.* **362**, 227–238.

Dixon, R. A. F., Kobilka, B. K., Strader, D. J., Benovic, J. L., Dohlman, H. G., Frielle, T., Bolanowski, M. A., Bennett, C. D., Rands, E., Diehl, R. E., Mumford, R. A., Slater, E. E., Sigal, I. S., Caron, M. G., Lefkowitz, R. J., and Strader, C. D. (1986) Cloning of the gene and cDNA for mammalian beta-adrenergic receptor and homology with rhodopsin. *Nature* **321**, 75–79.

Doods, H. I., Mathy, M. J., Dividesko, D., van Charldorp, K. J., de-Jonge, A., and van Zweiten, P. A. (1987) Selectivity of muscarinic antagonists in radioligand and in vivo experiments for the putative M_1, M_2 and M_3 receptors. *J. Pharmacol. Exp. Ther.* **242**, 257–262.

Dunlap, J. and Brown, J. H. (1983) Heterogeneity of binding sites on cardiac muscarinic receptors induced by the neuromuscular blocking agents gallamine and pancuronium. *Mol. Pharmacol.* **24**, 15–22.

Evans, R. A., Watson, M., Yamamura, H. I., and Roeske, W. R. (1985) Differential Ontogeny of putative M_1 and M_2 muscarinic receptor binding sites in the murine cerebral cortex and heart. *J. Pharmacol. Exp. Ther.* **235**, 612–618.

Fibiger, H. C. (1982) The organization and some projections of cholinergic neurons of the mammalian forebrain. *Brain Res. Rev.* **4**, 327–388.

Fisher, A., Grunfeld, Y., Weinstock, M., Gitter, S., and Cohen, S. (1976a) A study of muscarinic receptor heterogeneity with weak antagonists. *Eur. J. Pharmacol.* **38**, 131–139.
Fisher, A., Weinstock, M., Gitter, S., and Cohen, S. (1976b) A new probe for heterogeneity in muscarinic receptors: 2-Methyl-spiro(1,3-dioxolane-3,4)quinuclidine. *Eur. J. Pharmacol.* **37**, 329–338.
Fuder, H. (1982) The affinity of pirenzepine and other antimuscarinic compounds for pre- and postsynaptic muscarine receptors of the isolated rabbit and rat heart. *Scand. J. Gastroenterol.*, **suppl. 72**, 79–85.
Fukuda, K., Higashida, H., Kubo, T., Maeda, A., Akiba, I., Bujo, H., Mishina, M., and Numa, S. (1988) Selective coupling with K^+ currents of muscarinic acetylcholine receptor subtypes in NG108-15 cells. *Nature* **335**, 355–358.
Fukuda, K., Kubo, T., Akiba, I., Maeda, A., Mishina, M., and Numa, S. (1987) Molecular distinction between muscarinic acetylcholine receptor subtypes. *Nature* **327**, 623–625.
Gardier, R. W., Tsevdos, E. J., Jackson, D. B., and Delaunois, A. L. (1978) Distinct muscarinic mediation of suspected dopaminergic activity in sympathetic ganglia. *Fed. Proc.* **37**, 2422–2248.
Giachetti, A., MIcheletti, R., and Montagna, E. (1986) Cardioselective profile of AF-DX 116, a muscarinic M_2-receptor antagonist. *Life Sci.* **38**, 1663–1672.
Goyal, R. K. and Rattan, S. (1978) Neurohormonal, hormonal, and drug receptors for the lower esophageal sphicter. Prog. Gastroenterol. **74**, 598–619.
Haga, K. and Haga, T. (1983) Affinity chromatography of the muscarinic acetylcholine receptor. *J. Biol. Chem.* **258**, 13575–13579.
Hammer, R. and Giachetti, A. (1982) Muscarinic receptor subtypes: M_1 and M_2 biochemical and functional characterization. *Life Sci.* **31**, 2991–2998.
Hammer, R., Berrie, C. P., Birdsall, J. J. M., Burgen, A. S. V., and Hulme, E. C. (1980) Pirenzepine distinguishes between different subclasses of muscarinic receptors. *Nature* **283**, 90–92.
Hammer, R., Giraldo, E., Schiavi, G. B., Monferini, E., and Ladinsky, H. (1986) Binding profile of a novel cardioselective muscarine receptor antagonist, AF-DX 116, to membranes of peripheral tissues and brain in the rat. *Life Sci.* **38**, 1653–1662.
Herron, G. S. and Schimerlik, M. I. (1983) Glycoprotein properties of the solubilized atrial muscarinic acetylcholine receptor. *J. Neurochem.* **41**, 1414–1420.
Hulme, E. C., Burgen, A. S. V., and Birdsall, N. J. M. (1976) Interactions of agonists and antagonists with the muscarinic receptor. *INSERM* **50**, 49–69.
Jacobs, S. and Cuatrecasas, P. (1976) The mobile receptor hypothesis and "cooperativity" of hormone binding. Application to insulin. *Biochim. Biophys. Acta.* **433**, 482–495.
Jarv, J., Hedlund, B., and Bartfai, T. (1979) Isomerization of the mus-

carinic receptor-antagonist complex. *J. Biol. Chem.* **254**, 5595–5598.

Kubo, T., Fukuda, K., Mikami, A., Maeda, A., Takahashi, H., Mishina, M., Haga, T., Haga, K., Ichiyama, A., Kangawa, K., Kojima, M., Matsuo, H., Hirose, T., and Numa, S. (1986a) Cloning, sequencing and expression of complementary DNA encoding the muscarinic acetylcholine receptor. *Nature* **323**, 411–416.

Kubo, T., Meada, A., Sugimoto, K., Akiba, I., Mikami, A., Takahashi, H., Haga, T., Haga, K., Ichiyama, A., Kangawa, M., Matsuo, H., Hirose, T., and Numa, S. (1986b) Primary structure of porcine cardiac muscarinic acetylcholine receptor deduced from the cDNA sequence. *FEBS* **209**, 367–372.

Lee, J. H. and El-Fakahany, E. E. (1985) Heterogeneity of binding of muscarinic receptor antagonists in rat brain homogenates. *J. Pharmacol. Exp.* **233**, 707–714.

Lee, W. and Wolfe, B. B. (1985) Ontogeny of muscarinic receptor binding sites and muscarinic receptor-mediated stimulation of phosphoinositide breakdown and inhibition of cyclic AMP accumulation in rat forebrain. *Soc. Neurosci. Abstr.* **11**, 95.

Lee, W. and Wolfe, B. B. (1986) Selective regulation of putative muscarinic receptor subtypes and their responsiveness with chronic atropine treatment. *Soc. Neurosci. Abstr.* **12**, 307.

Li, M. and Wolfe, B. B. (1988) Preferential coupling of subtypes of muscarinic receptors to specific second messenger systems: The M_1 receptor prefers phosphoinositide breakdown while the M_2 receptor prefers inhibition of adenylate cyclase. *Soc. Neurosci. Abstr.* **14**, 228.

Luthin, G. R. and Wolfe, B. B. (1984a) Comparison of (^{3}H)-pirenzepine and (^{3}H)-QNB binding to muscarinic cholinergic receptors in rat brain. *J. Pharmacol. Exp. Ther.* **228**, 648–655.

Luthin, G. R. and Wolfe, B. B. (1984b) (^{3}H)-Pirenzepine and (^{3}H)-QNB binding to brain muscarinic cholinergic receptors: Differences in receptor density are not explained by differences in receptor isomerization. *Mol. Pharmacol.* **26**, 164–169.

Luthin, G. R. and Wolfe, B. B. (1985) Characterization of (^{3}H)-pirenzepine binding to muscarinic cholinergic receptors solubilized from rat brain. *J. Pharmacol. Exp. Ther.* **234**, 37–44.

Luthin, G. R., Harkness, J., Artymyshyn, R. P., and Wolfe, B. B. (1988) Antibodies to a synthetic peptide can be used to distinguish between mucscarinic acetylcholine receptor binding sites in brain and heart. *Mol. Pharmacol.* **34**, 327–333.

Martin, M. W., Evans, T., Smith, M. M., and Harden, T. K. (1984) Guanine nucleotide-insensitive binding of a tritiated agonist to muscarinic acetylcholine receptors of rat brain. *Fed. Proc.* **43**, 567.

Mei, L., Lai, J., Roeske, W. R., Fraser, C. M., Venter, J. C., and Yamamura, H. I. (1989) Pharmacological characterization of the M1 muscarinic receptors expressed in murine fibroblast B82 cells. *J. Pharmacol. Exp. Ther.* **248**, 661–670.

Minneman, K. P. and Molinoff, P. B. (1980) Classification and quantitation of *beta*-adrenergic receptor subtypes. *Biochem. Pharmacol.* **29**, 1317–1323.

Mutschler, E. and Lamrecht, G. (1984) Selective muscarinic agonists and antagonists in functional tests, in Subtypes of Muscarinic Receptors. *Trends Pharmacol. Sci. Supplement*, pp 39–44.

O'Donnell, S. R. and Wanstall, J. C. (1976) The contribution of extraneuronal uptake to the trachea-blood vessel selectivity of beta-adrenoceptor stimulants in vitro in guinea pigs. *Br. J. Pharmacol.* **57**, 369–373.

Peralta, E. G., Ashkenazi, A., Winslow, J. W., Ramachandran, J., and Capon, D. J. (1988) Differential regulation of PI hydrolysis and adenylyl cyclase by muscarinic receptor subtypes. *Nature* **334**, 434–437.

Peralta, E. G., Winslow, J. W., Peterson, G. L., Smith, D. H., Ashkenazi, A., Ramachandran, J., Schimerlik, M. I., and Capon, D. J. (1987a) Primary structure and biochemical properties of an M_2 muscarinic receptor. *Science* **236**, 600–605.

Peralta, E. G., Ashkenazi, A., Winslow, J. W., Smith, D. H., Ramachandran, J., and Capon, D. J. (1987b) Distinct primary structures, ligand-binding properties and tissue-specific expression of four human muscarinic acetylcholine receptors. *EMBO J.* **6**, 3923–3929.

Peterson, G. L., Herron, G. S., Yamaki, M., Fullerton, D. S., and Schimerlik, M. J. (1984) Purification of the muscarinic acetylcholine receptor from porcine atria. *Proc. Nat. Acad. Sci.* **81**, 4993–4997.

Potter, L. T., Flynn, D. D., Hanchett, H. E., Kalinoski, D. L., Luber-Narod, J., and Mash, D. C. (1984) Independent M_1 and M_2 receptors: Ligands, autoradiography and functions, in Subtypes of Muscarinic Receptors. *Trends Pharmacol. Sci.* **Supplement** pp22–31.

Raiteri, M., Leardi, R., and Marchi, M. (1984) Heterogeneity of presynaptic muscarinic receptors regulating neurotransmitter release in rat brain. *J. Pharmacol. Exp. Ther.* **228**, 209–214.

Roeske, W. R. and Venter, J. C. (1984) The differential loss (^{3}H)-pirenzepine *vs* (^{3}H)-QNB binding to soluble rat brain muscarinic receptors indicates that pirenzepine binds to an allosteric state of the muscarinic receptor. *Biochem. Biophys. Res. Comm.* **118**, 950–957.

Rosenberger, L. B., Yamamura, H. I., and Roeske, W. R. (1980) Cardiac muscarinic cholinergic receptor binding is regulated by Na^+ and guanyl nucleotides. *J. Biol. Chem.* **255**, 820–823.

Shapiro, R. A., Scherer, N. M., Habecker, B. A., Subers, E. M., and Nathanson, N. M. (1988) Isolation, sequence, and functional expression of the mouse M1 muscarinic acetylcholine receptor gene. *J. Biol. Chem.* **263**, 18397–18403.

Snyder, S. H., Chang, K. J., Kuhar, M. J., and Yamamura, H. I. (1975) Biochemical identification of the mammalian muscarinic cholinergic receptor. *Fed. Proc.* **34**, 1915–1921.

Stein, R., Pinkas-Kramarski, R., and Sokolovsky, M. (1988) Cloned M1 muscarinic receptors mediate both adenylate cyclase inhibition and phosphoinositide turnover. *EMBO J.* **7**, 3031–3035.

Stockton, J. M., Birdsall, N. J. M., Burgen, A. S. V., and Hulme, E. C. (1983) Modification of the binding properties of muscarinic receptors by gallamine. *Mol. Pharmacol.* **23**, 551–557.

Taylor, P. (1985) *The Pharmacological Basis of Therapeutics* (Gilman, A. G., Goodman, L. S., Rall, T. W., and Murad, F., eds.), Macmillian, New York, pp 244–225.

Waelbroeck, M., Gillard, M., Robberecht, P., and Christophe, J. (1987) Muscarinic receptor heterogeneity in rat central nervous system. I. Binding of four selective antagonists to three muscarinic receptor subclasses: A comparison with M_2 cardiac muscarinic receptors of the C type. *Mol. Pharmacol.* **32**, 91–99.

Wamsley, J. K., Gehert, D. R., Roeske, W. R., and Yamamura, H. I. (1984) Muscarinic antagonist binding site heterogeneity as evidenced by autoradiography after direct labeling with (^{3}H)-QNB and (^{3}H)-pirenzepine. *Life Sci.* **34**, 1395–1402.

Watson, M., Roeske, W. R., and Yamamura, H. I. (^{3}H)-Pirenzepine selectively identifies a high-affinity population of muscarinic cholinergic receptors in the rat cerebral cortex. *Life Sci.* **31**, 2019–2023.

Watson, M., Yamamura, H. I., and Roeske, W. R. (1983) A unique regulatory profile and regional distribution of (^{3}H)-pirenzepine binding in the rat provide evidence for distinct M_1 and M_2 muscarinic receptor subtypes. *Life Sci.* **32**, 3001–3011.

Watson, M., Roeske, W. R., Johnson, P. C., and Yamamura, H. I. (1984) (^{3}H)-Pirenzepine identifies putative M_1 muscarinic receptors in human stellate ganglia. *Brain Res.* **290**, 179–182.

Yamamura, H. I., Kuhar, J. J., Greenberg, D., and Snyder, S. H. (1974) Muscarinic cholinergic receptor binding: Regional distribution in monkey brain. *Brain Res.* **66**, 541–546.

Yamamura, H. I., Wamsley, J. K., Deshmukh, P., and Roeske, W. R. (1983) Differential light microscopic autoradiographic localization of muscarinic cholinergic receptors in the brainstem and spinal cord of the rat using (^{3}H)-pirenzepine. *Eur. J. Pharmacol.* **91**, 147–149.

5

Structural Determinants of Muscarinic Agonist Activity

Björn Ringdahl

1. Introduction

Traditionally, structure–activity relationships (SAR) among muscarinic cholinergic agonists have been derived from dose–effect studies on isolated tissues or in whole animals (Barlow, 1964). In general, in vitro assays are preferable to in vivo assays for studying relations between chemical structure and actual effect (pharmacodynamics). The interpretation of SAR obtained in vivo is likely to be complicated by distributional and pharmacokinetic factors.

Conventional dose–effect studies measure the physiological effect that is the final outcome of a sequence of events, among which initial binding of the agonist to the receptor is the first. Although such studies define agonist potency, a parameter of great importance, they usually do not provide detailed information on molecular mechanisms of action. However, they furnish the necessary base for more sophisticated structure–activity studies. These take into account the notion that agonist potency is determined by two drug-dependent parameters: affinity and intrinsic efficacy (Stephenson, 1956; Kenakin, 1984). Affinity, as reflected by the dissociation constant (K_D) of the agonist–receptor complex, is a measure of the ability of the agonist to bind to the receptor. Intrinsic efficacy, as reflected by the relative receptor occupancy required for a given response, is related to the ability of the agonist–receptor complex to elicit a cellular response.

Because of the existence in many tissues of a receptor reserve for muscarinic agonists, apparent affinity constants (ED_{50} values), determined from dose-response curves, generally are poor estimates of agonist affinity (Stephenson, 1956; Furchgott and Bursztyn, 1967; Ringdahl, 1984a). Comparison of agonist potencies, therefore, may give misleading information on the structural requirements for binding to mAChRs. Furthermore, structural changes in a series of agonists may affect affinity and intrinsic efficacy differently (Stephenson, 1956; Ringdahl, 1985a). Consequently, attempts to relate chemical structure to activity should take into account separate relations between structure and binding and between structure and receptor activation (efficacy). Fortunately, pharmacological methods are available that allow discrete assessment of agonist affinity and intrinsic efficacy in isolated tissues (Furchgott, 1966; Furchgott and Bursztyn, 1967; Ruffolo, 1982; Kenakin, 1984, 1985).

The development of radioligand binding assays has provided another approach to the study of SAR of muscarinic agonists (Yamamura and Snyder, 1974). Such assays may provide detailed information on structure–affinity relationships and also appear useful in assessment of agonist efficacy (Birdsall et al., 1978a,b; Freedman et al., 1988). Of particular importance is the ability of the binding techniques to measure affinity for central mAChRs (Birdsall et al., 1978a; Sokolovsky, 1984). However, the relationship between binding parameters and pharmacological potencies of muscarinic agonists is not yet fully understood (Birdsall and Hulme, 1976; Beld et al., 1980; Jim et al., 1982).

There is growing evidence of subtypes of mAChRs in both the central nervous system, and in peripheral neurons and effector organs (Levine et al., 1986, 1988; Mitchelson, 1988). This apparent receptor heterogeneity may complicate the interpretation of structure–activity studies on muscarinic agonists inasmuch as each receptor subtype presumably has its own structural requirements. Another complication arises from the fact that some muscarinic agents may act indirectly via the release of ACh from the nerve terminal. Furthermore, muscarinic agonists are known to activate not only postsynaptic mAChRs, but also inhibitory mAChRs that are located presynaptically and that modulate the release of ACh. However, pre- and postsynaptic mAChRs appear to have similar structural specificity (Kilbinger et al., 1984; Kilbinger and Wessler, 1980).

This chapter reviews results from recent investigations of SAR among muscarinic agonists. Special emphasis will be placed on structural determinants of agonist affinity and intrinsic efficacy, on partial agonists, and on tertiary amines capable of passing the blood–brain barrier. The isolated guinea-pig ileum has been a valuable preparation for assessing muscarinic agonist potency and, thus, for elucidating SAR. Published potency data are often expressed as pD_2 values (negative logarithm of molar concentration eliciting half-maximal contractile response) or as equipotent molar ratios (EPMR) relative to a standard agonist. The guinea-pig ileum is also the most widely used preparation for estimation of K_D values and of relative efficacies of muscarinic agonists. Unless otherwise stated, potency, affinity, and efficacy data referred to in this chapter were obtained from contractile responses on the isolated guinea-pig ileum. Whenever possible, comparisons will be made between these pharmacologically determined drug parameters, and results obtained from radioligand binding and other biochemical studies directed toward the activation of coupling mechanisms such as the PI response. The older literature on SAR among muscarinic agonists, which deals mostly with quaternary ammonium compounds, has been reviewed extensively (Barlow, 1964; Brimblecombe, 1974; Casy, 1975; Triggle, 1976, 1984; Cavallito, 1980; Cannon, 1981).

Many compounds have muscarinic actions. Although muscarinic agonists may differ substantially in chemical structure, they all are fairly small molecules. Common structural features include:

1. an ammonium group (tertiary or quaternary) or a similarly charged group (e.g., sulfonium)
2. an ester, ether, or similar function of high electron density (e.g., triple bond) and
3. a methyl or methylene group corresponding to the acetyl methyl group of ACh.

Muscarinic agonists are often classified as analogs of certain well-known parent compounds (ACh, muscarine, arecoline, oxotremorine, and so on). In general, muscarinic agonists show similar pharmacological effects, but there are important differences among and within the various classes, e.g., with regard to nicotinic actions, susceptibility to hydrolysis by cholinesterases, and ability to produce central effects on systemic administration.

2. Analogs of Acetylcholine

ACh and some of its analogs have pronounced nicotinic actions and many are hydrolysed by tissue cholinesterases. The ACh molecule is conformationally mobile, and different conformations may be involved in its interaction with nicotinic and various muscarinic receptors and with the active site of cholinesterases (Martin-Smith and Stenlake, 1967). The extensive literature on the solid-state and solution conformation of ACh has been reviewed by Casy (1975). Most studies have focused on the torsional angle of the NCCO grouping, which defines the relative disposition of the acetyl and trimethylammonium groups. ACh appears to exist predominantly as the *synclinal (gauche)* conformer both in the crystal and in solution (Fig. 1). Similar *synclinal* conformations were preferred by several analogs of ACh containing the NCCO fragment (Casy, 1975; Partington et al., 1972). However, the energy barriers for interconversion of the various rotamers of ACh-like compounds are relatively small (Makriyannis et al., 1979), and ligand binding may very well occur in a conformation that differs from that preferred in solution or in the crystal. Thus, ACh has been suggested to bind to mAChRs in a transoid *anticlinal/antiplanar* conformation (Fig. 1) (Casy, 1975; Mutschler and Lambrecht, 1983).

2.1. Conformationally Restricted Analogs of Acetylcholine

Evidence for a *trans* orientation of the trimethylammonium and ester groups in the receptor-bound conformation of ACh comes mainly from studies on rigid and semirigid analogs. Perhaps the most conclusive study is that of Cannon and coworkers (Chiou et al., 1969; Armstrong and Cannon, 1970) on the *trans* and *cis* isomers of 2-acetoxycyclopropyltrimethylammonium iodide (ACTM). They showed that the (1*S*,2*S*)-isomer of *trans*-ACTM was equipotent with ACh as a muscarinic agonist, whereas (1*R*,2*R*)-*trans*-ACTM and (±)-*cis*-ACTM were virtually inactive (Fig. 2). *Trans*-ACTM was found to have a near *anticlinal* arrangement of the acetyl and trimethylammonium groups (NCCO torsion angle 137°) in the solid state (Chothia and Pauling, 1970). This conformation probably is close to that of the receptor-bound molecule because of the rigidity of *trans*-ACTM. The corresponding cyclobutane (Cannon et al., 1975, 1982),

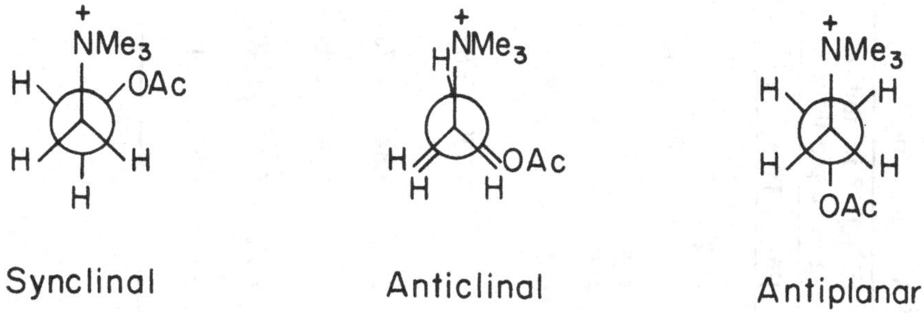

Synclinal Anticlinal Antiplanar

Fig. 1. Conformations of acetylcholine.

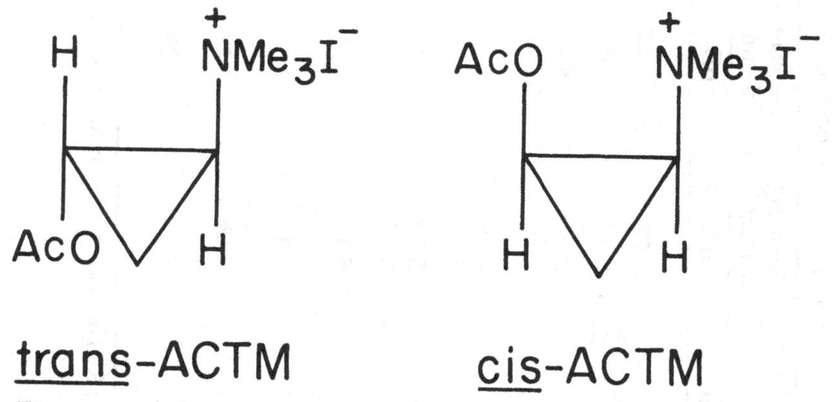

Fig. 2. *Trans* and *cis* isomers of 2-acetoxycyclopropyltrimethylammonium iodide (ACTM).

cyclopentane, and cyclohexane (Lambrecht, 1976a) analogs of ACh showed only weak muscarinic activity despite the transoid arrangement of the acetyl and trimethylammonium groups in the *trans* isomers of the former compounds. Presumably, the added methylene groups introduced changes in the steric or electronic properties as compared to ACh.

More recent studies have focused on heterocyclic, semirigid analogs of ACh (Höltje et al., 1979; Mutschler and Lambrecht, 1983). 3-Acetoxyquinuclidine (*1*), also referred to as aceclidine, is of particular interest because it is a tertiary amine having a more or less fixed *anticlinal* arrangement of the acetyl group and the amino group (Casy, 1975). The *S*-enantiomer of compound *1* was a potent muscarinic agonist in vitro (Table 1) and in vivo (Ringdahl et al., 1982a). Theoretical calculations showed that the "interaction pharmacophore" of (*S*)-*1*, defined by its electrostatic potential fields, was compatible with the requirements for ACh-

Table 1
Muscarinic Activity and Receptor Binding of Semirigid Heterocyclic Analogs of Acetylcholine[a]

Compound	Absolute configuration	Guinea-pig atrium[b]	Guinea-pig ileum[c]	[³H]QNB inhibition guinea-pig ileum		
		pD_2	pD_2	pK_i	pK_H	pK_L
1	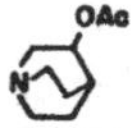 *R*	5.02	5.47	4.92[d]		
	S	6.10	6.62		6.47[d]	5.18[d]
2	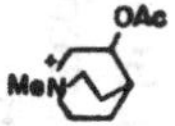 *R*	3.97	4.23	4.37[d]		
	S	3.56[e]	<3.30[e]	4.31[d]		
3	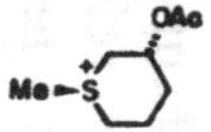 1*R*,3*R*	6.50	6.56	5.81	7.27	5.03
	1*S*,3*S*	4.85	4.84			
4	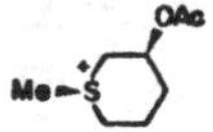 1*R*,3*S*	3.30	4.50	4.15[f]	5.03[f]	3.85[f]
	1*S*,3*R*	3.37	4.45			
5	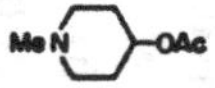	5.15	5.60	5.50	5.84	4.70
6	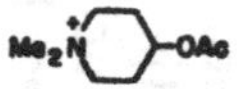	5.13	5.81	5.31	5.85	4.32
7	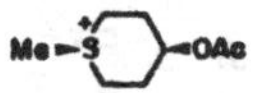	7.68	7.59	6.71	7.09	5.40
8	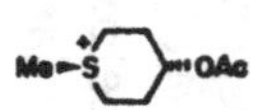	5.72	6.20	5.65	5.94	4.45

[a]Höltje et al., 1978; Lambrecht, 1981; Jim et al., 1982; Ringdahl et al., 1982a; Mutschler and Lambrecht, 1983, 1984.
[b]Negative inotropic effect.
[c]Contractile response.
[d]Rat brain stem.
[e]Partial agonist.
[f]Data for racemate.

like activity (Weinstein et al., 1975). *N*-Methylation of *1* to give *2* reduced muscarinic activity (Cho et al., 1972) and receptor binding affinity (Ringdahl et al., 1982a; Ehlert and Jenden, 1984), and also caused a reversal of stereoselectivity (Barlow and Casy, 1975). In contrast, *N*-methylation of flexible aminoacetates related to ACh increased muscarinic activity (Cho et al., 1972) and binding affinity (Ehlert and Jenden, 1984).

Compound (*S*)-*1* has been a useful model for determining the active conformations of other heterocyclic analogs of ACh. Thus, the low muscarinic activity of 3-acetoxy-*N*-methylpiperidine was ascribed to the fact that it must attain an energetically unfavored boat conformation (Fig. 3) in order to interact with the mAChR in a manner analogous to that of (*S*)-*1* (Lambrecht, 1976b; Mutschler and Lambrecht, 1983). In contrast to 3-acetoxy-*N*-methylpiperidine, which existed in solution as a mixture of interconvertible *trans* and *cis* chair conformations (Lambrecht, 1976b), the *trans* (*3*) and *cis* (*4*) isomers of its sulfonium analog could be readily isolated. Compound *3* was about 300-fold more potent than 3-acetoxy-*N*-methylpiperidine as a muscarinic agonist, whereas *4* was much less potent (Lambrecht, 1977). The high potency of *1* and *3* suggested that, in addition to an *anticlinal/antiplanar* arrangement between the hetero atom and the acetoxy group, an axial methyl group and an equatorial acetoxy group were required for high muscarinic activity among these heterocyclic ACh analogs (Fig. 3). In the case of 3-acetoxy-*N*-methylpiperidine and *4*, these "muscarinic-essential" conformations were of high energy and were present only in small amounts ($< 0.3\%$) in solution (Lambrecht, 1977; Mutschler and Lambrecht, 1983). The enantiomers of *3* and *4* have been synthesized and their absolute configurations established (Lambrecht, 1981; Jensen, 1981; Mutschler and Lambrecht, 1984). Pharmacological data for the enantiomers of *3* and *4* are summarized in Table 1.

In contrast to 4-acetoxy-*N*-methylpiperidine (*5*) and its methiodide (*6*), which were moderately potent muscarinic agents (Table 1), *cis*-4-acetoxy-1-methylthianium (*7*) was equipotent with ACh (Höltje et al., 1978). The high potency of *7* was ascribed to its ability to assume a chair conformation with an axial methyl group and an equatorial acetoxy group similar to the "muscarinic essential" chair conformation depicted in Fig. 3. The *trans* isomer *8* and the piperidine *5* were able to adopt such a conformation only at considerable energy expense and, thus, were about

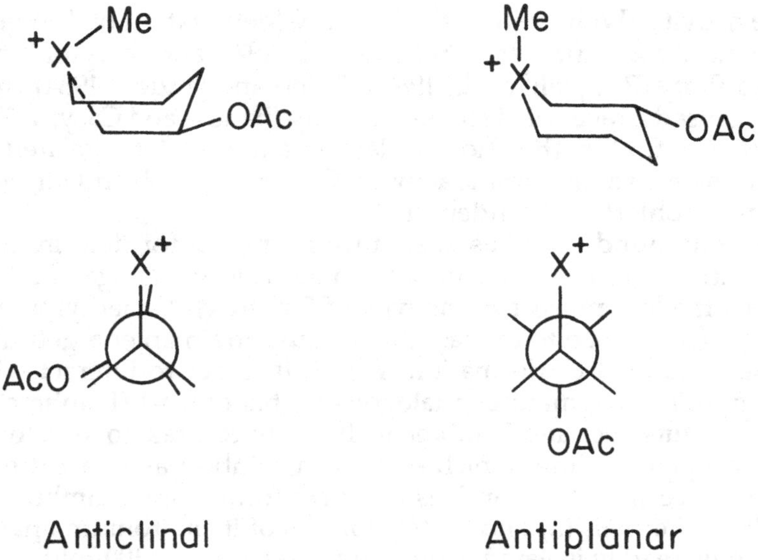

Anticlinal Antiplanar

Fig. 3. "Muscarinic essential" chair and boat conformations of piperidine and thiane analogs of acetylcholine (X = N or S).

100-fold less potent than *7*. The lower potency of *6*, which was able to adopt the required conformation, was ascribed to steric hindrance by the additional *N*-methyl group (Höltje et al., 1978, 1979). Compounds *7* and *8* also have been studied by X-ray crystallography (Jensen, 1979) and theoretical calculations (Höltje et al., 1978, 1979). 4-Acetoxyquinuclidine and its methiodide were found (Barlow and Kitchen, 1982) to be weak muscarinic agonists. With the exception of 4, the compounds in Table 1 showed only weak nicotinic actions as measured on the rectus abdominis muscle of the frog (Barlow and Casy, 1975; Höltje et al., 1979; Lambrecht, 1981).

2.2. Affinity and Efficacy of Acetylcholine Analogs

The dissociation constants of ACh and its analogs at mAChRs have been estimated in both pharmacological (Table 2) and radioligand binding studies (Tables 1 and 3). The sulfonium salt *7* was reported to have the same affinity as ACh for mAChRs in guinea-pig atrium, whereas its intrinsic efficacy exceeded that of ACh (Lambrecht, 1979). Its *trans* isomer (*8*) retained high intrinsic efficacy, but had greatly reduced affinity for atrial recep-

Table 2
Dissociation Constants (K_D) and Relative Efficacies (e_r) of Acetylcholine Analogs at Muscarinic Receptors in Isolated Tissues

Compound	Guinea-pig ileum[a]			Rabbit fundus[b]		Rat heart[c]		
	pD_2	pK_D	e_r	pK_D	e_r	pD_2	pK_D	e_r
Acetylcholine	7.44	5.77	1.0	5.68	1.0			
(*S*)-Methacholine	7.37	5.70	0.60			7.00	5.60	1.0
(*R*)-Methacholine	4.50	3.21	0.28			4.19	3.36	0.23
(±)-Methacholine				5.61	0.73	6.59	5.40	
Carbachol	7.10	4.79		4.80	1.05			
(*S*)-Bethanechol	6.33							
(*R*)-Bethanechol	3.37							

[a]Ringdahl, 1984a, 1986; Schwörer et al., 1985.
[b]Furchgott and Bursztyn, 1967.
[c]Fuder and Jung, 1985.

Table 3
Muscarinic Receptor Binding Parameters of Acetylcholine Analogs

Compound	Rat cerebral cortex [^{3}H]PrBCh inhibition[a]			Rat forebrain [^{3}H]CD inhibition[b]	Guinea-pig cerebral cortex [^{3}H]QNB inhibition[c]		
	pK_H	pK_L	K_L/K_H	pK_i	pK_H	pK_L	K_L/K_H
Acetylcholine	7.08	5.08	100	8.06	7.40	5.23	148
(*S*)-Methacholine	6.40	4.50	80	6.96			
(*R*)-Methacholine	4.03	3.00	11	4.24			
(±)-Methacholine					6.77	4.70	113
Carbachol	6.33	4.05	195	7.77	5.85	4.12	54
(±)-Bethanechol						4.00[d]	

[a]Birdsall et al., 1978a.
[b]Ehlert et al., 1980.
[c]Fisher et al., 1983.
[d]Only one binding site was observed.

tors. The receptor binding data in Table 1 are consistent with these observations, since *7* had greater potency than *8* in inhibiting [^{3}H]QNB binding in homogenates of the guinea-pig ileum, and since the two compounds had similar ratio of low- and high-affinity dissociation constants (K_L/K_H). This ratio has been suggested to reflect agonist efficacy (Birdsall et al., 1978a).

Carbachol (carbamoylcholine, CCh) had only about one-tenth of the affinity of ACh for mAChRs in the guinea-pig ileum

(Ringdahl, 1984a, 1986) and rabbit fundus (Furchgott and Bursztyn, 1967) as determined after fractional receptor inactivation with an irreversible antagonist (Table 2). Higher affinity of ACh as compared to CCh also was observed in ^{3}H-antagonist inhibition studies on rat (Birdsall et al., 1978a) and guinea-pig cerebral cortex (Fisher et al., 1983). The relative efficacies of ACh and CCh, determined on rabbit fundus (Furchgott and Bursztyn, 1967), appeared to be similar, as was their K_L/K_H ratio obtained from inhibition of ^{3}H-antagonist binding (Table 3). Their relative potencies in stimulating PI turnover in guinea-pig cerebral cortex also agreed, suggesting similar intrinsic efficacy (Fisher et al., 1983).

The affinity of (*S*)-methacholine (acetyl-β-methylcholine) for mAChRs in the guinea-pig ileum was similar to that of ACh. This observation suggested that the methyl group at the chiral center of (*S*)-methacholine had no apparent effect on the binding to the receptor. The 300-fold lower affinity of (*R*)-methacholine was explained by the difficulty with which it adopted the *anticlinal* conformation necessary for binding to mAChRs (Ringdahl, 1986). The dissociation constants of (*S*)- and (*R*)-methacholine also have been measured at presynaptic mAChRs in rat heart (Fuder and Jung, 1985) and that of (±)-methacholine on rabbit fundus (Furchgott and Bursztyn, 1967). The results agreed well with those obtained on the guinea-pig ileum (Table 2). Methacholine appeared to be slightly less potent than ACh in inhibiting ^{3}H-antagonist (Birdsall et al., 1978a; Fisher et al., 1983) and ^{3}H-agonist binding (Ehlert et al., 1980) in brain tissue. The intrinsic efficacy of methacholine as measured in functional studies on whole tissue (Furchgott and Bursztyn, 1967; Ringdahl, 1986), by the K_L/K_H ratio (Birdsall et al., 1978a; Fisher et al., 1983) and by ability to stimulate the PI response (Fisher et al., 1983), was similar to or slightly lower than that of ACh.

The enantiomers of bethanechol (carbamoyl-β-methylcholine), which differed 700- to 900-fold in muscarinic activity (De Micheli et al., 1983; Schwörer et al., 1985), were decidedly less potent than those of methacholine and were also less effective in inhibiting [^{3}H]QNB binding (Bühl et al., 1987). The intrinsic efficacy of bethanechol appeared to be somewhat lower than that of ACh, since it displayed a rather low K_L/K_H ratio (Fisher et al., 1983; Bühl et al., 1987) and since it was less effective than ACh in stimulating the PI response (Fisher et al., 1983). A hydrazinium analog of ACh, 1,1-dimethyl-1-(2-acetoxyethyl)-hydrazinium perchlorate, had 20-fold lower affinity than ACh for ileal mAChRs, whereas its intrinsic efficacy approached that of ACh (Newton et al., 1985).

2.3. Molecular Models for Acetylcholine-Like Ligands

New molecular models for the recognition of ACh-like compounds by the mAChR have been proposed. Schulman et al. (1983) employed quantum mechanical calculations to derive a theoretical model that predicted the pharmacologically active conformations of ACh and related agonists. With the exception of ACTM, these differed substantially from the conformations found in the crystal. The muscarinic pharmacophore (Fig. 4) was defined by the distance (PQ) between an anionic site P and a cationic site Q on the receptor, and the interaction dihedral angle (PNOQ), defined as the angle between the planes NPO and POQ. This angle describes the relative orientation of the drug and the receptor site. Calculations showed P to lie on the threefold axis of the trimethylammonium group at a distance of 30 nm from the nitrogen atom. The optimal distance between the cationic site Q and the ether oxygen of ACh was 12 nm. Since P and Q are receptor sites, it was assumed that the distance PQ was a constant property of the receptor in its active conformation. By applying the constraint of a common PQ distance for a series of agonists, the muscarinic pharmacophore was shown to correspond to PNOQ values of between 100 and 117° and to PQ values of 66–68 nm. The model was claimed to explain the pattern of stereoselectivity in agonists related to aceclidine. It also furnished a geometric criterion for partial agonism and antagonism, i.e., PC distances larger than 85 nm (Fig. 4). The approach of Schulman et al. (1983), referred to as "drug-inferred pharmacophore analysis" (Schulman and Disch, 1986), has been praised (Tollenaere, 1984), but also criticized (Kokkinidis and Gieren, 1984; Snyder, 1985).

Gieren and Kokkinidis (1986) interpreted crystal structures of several halide salts of agonists related to ACh in terms of agonist–receptor interactions. Since in these salts there was a stereoselective arrangement of the anions with respect to the cations, the anions in the crystal structures were considered as a model of the anionic site of the receptor. So-called "activity triangles," formed by the nitrogen of the ammonium group, a

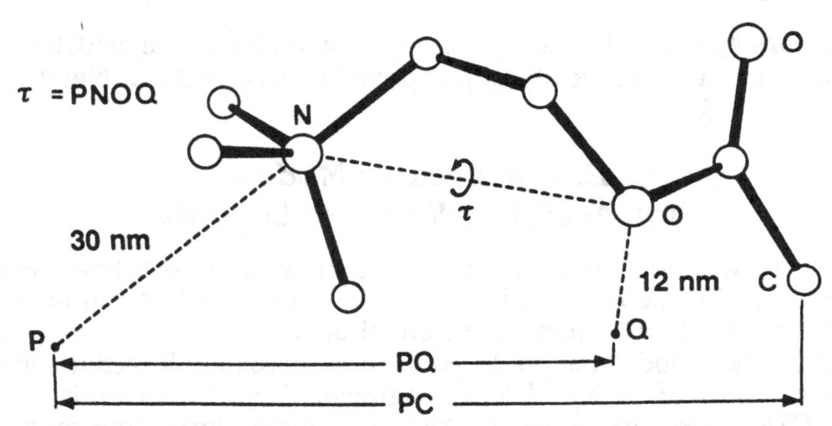

Fig. 4. Proposed mode of interaction of acetylcholine with an anionic site P and a cationic site Q on the muscarinic receptor.

second polar group of the agonist and the counterion, exhibited characteristic geometries that were associated with muscarinic and nicotinic activity. It was suggested that SAR of cholinergic stimulants do not depend on the conformational details of the agonist, but primarily on the relative positions of the polar centers with respect to the anionic binding site of the receptor.

3. Analogs of Muscarine

The early literature on the chemistry and pharmacology of muscarine, muscarone, and related compounds has been reviewed by Waser (1961). Muscarine shows a high degree of stereoselectivity toward the mAChR; the enantiomer of the naturally occurring L-(+)-muscarine (Fig. 5) is virtually inactive, as are its three diastereomers (*epi*muscarine, *allo*muscarine, and *epiallo*muscarine). Oxidation of muscarine to muscarone results in a slight increase of muscarinic activity, but the stereoisomers of muscarone differ much less in potency than those of muscarine. In contrast to muscarine, which is a very weak nicotinic agent, muscarone has considerable nicotinic potency. Methylfurthrethonium (*9*), which is an aromatic analog of muscarine and muscarone, is a very potent muscarinic agent and so is the dioxolane (*10*) (Fig. 5). Comparative studies of the muscarinic activities of muscarine, muscarone, *9*, *10*, and related compounds in several isolated tissues have recently been described (Grana et al., 1986). The muscarinic activity of *10* resides mainly in the

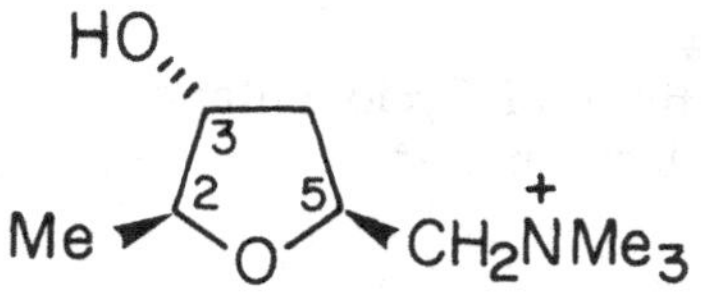

(2S,3R,5S)-Muscarine

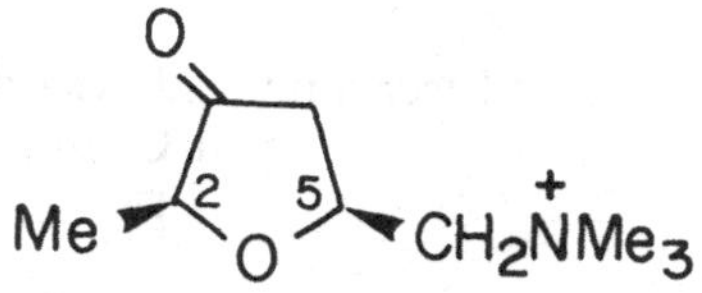

(2S,5S)-Muscarone

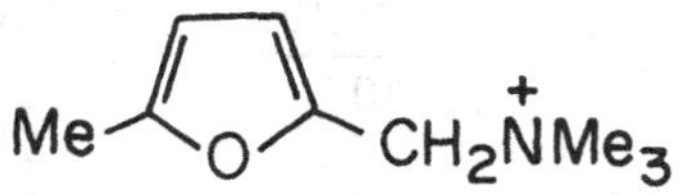

9

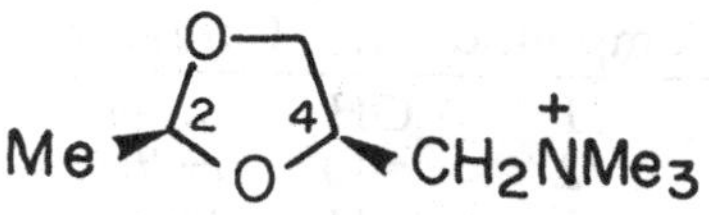

(2S,4R)-10

Fig. 5. Chemical structures of L-(+)-muscarine, L-(+)-muscarone, and related compounds.

cis isomer whose (2*S*,4*R*)-enantiomer is about 100-fold more potent than the (2*R*,4*S*)-enantiomer. The more potent stereoisomers of muscarine, *10*, and methacholine are configurationally related, suggesting similar modes of binding to the mAChR. However, the apparent inversion of stereoselectivity on conversion of muscarine to muscarone suggests that alternative modes of binding are possible (Triggle, 1976). In order to define further the structural and stereochemical requirements for interaction of muscarine-like compounds with the mAChR, several pentatomic, cyclic compounds having structural resemblance to muscarine, muscarone, and *10* have been investigated. These studies are summarized in recent reviews (Gualtieri et al., 1979, 1985; Timoney, 1983; Triggle, 1984). ACh was the standard agonist employed in most of these studies, which were performed in the absence of a cholinesterase inhibitor. Since analogs of muscarine are not hydrolyzed by tissue cholinesterases (Waser and Hopff, 1979), their relative potencies may therefore be overestimated.

3.1. Cyclopentane and Cyclopentene Derivatives

Replacement of the ether oxygen of muscarine by an isosteric methylene group to give desethermuscarine *11* (Table 4) was accompanied by a relatively small decrease in muscarinic activity

Table 4
Muscarinic and Nicotinic Activity of Cyclopentane and Cyclopentene Derivatives[a]

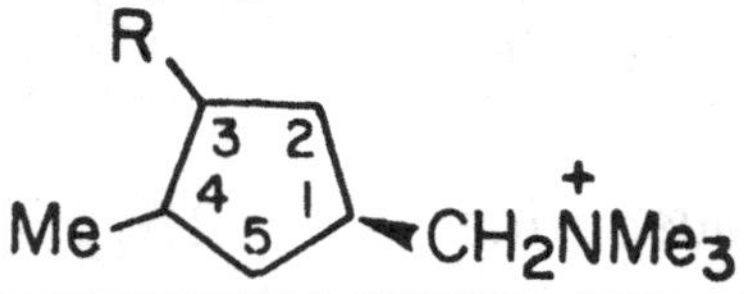

Compound	R	Configuration	Position of double bond	Guinea-pig ileum EPMR[b]	Frog rectus abdominis EPMR[b]
11	OH	*3t, 4c, 1r*		10	75
12	=O	*cis/trans*		0.5	2.8
13	OH	*trans*	4,5	50	>500
14	=O		4,5	5.9	0.7
15	H		4,5	26.6	3.1
16	H		1,5	19.3	5.0
17	H		3,4	11.1	12.8
18	H		1,2	36.7	3.1

[a]Angeli et al., 1981; Gualtieri et al., 1985.
[b]Equipotent molar ratio relative to ACh measured from contractile responses in the absence of a cholinesterase inhibitor.

(Givens and Rademacher, 1974). Alkylation or acetylation of the hydroxyl group of *11* did not dramatically affect its pharmacological profile. Hence the subsite on the receptor that interacts with the hydroxyl group of muscarine was also able to accommodate larger groups (Gualtieri et al., 1985). The cyclopentane analog (*12*) of muscarone (desethermuscarone) was equipotent with ACh, but somewhat less potent than muscarone on the guinea-pig ileum (Gualtieri et al., 1974; Givens and Rademacher, 1974). These results and those obtained with *11* showed that the ether oxygen of muscarine and muscarone was not necessary for muscarinic activity.

Introduction of a double bond in *11* and *12*, to give *13* and *14*, lowered muscarinic activity (Angeli et al., 1981). A comparison of the potencies of *13* and *14* with that of their nonoxygenated analog *15* showed that the carbonyl oxygen of *14* increased both muscarinic and nicotinic activity, the latter estimated on the rectus abdominis of the frog. The hydroxyl group of *13* more or less abolished nicotinic activity. Compounds *16–18*, which differ from *15* in the location of the ring double bond, retained considerable muscarinic and nicotinic activity (Gualtieri et al., 1985).

Table 5
Muscarinic and Nicotinic Activity of Tetrahydrofurane and Tetrahydrothiophene Derivatives[a]

X 3 2 4 5 1 Me Y $CH_2\overset{+}{N}Me_3$

Compound	X	Y	Configuration	Guinea-pig ileum EPMR[b]	Frog rectus abdominis EPMR[b]
10	O	O	*cis*	0.7	3.3
19	CH_2	O	*cis*	17.5	38.8
20	O	CH_2	*cis*	22.8	0.5
21	O	CHOH	*4c,5t,1r*	>1000	>500
22	CHOH	S	*3t,4c,1r*	216	>300

[a]Melchiorre et al., 1978; Pigini et al., 1980; Waser and Hopff, 1979.
[b]Equipotent molar ratio relative to ACh measured from contractile responses in the absence of a cholinesterase inhibitor.

3.2. Tetrahydrofurane and Tetrahydrothiophene Derivatives

Isosteric substitution of either oxygen atom of *10* by a methylene group, to give *19* and *20* (Table 5), caused a moderate decrease in potency on the ileum. Consequently, only one of the two oxygens of *10* appeared to be necessary for muscarinic activity. However, the importance of the two oxygens of *10* for nicotinic activity differed markedly, since *19* had 10-fold lower potency than *10* on the rectus abdominis muscle of the frog, whereas *20* was about five-fold more potent than *10* (Melchiorre et al., 1978).

In order to elucidate further the role of the oxygen atoms of *10* and muscarine, Pigini et al. (1980) synthesized compound *21*, which has the oxygenated functions inverted with respect to muscarine (isomuscarine). Compound *21* and its *epi-*, *allo-*, and *epiallo*-isomers were inactive, suggesting that a dipole–dipole interaction rather than a hydrogen bond is involved in the interaction of the ether oxygen of muscarine with the mAChR. Replacement of the ether oxygen of muscarine by a sulfur atom (*22*) lowered muscarinic activity considerably (Waser and Hopff, 1979). These results and those obtained with isomuscarine showed that the receptor site interacting with the ether oxygen of muscarine has limited size and very stringent electronic requirements. The only group that can replace this oxygen without substantially lowering muscarinic activity is a CH_2 group (Gualtieri et al., 1985).

3.3. Oxathiolanes

Elferink and Salemink (1975) showed that a mixture of the oxathiolanes *24* and *25* (Table 6) produced potent muscarinic and nicotinic actions. Pigini et al. (1981) later studied the racemic *cis* and *trans* isomers (*24* and *25*) as well as their sulfoxide and sulfone derivatives. Recently, the synthesis of the enantiomers of *24* and *25* (Teodori et al., 1986) and those of the sulfoxide *26* (Teodori et al., 1987) was reported. In Table 6, pharmacological data for the enantiomers of compounds *24-26* are compared with those of the enantiomers of the *cis* dioxolane *10* and of its *trans* isomer *23*.

In general, the muscarinic actions of the enantiomers of *24* and *25* paralleled those of the corresponding dioxolanes (Teodori et al., 1986). The more potent enantiomer of the *cis* isomer (*24*) was configurationally related to the more potent enantiomers of muscarine and *10*. Furthermore, the high enantiomeric potency ratio of *24* and the low ratio of *25* agreed with similar observations made among the muscarines and dioxolanes. Therefore, (2*R*,4*R*)-*24* was suggested to recognize the same binding site as that identified by more classical muscarinic agonists. The high potency of *24* appeared inconsistent with the proposed existence of a hydrogen bond between the mAChR and the carbonyl oxygen of muscarone, the hydroxyl group of muscarine and the oxygen in position 1 of *10*. A dipole–dipole interaction was believed to provide a better explanation of the high potency of *24*. The oxathiolanes *24* and *25* were quite potent on the rectus abdominis (Pigini et al., 1981; Teodori et al., 1986). Thus, although the stereochemistry and enantioselectivity of *24* make it a close analog of muscarine, its high potency at both muscarinic and nicotinic receptors resembles the behavior of muscarone. The conformational properties of *24* in the crystal and in solution did not differ substantially from those of muscarine and related compounds (Gualtieri et al., 1985).

The sulfoxide *26* is related configurationally to *24* (Teodori et al., 1987). The $S^{+}—O^{-}$ group of *26* appeared to interact with the mAChR in the same way as the CHOH group of muscarine,

Table 6
Muscarinic and Nicotinic Activity of Dioxolane and Oxathiolane Derivatives[a]

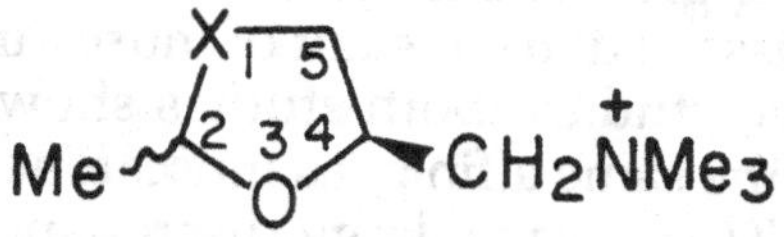

Compound	X	Configuration	Guinea-pig ileum		Frog rectus abdominis
			pD_2	EPMR[b]	EPMR[c]
10	O	*cis*(2*S*,4*R*)	8.64	0.2	
		cis(2*R*,4*S*)	6.62	12	
23	O	*trans*(2*S*,4*S*)		4	
		trans(2*R*,4*R*)		2	
24	S	*cis*(2*R*,4*R*)	8.06		1.2
		cis(2*S*,4*S*)	5.82		24
25	S	*trans*(2*R*,4*S*)	6.21		4.5
		trans(2*S*,4*R*)	6.38		2.8
26	SO	1*R*,2*R*,4*R*	6.95		33
		1*S*,2*S*,4*S*	4.83		120

[a]Belleau and Lavoie, 1968; Chang and Triggle, 1973; Teodori et al., 1986, 1987.
[b]Equipotent molar ratio relative to ACh.
[c]Equipotent molar ratio relative to CCh.

as suggested by the structural resemblance, similar configurational requirements and near equipotency of *26* and muscarine (Gualtieri et al., 1985). The effects of structural modifications of the trimethylammonium group of *24* and *26* were similar to those observed for *10* (Angeli et al., 1984). Remarkably, the tertiary *N*-desmethyl analog of *24* was reported to be equipotent with *24* on the rat jejunum. As shown previously for muscarine, muscarone, *9*, and *10* (Waser, 1961; Triggle, 1976), the methyl group attached to the ring was also important for muscarinic activity among oxathiolane derivatives. The hydrophobic interaction of this methyl group with the receptor was suggested to be more important than electrostatic interactions of the oxygenated functions of compounds related to muscarine (Gualtieri et al., 1985).

3.4. Affinity and Efficacy of Muscarine Analogs

Angeli et al. (1985) and Grana et al. (1987) determined dissociation constants and relative efficacies of compounds related to muscarine at mAChRs in the guinea-pig ileum. The higher potency of muscarone compared to muscarine appeared to arise primarily from the greater affinity of muscarone. However, the estimate of the dissociation constant of muscarine differed almost 20-fold in the two studies. Both studies showed the dioxolane *10* to have about the same affinity as muscarone, but to be slightly less efficacious. There was a large discrepancy in the estimate of the intrinsic efficacy of the oxathiolane *24*. Angeli et al. (1985) found *24* to be 20-fold more efficacious than *10*, whereas Grana et al. (1987) reported similar efficacies for *10* and *24*. The sulfoxide *26* was claimed to be even more efficacious than *24*. On the basis of their results, Angeli et al. (1985) concluded that a strong dipole ($S^{+}—O^{-}$) or a polarizable atom (S) in the place of the CHOH group of muscarine favored high efficacy. Since deoxamuscarine (*11*) had lower affinity than muscarine, it was suggested that the ether oxygen of muscarine increases affinity, but contributes little to efficacy (Angeli et al., 1985). In view of the above discrepancies, and in view of the 10-fold difference in the relative efficacy of CCh between the two studies, these suggestions of Angeli et al. appear to need confirmation. Recently, potencies, affinities, and relative efficacies of the enantiomers of compounds *24–26* were estimated in guinea-pig ileum and heart and in rat urinary bladder (Angeli et al., 1988).

4. Analogs of Arecoline

The cholinergic actions of arecoline (arecaidine methyl ester) have been known for almost a century. Arecoline is a tertiary amine that is mostly protonated at physiological pH (Burgen, 1965; Hanin et al., 1966). It produces a variety of muscarinic effects, including tremor, analgesia, and hypothermia, when injected to experimental animals (Palacios et al., 1986). However, these effects are generally short-lasting. The muscarinic actions of arecoline, at least in vitro, appear to be the result of direct stimulation of mAChRs. The nicotinic actions of arecoline are weak, and it is not a substrate for acetylcholinesterase (Mutschler et al., 1983). Extensive structural modifications of arecoline have been carried out mainly by Mutschler and coworkers (Hultzsch et al., 1971; Mutschler and Hultzsch, 1973; Lambrecht and Mutschler, 1973; Moser et al., 1983).

Table 7
Muscarinic Activity of Arecaidine Esters
as Measured from Contractile Responses[a]

COOR
N
Me

Compound	R	Guinea-pig ileum	
		ia[b]	pD_2
Arecoline	CH_3	1.0	7.2
27	C_2H_5	1.0[c]	7.8
28	C_3H_7	0.5[c]	5.7
29	CH_2CH^{8} CH_2	0.9	6.6
30	CH_2C^{0} CH	1.0	8.6
31	CH_2CH^{8} $CHCH_3$	0.9	7.5
32	CH_2C^{0} CCH_3	1.0	8.1

[a]Gloge et al., 1966; Mutschler and Hultzsch, 1973.
[b]Intrinsic activity compared to ACh.
[c]Intrinsic activity compared to arecoline.

4.1. Tetrahydropyridine Analogs

Replacement of the *O*-methyl group of arecoline by an ethyl group, to give *27*, increased muscarinic activity on the guinea-pig ileum (Table 7). In contrast, arecaidine propyl ester (*28*) and higher homologs were weak partial agonists (Gloge et al., 1966). Introduction of multiple bonds in the propyl group of *28* had pronounced effects on muscarinic activity. The allyl ester (*29*) was 10-fold more potent than *28*. The propargyl ester (*30*) surpassed ACh in potency. Since there were only small differences in charge distribution of compounds *28–30*, the potent muscarinic actions of *30* apparently were the result of additional binding to the receptor by the triple bond (Mutschler and Hultzsch, 1973). The crotyl (*31*) and butynyl esters (*32*) also were potent agonists. Methylation of the basic nitrogen of arecoline and of *27–32* reduced both potency and intrinsic activity at ileal mAChRs (Gloge et al., 1966; Mutschler and Hultzsch, 1973), but increased nicotinic activity (Christiansen et al., 1967). Hydrogenation of the ring double bond also yielded weak partial agonists.

Isoarecoline (isoarecaidine methyl ester, *35*) and its corresponding ethyl, propyl, and propargyl esters were weak muscarinic agonists on the rat ileum. However, *N*-methylation of isoarecaidine esters resulted in a 50-fold increase in potency (Lambrecht and Mutschler, 1973; Wess et al., 1987). The increase in muscarinic activity on methylation of *35*, to give *N*-methylisoarecoline *36* (Table 8), was in sharp contrast to the decrease in potency observed on *N*-methylation of other tertiary amines containing an endocyclic nitrogen atom, e.g., arecoline, aceclidine, and oxotremorine (Hanin et al., 1966). Hydrogenation of the double bond of *35* and *36* was accompanied by a loss of potency that, however, was less dramatic than in the arecoline series (Lambrecht and Mutschler, 1973).

4.2. Dihydrothiopyranes

The sulfonium analog (*34*) of arecoline was equipotent with arecoline on the guinea-pig ileum and considerably more potent than *N*-methylarecoline (*33*) (Table 8). Quantum chemical calculations revealed that *34* preferred a half-chair conformation with an axial *S*-methyl group, whereas in arecoline, the *N*-methyl group was equatorially oriented. Thus, high potency among arecoline analogs did not require a specific orientation (axial or equatorial) of the methyl group on the onium center. Comparison of charge distribution and ring geometries of arecoline, *33*, and *34* showed that the low potency of *33*, which has a methyl group in both the axial and equatorial positions, could best be explained by steric hindrance by the methyl group introduced by *N*-methylation (Mutschler et al., 1983). Similar calculations on *35*, *36*, and their sulfonium analog (*37*) revealed that in *35* the *N*-methyl group was equatorial and that *37* had an axial *S*-methyl group. The low potency of *35* and the high potency of *37* (Table 8) suggested that, among isoarecoline derivatives, an axial methyl group was required for muscarinic activity. However, an additional methyl group (equatorial) did not interfere sterically with the drug–receptor interaction as shown by the increase in potency on *N*-methylation of *35* to give *36* (Holtje et al., 1983). Compounds *34*, *36*, and *37* were quite potent in stimulating contractions of the rectus abdominis muscle of the frog, but produced a submaximal response in comparison to ACh. None of the compounds in Table 8 were substrates for acetylcholinesterase. Instead, they were competitive inhibitors of this enzyme (Mutschler et al., 1983; Höltje et al., 1983).

Table 8
Muscarinic and Nicotinic Activity of Tetrahydropyridine and Dihydrothiopyrane Derivatives[a]

Compound		Guinea-pig ileum EPMR[b]	Frog rectus abdominis EPMR[b]
Arecoline	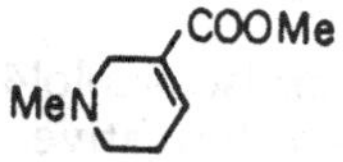 MeN (tetrahydropyridine ring), COOMe	9.4	401
33	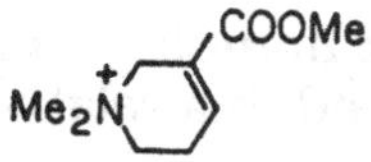 $Me_2\overset{+}{N}$ (ring), COOMe	89	15
34	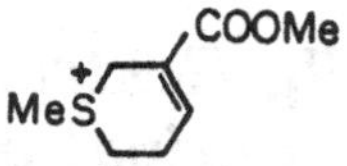 $Me\overset{+}{S}$ (ring), COOMe	12.2	4.1
35	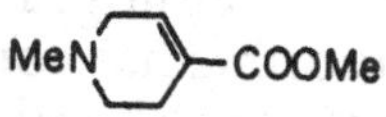 MeN (ring)–COOMe	265	150
36	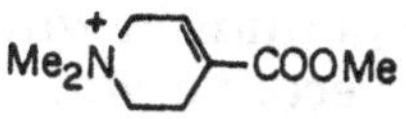 $Me_2\overset{+}{N}$ (ring)–COOMe	15	5.2
37	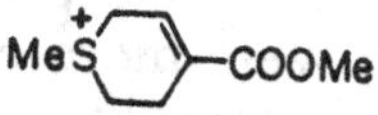 $Me\overset{+}{S}$ (ring)–COOMe	20	4.8

[a]Mutschler et al., 1983; Höltje et al., 1983.
[b]Equipotent molar ratios (EPMR) were measured relative to ACh and in the presence of physostigmine.

4.3. 3-Pyrroline Derivatives

The methyl, ethyl, and propyl esters of *N*-methyl-3-pyrroline-3-carboxylic acid (*38*, Fig. 6) were about 10-fold less potent on the guinea-pig ileum than the corresponding tetrahydropyridines (Section 4.1.). However, further increase in the size of the ester moiety was associated with a substantial increase in spasmogenic activity compared with the propyl derivative. *N*-Methylation generally had relatively little effect on potency, but quantitative comparison between tertiary amines and quaternary ammonium compounds was difficult because of nicotine-like effects of the

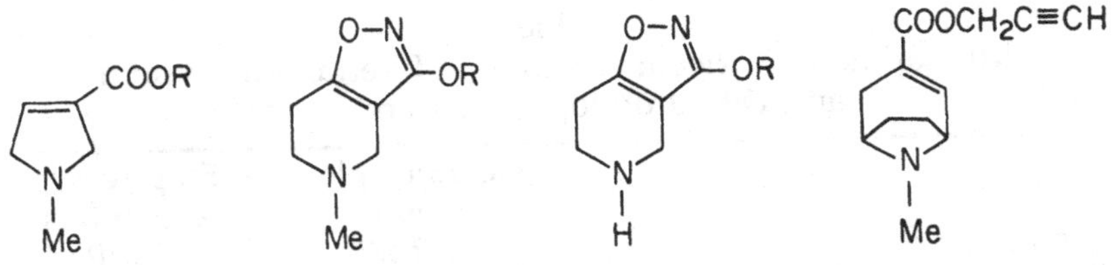

COOR 38 · 39 · 40 · 41

Fig. 6. 3-Pyrroline (*38*), 4,5,6,7-tetrahydroisoxazolo[4,5-c]pyridin-3-ol (*39* and *40*) and 2,3-dehydrotropane (*41*) derivatives of arecoline.

latter at ganglia, especially of higher homologs. Hydrogenation of the ring double bond of *38* resulted in a relatively small loss of potency (Hultzsch et al., 1971).

4.4. Conformationally Restricted Tetrahydropyridine Derivatives

Arecoline may be regarded as a semirigid analog of the reversed ester of ACh. However, arecoline has considerable conformational flexibility, especially around the bond connecting the ester moiety to the ring (Mutschler et al., 1983). Sauerberg et al. (1986a) described a series of compounds in which the methyl ester group of arecoline and norarecoline was replaced by a 3-alkoxyisoxazole moiety, which is resistant to hydrolysis. The resulting compounds (*39* and *40*, Fig. 6) had pK_a values close to physiological pH and in general were more lipophilic than arecoline. They were less potent than arecoline on the guinea-pig ileum, but appeared to exceed arecoline in central muscarinic activity as measured by anticonvulsant activity in mice. The propargyl derivative of *40* proved to be the most potent in this regard (Sauerberg et al., 1986a,b). Recently, some sulfonium analogs of *39* were described (Sauerberg et al., 1988).

The tertiary 2,3-dehydrotropane derivative *41* (Fig. 6) is a semirigid analog of isoarecaidine propargyl ester. Compound *41* was a partial muscarinic agonist on the rat atrium, with greater potency on force than on rate of contraction. On the rat ileum, *41* was a competitive antagonist. The affinity of *41* was claimed to be greater for muscarinic receptors in the ileum and myocardium than for those present in the pacemaker cells of the heart (Mutschler and Lambrecht, 1984; Lambrecht and Mutschler, 1985).

5. Analogs of Oxotremorine

Oxotremorine (Oxo) is a tertiary amine that is extensively protonated at physiological pH. It is a very potent muscarinic agonist when tested in vivo or on isolated organs (Cho et al., 1962; George et al., 1962). The nicotinic effects of Oxo are weak and its muscarinic actions are not potentiated by cholinesterases, nor does Oxo inhibit cholinesterases (Cho et al., 1962). It is generally accepted that Oxo is a directly acting muscarinic agent that is sufficiently nonpolar to penetrate the blood–brain barrier and exert effects on mAChRs in the brain. Evidence in favor of a direct action of Oxo in the central and peripheral nervous systems has been reviewed (Ringdahl and Jenden, 1983a). The structural requirements for muscarinic activity among congeners of Oxo are quite strict, because structural changes often result in antagonists, some of which are potent and highly selective in their central actions (Dahlbom, 1981; Ringdahl and Jenden, 1983a; Dahlbom et al., 1986).

5.1. Structure–Affinity and Structure–Efficacy Relationships

5.1.1. Modifications of the Amino Group

Replacement of the pyrrolidine ring of Oxo by larger cyclic amines such as piperidine and perhydroazepine yielded antagonists (Dahlbom, 1981). In contrast, substitution of the smaller azetidine ring for the pyrrolidine ring to give *42* enhanced muscarinic activity (Resul et al., 1980, 1982a). On the guinea-pig ileum, the greater muscarinic activity of *42* as compared to Oxo was the result of higher intrinsic efficacy of *42* (Table 9). The affinity of *42* for ileal mAChRs was less than that of Oxo (Ringdahl, 1985a). The dimethylamino analog (*43*) had about one-fifth the potency of Oxo, both as a stimulant of the ileum and as a tremorogenic agent (Resul et al., 1983). This reduction in potency was entirely the result of the loss of affinity, at least on the ileum, because *43* had six-fold greater intrinsic efficacy than Oxo (Ringdahl, 1985a). Stepwise replacement of the methyl groups of *43* by larger alkyl groups resulted in an attenuation of muscarinic activity. The methylethylamino derivative (*44*) was a full agonist on the ileum, whereas the diethylamino analog (*45*) was a partial agonist. More heavily substituted derivatives were

Table 9
Muscarinic Activity In Vitro of Tertiary Amines, Quaternary Ammonium, and Sulfonium Salts Related to Oxotremorine[a]

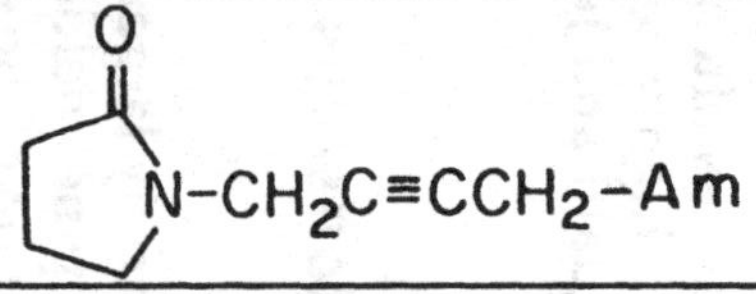

$$\text{(2-oxopyrrolidin-1-yl)}\,N\text{-}CH_2C{\equiv}CCH_2\text{-}Am$$

Compound	Am	Guinea-pig ileum				Guinea-pig urinary bladder			
		ia[b]	pD_2	pK_D	e_r[b]	ia[b]	pD_2	pK_D	e_r[b]
Oxo[c]	 [ring structure, N]	1.0	7.46	6.03	0.13	0.8	6.11	5.85	0.14
42[c]	 [ring structure, N]	1.0	7.96	5.86	0.60				
43[c]	$N(CH_3)_2$	1.0	6.72	4.64	0.85	1.0	5.64	4.46	1.1
44[c]	$NCH_3C_2H_5$	1.0	6.26	4.82	0.14				
45	$N(C_2H_5)_2$	0.7	4.48	4.74	0.008				
46	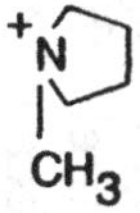 [ring structure, ^{+}N–CH_3]	1.0	6.50	5.17	0.11				
47	$\overset{+}{N}(CH_3)_3$	1.0	7.64	5.54	0.95	1.0	6.43	5.37	0.88

48	$\overset{+}{N}(CH_3)_2C_2H_5$	1.0	6.41	4.60	0.32				
49	$\overset{+}{N}(C_2H_5)_2CH_3$	1.0	4.95	4.28	0.028				
50	$\overset{+}{N}(C_2H_5)_3$	0		4.43					
51	S^{+} (ring)	1.0	6.60	5.89	0.054	0.5	5.92	5.96	0.057
52	$\overset{+}{S}(CH_3)_2$	1.0	7.54	5.82	0.41	1.0	6.70	5.89	0.48
CCh		1.0	7.00	4.81	1.0	1.0	6.08	4.77	1.0

[a]Ringdahl, 1985a, 1987a, 1988a.

[b]Intrinsic activities (ia) and relative efficacies (e_r) are given in relation to CCh.

[c]Oxo, *42*, *43*, and *44* had tremorogenic doses in mice (iv) of 0.5, 0.5, 2.7, and 5.3 μmol/kg, respectively (Resul et al., 1980, 1982a, 1983).

competitive antagonists (Resul et al., 1983). This decrease in muscarinic activity on structural modification of *43* arose because of a progressive loss of efficacy, rather than loss of affinity. Thus, compounds *43–45* had almost identical affinity, but their relative efficacies were within a range of 100-fold (Ringdahl, 1985a). Replacement of one methyl group of *43* by a hydrogen atom to give a secondary amine led to a 100-fold decrease of muscarinic activity (Resul et al., 1983).

The *N*-methyl derivative (*46*) of Oxo (Table 9) was about ten times less potent than Oxo on the ileum mainly because of loss of affinity on *N*-methylation. In contrast, *N*-methylation of the dimethylamino analog *43* to give *47*, often referred to as oxotremorine-*M* (Oxo-*M*), increased muscarinic activity, mainly as a result of increased affinity. Replacement of the methyl groups of *47* by ethyl groups (compounds *48–50*) reduced muscarinic activity, primarily because of a progressive decrease of efficacy with increasing substitution (Ringdahl, 1985a). The thiolanium (*51*) and dimethylsulfonium (*52*) derivatives were equipotent with the corresponding ammonium compounds (*46* and *47*, respectively), but had higher affinity and lower efficacy than the latter (Ringdahl, 1988a). Compounds *46–49*, *51*, and *52* elicited the same maximum response on the ileum, i.e., they had the same intrinsic activity (ia). Yet, their intrinsic efficacies varied over a range of 34-fold. These results clearly illustrate the inadequacy of the parameter intrinsic activity as a measure of receptor activating abilities of full agonists.

Among the tertiary amines, affinity did not seem to depend on the size of the amino group as evidenced by the similar affinities of compounds *43–45*. In addition, Oxo and *44* differed 20-fold in affinity despite the similar size of their amino groups. The affinities of the quaternary ammonium compounds also showed no clear dependence on molecular size. Efficacy, however, decreased in a regular fashion with increasing size of the amino or ammonium group, the latter estimated from apparent molal volumes. Quaternary derivatives, in spite of their larger size, generally had greater efficacies than their corresponding desmethyl analogs. For example, the tertiary amine *45* was a partial agonist on the ileum, whereas its *N*-methyl analog *49* was a full agonist. Thus molecular size was not the only factor determining efficacy (Ringdahl, 1985a).

Oxo, *43*, *47*, *51*, and *52* were 5- to 20-fold less potent on strips of the guinea-pig urinary bladder than on the ileum. Furthermore, Oxo and *51* behaved like partial agonists on the bladder. There was good agreement between dissociation constants and relative efficacies in the two tissues (Ringdahl, 1987a, 1988a). The trimethylammonium (*47*) and thiolanium (*51*) salts approached CCh in nicotinic activity on the rectus abdominis of the frog (Ringdahl, 1984b, 1988a).

5.1.2. Effects of Methyl Substitution

Addition of a methyl group at the 5-position of the lactam ring of Oxo abolished efficacy, but increased affinity for ileal mAChRs 10-fold (Ringdahl and Jenden, 1983b). The resulting compound (*53*) (Table 10) was more potent than atropine in inhibiting Oxo-induced tremors, but had only 1/20 of the mydriatic activity and 1/100 of the in vitro parasympatholytic activity of atropine (Amstutz et al., 1985). The azetidino (*54*) and dimethylamino (*55*) analogs of *53* were full or nearly full agonists on the ileum, but in contrast to their desmethyl analogs (*42* and *43*, Table 9), did not produce tremor. Compounds *54* and *55* had three- to five-fold greater affinity and about 25-fold lower efficacy than *42* and *43*. Consequently, the failure of *54* and *55* to produce tremor, in spite of their potent agonist effects on the ileum, was ascribed to their low intrinsic efficacies (Ringdahl, 1988b). Compounds *54* and *55* possessed affinity for those central receptors that mediate tremor as evidenced by their ability to block Oxo-induced tremor (Table 10). Furthermore, they were potent in inducing other central muscarinic effects, e.g., hypothermia and analgesia. These and similar observations suggested a need for high efficacy to induce a tremor response, i.e., a low receptor reserve for this response (Ringdahl, 1988b; Ringdahl et al., 1987). Although the 5-methyl-2-pyrrolidones *53–59*, on the average, had seven-fold higher affinity and 40-fold lower efficacy than the corresponding 2-pyrrolidones (Table 9), the effects of changing the structure of the amino group on affinity and efficacy were similar in the two series of compounds. A plot of the dissociation constants of compounds *53–59* vs those of the identically modified 2-pyrrolidones gave a linear regression (r = 0.85; $P < 0.05$) with a slope of unity. Thus, the incremental changes in affinity caused by identical modification of the amino group in the two series were the same. These observations suggested that a given amino group contributed to affinity by the same mechanism in both series of compounds, and that 2-pyrrolidones and 5-methyl-2-pyrrolidones had similar mode of binding to mAChRs. The

Table 10
Muscarinic and Antimuscarinic Activity In Vitro and Tremorolytic Activity in Mice of 5-Methyl-2-pyrrolidones Related to Oxotremorine[a]

5-methyl-2-pyrrolidone ring: $N{-}CH_2C{\equiv}CCH_2{-}Am$ (CH_3 at position 5)

Compound	Am	Guinea-pig ileum				Guinea-pig urinary bladder				Tremor blockade
		ia[b]	pD_2	pK_D	e_r[b]	ia[b]	pD_2	pK_D	e_r[b]	ID_{50}, μmol/kg
53	N	0		7.03		0		6.98		0.4
54	N	0.9	6.64	6.28	0.020	0.2	5.34	6.24	0.022	4.6
55	$N(CH_3)_2$	1.0	6.04	5.37	0.040	0.6	4.82	4.93	0.064	23
56	$NCH_3C_2H_5$	0		5.74		0		5.60		6.7
57	$N(C_2H_5)_2$	0		5.96		0		5.68		5.1
58	$\overset{+}{N}(CH_3)_3$	1.0	6.74	5.77	0.057	0.7	5.70	5.55	0.062	
59	$\overset{+}{N}(CH_3)_2C_2H_5$	0.3	5.49	5.31	0.0036	0		5.46		

[a]Ringdahl, 1988b.
[b]Intrinsic activities (ia) and relative efficacies (e_r) are given in relation to CCh.

good correlation between the dissociation constants of compounds *53–57* at receptors in the ileum and tremorolytic activity indicated similar structural demands for binding at ileal mAChRs and at those central mAChRs that mediate tremor (Ringdahl, 1988b).

None of compounds *53–59* elicited a full contractile response compared to CCh on the urinary bladder (Table 10). Compounds *54, 55,* and *58* were partial agonists, but were 10- to 20-fold less potent than on the ileum. In spite of the large differences in potency and maximal responses between the ileum and bladder, the dissociation constant of each compound showed good agreement (within threefold) in the two tissues (Ringdahl, 1988b).

Introduction of a methyl group adjacent to the carbonyl carbon (position 3′) of Oxo (Table 11) reduced affinity for mAChRs in the ileum 10-fold, but had no apparent effect on efficacy (Ringdahl and Jenden, 1983b). The resulting compound (*60*) was tremorogenic, although less potent than Oxo (Ringdahl et al., 1979). In contrast, methyl substitution in the 4′-position (*61*) and in the 5′-position (*53*) of the lactam ring abolished efficacy. Compound *61* had 10-fold *lower* affinity for ileal mAChRs than Oxo (Ringdahl and Jenden, 1983b), whereas the affinity of the *R*-enantiomer of *53* was about 15-fold *higher* than that of Oxo (Amstutz et al., 1985). (*R*)-*53* was half as potent as PZ in inhibiting [^{3}H]PZ binding in the rat cerebral cortex, but was eightfold more potent than PZ in inhibiting muscarine-induced depolarization of the isolated rat superior cervical ganglion. Furthermore, it had about 100-fold greater affinity for mAChRs in the ganglion than for those in the ileum (Amstutz et al., 1987).

Addition of a methyl group in the 1-position of the butynyl chain (*62*) also abolished efficacy and increased affinity (Lindgren et al., 1973; Dahlbom et al., 1974). The *R*-enantiomer of *62* had 15 times higher and the *S*-enantiomer 14 times lower affinity than Oxo for mAChRs in the ileum (Ringdahl and Jenden, 1983b; Ringdahl, 1984a). (*R*)-*62* was 10-fold more potent than atropine as a tremorolytic agent, but had only a small fraction of the peripheral parasympatholytic activity of atropine (Dahlbom et al., 1974). It appeared that the methyl groups of (*R*)-*62* and (*R*)-*53* participated in an additional interaction with the receptor, presumably at a common site (Ringdahl and Jenden, 1983b; Amstutz et al., 1985). The derivative having a methyl group in the 4-position of the butynyl chain (*63*) was a weak antagonist (Resul et al., 1979a) whose affinity was 40-fold lower than that of Oxo (Ringdahl and Jenden, 1983b).

Table 11
Muscarinic and Antimuscarinic Activity In Vitro
and Tremorolytic Activity in Mice
of Some Methyl-Substituted Oxotremorine Analogs[a]

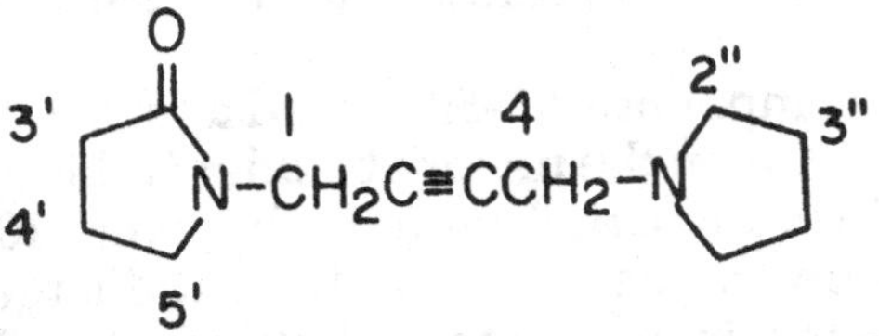

O 3' 4' 5' 1 4 2" 3"

$N{-}CH_2C{\equiv}CCH_2{-}N$

Compound	Position of CH_3	Guinea-pig ileum ia[b]	pD$_2$	pK$_D$	e_r[b]	Tremor blockade ID$_{50}$, μmol/kg
60	3′	1.0	6.30	4.94	0.12	8.8[c]
61	4′	0		5.00		37
(*R*)-*53*	5′	0		7.29		0.19
(*S*)-*53*	5′	0		6.01		4.2
(*R*)-*62*	1	0		7.34		0.26
(*S*)-*62*	1	0		5.02		20
63	4	0		4.43		11
(*S*)-*64*	2″	0.6	6.06	6.18	0.009	2.6
(*R*)-*64*	2″	0.7	5.55	5.30	0.014	56
(*S*)-*65*	3″	0		5.68		5.8
(*R*)-*65*	3″	0		5.69		6.5

[a]Ringdahl and Jenden, 1983b; Ringdahl, 1984a; Amstutz et al., 1985.
[b]Intrinsic activities (ia) and relative efficacies (e_r) are given in relation to CCh.
[c]Tremorogenic dose (Ringdahl et al., 1979).

Methyl substitution in the pyrrolidine ring also abolished the tremorogenic activity of Oxo and produced tremorolytic agents (Ringdahl and Dahlbom, 1978). The enantiomers of the 2-methylpyrrolidino analog (*64*) were partial agonists on the ileum, whereas those of the 3-methylpyrrolidino derivative (*65*) were competitive antagonists. Compounds *64* and *65* displayed little or no stereoselectivity in vitro (Ringdahl and Jenden, 1983b; Ringdahl, 1984a) or in vivo (Ringdahl and Dahlbom, 1978). Their dissociation constants at mAChRs in the ileum were similar to or only slightly lower than that of Oxo. In summary, methyl substitution in Oxo abolished or greatly reduced efficacy, except for substitution in the 3-position of the lactam ring and on the basic nitrogen (*46*). Affinity was increased by a methyl group adjacent to the lactam nitrogen and virtually unaltered by a methyl group in the pyrrolidine ring. A methyl group in all other positions, including the basic nitrogen, reduced affinity. All of the methyloxotremorines in Table 11 stimulated contractions of the frog rectus abdominis preparation but only at high concentrations (Ringdahl, 1984b).

5.1.3. N-Alkylcarboxamides

The *N*-methylacetamido analog (*66*) of Oxo (Table 12) had pharmacological properties that were quite similar to those of Oxo (Bebbington et al., 1966; Svensson et al., 1978). However, *66* was two- to threefold less potent than Oxo. On the guinea-pig ileum, this lower potency was entirely the result of lower affinity, since the intrinsic efficacy of *66* was slightly greater than that of Oxo (Ringdahl, 1984a,c). The effects of structural modifications of the amino group of *66* on potency, affinity, and efficacy were similar to those observed in the 2-pyrrolidone series (Table 9) (Resul et al., 1983; Ringdahl, 1984c). The trimethylammonium derivative (*68*) was more efficacious than CCh and is the most efficacious Oxo analog known. It required less than 0.5% receptor occupation in the guinea-pig ileum to produce a half-maximal response (Ringdahl, 1984c). Replacement of the *N*-methyl group of *66* by a hydrogen atom (*69*) abolished efficacy and caused a 15-fold reduction in affinity (Resul et al., 1979b; Ringdahl et al., 1982b). Substitution of an ethyl group for the *N*-methyl group of *66*, to give *70* (Svensson et al., 1978), also abolished efficacy, but did not have much effect on affinity (Ringdahl et al., 1982b).

The formamide analog (*71*) of *66* was a weak antagonist (Ringdahl et al., 1982b). Introduction of progressively larger alkyl groups in the acetyl moiety of *66* resulted in an attenuation of muscarinic activity. The propionic acid amide (*72*) retained considerable muscarinic activity in vivo (Svensson et al., 1978) and in vitro (Ringdahl et al., 1982b), whereas higher homologs were weak agonists (*73*) or antagonists (*74*). Fluorine substitution in the acetyl group of *66* was detrimental to muscarinic activity. The resulting compound (*75*) was a weak partial agonist (Nilsson et al., 1988). The β-lactam derivative (*76*), which may be regarded as a ring closed analog of *66*, had about one-third of its muscarinic activity in vivo and in vitro (Resul et al., 1981). Except for the trimethylammonium derivative *68*, the acetamides in Table 12 had weak nicotinic actions (Ringdahl, 1984b).

Introduction of a methyl group in position 1 of the butynyl chain of *66* (Table 13) abolished tremorogenic activity and yielded a potent tremorolytic agent (*77*), commonly referred to

Table 12
Muscarinic Activity In Vitro and Tremorogenic Activity in Mice of Carboxamides Related to Oxotremorine[a]

$$R-\overset{\overset{\displaystyle O}{\|}}{C}-\underset{\underset{\displaystyle R'}{|}}{N}-CH_2C{\equiv}CCH_2-Am$$

Compound	R	R′	Am	Guinea-pig ileum				Tremor ED_{50}, μmol/kg
				ia[b]	pD_2	pK_D	e_r[b]	
66[c]	CH_3	CH_3	[pyrrolidine ring, N]	1.0	7.16	5.66	0.16	1.3
67[c]	CH_3	CH_3	$N(CH_3)_2$	1.0	6.41	4.16	0.90	15
68[c]	CH_3	CH_3	$\overset{+}{N}(CH_3)_3$	1.0	7.43	5.00	1.32	
69	CH_3	H		0		4.50		
70	CH_3	C_2H_5		0		5.44		4.8[d]
71	H	CH_3		1.0	5.40			
72	C_2H_5	CH_3	[pyrrolidine ring, N]	1.0	6.38			2.1
73	C_3H_7	CH_3		1.0	5.18			
74	$(CH_3)_2CH$	CH_3		0		4.36		
75	CF_3	CH_3		0.16	4.77	4.81	0.003	
76	$-CH_2CH_2-$			1.0	6.66			3.0

[a]Svensson et al., 1978; Resul et al., 1981, 1983; Ringdahl et al., 1982b; Ringdahl, 1984c; Nilsson et al., 1988.
[b]Intrinsic activities (ia) and relative efficacies (e_r), are given in relation to CCh.
[c]Guinea-pig urinary bladder: *66*, ia 0.8; pD_2 5.51; pK_D 4.99; e_r 0.18. *67*, ia 1.0; pD_2 4.79; pK_D 3.61; e_r 1.51. *68*, ia 1.0; pD_2

Table 13
Muscarinic Activity In Vitro and Tremorolytic Activity in Mice of *N*-Methylcarboxamides Related to Oxotremorine[a]

$$R{-}C({=}O){-}N(CH_3){-}CH(CH_3)C{\equiv}CCH_2{-}Am$$

Compound	R	Am	Guinea-pig ileum				Guinea-pig urinary bladder				Tremor blockade
			ia[b]	pD_2	pK_D	e_r[b]	ia[b]	pD_2	pK_D	e_r[b]	ID_{50}, μmol/kg
(*R*)-77	CH_3	pyrrolidin-1-yl	0.8	7.03	6.82	0.013	<0.05[c]		6.85[c]		0.35
(*S*)-77			0.9	5.77	5.38	0.018					10
(*R*)-78	CH_3	$N(CH_3)_2$	1.0	6.60	5.19	0.13	0.9	5.26	4.93	0.21	12[d]
(*S*)-78			1.0	5.27	3.97	0.10	0.9	3.81	3.58	0.15	
(*R*)-79	CH_3	$\overset{+}{N}(CH_3)_3$	1.0	6.86	5.14	0.26	1.0[c]	5.13[c]	4.53[c]	0.49[c]	
(*S*)-79			1.0	5.34	4.22	0.069					
80	CF_3	pyrrolidin-1-yl	0		6.24						21
81	C_2H_5	pyrrolidin-1-yl	1.0	6.38			0.5	5.48	5.29	0.019	8.0

[a]Resul et al., 1982b; Dahlbom et al., 1982; Ringdahl, 1984a,c, 1987a; Ringdahl and Markowicz, 1987; Nilsson et al., 1988.
[b]Intrinsic activities (ia) and relative efficacies (e_r) are given in relation to CCh.
[c]Data for the racemate.
[d]Tremorogenic dose.

as BM 5. In the guinea-pig ileum, *77* was a partial agonist (Resul et al., 1982b), whose *R*-enantiomer (Dahlbom et al., 1982) had 15 times higher affinity and 12 times lower intrinsic efficacy than *66* (Ringdahl, 1984a). Replacement of the pyrrolidine ring of (*R*)-*77* by a dimethylamino [(*R*)-*78*] or a trimethylammonium [(*R*)-*79*] group (Dahlbom et al., 1982) gave the expected (cf. Table 12) increase in efficacy and decrease in affinity (Ringdahl, 1984c). Fluorine substitution in the acetyl group of *77*, to give *80*, decreased affinity for ileal mAChRs fourfold (Nilsson et al., 1988). In contrast to *77*, the propionyl derivative *81* was a full agonist on the ileum (Resul et al., 1982b), suggesting that substitution of an ethyl group for the acetyl methyl group of *77* increased efficacy. This suggestion was confirmed on the urinary bladder, where *77* was an antagonist and *81* a partial agonist (Ringdahl and Markowicz, 1987). The affinity of *81*, as estimated from its dissociation constant on the bladder or its tremorolytic potency, was lower than that of *77*. Among compounds *77–81*, only the trimethylammonium derivative (*79*) was a full agonist on the bladder, although it was much less potent than on the ileum (Ringdahl, 1987a; Ringdahl and Markowicz, 1987). Differences in potency between the enantiomers of *77–79* were mainly the result of differences in affinity for the receptor (Ringdahl, 1984a,c 1987a). The compounds in Table 13 had only weak nicotinic actions (Ringdahl, 1984b).

5.1.4. 4-Substituted N-(2-Butynyl)Succinimides

The succinimide analog (*82*) of Oxo was a full agonist on the guinea-pig ileum. It had about one-fifth of the affinity and intrinsic efficacy of Oxo (Ringdahl, 1987b). The effects of structural modifications of the amino group of *82* on affinity and efficacy at mAChRs in the guinea-pig ileum and urinary bladder (Table 14) were almost identical to those observed for the corresponding 2-pyrrolidones (Table 9). A plot of dissociation constants in one series vs those in the other yielded a highly significant ($r = 0.91$; $P < 0.001$) linear regression with a slope of unity. Relative efficacies in the two series were also highly correlated ($r = 0.95$; $P < 0.001$). From these observations, it was concluded that succinimides and 2-pyrrolidones modified at the 4-position of the butynyl chain bind to and activate mAChRs in a similar fashion (Ringdahl, 1987b). The succinimides, on the average, had 3.9-fold lower efficacy than the corresponding 2-pyrrolidones. This lower efficacy probably explains why only one of the succinimides (*83*) produced tremor (Karlen et al., 1970). Compound *83* had the highest efficacy but the lowest affinity among the tertiary amines in Table 14. These observations further emphasize high efficacy as a requirement for tremorogenic activity (Ringdahl et al., 1987).

5.1.5. Modifications in the 5-Position of the 2-Pyrrolidone Ring

Isosteric replacement of the CH_2 group in the 5-position of the lactam ring of Oxo by oxygen (*92*) or nitrogen (*93*) yielded muscarinic agonists that, like Oxo, were more potent on the rat ganglion than on the guinea-pig ileum (Table 15). The formyl derivative *94* was an antagonist and displayed greater selectivity than PZ for mAChRs in the ganglion as opposed to those present in the ileum. In this regard, it was more selective than (*R*)-*53* (Section 5.1.2.). In contrast to PZ, which was equipotent in antagonizing muscarine-induced depolarization of the rat ganglion and in inhibiting [^{3}H]PZ binding in the cortex, *94* and (*R*)-*53* were relatively less potent in inhibiting [^{3}H]PZ binding (Amstutz et al., 1987). These results suggest that cortical mAChRs labeled by low concentrations of [^{3}H]PZ (M1-receptors) are not identical to ganglionic mAChRs mediating depolarization.

5.1.6. Modifications of the Acetylenic Group

Replacement of the triple bond of Oxo by a double or single bond abolished activity (Bebbington et al., 1966). Similarly, an allenic group (Claesson et al., 1978) or a cyclopropane ring (Dahlbom et al., unpublished results) in place of the triple bond yielded inactive compounds.

5.1.7. Summary of Structure–Affinity and Structure–Efficacy Relationships

In general, a cyclic tertiary amino or cyclic sulfonium group at the "cationic head" of Oxo favored high affinity, whereas small acyclic tertiary amino, quaternary ammonium, and sulfonium groups favored high efficacy. Methyl substitution adjacent to the lactam nitrogen greatly increased affinity, but abolished efficacy. Replacement of the 2-pyrrolidone ring by an *N*-methylacetamide group decreased affinity, but slightly increased efficacy. All modifications of Oxo that led to higher affinity at the same time reduced or abolished efficacy. Structural alterations of Oxo that

Table 14
Muscarinic Activity In Vitro of Succinimides Related to Oxotremorine[a]

N-$CH_2C{\equiv}CCH_2$-Am (succinimide)

Compound	Am	Guinea-pig ileum				Guinea-pig urinary bladder			
		ia[b]	pD_2	pK_D	e_r[b]	ia[b]	pD_2	pK_D	e_r[b]
82	pyrrolidin-1-yl	1.0	6.00	5.27	0.029	0.3	4.73	5.06	0.025
83	$N(CH_3)_2$	1.0	5.61	3.68	0.37	1.0	4.23	3.55	0.40
84	$NCH_3C_2H_5$	1.0	4.72	4.20	0.025	0		4.01	
85	$N(C_2H_5)_2$	0		4.51					
86	$^+$N-CH_3 pyrrolidinium	1.0	5.30	4.50	0.042	0.2	4.46	4.60	0.019
87	$\overset{+}{N}(CH_3)_3$	1.0	6.85	4.78	0.65	1.0	5.83	4.81	0.68

88	$\overset{+}{N}(CH_3)_2C_2H_5$	1.0	5.72	4.25	0.15	0.8	4.47	4.18	0.14
89	$\overset{+}{N}(C_2H_5)_2CH_3$	0		4.11					
90	$\overset{+}{S}$ (five-membered ring)	0.5	5.10	5.12	0.0061	0		4.88	
91	$\overset{+}{S}(CH_3)_2$	1.0	6.39	5.16	0.11	1.0	5.77	5.23	0.15

[a] Ringdahl, 1987b, 1988a.
[b] Intrinsic activities (ia) and relative efficacies (e_r) are given in relation to CCh.

Table 15
Muscarinic Activity and Receptor Binding Affinity of Compounds *92–95*[a]

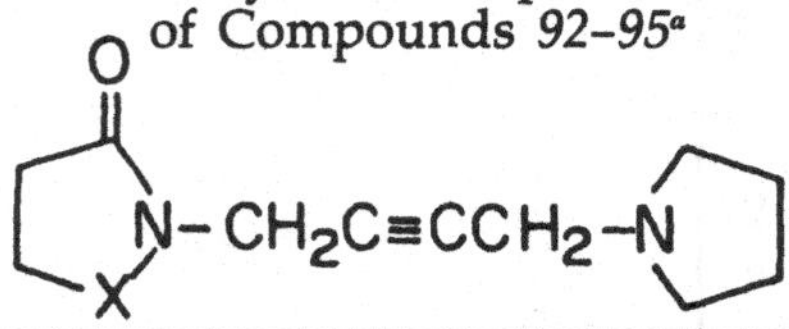

Compound	X	Guinea-pig ileum		Rat ganglion[b]		[³H]PZ inhibition[c]
		pD_2	pA_2	pD_2	pA_2	pIC_{50}
92	O	6.6		7.3		6.8
93	NH	5.6		6.3		6.2
94	NCHO		5.2		7.9	6.8
95	$NCOCH_3$		6.3		6.9	7.3
(*R*)-*53*			7.2		9.2	8.1
Oxo		7.1		8.0		6.9
PZ			6.8		8.3	8.4

[a]Amstutz et al., 1987.
[b]Isolated superior cervical ganglion (depolarization).
[c]Inhibition of [³H]pirenzepine binding (0.3 n*M*) in homogenates of the rat cerebral cortex.

decreased affinity had no consistent effect on efficacy. There was, therefore, no simple relationship between the affinity of Oxo analogs for the mAChR and their ability to activate the receptor. In fact, the absence of a correlation between affinities and relative efficacies suggested different structural requirements for these two parameters of drug action (Ringdahl, 1984a,c, 1985a,b, 1987b).

5.2. Relationship between Affinity and Efficacy and Biochemically Determined Drug Parameters

Fisher et al. (1984) studied the ability of some analogs of Oxo to inhibit [³H]QNB binding and to stimulate inositol phospholipid turnover in nerve-ending fractions from the guinea-pig cerebral cortex (Table 16). There was an excellent agreement between the low affinity dissociation constants (or the single affinity constants of compounds that displayed only one binding site) of the Oxo analogs studied and their pharmacologically determined dissociation constants (Fig. 7). Furthermore, agonists having high efficacy on the ileum displayed large K_L/K_H ratios. These results are in agreement with previous correlations of pharmacological re-

Table 16
Muscarinic Receptor Binding Parameters and Effects on Phosphatidylinositol Turnover of Oxotremorine Analogs[a]

Compound[b]	Guinea-pig cerebral cortex: [³H]QNB inhibition pK_H	pK_L	K_L/K_H	[³²P]PhI formation % Control	Relative efficacy
Oxo		6.48[c]		114	0.18
43	6.68	4.65	106	162	0.78
47	6.82	4.87	91	179	0.99
53		7.10[c]		103	0.038
66		5.60[c]		121	0.26
67	6.62	4.31	206	153	0.66
68	6.68	4.47	160	173	0.91
CCh	5.85	4.12	54	180	1.00

[a]Fisher et al., 1983, 1984.
[b]For chemical structures, *see* Tables 9, 10, and 12.
[c]Inhibition data were fitted to a one-site model.

sponses and ligand binding (Birdsall et al., 1978a,b). Figure 8 shows a comparison of the relative efficacies of these Oxo analogs measured pharmacologically on the ileum and from ability to stimulate PI turnover in guinea-pig brain. There was a close correlation between the two measures of agonist efficacy. Receptor binding parameters for sulfonium analogs of Oxo also agreed with pharmacologically determined affinities and efficacies (Ringdahl, 1988a).

5.3. Selectivity of Partial Agonists

One group of Oxo analogs (*43, 47, 52, 67, 68, 79, 83, 87,* and *91*) that were full agonists on the guinea-pig ileum also produced full responses on the urinary bladder. A second group of analogs (Oxo, *51, 55, 58, 66, 78, 81, 82, 86,* and *88*) were partial agonists on the bladder despite their full agonist effects on the ileum. A third group of compounds, (*54, 59, 77,* and *90*), which were partial agonists on the ileum, showed very weak or no responses on the bladder and were competitive antagonists in this tissue. Concentration–response curves for members of each of these groups are shown in Fig. 9. Analogs in the first group discriminated between the two tissues by being less potent (4- to 40-fold) on the bladder. Analogs in the latter groups also discriminated between the ileum and bladder in terms of maxi-

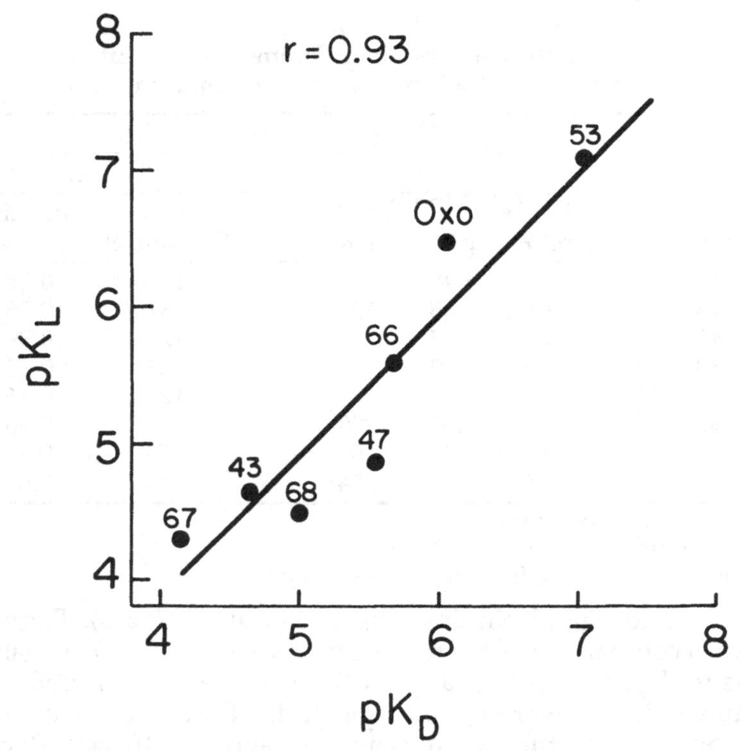

Fig. 7. Relationship between dissociation constants of oxotremorine analogs determined from contractile responses on the guinea-pig ileum (K_D) and from inhibition of [^{3}H]QNB binding in guinea-pig brain (K_L). K_D values are from Tables 9, 10, and 12, and K_L values from Table 16.

mum response obtainable. The ability to discriminate between the two tissues clearly was related to intrinsic efficacy as estimated in the ileum (Ringdahl, 1987a,b; Ringdahl and Markowicz, 1987). Compounds in the first group all were highly efficacious, whereas those in the second group, including Oxo itself, had intermediate intrinsic efficacy. The third group of compounds had low intrinsic efficacy. In spite of these striking differences in agonist potency and relative maximal responses between the ileum and bladder, no corresponding differences were observed between dissociation constants in the two tissues (Fig. 10), suggesting that mAChRs in the two tissues are pharmacologically similar. Furthermore, the relative efficacy of each agonist was similar in the two tissues (Fig. 11), confirming the homogeneity of mAChRs

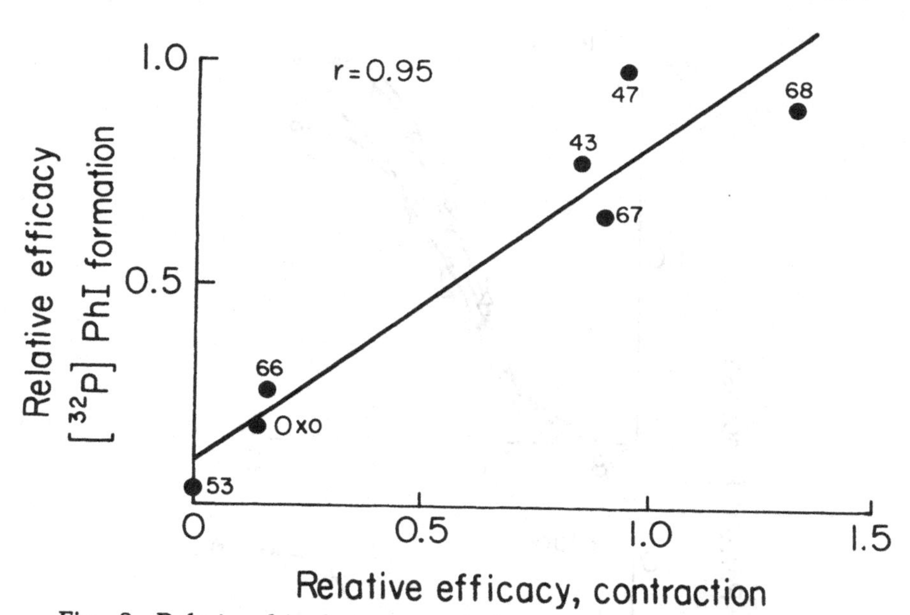

Fig. 8. Relationship between relative efficacies of oxotremorine analogs determined from contractile responses on the guinea-pig ileum (Tables 9, 10, and 12) and from ability to stimulate PI turnover in guinea-pig brain (Table 16). Efficacies are expressed relative to that of carbachol.

in the ileum and bladder with respect to analogs of Oxo. Thus, agonists of low intrinsic efficacy (partial agonists) stimulated responses on the ileum and blocked responses on the bladder, and therefore displayed tissue selectivity without discriminating between tissue receptors.

In contrast to these similarities in affinity and relative efficacy, the degree of receptor occupancy required for half-maximal response was considerably larger in the bladder than in the ileum. Muscarinic receptor occupancy therefore appeared to be less efficiently translated to a contractile response in the bladder. It was suggested (Ringdahl, 1987a; Ringdahl and Markowicz, 1987) that the observed selectivity for stimulating responses in the ileum (or blocking responses on the bladder) was derived from differences in tissue sensitivity resulting from a smaller effective receptor reserve in the bladder. The latter may be the result of low receptor density or less efficient coupling mechanisms. Thus, responses to agonists of low intrinsic efficacy were more affected by tissue differences in receptor reserve than were

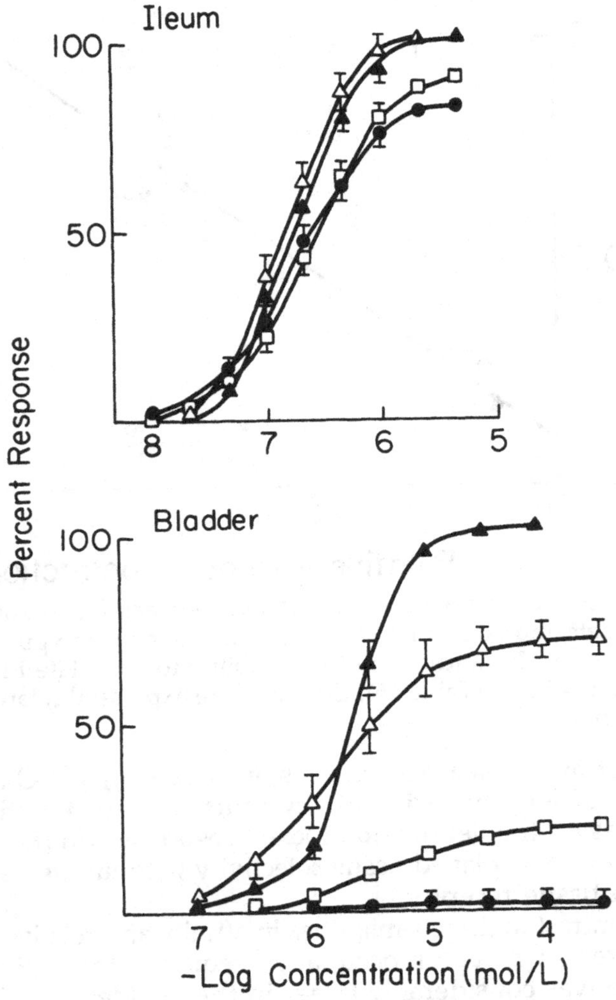

Fig. 9. Concentration–response curves of *43* (▲), *54* (□), *58* (△), and *77* (●) in strips of the guinea-pig ileum and urinary bladder. Responses are expressed relative to the maximum response elicited by carbachol.

responses to agonists of higher efficacy, as was expected on theoretical grounds (Kenakin, 1986).

Traditionally, drug selectivity is thought of in terms of receptor selectivity. The results obtained with these Oxo analogs suggested an alternate and virtually unexplored mechanism, based on tissue differences in receptor density and/or in the efficiency of the receptor–effector coupling, by which selectivity may be

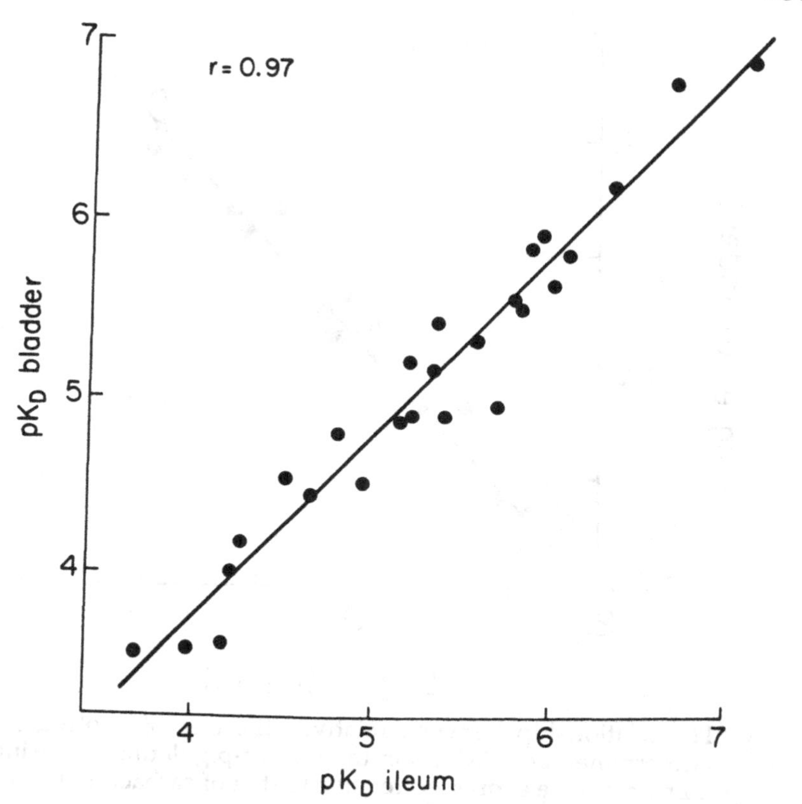

Fig. 10. Relationship between dissociation constants of oxotremorine analogs determined at mAChRs in the guinea-pig ileum and urinary bladder. Data are from Tables 9, 10, 12, 13, and 14.

achieved among muscarinic stimulants, even in the absence of distinct subtypes of mAChRs. They also showed the importance of carefully evaluating both the affinity and efficacy components of agonist potency before ascribing apparently selective actions to receptor-selective properties of an agonist (Ringdahl, 1987a).

Similar tissue selectivity was also demonstrated in vivo (Ringdahl et al., 1987; Ringdahl, 1988b). Tertiary amines, such as Oxo, *43*, and *78*, having relatively high intrinsic efficacy, induced salivation, analgesia, hypothermia, and tremor in mice. In contrast, compounds *54* and *55* having lower intrinsic efficacy differentiated between centrally mediated muscarinic effects as they, like Oxo, were potent in producing analgesia and hypothermia, but did not produce tremor. Instead, they antagonized Oxo-induced tremor. Compound *77* (BM 5) having still lower intrin-

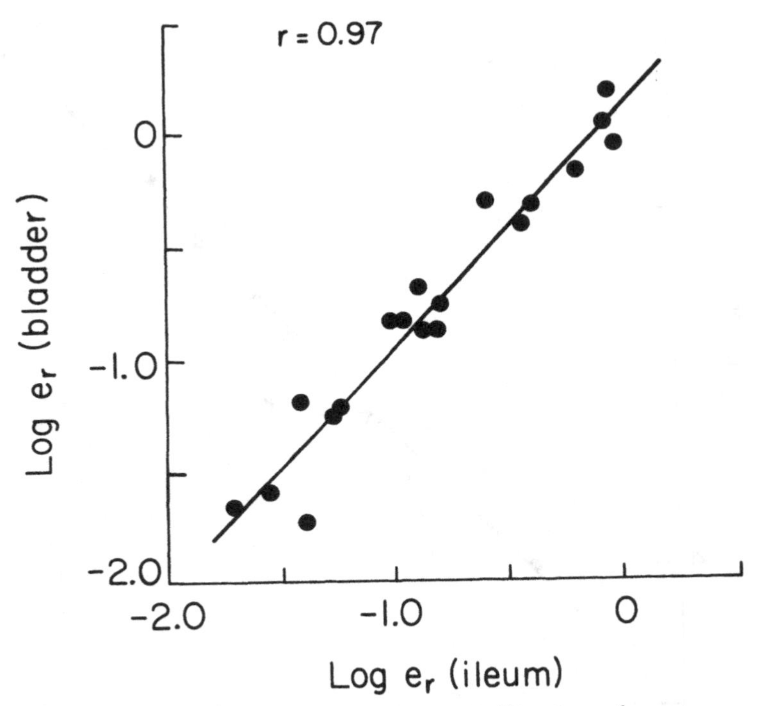

Fig. 11. Relationship between relative efficacies of oxotremorine analogs determined at mAChRs in the guinea-pig ileum and urinary bladder. Efficacies are expressed relative to that of carbachol. Data are from Tables 9, 10, 12, 13, and 14.

sic efficacy showed even better discrimination among the various responses, since it was potent in producing analgesia, but was a partial agonist with respect to hypothermia and an antagonist with respect to tremor. This compound under certain conditions discriminated between pre- and postsynaptically mediated muscarinic effects in vitro (Nordström et al., 1983) and in vivo (Casamenti et al., 1986) and showed regional differences in its effects on ACh levels in rat brain (Engström et al., 1987). The apparent selectivity of *54*, *55*, and *77* did not seem to be the result of any major differences in affinity for receptors at the various sites of action or to differences in drug distribution (Ringdahl et al., 1987). Rather, there appeared to be regional differences in receptor reserve for muscarinic agonists that affected in vivo responses of partial agonists in a qualitative manner. Responses to more efficacious agonists were affected only quantitatively by such differences. Efficacious agonists were shown to have a large receptor reserve with respect to salivation and analgesia, intermediate receptor reserve with regard to hypothermia, and a low receptor reserve for the tremor response (Ringdahl et al., 1987).

6. Atypical Muscarinic Agonists

6.1. Analogs of McN-A-343

McN-A-343 (Fig. 12) was first described by Roszkowski (1961). Like many other muscarinic agents, McN-A-343 is potent in stimulating mAChRs in autonomic ganglia. The unique feature of McN-A-343 is its relatively weak muscarinic actions outside ganglia, as for example in the isolated heart or on intestinal smooth muscle. The selective actions of McN-A-343 have been claimed (Hammer and Giachetti, 1982) to be mediated by a subtype of mAChRs (M1 receptors). Recent evidence suggests, however, that its selectivity can be equally well explained by a combination of low intrinsic efficacy and tissue differences in receptor reserve (Black and Shankley, 1985; Eglen et al., 1987).

The structural requirements for McN-A-343-like activity are very specific. Thus, *N*-demethylation of McN-A-343 abolished activity, whereas replacement of the trimethylammonium group by a triethylammonium group yielded an antagonist. A shift of the chlorine atom of the benzene ring from the 3- to the 4-position produced a threefold enhancement of ganglionic stimulant activity. The 2-chloro derivative had only one-tenth of the potency of McN-A-343. Saturation of the triple bond abolished activity (Roszkowski and Yelnosky, 1967). Nelson et al. (1973) reported that the *trans* olefinic derivative *96* (Fig. 12) had about half the potency of McN-A-343 as a ganglionic stimulant in the anesthetized cat, whereas the *cis* olefine *97* was much less active. Both *96* and *97* were very weak partial muscarinic agonists on the rabbit ileum. More or less identical results were obtained by measurements of ganglionic and end-organ responses to *96* and *97* in the same species, i.e., the rat (Mutschler and Lambrecht, 1984). The 4-chlorophenyl analog *98* was more potent than *96* as a ganglionic stimulant, but was less selective, whereas its *cis* isomer *99* was virtually inactive. The *trans* epoxide *100* approached McN-A-343 in potency at the ganglion, and also was very selective since it showed little or no muscarinic activity on smooth muscle. Its *cis* isomer *101* as well as the cyclopropane

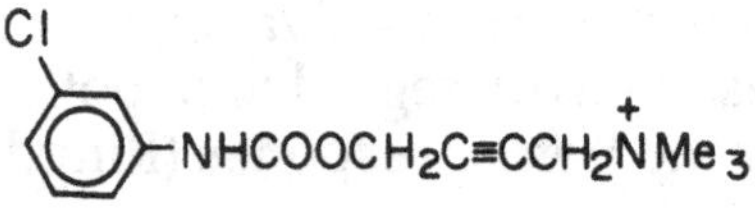

Cl

$NHCOOCH_2C{\equiv}CCH_2\overset{+}{N}Me_3$

McN-A-343

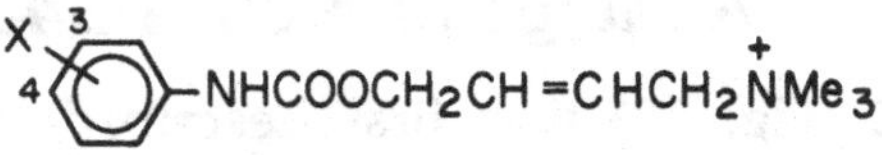

$NHCOOCH_2CH{=}CHCH_2\overset{+}{N}Me_3$

trans		cis	
96	X = 3-Cl	97	X = 3-Cl
98	X = 4-Cl	99	X = 4-Cl

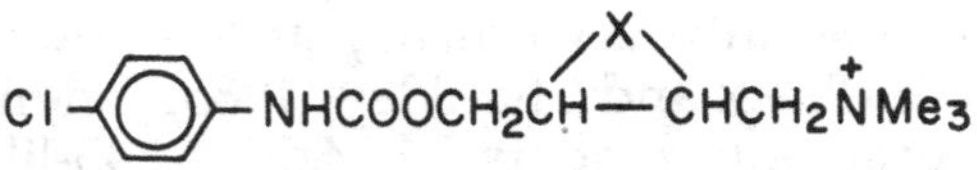

$Cl{-}C_6H_4{-}NHCOOCH_2CH{-}CHCH_2\overset{+}{N}Me_3$ (X bridging)

trans		cis	
100	X = O	101	X = O
102	X = CH_2	103	X = CH_2

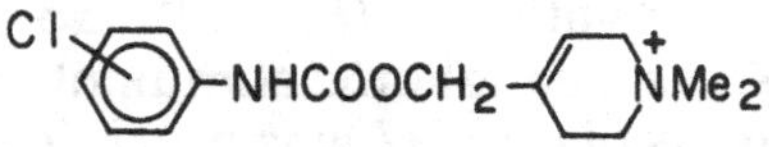

$NHCOOCH_2$ $\overset{+}{N}Me_2$

104

Fig. 12. Chemical structures of some analogs of McN-A-343.

derivatives *102* and *103* were inactive (Nelson et al., 1976). Some quaternary isoarecolinol derivatives (*104*) of *96* and *98* in which the conformational flexibility of the cationic head is reduced (Fig. 12) were about one-fourth to one-sixth as potent muscarinic ganglionic stimulants as McN-A-343. The corresponding tertiary amines as well as dihydroisoarecolinol and arecolinol derivatives were virtually inactive (Lambrecht et al., 1986). Collectively, the structure–activity studies on analogs of McN-A-343 showed that a quaternary nitrogen, unsaturation at C-2 of the amino alcohol moiety, and a distance of about 57 nm between the ether oxygen and the ammonium group are necessary for high muscarinic ganglion-stimulating activity (Nelson et al., 1976; Lambrecht et al., 1986). The compound AHR-602 (*N*-benzyl-3-pyrrolidyl acetate methobromide), despite its structural dissimilarity to McN-A-343, has pharmacological properties similar to those of McN-A-343 (Franko et al., 1963).

6.2. Analogs of RS 86

The spiro-succinimide RS 86 (Fig. 13) was somewhat more potent than arecoline, aceclidine, and pilocarpine in inducing hypothermia in mice, but was at least 10-fold less potent than Oxo. The tremorogenic potency of RS 86 was similar to those of arecoline and aceclidine and about 25-fold lower than that of Oxo. RS 86 also was more potent than aceclidine and pilocarpine in its behavioral effects as observed in an open field. Like Oxo, muscarine, and pilocarpine, RS 86 was more potent in depolarizing the rat superior cervical ganglion in vitro (M1-mediated effect) than in contracting the isolated guinea-pig ileum (M2-mediated effect). In contrast, aceclidine showed the reversed selectivity, and arecoline was almost equipotent in the two preparations (Table 17). These and similar results obtained in radioligand displacement studies (Palacios et al., 1986; Tonnaer et al., 1987) suggest that RS 86 has a certain degree of selectivity for M1 receptors when compared to aceclidine and arecoline. Several analogs of RS 86, modified in the *N*-substituents, succinimide ring, and in the piperidine ring, were found to be less potent than RS 86 (Bolliger et al., 1986).

6.3. Analogs of Pilocarpine

Pilocarpine also has been claimed to possess M1 selectivity (Caulfield and Stubley, 1982), and these claims are not contradicted by the data in Table 17. However, the compounds in Table 17 (with the exception of muscarine) are partial agonists at many sites of action (Fisher et al., 1983; Palacios et al., 1986; Freedman et al., 1988) and regional differences in muscarinic receptor reserve may influence potency both in vitro (Ringdahl, 1987a) and in vivo (Ringdahl et al., 1987). Some pilocarpic acid esters were recently described as prodrugs of pilocarpine. These

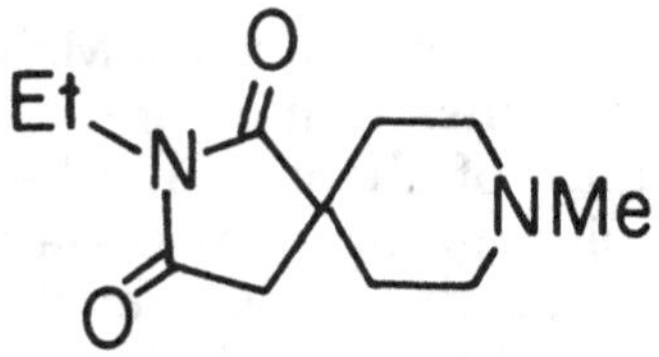

RS 86

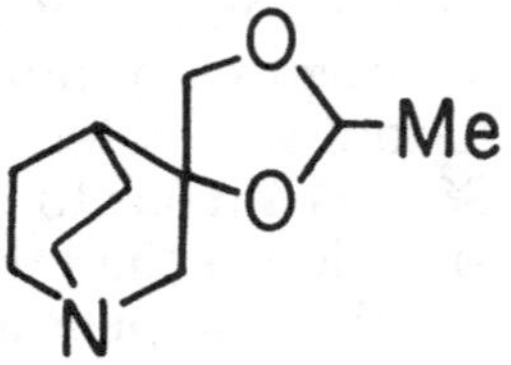

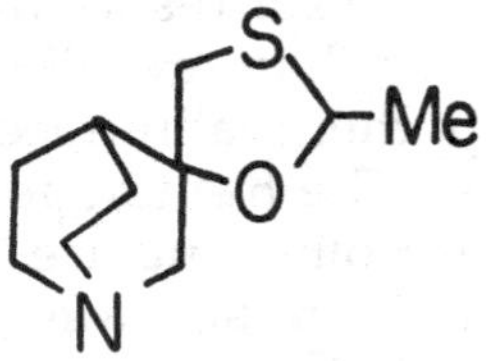

AF 30

AF 102

Fig. 13. Chemical structures of some rigid muscarinic agonists.

compounds had better bioavailability and longer duration of action than pilocarpine (Bundgaard et al., 1985).

6.4. Analogs of AF 30

AF 30 (Fig. 13) may be regarded as a rigid analog of aceclidine (Table 1). It was claimed to be 17-fold less potent than aceclidine on the guinea-pig ileum, but equipotent with aceclidine as a stimulant of the superior cervical ganglion of the cat and in inducing tremors in mice (Fisher et al., 1976). These results were confirmed in part (Palacios et al., 1986), but when the four separated isomers of AF 30 were investigated, the rank order of potency was ileum > ganglion > atrium for all isomers (Saunders et al., 1987). The isomers also failed to show appreciable M1 selectivity in radioligand binding assays. A sulfur analog (AF 102, Fig. 13) of AF 30 was recently described (Fisher et al., 1987). The *cis* isomer of AF 102 was reported to be an M1 selective agonist that improved cognitive impairment in AF 64A-treated rats without causing overt muscarinic effects. On the

Table 17
Effects of Muscarinic Agonists in the Isolated Guinea-Pig Ileum (Contraction) and Rat Superior Cervical Ganglion (Depolarization) and on [^{3}H]*cis*methyldioxolane (CD) and [^{3}H]Pirenzepine (PZ) Binding in Rat Cerebral Cortex[a]

Compound	Ileum		Ganglion		[^{3}H]CD[c]	[^{3}H]PZ[d]
	ia[b]	pD_2	ia[b]	pD_2	pIC_{50}	pIC_{50}
Oxo	1.0	7.1	0.63	8.0	8.55	6.95
RS 86	1.0	6.05	0.82	6.7	7.06	6.12
Muscarine	1.0	6.8	1.0	7.4		
Pilocarpine	0.84	5.9	1.34	6.5	7.10	6.02
Arecoline	1.0	6.4	0.89	6.6	7.70	5.76
Aceclidine	1.0	6.0	0.95	5.4	7.25	5.15
cis AF-30	1.0	4.8	0.56	6.0	7.05	5.80
McN-A-343	Inactive		1.79	6.3		

[a]Palacios et al., 1986.
[b]Intrinsic activity relative to muscarine.
[c][^{3}H]CD concentration was 1 n*M*.
[d][^{3}H]PZ concentration was 0.3 n*M*.

guinea-pig ileum, *cis* AF 102 was a weak muscarinic agonist. The relatively low muscarinic activity of compounds such as AF 30 and AF 102 may result from their rigidity. Saunders et al. (1988) have suggested that rigidity may well contribute to selectivity, but rigidity may also limit agonist efficacy, since subtle but complimentary conformational changes in both receptor and ligand may be necessary to induce a response. However, more flexible analogs of AF 30 also showed quite low muscarinic potency, although some were more efficacious than AF 30 (Saunders et al., 1988).

7. Haloalkylamine Derivatives of Muscarinic Agonists

Reactive 2-haloalkylamine moities have been introduced into the structure of many compounds to produce irreversibly acting analogs. This approach appears particularly relevant to muscarinic agonists because of the close structural similarity between the aziridinium ion derived from 2-haloethylamines and the trimethylammonium group of ACh.

7.1. Analogs of Acetylcholine

The 2-chloroethylamino analog of ACh (ACh mustard), in its cyclized aziridinium ion form, was a potent agonist at mAChRs in intestinal smooth muscle, but showed little (Hudgins and Stubbins, 1972) or no (Hirst and Jackson, 1972) irreversible binding as measured by inhibition of ACh-induced contractions. The surprisingly weak irreversible effects of ACh mustard in whole tissues were probably the result of its enzymatic hydrolysis (Robinson et al., 1975; Hudgins and Stubbins, 1975). However, in the presence of physostigmine, ACh mustard bound irreversibly to mAChRs in the guinea-pig small intestine as shown by persistent inhibition of the binding of [^{3}H]PrBCM (Robinson et al., 1975). ACh mustard also had pronounced nicotinic actions (Hudgins and Stubbins, 1972; Willcockson et al., 1981).

An *N*-ethyl analog of ACh mustard was recently shown to be a muscarinic agonist that binds irreversibly to cardiac mAChRs (Baker and Posner, 1986). *N*-Ethylacetylcholine mustard produced sustained inhibition of ACh release in rat cortical synaptosomes, suggesting that it acted as an irreversible agonist at presynaptic mAChRs (Meyer et al., 1987). A similar observation was made with an alkylating analog (BM 123A) of Oxo (Ehlert and Jenden, 1986). Neurochemical and behavioral effects of *N*-ethylacetylcholine mustard also have been reported (Pope et al., 1987). The 2-chloroethylamino derivatives of the *cis*-1,3-dioxolane *10* and the *cis*-1,3-oxathiolane *24* were poor muscarinic agonists on the rat jejunum and showed only weak irreversible effects (Cassinelli et al., 1987).

7.2. Analogs of Oxotremorine

A 2-chloroethylamino analog (BM 123) of Oxo, first described by György et al. (1971), has been studied extensively (Ringdahl and Jenden, 1987a). BM 123 cyclized quite slowly (Table 18) to the corresponding aziridinium ion (BM 123A). BM 123A was a potent agonist on the guinea-pig ileum (Table 18), whereas the parent compound appeared to be inactive. The nicotinic actions of BM 123A, as measured on the frog rectus, were weak (Ringdahl et al., 1984). In the rabbit heart, BM 123A was almost equipotent with Oxo-M as an inhibitor of heart rate (Ehlert, 1987). Furthermore, Oxo-M and BM 123A had similar affinity for mAChRs in the rat cortex and rabbit myocardium as measured

Table 18
Rates of Cyclization ($t_{1/2}$) of Haloalkylamines Related to Oxotremorine and Muscarinic Activity of the Corresponding Ammonium Ions[a]

$N\text{-}CH_2C{\equiv}CCH_2\text{-}N(CH_3)(CH_2)_nX \longrightarrow N\text{-}CH_2C{\equiv}CCH_2\text{-}N^+(CH_3)(CH_2)_n\ X^-$

Compound	n	X	$t_{1/2}$,[b] min	Guinea-pig ileum, pD_2
BM 123	2	Cl	36.5	7.49
BR 401	2	Br	0.8	7.91
104	3	Cl	436	6.92
105	3	Br	11.4	6.94
106	3	I	14.1	6.87
107	4	Cl	<0.4	6.27
(+)-BM 130			0.7[c]	6.78
(−)-BM 130				6.69

[a]Ringdahl et al., 1984, 1988, 1989a; Ringdahl and Jenden, 1987c.
[b]Half-lives were determined at 37°C and pH 7.0–7.3.
[c]Measured for the racemate.

from inhibition of [^{3}H]NMS binding at 0°C. The parent mustard (BM 123) was about 100-fold less potent than BM 123A in displacing [^{3}H]NMS both in the myocardium and cortex (Ehlert and Jenden, 1985; Ehlert, 1987).

Alkylation of mAChRs by BM 123A was demonstrated in the guinea-pig ileum (Ringdahl et al., 1984), rat cerebral cortex and heart (Ehlert and Jenden, 1986), as well as in the rabbit heart (Ehlert, 1987). The kinetics of alkylation of mAChRs by BM 123A were consistent with a model in which the aziridinium ion (A) rapidly forms a reversible complex (AR) with the receptor (R), which then undergoes a slower irreversible change to a covalent complex (AR*) as follows:

$$A + R \rightleftharpoons AR \rightarrow AR^*$$

When administered to mice (György et al., 1971; Ringdahl and Jenden, 1987b) and rats (Russell et al., 1986a), BM 123 produced signs typical of central and peripheral muscarinic stimulation, e.g., tremor and salivation (Table 19). These signs were followed by a sustained resistance to the effects of muscarinic agonists. In mice (Ehlert et al., 1984) and rats (Russell et al., 1986a)

Table 19
Muscarinic Activity in Mice
of Haloalkylamines Related to Oxotremorine[a]

	ED_{50}, μmol/kg			
	Salivation		Tremor	
Compound	iv	ip	iv	ip
BM 123	0.95	1.1	2.8	5.3
BR 401	0.068	0.13	0.14	1.5
(+)-BM 130	1.4	4.7	2.8	[b]
(−)-BM 130	1.3	6.2	3.2	[b]
104	22.4	29.3	[b]	[b]
105	2.9	6.0	6.3	15.6
106	4.5	10.4	8.9	29.7
107	1.1	4.2	3.6	[b]
Oxo	0.19	0.29	0.44	0.59

[a]Ringdahl and Jenden, 1987b; Ringdahl et al., 1988, 1989a.
[b]Inactive.

treated with BM 123, there was a long-lasting decrease in the number of [^{3}H]QNB binding sites in the brain. Apparently, the initial agonist effects observed with BM 123 were associated with the reversible complex AR, whereas the subsequent decreased sensitivity to muscarinic agonists was the result of depletion of functional receptors because of formation of the covalent complex AR* (Russell et al., 1986a).

The 2-bromoethylamino analog (BR 401) of BM 123 cyclized to the aziridinium ion about 50-fold faster than did BM 123 (Table 18). The higher peak concentration of the aziridinium ion that resulted was reflected in a greater muscarinic activity of BR 401 compared to BM 123 (Ringdahl and Jenden, 1987c). Cyclized BR 401 (BR 401A) was threefold more potent than Oxo on the guinea-pig ileum and is one of the most potent muscarinic agents known. Upon iv administration to mice, BR 401 was threefold more potent than Oxo and 20-fold more potent than BM 123 in producing tremor. However, after ip administration, BR 401 was less potent than Oxo and only 3.5-fold more potent than BM 123 (Table 19). Similarly, BR 401 given iv was twofold more potent than Oxo as an analgesic, but fourfold less potent when given ip. BR 401A did not cause tremor by either route of administration. These results showed that BR 401, when given iv, penetrated effectively into the central nervous system where it cyclized rapidly to the pharmacologically active aziridinium ion. In contrast, after ip administration, significant amounts of the aziridinium ion were formed before BR 401 reached its central site of action. BM 123 apparently reached the central nervous system effectively by both routes of administration, but cyclized too slowly to produce high levels of the aziridinium ion (Ringdahl and Jenden, 1987b). Recently, two 5-methyl-2-pyrrolidone analogs (EK 17 and BR 402) of BM 123 and BR 401 were described. EK 17 and BR 402, after cyclization to the corresponding aziridinium ions, behaved like partial agonists and were considerably more potent than BM 123A and BR 401A in alkylating mAChRs (Ringdahl et al., 1989b).

BM 130 (Fig. 14) cyclized to a bicyclic aziridinium ion (BM 130A) at a rate similar to that observed for BR 401 (Table 18). BM 130A was less potent than BR 401A as a muscarinic agonist in vitro, but equipotent with BR 401A as a nicotinic agent (Ringdahl et al., 1984). When administered iv to rats, BM 130 produced typical muscarinic effects that were followed by a long-lasting resistance to the central effects of Oxo (Russell et al., 1986b). BM 130A also caused irreversible reduction in the binding of [^{3}H]QNB in homogenates of the guinea-pig ileum (Ringdahl et al., 1984), and the rat cerebral cortex and heart (Ehlert and Jenden, 1986). The enantiomers of BM 130 were equipotent in producing central and peripheral muscarinic effects in mice (Table 19). Furthermore, the aziridinium ions [(+)- and (−)-BM 130A] had similar muscarinic activity in vitro (Table 18) and in vivo, and similar affinity for mAChRs in the rat cerebral cortex. However, (+)- and (−)-BM 130A alkylated cortical mAChRs at different rates. These results supported the hypothesis, outlined above for BM 123A, of two consecutive recognition processes in the covalent binding of BM 130A to mAChRs (Ringdahl et al., 1989a).

Although these 2-haloalkylamino analogs of Oxo were potent in eliciting central muscarinic effects, these were generally of quite short duration, presumably because of receptor alkylation by the reactive aziridinium ions. The 3- and 4-haloalkylamines *104–107* cyclized to stable (nonalkylating) azetidinium and pyrrolidinium ions (Table 18). These were potent muscarinic agonists on the guinea-pig ileum, whereas the parent haloalkylamines were inactive. The 3-chloropropylamine *104*, which cyclized very slowly (Table 18), had low muscarinic activity in mice (Table 19). The 3-bromo- and 3-iodopropylamines cyclized more quickly and were more potent. The rapidly cyclizing 4-chlorobutylamine *107* was potent when administered iv, but like BM 130, elicited no tremor after ip administration (Ringdahl et al., 1988).

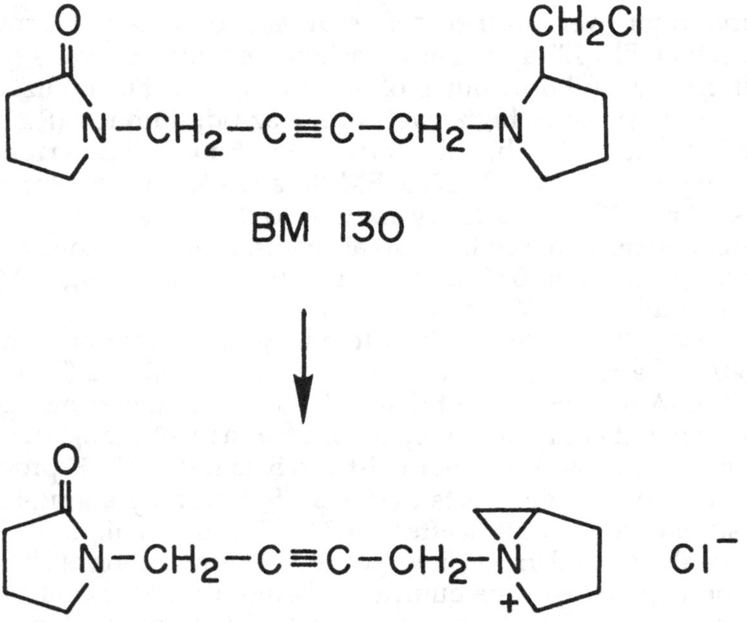

BM 130

BM 130A

Fig. 14. Chemical structures of BM 130 and its aziridinium ion (BM 130A).

In conclusion, tertiary haloalkylamines related to Oxo served in vivo as prodrugs of potent quaternary ammonium salts and also as carriers for the passage of such salts into the central nervous system. Central potency of these haloalkylamines was critically dependent on rate of cyclization and route of administration. The potency of rapidly cyclizing compounds appeared to be limited by cyclization in the periphery, especially after ip administration. For slowly cyclizing compounds, central potency apparently was limited by metabolism and elimination of the parent haloalkylamines. Haloalkylamines such as *104–107*, which cyclize to stable ammonium ions, may have long-lasting muscarinic actions selectively in the central nervous system because of restricted diffusion of charged compounds across the blood–brain barrier into the periphery (Ringdahl et al., 1988).

Acknowledgment

Part of the work described in this chapter was supported by grants from the United States Public Health Service (GM 37816 and

MH 17691). Part of the synthetic work on oxotremorine analogs described in this chapter was done in collaboration with Drs. Richard Dahlbom and Bahram Resul, University of Uppsala, Sweden. The author thanks Dr. Donald J. Jenden, UCLA, for helpful comments and suggestions. Excellent secretarial assistance by Holly Batal is gratefully acknowledged.

References

Amstutz, R., Ringdahl, B., Karlén, B., Roch, M., and Jenden, D. J. (1985) Stereoselectivity of muscarinic receptors in vivo and in vitro for oxotremorine analogues. *N*-[4-(Tertiary amino)-2-butynyl]-5-methyl-2-pyrrolidones. *J. Med. Chem.* **28**, 1760–1765.

Amstutz, R., Closse, A., and Gmelin, G. (1987) Die Position 5 im Oxotremoringeruest: Eine zentrale Stelle fuer die Steuerung der Aktivitaet am muskarinischen Rezeptor. *Helv. Chim. Acta* **70**, 2232–2244.

Angeli, P., Brasili, L., Giardina, D., and Melchiorre, C. (1981) Molecular requirements of the active site of cholinergic receptors XV: Synthesis and biological activity of 2,3-dehydrodeoxamuscarone and 2,3-dehydrodeoxamuscarines. *J. Pharmaceut. Sci.* **70**, 489–492.

Angeli, P., Brasili, L., Giannella, M., Gualtieri, F., Picchio, M. T., and Teodori, E. (1988) Chiral muscarinic agonists possessing a 1,3-oxathiolane nucleus: enantio- and tissue-selectivity on isolated preparations of guinea-pig ileum and atria and of rat urinary bladder. *Naunyn-Schmiedeberg's Arch. Pharmacol.* **337**, 241–245.

Angeli, P., Giannella, M., Pigini, M., Gualtieri, F., and Cingolani, M. L. (1984) Size of muscarinic receptor anionic binding site related to onium group of ligands with a 1,3-oxathiolane nucleus. *Eur. J. Med. Chem.* **19**, 495–500.

Angeli, P., Brasili, L., Giannella, M., Gualtieri, F., and Pigini, M. (1985) Affinity and efficacy correlate with chemical structure more than potency does in a series of pentatomic cyclic muscarinic agonists. *Br. J. Pharmacol.* **85**, 783–786.

Armstrong, P. D. and Cannon, J. G. (1970) Small ring analogs of acetylcholine. Synthesis and absolute configurations of cyclopropane derivatives. *J. Med. Chem.* **13**, 1037–1039.

Baker, S. P. and Posner, P. (1986) Irreversible binding of acetylethylcholine mustard to cardiac cholinergic muscarinic receptors. *Mol. Pharmacol.* **30**, 411–418.

Barlow, R. B. (1964) *Introduction to Chemical Pharmacology* (Methuen & Co., Ltd., London) pp. 185–240.

Barlow, R. B. and Casy, A. F. (1975) Inversion of stereospecificity by methylation of compounds acting at acetylcholine receptors. *Mol. Pharmacol.* **11**, 690–693.

Barlow, R. B. and Kitchen, R. (1982) The actions of some esters of 4-hydroxyquinuclidine on guinea-pig ileum, atria and rat fundus strip. *Br. J. Pharmacol.* **77**, 549–557.

Bebbington, A., Brimblecombe, R. W., and Shakeshaft, D. (1966) The central and peripheral activity of acetylenic amines related to oxotremorine. *Br. J. Pharmacol.* **26**, 56–67.

Beld, A. J., Klok, E. J., and Rodrigues de Miranda, J. F. (1980) Inhibition of [^{3}H]dexetimide binding by a homologous series of methylfurthrethonium analogues at the peripheral muscarinic receptor. *Biochem. Biophys. Res. Commun.* **97**, 430–436.

Belleau, B. and Lavoie, J. L. (1968) A biophysical basis of ligand-induced activation of excitable membranes and associated enzymes. A thermodynamic study using acetylcholinesterase as a model receptor. *Can. J. Biochem.* **46**, 1397–1409.

Birdsall, N. J. M. and Hulme, E. C. (1976) Biochemical studies on muscarinic acetylcholine receptors. *J. Neurochem.* **27**, 7–16.

Birdsall, N. J. M., Burgen, A. S. V., and Hulme, E. C. (1978a) The binding of agonists to brain muscarinic receptors. *Mol. Pharmacol.* **14**, 723–736.

Birdsall, N. J. M., Burgen, A. S. V., and Hulme, E. C. (1978b) Correlation between the binding properties and pharmacological responses of muscarinic receptors, in *Cholinergic Mechanisms and Psychopharmacology* (Jenden, D. J., ed), Plenum, New York, pp. 25–33.

Black, J. W. and Shankley, N. P. (1985) Pharmacological analysis of the muscarinic receptors involved when McN-A-343 stimulates acid secretion in the mouse isolated stomach. *Br. J. Pharmacol.* **86**, 609–617.

Bolliger, G., Palacios, J. M., Closse, A., Gmelin, G., and Malanowski, J. (1986) Structure–activity relationships of RS 86 analogues. *Adv. Behav. Biol.* **29**, 585–592.

Brimblecombe, R. W. (1974) *Drug Actions on Cholinergic Systems* (University Park Press, Baltimore).

Bühl, T., Rodrigues de Miranda, J. F., Beld, A. J., Lambrecht, G., and Mutschler, E. (1987) Binding of the S(+)- and R(−)-enantiomers of bethanechol to muscarinic receptors in jejunal smooth muscle, nasal mucosa, atrial and ventricular myocardium of the rat. *Eur. J. Pharmacol.* **140**, 221–226.

Bundgaard, H., Falch, E., Larsen, C., Mosher, G. L., and Mikkelson, T. J. (1985) Pilocarpic acid esters as novel sequentially labile pilocarpine prodrugs for improved ocular delivery. *J. Med. Chem.* **28**, 979–981.

Burgen, A. S. V. (1965) The role of ionic interaction at the muscarinic receptor. *Br. J. Pharmacol. Chemother.* **25**, 4–17.

Cannon, J. G. (1981) Cholinergics, in *Burger's Medicinal Chemistry*, Vol. 3, 4th Ed. (Wolf, M., ed). John Wiley, New York, pp. 339–360.

Cannon, J. G., Crockatt, D. M., Long, J. P., and Maixner, W. (1982) (±)-*cis*-2-Acetoxycyclobutyltrimethylammonium iodide: A semirigid analogue of acetylcholine. *J. Med. Chem.* **25**, 1091–1094.

Cannon, J. G., Lee, T., Sankaran, V., and Long, J. P. (1975) *trans*-2-Acetoxycyclobutyltrimethylammonium iodide, a cyclobutane analog of "*trans*-ACTM." *J. Med. Chem.* **18**, 1027–1028.

Casamenti, F., Cosi, C., and Pepeu, G. (1986) Effect of BM-5, a presynaptic antagonist-postsynaptic agonist, on cortical acetylcholine release. *Eur. J. Pharmacol.* **122**, 288–290.

Cassinelli, A., Angeli, P., Giannella, M., and Gualtieri, F. (1987) β-Haloalkylamine derivatives with 1,3-oxathiolane or 1,3-dioxolane nuclei. *Eur. J. Med. Chem.* **22**, 5–10.

Casy, A. F. (1975) Stereochemical aspects of parasympathomimetics and their antagonists: recent developments. *Prog. Med. Chem.* **11**, 1–65.

Caulfield, M. P. and Stubley, J. K. (1982) Pilocarpine selectively stimulates muscarinic receptors in rat sympathetic ganglia. *Br. J. Pharmacol.* **76**, 216P.

Cavallito, C. J. (1980) Quaternary ammonium salts—advances in chemistry and pharmacology since 1960. *Prog. Drug Res.* **24**, 267–373.

Chang, K.-J. and Triggle, D. J. (1973) Stereoselectivity of cholinergic activity in a series of 1,3-dioxolanes. *J. Med. Chem.* **16**, 718–720.

Chiou, C. Y., Long, J. P., Cannon, J. G., and Armstrong, P. D. (1969) The cholinergic effects and rates of hydrolysis of conformationally rigid analogs of acetylcholine. *J. Pharmacol. Exp. Ther.* **166**, 243–248.

Cho, A. K., Haslett, W. L., and Jenden, D. J. (1962) The peripheral actions of oxotremorine, a metabolite of tremorine. *J. Pharmacol. Exp. Ther.* **138**, 249–257.

Cho, A. K., Jenden, D. J., and Lamb, S. I. (1972) Rates of alkaline hydrolysis and muscarinic activity of some aminoacetates and their quaternary ammonium analogs. *J. Med. Chem.* **15**, 391–394.

Chothia, C. and Pauling, P. (1970) Absolute configuration of cholinergic molecules; the crystal structure of (+)-*trans*-2-acetoxycyclopropyl trimethylammonium iodide. *Nature (London)* **226**, 541–542.

Christiansen, A., Lüllmann, H., and Mutschler, E. (1967) Investigations on structure–activity relationships of arecaidine-derivatives on the guinea pig isolated atria. *Eur. J. Pharmacol.* **1**, 81–85.

Claesson, A., Asell, A., Björkman, S., and Olsson, L.-I. (1978) An allenic oxotremorine analogue. *Acta Pharm. Suec.* **15**, 105–110.

Dahlbom, R. (1981) Structural and steric aspects of compounds related to oxotremorine, in *Cholinergic Mechanisms* (Pepeu, G. and Ladinsky, H., eds.), Plenum, New York, pp. 621–638.

Dahlbom, R., Lindquist, A., Lindgren, S., Svensson, U., Ringdahl, B., and Blair, M. R., Jr. (1974) Stereospecificity of oxotremorine antagonists. *Experientia* **30**, 1165–1166.

Dahlbom, R., Jenden, D. J., Resul, B., and Ringdahl, B. (1982) Stereochemical requirements for central and peripheral muscarinic and antimuscarinic activity of some acetylenic compounds related to oxotremorine. *Br. J. Pharmacol.* **76**, 299–304.

Dahlbom, R., Ringdahl, B., Resul, B., and Jenden, D. J. (1986) Stereoselectivity of some muscarinic and antimuscarinic agents related to oxotremorine, in *Dynamics of Cholinergic Function* (Hanin, I., ed.), Plenum, New York, pp. 385–393.

De Micheli, C., De Amici, M., Pratesi, P., Grana, E., and Santagostino Barbone, M. G. (1983) Relative potencies of the stereoisomers of bethanechol on muscarinic receptor. *Farmaco, Ed. Sci.* **38**, 514–520.

Eglen, R. M., Kenny, B. A., Michel, A. D., and Whiting, R. L. (1987) Muscarinic activity of McN-A-343 and its value in muscarinic receptor classification. *Br. J. Pharmacol.* **90**, 693–700.

Ehlert, F. J. (1987) Effects of an alkylating derivative of oxotremorine (BM 123A) on heart rate, muscarinic receptors and adenylate cyclase activity in the rabbit myocardium. *J. Pharmacol. Exp. Ther.* **241**, 804–811.

Ehlert, F. J. and Jenden, D. J. (1984) Comparison of the muscarinic receptor binding activity of some tertiary amines and their quaternary ammonium analogues. *Mol. Pharmacol.* **25**, 46–50.

Ehlert, F. J. and Jenden, D. J. (1985) The binding of a 2-chloroethylamine derivative of oxotremorine (BM 123) to muscarinic receptors in the rat cerebral cortex. *Mol. Pharmacol.* **28**, 107–119.

Ehlert, F. J. and Jenden, D. J. (1986) Alkylating derivatives of oxotremorine have irreversible actions on muscarinic receptors, in *Dynamics of Cholinergic Function* (Hanin, I., ed.). Plenum, New York, pp. 1177–1185.

Ehlert, F. J., Dumont, Y., Roeske, W. R., and Yamamura, H. I. (1980) Muscarinic receptor binding in rat brain using the agonist [^{3}H]cismethyldioxolane. *Life Sci.* **26**, 961–967.

Ehlert, F. J., Jenden, D. J., and Ringdahl, B. (1984) An alkylating derivative of oxotremorine interacts irreversibly with the muscarinic receptor. *Life Sci.* **34**, 985–991.

Elferink, J. G. R. and Salemink, C. A. (1975) Relation between structure and cholinergic activity. II. Synthesis and muscarinic activity of some sulfur analogs of 2-methyl-4-trimethylammoniomethyl-1,3-dioxolane-iodide. *Arzneim.-Forsch. (Drug Res.)* **25**, 1702–1705.

Engström, C., Undén, A., Ladinsky, H., Consolo, S.,, and Bartfai, T. (1987) BM-5, a centrally active partial muscarinic agonist with low tremorogenic activity. In vivo and in vitro studies. *Psychopharmacol.* **91**, 161–167.

Fisher, A., Weinstock, M., Gitter, S., and Cohen, S. (1976) A new probe for heterogeneity in muscarinic receptors: 2-methyl-spiro-(1,3-dioxolane-4,3′)-quinuclidine. *Eur. J. Pharmacol.* **37**, 329–338.

Fisher, S. K., Figueiredo, J. C., and Bartus, R. T. (1984) Differential

stimulation of inositol phospholipid turnover in brain by analogs of oxotremorine. *J. Neurochem.* **43**, 1171–1179.
Fisher, S. K., Klinger, P. D., and Agranoff, B. W. (1983) Muscarinic agonist binding and phospholipid turnover in brain. *J. Biol. Chem.* **258**, 7358–7363.
Fisher, A., Heldman, E., Brandeis, R., Pittel, Z., Dachir, S., Levy, A., and Karton, I. (1987) Restoration of cognitive functions in an animal model of Alzheimer's disease, in *Cellular and Molecular Basis of Cholinergic Function* (Dowdall, M. J. and Hawthorne, J. N., eds.), Ellis Horwood, Chichester, pp. 913–927.
Franko, B. V., Ward, J. W., and Alphin, R. S. (1963) Pharmacological studies of N-benzyl-3-pyrrolidyl acetate methobromide (AHR-602), a ganglion stimulating agent. *J. Pharmacol. Exp. Ther.* **139**, 25–30.
Freedman, S. B., Harley, E. A., and Iversen, L. L. (1988) Relative affinities of drugs acting at cholinoceptors in displacing agonist and antagonist radioligands: the NMS/Oxo-*M* ratio as an index of efficacy at cortical muscarinic receptors. *Br. J. Pharmacol.* **93**, 437–445.
Fuder, H. and Jung, B. (1985) Affinity and efficacy of racemic, (+)- and (–)-methacholine in muscarinic inhibition of [^{3}H]-noradrenaline release. *Br. J. Pharmacol.* **84**, 477–487.
Furchgott, R. F. (1966) The use of β-haloalkylamines in the differentiation of receptors and in the determination of dissociation constants of receptor-agonist complexes. *Adv. Drug Res.* **3**, 21–55.
Furchgott, R. F. and Bursztyn, P. (1967) Comparison of dissociation constants and of relative efficacies of selected agonists acting on parasympathetic receptors. *Ann. N.Y. Acad. Sci.* **144**, 882–899.
George, R., Haslett, W. L., and Jenden, D. J. (1962) The central action of a metabolite of tremorine. *Life Sci.* 361–363.
Gieren, A. and Kokkinidis, M. (1986) Structure investigations of agonists of the natural neurotransmitter acetylcholine, V[1]. Structure—activity correlations for cholinergic stimulants derived from crystal structures of their halides. *Z. Naturforsch.* **41c**, 627–640.
Givens, R. S. and Rademacher, D. R. (1974) Further studies on carbocyclic analogs of muscarine. Oxidation of desethermuscarine to desethermuscarone. *J. Med. Chem.* **17**, 457–459.
Gloge, H., Lüllmann, H., and Mutschler, E. (1966) The action of tertiary and quaternary arecaidine and dihydroarecaidine esters on the guinea pig isolated ileum. *Br. J. Pharmacol. Chemother.* **27**, 185–195.
Grana, E., Lucchelli, A., Zonta, F., Santagostino-Barbone, M. G., and D'Agostino, G. (1986) Comparative studies of the postjunctional activities of some very potent muscarinic agonists. *Naunyn-Schmiedeberg's Arch. Pharmacol.* **332**, 213–218.
Grana, E., Lucchelli, A., Zonta, F., and Boselli, C. (1987) Determination of dissociation constants and relative efficacies of some potent muscarinic agonists at postjunctional muscarinic receptors. *Naunyn-Schmiedeberg's Arch. Pharmacol.* **335**, 8–11.

Gualtieri, F., Angeli, P., Brasili, L., Giannella, M., and Pigini, M. (1985) Pentatomic cyclic compounds as probes for the cholinergic receptor, in *VIIIth International Symposium on Medicinal Chemistry*, Proceedings, Vol. 2 (Dahlbom, R. and Nilsson, J. L. G., eds.), Swedish Pharmaceutical, Stockholm, pp. 404–420.

Gualtieri, F., Angeli, P., Giannella, M., Melchiorre, C., and Pigini, M. (1979) Pentatomic cyclic compounds as tools for ACh receptor topography, in *Recent Advances in Receptor Chemistry* (Gualtieri, F., Giannella, M., and Melchiorre, C., eds.), Elsevier/North-Holland, Amsterdam, pp. 267–279.

Gualtieri, F., Giannella, M., Melchiorre, C., and Pigini, M. (1974) A cyclopentane analog of muscarone. *J. Med. Chem.* **17**, 455–457.

György, L., Gellen, B., Doda, M., and Sterk, L. (1971) Pharmacology of 1-(2-oxo-1-pyrrolidino)-4-(2-chloroethylmethylamino)but-2-yne HCl (DSO-16). I. Its central cholinomimetic and cholinolytic effects on mice. *Acta Physiol. Acad. Sci. Hung.* **40**, 373–379.

Hammer, R. and Giachetti, A. (1982) Muscarinic receptor subtypes: M1 and M2 biochemical and functional characterization. *Life Sci.* **31**, 2991–2998.

Hanin, I., Jenden, D. J., and Cho, A. K. (1966) The influence of pH on the muscarinic action of oxotremorine, arecoline, pilocarpine, and their quaternary ammonium analogs. *Mol. Pharmacol.* **2**, 352–359.

Hirst, M. and Jackson, C. H. (1972) The conversion of methyl-2-acetoxyethyl-2′-chloroethylamine to an acetylcholine-like aziridinium ion and its action on the isolated guinea pig ileum. *Can. J. Physiol. Pharmacol.* **50**, 798-808.

Höltje, H.-D., Jensen, B., and Lambrecht, G. (1978) Cyclic acetylcholine analogues. IV. 4-Acetoxy-piperidines and -thiacyclohexanes. *Eur. J. Med. Chem.* **13**, 453-463.

Höltje, H.-D., Jensen, B., and Lambrecht, G. (1979) Bioisosteric substitution of the ammonium group by the sulfonium group as a tool to determine the active conformation of heterocyclic acetylcholine analogues on the muscarinic receptor. A pharmacological and stereochemical study, in *Recent Advances in Receptor Chemistry* (Gualtieri, F., Giannella, M., and Melchiorre, C., eds.), Elsevier/North-Holland, Amsterdam, pp. 281-302.

Höltje, H.-D., Lambrecht, G., Moser, U., and Mutschler, E. (1983) Struktur- und Konformations-Wirkungs-Beziehungen heterozyklischer Acetylcholin-Analoga. 16. Mitt.: Konformatives Verhalten und cholinerge Eigenschaften von Isoarecaidin- und Sulfoisoarecaidin-Estern. *Arzneim.-Forsch./Drug Res.* **33**, 190–197.

Hudgins, P. M. and Stubbins, J. F. (1972) A comparison of the action of acetylcholine and acetylcholine mustard (chloroethylmethylaminoethyl acetate) on muscarinic and nicotinic receptors. *J. Pharmacol. Exp. Ther.* **182**, 303–311.

Hudgins, P. M. and Stubbins, J. F. (1975) Interactions of acetylcholine mustard with acetylcholinesterase. *J. Pharmaceut. Sci.* **64**, 1419–1421.

Hultzsch, K., Moser, U., Back, W., and Mutschler, E. (1971) Struktur-Wirkungs-Beziehungen von *N*-Methyl-3-pyrrolin- und *N*-Methyl-pyrrolidin-3-carbonsaureestern. *Arzneim.-Forsch./Drug Res.* **21**, 1979–1984.

Jensen, B. (1979) The crystal structures of *cis*- and *trans*-4-acetoxy-1-methylthiacyclohexane iodides. *Acta Chem. Scand. B* **33**, 359–364.

Jensen, B. (1981) The crystal structures of (±)-*cis*- and (−)-(1*S*,3*S*)-*trans*-3-acetoxy-1-methylthiane perchlorates. *Acta Chem. Scand. B.* **35**, 607–612.

Jim, K., Bolger, G. T., Triggle, D. J., and Lambrecht, G. (1982) Muscarinic receptors in guinea pig ileum. A study by agonist—^{3}H-labelled antagonist competition. *Can. J. Physiol. Pharmacol.* **60**, 1707–1714.

Karlén, B., Lindeke, B., Lindgren, S., Svensson, K.-G., Dahlbom, R., Jenden, D. J., and Giering, J. E. (1970) Acetylene compounds of potential pharmacological value. XIV. *N*-(*t*-Aminoalkynyl)-substituted succinimides and maleimides. A class of central anticholinergic agents. *J. Med. Chem.* **13**, 651–657.

Kenakin, T. P. (1984) The classification of drugs and drug receptors in isolated tissues. *Pharmacol. Rev.* **36**, 165–222.

Kenakin, T. P. (1985) The quantification of relative efficacy of agonists. *J. Pharmacol. Methods* **13**, 281–308.

Kenakin, T. P. (1986) Tissue and receptor selectivity: similarities and differences. *Adv. Drug Res.* **15**, 71–109.

Kilbinger, H. and Wessler, I. (1980) Pre- and postsynaptic effects of muscarinic agonists in the guinea-pig ileum. *Naunyn-Schmiedeberg's Arch. Pharmacol.* **314**, 259–266.

Kilbinger, H., Halim, S., Lambrecht, G., Weiler, W., and Wessler, I. (1984) Comparison of affinities of muscarinic antagonists to pre- and postjunctional receptors in the guinea-pig ileum. *Eur. J. Pharmacol.* **103**, 313–320.

Kokkinidis, M. and Gieren, A. (1984) Cholinergic neurotransmitter-receptor interactions. *Trends Pharmacol. Sci.* **5**, 369–370.

Lambrecht, G. (1976a) Struktur-Wirkungs-Beziehungen cyclischer Acetylcholinanaloga. *Pharmazie* **31**, 209–215.

Lambrecht, G. (1976b) Struktur- und Konformations-Wirkungs-Beziehungen heterocyclischer Acetylcholinanaloga. I. Muskarinwirkung enantiomerer 3-Acetoxychinuclidine und 3-Acetoxypiperidine. *Eur. J. Med. Chem.* **11**, 461–466.

Lambrecht, G. (1977) Struktur- und Konformations-Wirkungs-Beziehungen heterocyclischer Acetylcholinanaloga. II. Cholinerge Wirkung isomerer 3-Acetoxythiacyclohexane. *Eur. J. Med. Chem.* **12**, 41–47.

Lambrecht, G. (1979) Dissociation constants of 4-acetoxy-piperidines and -thiacyclohexanes at the muscarinic receptor. *Experientia* **35**, 1212–1213.

Lambrecht, G. (1981) Struktur- und Konformations-Wirkungs-Beziehungen heterozyklischer Acetylcholin-Analoga. 12. Mitteilung: Synthese und cholinerge Eigenschaften stereoisomerer 3-Acetoxythiacyclohexane. *Arzneim.-Forsch./Drug Res.* **31**, 634–640.

Lambrecht, G. and Mutschler, E. (1973) Struktur-Wirkungs-Beziehungen von Isoarecaidinestern. *Arzneim.-Forsch./Drug Res.* **23**, 1427–1431.

Lambrecht, G. and Mutschler, E. (1985) Heterogeneity in muscarinic receptors. Evidence from pharmacological studies with agonists and antagonists, in *VIIIth International Symposium on Medicinal Chemistry*. Proceedings, Vol. 2 (Dahlbom, R. and Nilsson, J. L. G., eds.) Swedish Pharmaceutical, Stockholm, pp. 379–390.

Lambrecht, G., Moser, U., Mutschler, E., Walther, G., and Wess, J. (1986) Muscarinic ganglionic stimulants: conformationally restrained analogues related to [4-[[*N*-(3-chlorophenyl)carbamoyl]oxy]-2-butynyl]trimethylammonium chloride. *J. Med. Chem.* **29**, 1309–1311.

Levine, R. R., Birdsall, N. J. M., Giachetti, A., Hammer, R., Iversen, L. L., Jenden, D. J., and North, R. A. (eds.) (1986) Subtypes of Muscarinic Receptors II. *Trends Pharmacol. Sci.*, Supplement.

Levine, R. R., Birdsall, N. J. M., North, R. A., Holman, M., Watanabe, A., and Iversen, L. L. (eds.) (1988) Subtypes of Muscarinic Receptors III. *Trends Pharmacol. Sci.*, Supplement.

Lindgren, S., Lindquist, A., Lindeke, B., Svensson, U., Karlén, B., Dahlbom, R., and Blair, M. R., Jr. (1973) Acetylene compounds of potential pharmacological value. XVIII. *N*-(*t*-Aminoalkynyl)-substituted 2-pyrrolidones, a new series of potent oxotremorine antagonists. *Acta Pharm. Suec.* **10**, 435–440.

Makriyannis, A., Theard, J. M., and Mautner, H. G. (1979) A direct approach to the determination of the rotational barriers of the acetylcholine and choline molecules. *Biochem. Pharmacol.* **28**, 1911–1915.

Martin-Smith, M. and Stenlake, J. B. (1967) The possible role of conformational isomerism in the biological actions of acetylcholine. *J. Pharm. Pharmacol.* **19**, 561–589.

Melchiorre, C., Angeli, P., Giannella, M., Gualtieri, F., Pigini, M., Cingolani, M. L., Gamba, G., Leone, L., Pigini, P., and Re, L. (1978) Molecular requirements at the cholinergic receptors. Importance of ether oxygens in the dioxolane series. *Eur. J. Med. Chem.* **13**, 357–361.

Meyer, E. M., Otero, D. H., Morgan, E., Marchand, S., and Baker, S. P. (1987) Effects of acetylethylcholine mustard on [^{3}H]quinuclidinyl benzilate binding and acetylcholine release in rat brain synaptosomes. *J. Neurochem.* **48**, 477–482.

Mitchelson, F. (1988) Muscarinic receptor differentiation. *Pharmacol. Ther.* **37**, 357–423.

Moser, U., Lambrecht, G., Mutschler, E., and Sombroek, J. (1983) Struktur- und Konformations-Wirkungs-Beziehungen heterocyclischer Acetylcholinanaloga. 18. Mitt. Synthese und muskarinartige

Wirkung von Estern des Sulfoarecaidins und Sulfoisoarecaidins. *Arch. Pharm. (Weinheim)* **316**, 670–677.

Mutschler, E. and Hultzsch, K. (1973) Über Struktur-Wirkungs-Beziehungen von ungesattigten Estern des Arecaidins und Dihydroarecaidins. *Arzneim.-Forsch./Drug Res.* **23**, 732–737.

Mutschler, E. and Lambrecht, G. (1983) Stereoselectivity and conformation: flexible and rigid compounds, in *Stereochemistry and Biological Activity of Drugs* (Ariens, E. J., Soudijn, W., and Timmermans, P. B. M. W. M., eds.). Blackwell Scientific, Oxford, pp. 63–79.

Mutschler, E. and Lambrecht, G. (1984) Selective muscarinic agonists and antagonists in functional tests. *Trends Pharmacol. Sci.* Suppl. **5**, 39–44.

Mutschler, E., Höltje, H.-D., Lambrecht, G., and Moser, U. (1983) Struktur- und Konformations-Wirkungs-Beziehungen heterozyklischer Acetylcholin-Analoga. 17. Mitt.: Konformatives Verhalten und cholinerge Eigenschaften von Arecaidin- und Sulfoarecaidin-Estern. *Arzneim.-Forsch./Drug Res.* **33**, 806–812.

Nelson, W. L., Freeman, D. S., and Vincenzi, F. F. (1976) Stereochemical analogs of a muscarinic, ganglionic stimulant. 2. *Cis* and *trans* olefinic, epoxide, and cyclopropane analogs related to 4-[*N*-(3-chlorophenyl)carbamoyloxy]-2-butynyltrimethylammonium chloride (McN-A-343). *J. Med. Chem.* **19**, 153–158.

Nelson, W. L., Freeman, D. S., Wilkinson, P. D., and Vincenzi, F. F. (1973) Stereochemical analogs of a muscarinic, ganglionic stimulant. *cis*- and *trans*-4-[*N*-(3-chlorophenyl)carbamoyloxy]-2-butenyltrimethylammonium iodides. *J. Med. Chem.* **16**, 506–510.

Newton, M. W., Ringdahl, B., and Jenden, D. J. (1985) Acetyl-*N*-aminodeanol: a cholinergic false transmitter in rat phrenic nerve-diaphragm and guinea-pig myenteric plexus preparations. *J. Pharmacol. Exp. Ther.* **235**, 147–156.

Nilsson, B. M., Ringdahl, B., and Hacksell, U. (1988) Derivatives of the muscarinic agent *N*-methyl-*N*-(1-methyl-4-pyrrolidino-2-butynyl)acetamide (BM 5). *J. Med. Chem.* **31**, 577–582.

Nördstrom, O., Alberts, P., Westlind, A., Undén, A., and Bartfai, T. (1983) Presynaptic antagonist-postsynaptic agonist at muscarinic cholinergic synapses. *N*-Methyl-*N*-(1-methyl-4-pyrrolidino-2-butynyl)acetamide. *Mol. Pharmacol.* **24**, 1–5.

Palacios, J. M., Bolliger, G., Closse, A., Enz, A., Gmelin, G., and Malanowski, J. (1986) The pharmacological assessment of RS 86 (2-ethyl-8-methyl-2,8-diazaspiro-[4,5]-decan-1,3-dion hydrobromide). A potent, specific muscarinic acetylcholine receptor agonist. *Eur. J. Pharmacol.* **125**, 45–62.

Partington, P., Feeney, J., and Burgen, A. S. V. (1972) The conformation of acetylcholine and related compounds in aqueous solution as studied by nuclear magnetic resonance spectroscopy. *Mol. Pharmacol.* **8**, 269–277.

Pigini, M., Ballini, R., Gualtieri, F., and Brasili, L. (1980) Molecular requirements of the active sites of the cholinergic receptors. XI. Synthesis and biological evaluation of isomuscarine and its isomers. *Farmaco, Ed. Sci.* **35**, 167–180.

Pigini, M., Brasili, L., Giannella, M., and Gualtieri, F. (1981) Molecular requirements of the active sites of the cholinergic receptors. XVII. 1,3-oxathiolane nucleus as a carrier of the active moieties. *Eur. J. Med. Chem.* **16**, 415–419.

Pope, C. N., Ho, B. T., and Wright, A. A. (1987) Neurochemical and behavioral effects of *N*-ethyl-acetylcholine aziridinium chloride in mice. *Pharmacol. Biochem. Behav.* **26**, 365–371.

Resul, B., Ringdahl, B., and Dahlbom, R. (1979a) Acetylene compounds of potential pharmacological value. XXX. Synthesis and pharmacological properties of *N*-(4-pyrrolidino-2-pentynyl)-substituted 2-pyrrolidone and succinimide. *Acta Pharm. Suec.* **16**, 161–165.

Resul, B., Ringdahl, B., Hacksell, U., Svensson, U., and Dahlbom, R. (1979b) Acetylene compounds of potential pharmacological value. XXXI. Studies on *N*-(4-tert-amino-2-butynyl)carboxamides as muscarinic agonists. *Acta Pharm. Suec.* **16**, 225–232.

Resul, B., Ringdahl, B., and Dahlbom, R. (1981) Acetylene compounds of potential pharmacological value. XXXV. The β-lactam analogue of oxotremorine. *Eur. J. Med. Chem.* **16**, 379–381.

Resul, B., Lewander, T., Zetterström, T., Ringdahl, B., Muhi-Eldeen, Z., and Dahlbom, R. (1980) Synthesis and pharmacological properties of *N*-[4-(1-azetidinyl)-2-butynyl]-2-pyrrolidone, a highly potent oxotremorine-like agent. *J. Pharm. Pharmacol.* **32**, 439–440.

Resul, B., Lewander, T., Ringdahl, B., Zetterström, T., and Dahlbom, R. (1982a) The pharmacological assessment of a new, potent oxotremorine analogue in mice and rats. *Eur. J. Pharmacol.* **80**, 209–215.

Resul, B., Dahlbom, R., Ringdahl, B., and Jenden, D. J. (1982b) *N*-Alkyl-*N*-(4-tert-amino-1-methyl-2-butynyl)carboxamides, a new class of potent oxotremorine antagonists. *Eur. J. Med. Chem.* **17**, 317–322.

Resul, B., Ringdahl, B., Dahlbom, R., and Jenden, D. J. (1983) Muscarinic activity of some secondary and tertiary amines and quaternary ammonium salts structurally related to oxotremorine. *Eur. J. Pharmacol.* **87**, 387–396.

Ringdahl, B. (1984a) Determination of dissociation constants and relative efficacies of oxotremorine analogs at muscarinic receptors in the guinea pig ileum by pharmacological procedures. *J. Pharmacol. Exp. Ther.* **229**, 199–206.

Ringdahl, B. (1984b) A comparison of the stimulant activities of oxotremorine analogues on the frog rectus abdominis and the guinea pig ileum. *Eur. J. Pharmacol.* **99**, 177–184.

Ringdahl, B. (1984c) Muscarinic receptor occupation and receptor activation in the guinea-pig ileum by some acetamides related to oxotremorine. *Br. J. Pharmacol.* **82**, 269–274.

Ringdahl, B. (1985a) Structural requirements for muscarinic receptor occupation and receptor activation by oxotremorine analogs in the guinea pig ileum. *J. Pharmacol. Exp. Ther.* **232**, 67–73.

Ringdahl, B. (1985b) Stereoselectivity, affinity and efficacy of oxotremorine analogues for muscarinic cholinergic receptors, in *VIIIth International Symposium on Medicinal Chemistry*, Proceedings, Vol. 2 (Dahlbom, R. and Nilsson, J. L. G., eds.). Swedish Pharmaceutical, Stockholm, pp. 391–403.

Ringdahl, B. (1986) Dissociation constants and relative efficacies of acetylcholine, (+)- and (−)-methacholine at muscarinic receptors in the guinea-pig ileum. *Br. J. Pharmacol.* **89**, 7–13.

Ringdahl, B. (1987a) Selectivity of partial agonists related to oxotremorine based on differences in muscarinic receptor reserve between the guinea pig ileum and urinary bladder. *Mol. Pharmacol.* **31**, 351–356.

Ringdahl, B. (1987b) Structural requirements for affinity and efficacy of *N*-(4-amino-2-butynyl)succinimides at muscarinic receptors in the guinea-pig ileum and urinary bladder. *Eur. J. Pharmacol.* **140**, 13–23.

Ringdahl, B. (1988a) Dimethylsulfonium and thiolanium analogues of the muscarinic agent oxotremorine. *J. Med. Chem.* **31**, 164–168.

Ringdahl, B. (1988b) 5-Methyl-2-pyrrolidone analogues of oxotremorine as selective muscarinic agonists. *J. Med. Chem.* **31**, 683–688.

Ringdahl, B. and Dahlbom, R. (1978) Stereoselectivity of oxotremorine antagonists containing a chiral pyrrolidine group. *Experientia* **34**, 1334–1335.

Ringdahl, B. and Jenden, D. J. (1983a) Pharmacological properties of oxotremorine and its analogs. *Life Sci.* **32**, 2401–2413.

Ringdahl, B. and Jenden, D. J. (1983b) Affinity, efficacy, and stereoselectivity of oxotremorine analogues for muscarinic receptors in the isolated guinea pig ileum. *Mol. Pharmacol.* **23**, 17–25.

Ringdahl, B. and Jenden, D. J. (1987a) Properties of a new alkylating ligand for muscarinic receptors, in *Cellular and Molecular Basis of Cholinergic Function* (Dowdall, M. J. and Hawthorne, J. N., eds.), Ellis Horwood, Chichester, pp. 120–130.

Ringdahl, B. and Jenden, D. J. (1987b) Muscarinic actions of an *N*-methyl-*N*-2-bromoethylamino analog of oxotremorine (BR 401) in the mouse. *J. Pharmacol. Exp. Ther.* **240**, 370–375.

Ringdahl, B. and Jenden, D. J. (1987c) Kinetics of solvolysis and muscarinic actions of an *N*-methyl-*N*-(2-bromoethyl)amino analogue of oxotremorine. *J. Med. Chem.* **30**, 852–854.

Ringdahl, B. and Markowicz, M. E. (1987) Muscarinic and antimuscarinic activity of acetamides related to oxotremorine in the guinea pig urinary bladder. *J. Pharmacol. Exp. Ther.* **240**, 789–794.

Ringdahl, B., Muhi-Eldeen, Z., Ljunggren, C., Karlén, B., Resul, B., Dahlbom, R., and Jenden, D. J. (1979) Acetylene compounds of potential pharmacological value. XXVIII. Oxotremorine analogues substituted with a methyl group in the lactam ring. *Acta Pharm. Suec.* **16**, 89–94.

Ringdahl, B., Ehlert, F. J., and Jenden, D. J. (1982a) Muscarinic activity and receptor binding of the enantiomers of aceclidine and its methiodide. *Mol. Pharmacol.* **21**, 594–599.

Ringdahl, B., Resul, B., Jenden, D. J., and Dahlbom, R. (1982b) Muscarinic activity in the isolated guinea pig ileum of some caboxamides related to oxotremorine. *Eur. J. Pharmacol.* **85**, 79–83.

Ringdahl, B., Resul, B., Ehlert, F. J., Jenden, D. J., and Dahlbom, R. (1984) The conversion of 2-chloroalkylamine analogues of oxotremorine to aziridinium ions and their interactions with muscarinic receptors in the guinea pig ileum. *Mol. Pharmacol.* **26**, 170–179.

Ringdahl, B., Roch, M., and Jenden, D. J. (1987) Regional differences in receptor reserve for analogs of oxotremorine in vivo: implications for development of selective muscarinic agonists. *J. Pharmacol. Exp. Ther.* **242**, 464–471.

Ringdahl, B., Roch, M., and Jenden, D. J. (1988) Tertiary 3- and 4-haloalkylamine analogues of oxotremorine as prodrugs of potent muscarinic agonists. *J. Med. Chem.* **31**, 160–164.

Ringdahl, B., Katz, E. D., Roch, M., and Jenden, D. J. (1989a) Muscarinic actions and receptor binding of the enantiomers of BM 130, an alkylating analog of oxotremorine. *J. Pharmacol. Exp. Ther.*, in press.

Ringdahl, B., Roch, M., Katz, E. D., and Frankland, M. C. (1989b) Alkylating partial muscarinic agonists related to oxotremorine. N-[4-[(2-haloethyl)methylamino]-2-butynyl]-5-methyl-2-pyrrolidones. *J. Med. Chem.*, in press.

Robinson, D. A., Taylor, J. G., and Young, J. M. (1975) The irreversible binding of acetylcholine mustard to muscarinic receptors in intestinal smooth muscle of the guinea-pig. *Br. J. Pharmacol.* **53**, 363–370.

Roszkowski, A. P. (1961) An unusual type of sympathetic ganglionic stimulant. *J. Pharmacol. Exp. Ther.* **132**, 156–170.

Roszkowski, A. P. and Yelnosky, J. (1967) Structure–activity relationships among a series of acetylenic carbamates related to McN-A-343. *J. Pharmacol. Exp. Ther.* **156**, 238–245.

Ruffolo, R. R., Jr. (1982) Important concepts of receptor theory. *J. Auton. Pharmacol.* **2**, 277–295.

Russell, R. W., Smith, C. A., Booth, R. A., Jenden, D. J., and Waite, J. J. (1986a) Behavioral and physiological effects associated with changes in muscarinic receptors following administration of an irreversible cholinergic agonist (BM 123). *Psychopharmacol.* **90**, 308–315.

Russell, R. W., Crocker, A. D., Booth, R. A., and Jenden, D. J. (1986b) Behavioral and physiological effects of an aziridinium analog of oxotremorine (BM 130). *Psychopharmacol.* **88**, 24–32.

Saunders, J., Showell, G. A., Baker, R., Freedman, S. B., Hill, D.,

McKnight, A., Newberry, N., Salamone, J. D., Hirschfield, J., and Springer, J. P. (1987) Synthesis and characterization of all four isomers of the muscarinic agonist 2′-methylspiro[1-azabicyclo[2.2.2]-octane-3,4′-[1,3]dioxolane]. *J. Med. Chem.* **30**, 969–975.

Saunders, J., Showell, G. A., Snow, R. J., Baker, R., Harley, E. A., and Freedman, S. B. (1988) 2-Methyl-1,3-dioxaazaspiro[4,5]decanes as novel muscarinic cholinergic agonists. *J. Med. Chem.* **31**, 486–491.

Sauerberg, P., Larsen, J.-J., Falch, E., and Krogsgaard-Larsen, P. (1986a) A novel class of conformationally restricted heterocyclic muscarinic agonists. *J. Med. Chem.* **29**, 1004–1009.

Sauerberg, P., Fjälland, B., Larsen, J.-J., Bach-Lauritsen, T., Falch, E., and Krogsgaard-Larsen, P. (1986b) Pharmacological profile of a novel class of muscarinic acetylcholine receptor agonists. *Eur. J. Pharmacol.* **130**, 125–131.

Sauerberg, P., Falch, E., Meier, E., Lemböl, H. L., and Krogsgaard-Larsen, P. (1988) Heterocyclic muscarinic agonists. Synthesis and biological activity of some bicyclic sulfonium arecoline bioisosteres. *J. Med. Chem.* **31**, 1312–1316.

Schulman, J. M. and Disch, R. L. (1986) On the recognition of cholinergic agonists by the muscarinic receptor. *Trends Pharmacol. Sci.* **7**, 263–264.

Schulman, J. M., Sabio, M. L., and Disch, R. L. (1983) Recognition of cholinergic agonists by the muscarinic receptor. 1. Acetylcholine and other agonists with the NCCOCC backbone. *J. Med. Chem.* **26**, 817–823.

Schwörer, H., Lambrecht, G., Mutschler, E., and Kilbinger, H. (1985) The effects of racemic bethanechol and its (R)- and (S)-enantiomers on pre- and postjunctional muscarine receptors in the guinea-pig ileum. *Naunyn-Schmiedeberg's Arch. Pharmacol.* **331**, 307–310.

Snyder, J. P. (1985) Molecular models for muscarinic receptors. *Trends Pharmacol. Sci.* **6**, 464–466.

Sokolovsky, M. (1984) Muscarinic receptors in the central nervous system. *Internat. Rev. Neurobiol.* **25**, 139–183.

Stephenson, R. P. (1956) A modification of receptor theory. *Br. J. Pharmacol. Chemother.* **11**, 379–393.

Svensson, U., Hacksell, U., and Dahlbom, R. (1978) Acetylene compounds of potential pharmacological value. XXV. *N*-(4-Pyrrolidino-2-butynyl)-*N*-alkylcarboxamides. *Acta Pharm. Suec.* **15**, 67–70.

Teodori, E., Gualtieri, F., Angeli, P., Brasili, L., and Giannella, M. (1987) Resolution, absolute configuration, and cholinergic enantioselectivity of (−)- and (+)-*c*-2-methyl-*r*-5-[(dimethylamino)methyl]-1,3-oxathiolane *t*-3-oxide methiodide and related sulfones. *J. Med. Chem.* **30**, 1934–1938.

Teodori, E., Gualtieri, F., Angeli, P., Brasili, L., Giannella, M., and Pigini, M. (1986) Resolution, absolute configuration, and cholinergic enantioselectivity of (+)- and (−)-*cis*-2-methyl-5-[(dimethylamino)-

methyl]-1,3-oxathiolane methiodide. *J. Med. Chem.* **29**, 1610–1615.

Timoney, R. F. (1983) Structural requirements of muscarine, 1,3-dioxolane and their analogues for agonist activity at cholinergic receptors. *Internat. J. Pharmaceut.* **15**, 223–234.

Tollenaere, J. P. (1984) Muscarinic pharmacophore identification. *Trends Pharmacol. Sci.* **5**, 85–86.

Tonnaer, J. A. D. M., van Vugt, M. A., de Boer, Th., and de Graaf, J. S. (1987) Differential interactions of muscarinic drugs with binding sites of [^{3}H]pirenzepine and [^{3}H]quinuclidinyl benzilate in rat brain tissue. *Life Sci.* **40**, 1981–1987.

Triggle, D. J. (1976) Structure–activity relationships: chemical constitution and biological activity, in *Chemical Pharmacology of the Synapse* (Triggle, D. J. and Triggle, C. R., eds.). Academic, New York, pp. 233–430.

Triggle, D. J. (1984) Muscarinic cholinergic drugs: effects of stereoisomers, in *Handbook of Stereoisomers: Drugs in Psychopharmacology* (Smith, D. F., ed.). CRC Press, Boca Raton, pp. 31–66.

Waser, P. G. (1961) Chemistry and pharmacology of muscarine, muscarone and related compounds. *Pharmacol. Rev.* **13**, 465–515.

Waser, P. G. and Hopff, W. (1979) Structure and activity of some new muscarine derivatives and acetylcholine analogues, in *Advances in Pharmacology and Therapeutics*, Vol. 3 (Stoclet, J. C., ed.). Pergamon, Oxford, pp. 261–269.

Weinstein, H., Maayani, S., Srebrenik, S., Cohen, S., and Sokolovsky, M. (1975) A theoretical and experimental study of the semirigid cholinergic agonist 3-acetoxyquinuclidine. *Mol. Pharmacol.* **11**, 671–689.

Wess, J., Lambrecht, G., Moser, U., and Mutschler, E. (1987) Stimulation of ganglionic muscarinic M_1 receptors by a series of tertiary arecaidine and isoarecaidine esters in the pithed rat. *Eur. J. Pharmacol.* **134**, 61–67.

Willcockson, W. S., Kahlid, M., Ahmed, A. E., and Hillman, G. R. (1981) Effects of acetylcholine mustard analogs on schistosome and vertebrate neuromuscular preparations. *J. Pharmacol. Exp. Ther.* **218**, 330–336.

Yamamura, H. I. and Snyder, S. H. (1974) Muscarinic cholinergic binding in rat brain. *Proc. Nat. Acad. Sci.* **71**, 1725–1729.

SECTION 3

Biochemical Effectors Coupled to Muscarinic Receptors

6

Muscarinic Cholinergic Receptor-Mediated Regulation of Cyclic AMP Metabolism

T. Kendall Harden

1. Introduction

The role of cyclic AMP as a mediator in multifarious physiological processes is well established (Robison et al., 1971). Although its direct association with mAChR-mediated responses has been unambiguously demonstrated in only a few cases, the central role of cyclic AMP as a regulatory substance, together with the widespread observation of marked effects of mAChR agonists on cyclic AMP levels, makes it essentially irrefutable that physiological effects mediated through mAChR in many target tissues involve, at least in part, cyclic AMP.

Activation of mAChR results in attenuation of intracellular cyclic AMP levels. The most widely accepted mechanism for hormonally mediated reduction of cyclic nucleotide levels involves inhibition of adenylate cyclase, and such a response to mAChR stimulation has been widely reported. However, activation of mAChR also increases the rate of cyclic AMP degradation by cyclic AMP phosphodiesterase. This mechanism has been less prominently recognized and has been studied in only a few tissues. Nevertheless, activation of a Ca^{2+}-calmodulin-regulated phosphodiesterase may accompany mAChR-mediated elevation of cytoplasmic Ca^{2+} levels in many cells and likely represents an important regulatory mechanism. The details of both mechanisms of lowering cyclic AMP levels are reviewed below. It should be emphasized that, although aspects of this literature most closely related to the mAChR will be considered, as far as is known, there is nothing fundamentally different between the association of mAChR and other receptors with these two mechanisms. For example, mAChR and α-2 adrenergic receptors likely inhibit adenylate cyclase through a common mechanism and muscarinic receptors, and α-1 adrenergic receptors activate phosphodiesterase through a common mechanism.

2. Regulation of Cyclic AMP Synthesis

2.1. General Model for Activation of Adenylate Cyclase

Appreciation of the existence and function of guanine nucleotide regulatory proteins (G-proteins) has been key to the development of understanding of the mechanisms whereby hormones and neurotransmitters activate and inhibit adenylate cyclase. Rodbell and coworkers (Rodbell et al., 1971, 1975; Rodbell, 1975) first recognized that the stimulatory effects of glucagon on adenylate cyclase activity in liver cell membranes was dependent on the presence of GTP. Observation of biphasic effects of guanine nucleotides by Rodbell and associates (Harwood et al., 1973; Rodbell, 1975; Yamamura et al., 1977; Cooper et al., 1979) also presaged the existence of two G-proteins in association with adenylate cyclase, one possessing stimulatory and the other inhibitory properties.

The early history of the delineation of the components of the adenylate cyclase response system mainly centered on the β-adrenergic receptor-regulated enzyme. Among the key advances were:

1. The development of successful radioligand binding assays for the β-receptor in the laboratories of Lefkowitz (Lefkowitz et al., 1974), Aurbach (Aurbach et al., 1974), and Levitzki (Levitzki et al., 1974)
2. The development of S49 lymphoma cell mutants that were deficient in various adenylate cyclase activities (Bourne et al., 1975; Haga et al., 1977; Johnson et al., 1980)
3. Pfeuffer's partial resolution by GTP-Sepharose chromatography of a protein fraction that conferred guanine nucleotide regulatory properties on the enzyme (Pfeuffer and Helmreich, 1975; Pfeuffer, 1977)

4. Development of understanding of the reasons for the quasi-irreversible activation of adenylate cyclase catalyzed by cholera toxin (Cassel and Selinger, 1977; Moss and Vaughan, 1977; Gill and Meren, 1978; Cassel and Pfeuffer, 1978)
5. The work of Cassel and Selinger suggesting that a GTPase activity and hormonally catalyzed exchange of GTP for GDP were involved in an adenylate cyclase regulatory cycle (Cassel et al., 1977; Cassel and Selinger, 1977, 1978)
6. The work of Gilman, Ross, and coworkers (Ross and Gilman, 1977; Ross et al., 1978; Ross and Gilman, 1980) indicating that a protein extracted (and later purified; Northup et al., 1980; Sternweis et al., 1981) from plasma membranes would confer guanine nucleotide-dependent stimulation of adenylate cyclase by β-adrenergic receptor agonists in membranes devoid of this activity.

These developments, together with work from several other laboratories, led by 1980 to the description of a general model for activation of adenylate cyclase that is summarized in Fig. 1.

It is proposed that the binding of a stimulatory agonist to a receptor, e.g., the β-adrenergic receptor, results in an association of the agonist–receptor complex with the stimulatory G-protein, G_s, with a consequential exchange of GTP for GDP. The GTP-liganded G-protein (or more accurately, an activated subunit; *see below*) then activates the adenylate cyclase catalytic subunit. Steady-state production of cyclic AMP in the presence of hormone is governed then by the lifetime (which is long) of the GTP-activated G-protein and by a GTPase activity of the G_s protein. As a consequence of the "off reaction" catalyzed by this GTPase, activity is reduced to the ground state; in the continued presence of hormone, another activation cycle is initiated in the "on reaction" catalyzed as described above by activated receptor. Activation of adenylate cyclase by cholera toxin is explained by the fact that cholera toxin is an enzyme that, in the presence of NAD, ADP-ribosylates G_s (Cassel and Pfeuffer, 1978; Gill and Meren, 1978). This covalent modification inactivates the GTPase activity associated with G_s and, thus, leads to a quasi-irreversibly activated catalyst.

The view of hormonal activation of adenylate cyclase depicted in Fig. 1 was modified considerably as a consequence of two realizations. First, the discovery that G_s was a heterotrimeric protein (Northup et al., 1980; Sternweis et al., 1981) led to a more complex scheme for explanation of its mechanism of activation. Moreover, as this simplified model for activation came to fruition in the late 1970s, it became clear that a large family of receptors inhibited rather than activated adenylate cyclase.

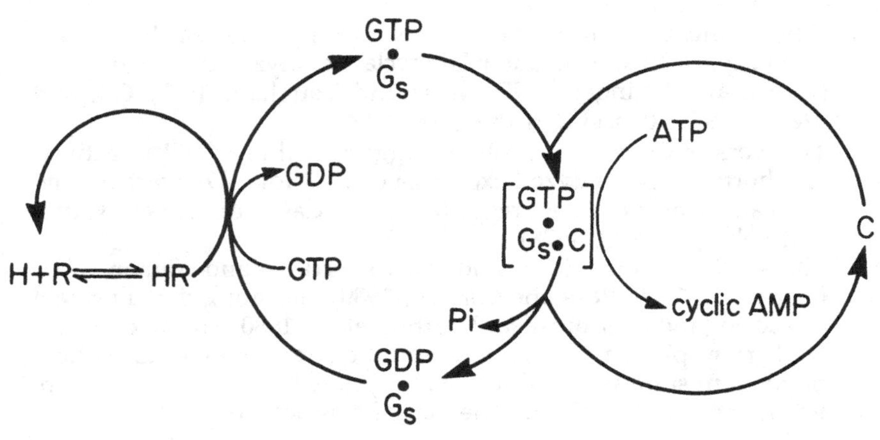

Fig. 1. General model for hormonal activation of adenylate cyclase. This model describes the hypothetical hormone (H)-induced interactions of a cell surface receptor (R), the stimulatory guanine nucleotide regulatory protein (G_s), the catalytic protein (C), and the guanine nucleotides GTP and GDP. It is proposed that H-occupied R forms a ternary complex with G_s and, thus, accelerates the exchange of GTP for GDP on G_s. Formation H·R·G_s is considered to be rate-limiting in the activation process, with G_s·GTP rapidly interacting with C to form the enzymically active complex, GTP·G_s·C. The lifetime of the active complex is determined by the activity of a GTPase, which hydrolyzes the bound GTP, releasing P_i, G_s·GDP, and the inactive form of C. This simple model does not take into account the GTP-promoted dissociation of the G_s heterotrimer. Also, activation and deactivation of C may not actually involve changes in its physical association with G_s. Finally, the number of partial reactions that constitute the catalytic cycle are oversimplified in this model. For example, acceleration of guanine nucleotide exchange by activated hormone receptors likely involves more steps than are implied in this general model.

2.2. Receptor-Mediated Inhibition of Adenylate Cyclase

The mAChR shares with several receptors—e.g., α-2 adrenergic, opiate, A_1-adenosine, somatostatin, angiotensin II, and ADP—the capacity to inhibit adenylate cyclase (Jakobs, 1979; Rodbell, 1980; Limbird, 1981; Cooper, 1982; Jakobs et al., 1984, 1985). As mentioned previously, there is no confirmed reason to expect that the mechanism whereby mAChR regulate the enzyme differs from that of other hormones and neurotransmitters, although very little information is available that directly addresses this possibility. Therefore, an essentially generic discussion of receptor-mediated inhibition of adenylate cyclase follows. Specific properties of mAChR-regulated adenylate cyclase are considered in a subsequent section.

Based on knowledge that had already accumlated on the β-adrenergic and glucagon receptor-stimulated adenylate cyclase, it was not surprising that, upon clear documentation of inhibitory effects of hormones on adenylate cyclase, a G-protein was immediately implicated. Receptor-mediated inhibition of enzyme activity required the presence of GTP (Jakobs et al., 1978; Londos et al., 1978; Jakobs et al., 1979), and high affinity binding of agonists to inhibitory receptors was shown to be reduced by guanine nucleotides (Berrie et al., 1979; Rosenberger et al., 1980; Hoffman et al., 1980; Smith and Limbird, 1981). Whether this putative G-protein was G_s or another G-protein was a subject of debate. However, indirect evidence suggested that functionally unique proteins were involved. First, the biphasic effects of GTP on adenylate cyclase reported by Rodbell and coworkers (Harwood et al., 1973; Rodbell, 1975; Yamamura et al., 1977; Cooper et al., 1979) were consistent with the idea that the catalytic protein was under dual control. The effects of guanine nucleotides on agonist binding to inhibitory receptors and coupling of inhibitory receptors to adenylate cyclase expressed differential sensitivities to inactivation by various reagents when compared to analogous activities of stimulatory receptors (Hoffman et al., 1981; Jakobs et al., 1982; Harden et al., 1982). Pseudohypoparathyroid patients are functionally deficient in G_s (Farfel et al. 1981), yet expressed normal inhibitory responses to α-2 adrenergic receptor stimulation (Motulsky et al., 1982). Finally, cys$^-$ S49 lymphoma cells lack functional G_s, but adenylate cyclase activity in membranes from these cells was markedly inhibited by guanine nucleotides and by somatostatin in a GTP-dependent manner (Hildebrandt et al., 1982, 1983; Jakobs et al., 1983). These observations allowed a strong circumstantial argument to be made in favor of the existence of a unique protein involved in coupling inhibitory receptors to adenylate cyclase. However, it was the pioneering work of Ui and coworkers that was key to arrival at unambiguous conclusions regarding this question.

Out of a series of experiments designed to examine the mechanisms of action of a toxin isolated from *B. pertussis*, it became clear to Ui and associates that this material had profound effects on adenylate cyclase. Most notably, the pertussis toxin (termed islet activating protein, IAP, by Ui and coworkers) blocked the inhibitory effects of α-2 adrenergic receptors on adenylate cyclase activity in pancreatic islets (Katada and Ui, 1980; Katada and Ui, 1981). Like cholera toxin, pertussis toxin catalyzed the ADP-ribosylation of a plasma membrane protein. The principal pertussis toxin substrate was shown to be a 41-kdalton protein, the extent of ADP-ribosylation of which was shown to be directly correlated with the extent of modification of adenylate cyclase activities (Katada and Ui, 1982a), including the blockade of inhibitory effects of muscarinic (Hazeki and Ui, 1981; Kurose et al., 1983; Hughes et al., 1984) and a variety of other inhibitory receptors (Ui, 1984). The 41-kdalton substrate was shown to copurify with a GTP-binding activity (Bokoch et al., 1983; Codina et al., 1983) that reconstituted inhibitory effects of guanine nucleotides and hormone receptors on adenylate cyclase activity and guanine-nucleotide-sensitive binding of agonists to inhibitory receptors (Bokoch et al., 1984; Katada et al., 1984a,b,c). This protein was named G_i to distinguish it from the stimulatory G-protein, G_s.

A discussion of the mechanism whereby mAChR and other receptors inhibit adenylate cyclase requires a consideration of the oligomeric structure of both G_s and G_i. G_s was initially purified as an apparently oligomeric protein consisting of 45-kdalton (and 52-kdalton depending on the source) and 35-kdalton proteins (Northup et al., 1980; Sternweis et al., 1981). The 45-kdalton protein was shown to be the GTP-binding α subunit based on its capacity to be selectively ADP-ribosylated by cholera toxin (Northup et al., 1980; Sternweis et al., 1981) and labeled by [^{32}P]8-azido-GTP (Northup et al., 1982). Results from a variety of model systems indicate that it is the guanine nucleotide liganded α subunit that activates the enzyme, and thus, activation of the protein by GTP would proceed with dissociation of subunits as follows:

$$\alpha\beta + GTP \rightarrow GTP{\cdot}\alpha + \beta$$

Based on data from experiments with resolved subunits and principally using reconstituted model preparations, Gilman and coworkers (Katada et al., 1984a,b,c; Gilman, 1984, 1987) have concluded that the β subunit released from the dissociation reaction depicted above serves as a deactivator of the enzyme in that it promotes formation of the inactive oligomeric form of G_s.

G_i also possesses an oligomeric structure similar to that for G_s. That is, this G-protein consists of an α subunit of 41-kdaltons and a β subunits (35-kdaltons) that is apparently identical to that for G_s. (Bokoch et al., 1983, 1984; Codina et al., 1983). Both G_s and G_i are actually heterotrimers in that a γ subunit of 5–10 kdaltons also copurifies with the holo-protein and is found tightly associated with the β subunit under activating conditions (Hildebrandt et al., 1984).

Activation of G_i with a stable GTP analog such as GTPγS results in dissociation of the protein as was described above for G_s. (Bokoch et al., 1984; Katada et al., 1984a,b,c). This observation, together with the fact that G_i is found in tissues at concentrations in excess of those for G_s, suggests that activation of inhibitory receptors could result in inhibition of adenylate cyclase as a consequence of stimulated dissociation of G_i, which in turn would promote formation of the inactive heterotrimeric form of G_s. Several observations support this model. Reconstitution of wild-type S49 lymphoma cell or platelet membranes with the subunits resolved from GTPγS-activated G_i revealed that the β subunit produced marked inhibitory effects on basal adenylate cyclase activity; only small inhibitory effects of GTPγS$\cdot\alpha_i$ were observed (Katada et al., 1984a). Thus, based on the structure of G_i, its relative abundance in the plasma membrane, its activation-promoted dissociation, and the inhibitory effects of the β subunit protein on adenylate cyclase activity, the model for receptor-mediated inhibition of adenylate cyclase depicted in Fig. 2 has been proposed (Katada et al., 1984a,b,c; Gilman, 1984, 1987). An agonist-occupied inhibitory receptor, e.g., the mAChR, promotes the exchange of GTP for GDP on G_i with the consequential dissociation of the protein into GTP-liganded α subunit + free $\beta\gamma$. Active α subunits then would be driven to the inactive oligomeric state by released $\beta\gamma$ according to mass action.

The proposal that inhibitory receptors would modify adenylate cyclase at the level of the G-protein, i. e., by promoting formation of the undissociated inactive state of G_s, rather than at the level of the adenylate cyclase catalyst, has not been proven unambiguously and, indeed, competes with another plausible mechanism. Birnbaumer and colleagues have suggested that the evidence in support of a mediatory role of β_i is weak and have porposed that the inhibitory activity of G_i resides in the capacity of α_i to directly inhibit the catalytic protein, rather than to modify directly any function of G_s (Hildebrandt et al., 1984; Birnbaumer

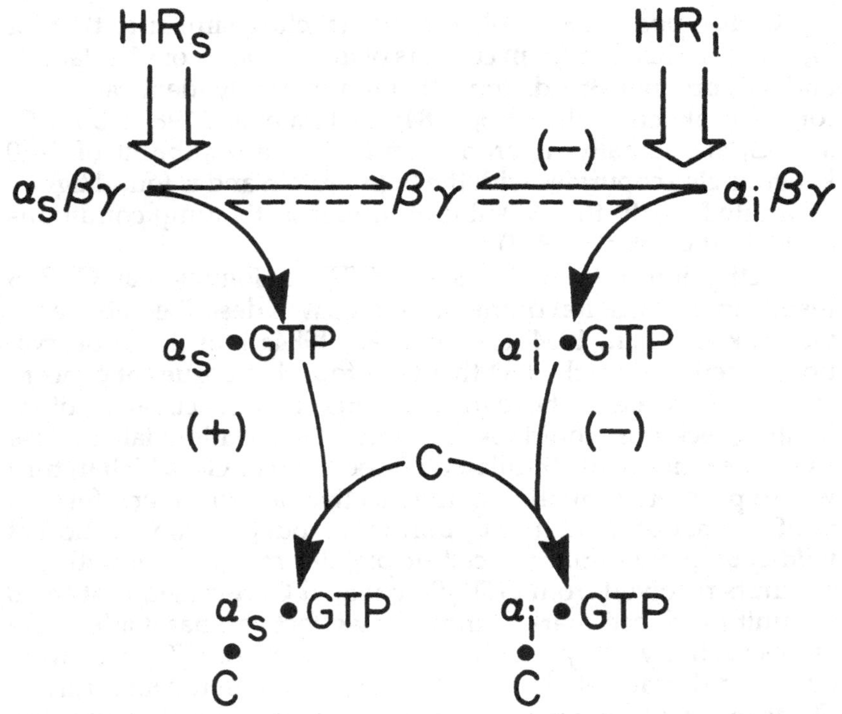

Fig. 2. Concensus model for hormonal activation and inhibition of adenylate cyclase. The stimulatory (R_s) and inhibitory (R_i) receptor-mediated regulation of adenylate cyclase through G_s ($\alpha_s\beta\gamma$) and $G_i(\alpha_i\beta\gamma)$ is depicted in this model. Occupation of R_s by H_s results in activation/dissociation of $\alpha_s\beta\gamma$ into GTP·α_s +$\beta\gamma$. GTP·α_s then interacts with C to form the active enzymic species (GTP·α_s·C). Occupation of R_i by H_i results in activation/dissociation of $\alpha_i\beta\gamma$ into GTP·α_i + $\beta\gamma$. The resultant inhibition of adenylate cyclase has been proposed to occur by two mechanisms. First, since G_i is usually present in excess concentration over G_s, it has been proposed that the HR_i-stimulated increase in free $\beta\gamma$ favors formation of the inactive form of G_s because of mass-action considerations. An opposing model proposes that inhibition of adenylate cyclase occurs through direct inhibition of C by α_i·GTP, i.e., GTP·α_i·C represents a form of C expressing relatively poor catalytic activity.

et al., 1985; Birnbaumer, 1987). The evidence in favor of this second model is as follows.

First, support for the model of Gilman and coworkers has been obtained principally in studies in which reconstituted $\beta\gamma$ itself has been shown to inhibit G_s-stimulated adenylate cyclase activity, rather than in experiments in which activation of a receptor inhibits activity through a mediatory role of $\beta\gamma$. Katada et al. (1984a) found in their initial studies that, although the greatest inhibition of platelet membrane adenylate cyclase occurred with resolved β subunit, GTPγS·α_i also inhibited activity. The activity of α_i was even more pronounced in experiments with cyc$^-$ membranes. Here the β subunit had no effect (G_s is not present in these cells), and GTPγS·α_i markedly inhibited enzyme activity (Katada et al., 1984b; Hildebrandt et al., 1984). Moreover, Hildebrandt et al. (1984) have reported that activated G_i inhibits adenylate cyclase in a manner that is noncompetitive with G_s. Since the β-mediated inhibition model would predict that this would not be the case, these workers proposed that their results were consistent with the existence of two sites on the catalytic protein: one that mediates activation of the enzyme by α_s and the other that mediates inhibition of the enzyme by α_i. This model is depicted in the right side of Fig. 2. Resolution of the relative validities of the Gilman vs Birnbaumer models requires additional work, with particular emphasis on the properties of receptor-stimulated dissociation of G-proteins in their membrane environment. The details of hormonal regulation of adenylate cyclase have been reviewed comprehensively in several excellent recent reviews (Birnbaumer et al., 1985; Stryer and Bourne, 1986; Gilman, 1987).

2.3. Observations on mAChR-Mediated Inhibition of Adenylate Cyclase

The initial report of receptor-mediated inhibiton of adenylate cyclase was made in 1962 by Murad and coworkers (Murad et al., 1962) in their ground-breaking studies of adenylate cyclase. It was shown that adenylate cyclase activity measured in cardiac membranes from several species was inhibited by cholinergic agonists, and that the effect involved an mAChR in that atropine completely blocked the inhibitory effect of the agonists. Probably because of the lack of a sufficiently sensitive assay for the enzyme as well as the lack of appreciation of the role of GTP in receptor-mediated regulation of adenylate cyclase activity, the question of mAChR-mediated regulation of adenylate cyclase was left essentially unexplored for 15 years.

During the 1970s, however, several observations were made with intact tissues that supported the idea that mAChR-mediated inhibition of adenylate cyclase may occur. Thus, mAChR agonists were shown to attenuate receptor-stimulated cyclic AMP accumulation in a variety of cardiac preparations from several species (Gardner and Allen, 1972; Kuo et al., 1972; George et al., 1973; Brown, 1979; Brown et al., 1979), in rat uterus (Triner et al., 1972), in rat lung slices (Kuo and Kuo, 1973), in guinea-pig ileum (Lee et al., 1972), in bovine tracheal smooth muscle (Lohmann et al., 1977), in bovine superior cervical ganglion (Kebabian et al., 1975), and in dog, pig, and bovine thyroid slices (Champion et al., 1974; Erneux et al., 1977). Attenuating effects of mAChR agonists occurred with a variety of stimulatory agonists, e.g., isoproterenol, norepinephrine, epinephrine, PGE_1, glucagon, dopamine, thyrotropin, and adenosine, and in several cases occurred in the presence of a phosphodiesterase inhibitor.

The question of whether mAChR-mediated inhibition of cyclic AMP accumulation occurred as a consequence of inhibition of adenylate cyclase was addressed directly in a series of experiments carried out in the late 1970s. These studies were in great part directed by the seminal work of Jakobs, Schultz, and coworkers in Heidelberg during this period (*see* Jakobs, 1979; Jakobs et al., 1984, 1985). Based on their initial data on α-2 receptor-mediated inhibition of adenylate cyclase in platelet membranes (Jakobs et al., 1976, 1978) and their subsequent data on the inhibition of adenylate cyclase by multifarious receptors in a variety of tissues, a relatively consistent picture of mAChR-mediated inhibition of adenylate cyclase was soon revealed. Thus, Blume and coworkers working with NG108-15 neuroblastoma × glioma cells (Lichtshtein et al., 1979), and Jakobs et al. (1979) and Watanabe et al. (1978) using dog heart membranes established that activation of mAChR resulted in an inhibition of both basal and hormone-stimulated adenylate cyclase activity, and that the inhibition was dependent on the presence of GTP. As opposed to the general observation in intact cells where hormonally mediated inhibition of receptor-activated cyclic AMP accumulation is usually greater than inhibition of basal cyclic AMP levels, the mAChR-mediated inhibition of basal adenylate cyclase activity in membrane preparations was greater than that observed with hormone stimulated activity. The reasons for these differences are not completely clear, although arguments based on one or both of the models described above for inhibition of adenylate cyclase can be marshaled. It has been reported for both mAChR and other inhibitiory hormone receptors that at least

10-fold higher concentrations of GTP are required for half-maximal inhibition of adenylate cyclase compared to the half-maximal concentration of nucleotide necessary for activation of adenylate cyclase by stimulatory hormones. This may in great part explain why receptor-mediated activation of adenylate cyclase was readily observed in early experiments with crude membrane preparations, whereas receptor-mediated inhibition of the enzyme required more careful provision and maintenance of exogenously added GTP. As with activation of adenylate cyclase, GTP-supported receptor-mediated inhibition of adenylate cyclase can be blocked by the stable analog of GDP, GDPβS. Although it was initially reported that receptor-mediated inhibition of adenylate cyclase did not occur in the presence of stable analogs of GTP, it is now clear that these results were to a great degree misleading. That is, under appropriate assay conditions, i.e., during stimulation of the enzyme with the diterpene activator, forskolin, stable analogs such as GppNHp inhibit activity, and the rate of occurrence and extent of inhibition can be accelerated or potentiated by mAChR and other inhibitory receptor agonists (*see* Jakobs et al., 1984, 1985).

As with stimulatory hormone receptors, activation of mAChR and other inhibitory receptors results in activation of a membrane GTPase. Since the GTPase associated with G_s was know to activated adenylate cyclase, it was proposed that inhibitory receptors produced their effects on adenylate cyclase activity by receptors produced their effects on adenylate cyclase activity by promoting GTP hydrolysis by G_s (Aktories and Jakobs, 1981; Koski and Klee, 1981). However, several observations suggested that this was not the case. Inhibitory receptor-mediated activation of GTPase additive with that for stimulatory receptors, is unaffected by cholera toxin (Aktories et al., 1982), and is blocked by pertussis toxin (Aktories et al., 1983a,b; Burns et al., 1983). Inhibitory receptor-mediated activation of GTPase was shown to occur in cyc^- membranes (Aktories et al., 1983b), which are devoid of G_s, and the purified G_i protein has been shown to possess GTPase activity that can be stimulated by reconstituted hormone receptors in model phospholipid vesicles (Asano et al., 1984). In spite of these observations suggesting that the GTPase stimulated by inhibitory receptors is the one associated with G_i, mAChR-mediated increases in the rate of inactivation of the activated adenylate cyclase in both rat (Smith and Harden, 1985) and dog (Fleming et al., 1987) heart membranes have been reported. Several possible explanations have been provided, although no unequivocal explanation can be marshaled in face of the incomplete understanding of the mechanism of receptor-mediated inhibition of the enzyme that was discussed in detail in the section above on mechanism.

In some tissues, monovalent cations amplify the inhibitory effects observed with hormone receptors, including the mAChR (*see* Limbird, 1981; Cooper, 1982; Jakobs et al., 1985). The order of potency usually observed is $Na^+ > Li^+ > K^+$. One effect of monovalent cations is to counteract an inhibition of adenylate cyclase produced by GTP alone, thus allowing the observation of GTP-dependent inhibition of the enzyme by inhibitory receptors (Londos et al., 1978; Jakobs et al., 1985). The site of action of monovalent cations is not completely clear, although involvement of more than one site is possible. As is discussed in detail in Chapter 2 of this volume, monovalent cations produce effects on agonist binding to mAChR and other inhibitory receptors that are similar to, and additive with, the effects of guanine nucleotides (Hoffman and Lefkowitz, 1980).

At least three divalent cation binding sites (one on G_i, one on G_s, and one on the catalytic unit) are involved in the hormone receptor-regulated adenylate cyclase (Jakobs et al., 1983; Birnbaumer et al., 1985; Jakobs et al., 1985). Mg^{2+} allosterically activates adenylate cyclase, and stimulatory hormones lower the $K_{0.5}$ for activation of the enzyme by Mg^{2+} (Iyengar and Birnbaumer, 1982; Iyengar, 1981). Activation of mAChR in rat heart membranes has been shown to increase the $K_{0.5}$ for activation of the enzyme by Mg^{2+} (Smith and Harden, 1985). The percent of inhibition by mAChR of both basal and β-receptor-stimulated adenylate cyclase activity was greater at low, e.g., 0.2 m*M*, than at high, e.g., >2 m*M* Mg^{2+} concentrations. The apparent decrease in the "affinity" of the adenylate cyclase system for Mg^{2+} in the presence of mAChR agonists likely has physiological significance in that this change occurs over a concentration range of Mg^{2+} that approximates that of intracellular-free Mg^{2+}.

3. Receptor-Mediated Activation of Phosphodiesterase

3.1. General Properties

Hormonal regulation of cyclic AMP accumulation usually has been considered as a phenomenon occurring at the level of second-messenger synthesis. However, cyclic AMP levels also could be modified by changes in the amount of egress of nucleotide or by modifications in the rate of its degradation. To our knowledge, regulation of egress by mAChR has not been reported, although activation of a prostaglandin receptor in several target cells has been shown to modify egress (Heasley et al., 1985). On the other hand, regulation of cyclic AMP phosphodiesterase by mAChR has been shown to be a prominent effect in several tissues, and our prejudice is that this phenomenon, which has been studied in only a limited way, is of considerable physiological importance.

Cyclic nucleotide phosphodiesterase exists in several forms differing in their

1. Selectivity for cyclic AMP vs cyclic GMP
2. Relative K_m and V_{max} for cyclic AMP
3. Regulation by various activators, e.g., Ca^{2+}-calmodulin, cyclic GMP and
4. Cellular and subcellular distribution.

The properties of cyclic nucleotide phosphodiesterase isozymes have been considered in detail in several excellent reviews (Wells and Hardman, 1977; Beavo et al., 1982; Strada et al., 1984) and will be considered here only as they relate to the mAChR.

3.2. Phosphodiesterase in Thyroid Slices

The first hint that cyclic nucleotide phosphodiesterase might be involved in mAChR action came in the studies by Dumont and coworkers with dog thyroid slices. Thyrotropin (TSH) is an effective activator of adenylate cyclase in this tissue, and the elevation of cyclic AMP levels by TSH was shown to be depressed by mAChR agonists (Van Sande et al., 1977). Carbachol attenuated cyclic AMP accumulation, irrespective of whether it was added at the time of TSH addition or 40 min subsequent to TSH addition. The phosphodiesterase inhibitor, isobutylmethylxanthine, greatly potentiated TSH-stimulated cyclic AMP accumulation, but completely antagonized the inhibitory effects of carbachol. In contrast, the phosphodiesterase inhibitor, Ro 20-1724, potentiated TSH-stimulated cyclic AMP accumulation, but had no effect on the mAChR-mediated response. Since isobutylmethylxanthine, but not Ro 20-1724, inhibited cyclic GMP-stimulated phosphodiesterase, it was suggested that the effects of carbachol might occur through activation of phosphodiesterase subsequent to mAChR-stimulated increases in cyclic GMP. The rate of disappearance of cyclic AMP was measured in TSH-preactivated cells to which trypsin was added to decrease stimulatory hormone levels rapidly. This rate was increased by carbachol, again suggesting that phosphodiesterase was activated by mAChR stimulation.

Subsequent work by Dumont and coworkers suggested that effects on intracellular Ca^{2+} were responsible for the mAChR-mediated effects in thyroid slices (Van Sande et al., 1979; Decoster et al., 1980). The effects of carbachol disappeared in calcium-depleted slices or in the presence of calcium antagonists, and the divalent ion ionophore, A23187, mimicked the effect of mAChR on cyclic GMP levels. Since the stimulatory effects of mAChR agonists were not observed in the absence of Ca^{2+} and were mimicked by A23187, it was not possible to conclude whether the effects of mAChR stimulation were mediated through a Ca^{2+}-calmodulin- or a cyclic GMP-stimulated phosphodiesterase. However, based on the fact that selective inhibitors of calmodulin-sensitive phosphodiesterase blocked mAChR-mediated effects on cyclic AMP levels, it was concluded that cyclic AMP levels were decreased through a mAChR-stimulated increase in cytoplasmic Ca^{2+}, which in turn activates a Ca^{2+}-calmodulin-stimulated phosphodiesterase (Miot et al., 1984).

3.3. Phosphodiesterase in Astrocytoma Cells

Gross and Clark reported in 1977 that activation of a receptor expressing the general pharmacological properties of a mAChR resulted in a decrease in cyclic AMP accumulation in 1321N1 human astrocytoma cells. Meeker and Harden (1982) later used [^{3}H]quinuclidinyl benzilate ([^{3}H]QNB) to demonstrate directly the presence of mAChR on 1321N1 cells and extended the original observations of Gross and Clark on mAChR-stimulated cyclic AMP accumulation. Activation of the mAChR was shown to inhibit the accumlation of cyclic AMP stimulated by isoproterenol, PGE_1, or by forskolin. Inhibition could be observed either as an attenuation in cyclic AMP accumulation by addition of a mAChR agonist at the time of stimulatory hormone addition or as a rapid decrease in the level of accumulated cyclic AMP if the mAChR agonist was added 5 min after challenge by stimulatory agonist. Three observations were consistent with the effect of mAChR occurring at the level of degradation, rather than synthesis of cyclic AMP

1. No decrease in cyclic AMP accumulation occurred during mAChR stimulation in the presence of a phosphodiesterase inhibitor; indeed, mAChR-stimulated increases in cyclic AMP levels occurred (*see below*)
2. The time to reach maximal levels of cyclic AMP accumulation in the presence of isoproterenol was greatly decreased by mAChR stimulation
3. No mAChR-mediated inhibition of adenylate cyclase activity occurred in cell-free preparations from 1321N1 cells under conditions that were optimal for measuring inhibitory activity in membranes from other tissues, e.g., rat heart (Harden et al., 1982; Meeker and Harden, 1982) or NG108-15 neuroblastoma × glioma cells (Smith and Harden, 1984).

Using an assay previously developed to measure cyclic AMP degradation in intact 1321N1 cells (Su et al., 1976), Meeker and Harden (1982, 1983) directly demonstrated mAChR-mediated stimulation of phosphodiesterase activity. Since activation of β-adrenergic receptors results in up to 100-fold elevations of cyclic AMP accumulation in 1321N1 cells, propranolol can be added to inhibit adenylate cyclase in isoproterenol-pretreated cells, and the consequential disappearance of cyclic AMP is solely a function of phosphodiesterase activity. Coaddition of carbachol with propranolol was shown by Meeker and Harden (1982) to increase the cyclic AMP degradation rate constant from 0.3 min^{-1} to 1.0 min^{-1} (Fig. 3). Assuming that mAChR stimulation in 1321N1 cells has no effect on cyclic AMP synthesis, this magnitude of increase in cyclic nucleotide degradation can account quantitatively for the 60–80% reduction in cyclic AMP accumulation, and the approximately threefold decrease in the time required to reach maximal β-adrenergic receptor-stimulated accumulation of cyclic AMP.

3.4. Mechanism of Activation of Phosphodiesterase

Receptor-mediated regulation of phosphodiesterase in 1321N1 cells and thyroid slices was not without biological precedence. That is, the receptor for light, rhodopsin, in retinal rod outer segment was known to regulate a cyclic GMP phosphodiesterase (Stryer et al., 1981; Stryer and Bourne, 1986). This occurred through a transmembrane coupling function of the G-protein, transducin, by steps completely analogous to those described above for hormonal activation of adenylate cyclase. Thus,

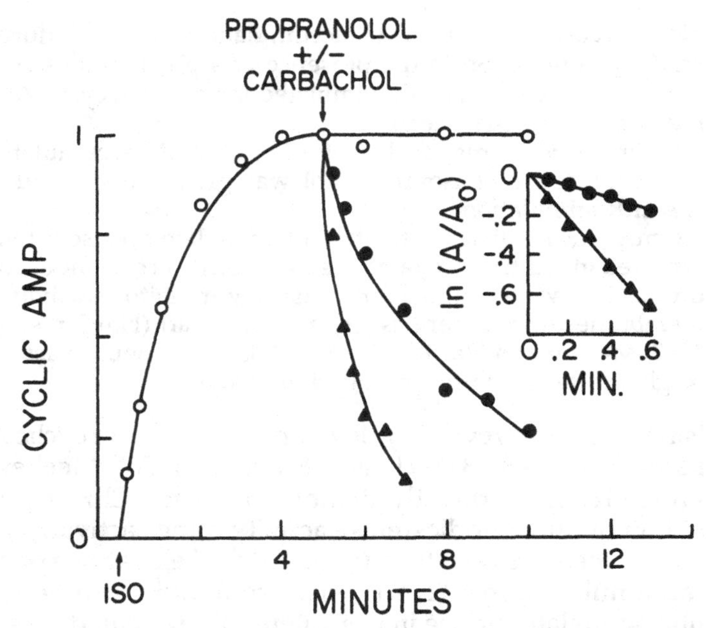

Fig. 3. Muscarinic receptor-mediated activation of phosphodiesterase. The data were obtained from 1321N1 human astrocytoma cells; specific methodology is discussed in Meeker and Harden (1982). Intracellular cyclic AMP levels were measured (units on ordinate are arbitrary) following challenge of cells with the β-adrenergic receptor agonist, isoproterenol (ISO). After 5-min incubation with ISO, the β-receptor antagonist propranolol (-●-) or propranolol in combination with the muscarinic receptor agonist carbachol (-▲-) was added and cyclic AMP levels measured as a function of time of continued incubation. Under these conditions, propranolol completely blocks ISO-stimulated cyclic AMP synthesis, there is essentially no basal adenylate cyclase activity, and there is no efflux of cyclic AMP from the cells. Therefore, changes in cyclic AMP levels are solely a reflection of the activity of cyclic AMP phosphodiesterase. The data from a similar experiment in which propranolol alone (-●-) or propranolol plus carbachol (-▲-) was added after isoproterenol prechallenge are presented as a first-order rate plot in the inset. The first-order rate constants for degradation were 0.3 min^{-1} and 1.0 min^{-1}, for the conditions without and with carbachol, respectively.

it was important to determine whether mAChR-stimulated activation of phosphodiesterase could be observed in cell-free preparations. Since mAChR previously had been shown to be coupled to a G-protein, and that protein tentatively identified as G_i, it also was of interest to determine whether the 1321N1 cell mAChR used a G-protein to couple to phosphodiesterase rather than to adenylate cyclase.

Meeker and Harden (1982) were unable to reproduce in membrane preparations the effects of mAChR stimulation on phosphodiesterase activity observed with intact cells. This suggested that the activation of phosphodiesterase required more than a transmembrane interaction of proteins and that a second-messenger-mediated activation of a soluble phosphodiesterase possibly was involved. The idea of coupling of the 1321N1 cell mAChR to mechanisms not involving G_i was further supported in experiments using pertussis toxin. Hughes et al. (1984) reported that two mechanisms of attenuation of cyclic AMP accumulation by mAChR could be distinguished clearly on the basis of their sensitivity to pertussis toxin. In NG108-15 neuroblastoma × glioma cells, pertussis toxin pretreatment reversed the inhibitory effects of carbachol on cyclic AMP accumulation with a concentration–effect curve similar to that for toxin-catalyzed ADP ribosylation of a 42-kdalton substrate. In contrast, pertussis toxin at concentrations four orders of magnitude greater than those necessary to fully ADP-ribosylate G_i had no effect on mAChR-mediated activation of phosphodiesterase in 1321N1 cells. Thus, not only did mAChR action in 1321N1 cells not involve a direct interaction with adenylate cyclase, it also apparently did not involve a G-protein with which mAChR had been shown previously to interact. Pertussis toxin also discriminates between the α-2 adrenergic receptor- and mAChR-mediated attenuation of cyclic AMP accumulation in dog thyroid slices; under conditions where toxin completely blocked the effects of norepinephrine, no effect on the action of carbachol was observed (Cochaux et al., 1985).

The initial work of Gross and Clark (1977) indicated that Ca^{2+} plays a role in mAChR action in 1321N1 cells in that incubation of cells in Ca^{2+}-free medium eliminated the inhibitory effect of carbachol on cyclic AMP accumulation, but had no effect on isoproterenol-stimulated cyclic AMP accumulation. This observation was extended by Meeker and Harden (1982). It was shown that the divalent cation ionophore A23187 decreases isoproterenol-stimulated cyclic AMP accumulation and that the phosphodiesterase inhibitor isobutylmethyl xanthine antagonizes the effects of ionophore. Moreover, in the absence of extracellular Ca^{2+} or in the presence of blockers of transmembrane movement of Ca^{2+}, the effects of mAChR stimulation were not observed.

Definition of the exact role of Ca^{2+} in mAChR action in 1321N1 cells proceeded on two fronts, one involving delineation of the probable isozyme form of phosphodiesterase that is involved in mAChR-mediated increases in cyclic AMP degradation, and the other involving clear definition of the primary second-messenger response system that is under control by mAChR in these cells.

Tanner et al. (1986) compared the relative effects of a series of selective phosphodiesterase inhibitors on intact cell responses to mAChR and β-adrenergic receptor stimulation to the relative effects of these drugs on the activities of three forms of phosphodiesterase purified from 1321N1 cells. Three antagonists were identified that selectively inhibited a Ca^{2+}-calmodulin-stimulated form of phosphodiesterase activity. Similar to the observations made previously in dog thyroid slices, these antagonists also were the most effective blockers of mAChR-stimulated activation of phosphodiesterase in intact cells. Two drugs that enhanced β-adrenergic receptor-stimulated cyclic AMP accumulation did not inhibit Ca^{2+}-calmodulin-regulated phosphodiesterase activity and had no effect on mAChR action in intact 1321N1 cells. Taken together, these pharmacological experiments suggested that it is a Ca^{2+}-calmodulin-regulated form of phosphodiesterase that is activated by mAChR in 1321N1 cells.

Receptor-stimulated hydrolysis of phosphatidylinositol 4,5-bisphosphate results in production of the second-messenger $Ins(1,4,5)P_3$ and the subsequent release of Ca^{2+} from intracellular stores (Berridge and Irvine, 1984). As is reviewed in detail in Chapter 7 of this volume, it recently has become clear that inositol phosphate/Ca^{2+} signaling accounts for mAChR action in a variety of tissues. Thus, based on the implied role of Ca^{2+} in mAChR-mediated activation of phosphodiesterase in 1321N1 cells, it was not surprising that activation of mAChR in 1321N1 cells also results in a marked increase in phosphoinositide hydrolysis measured as the accumulation of total inositol phosphates (Masters et al., 1984) or as an increase in Ins $(1,4,5)P_3$ (Nakahata and Harden, 1987). Moreover, elevation of cytoplasmic Ca^{2+} in response to mAChR agonists has been measured indirectly by prelabeling of Ca^{2+} stores with $^{45}Ca^{2+}$ (Masters et al., 1984) or directly with the fluorescent Ca^{2+} indicators, quin-2 (Hepler et al., 1987) or fura-2 (McDonough et al., 1987). Mobilization of intracellular stores of Ca^{2+} accounts for most, but probably not all (McDonough et al., 1987), of the increase in cytoplasmic Ca^{2+} in response to mAChR stimulation, suggesting that a mAChR-operated Ca^{2+} channel also may exist in 1321N1 cells.

The time course of Ins(1,4,5)P_3 formation and Ca^{2+} mobilization can account for the effects of mAChR on cyclic AMP degradation. Some discrepancies occurred, however, when the concentration dependence and maximal intrinsic activities of mAChR agonists were compared between the inositol phosphate and cyclic AMP responses. For example, oxotremorine which is a full agonist under most conditions for activation of phosphodiesterase (Meeker and Harden, 1982, 1983) exhibited a maximal intrinsic activity that was only 10% of that of carbachol for stimulation of inositol phosphate accumulation (Evans et al., 1985a). Harden et al. (1986a) illustrated that these discrepancies could be accounted for by "receptor reserve" in the activation of phosphodiesterase by Ca^{2+}. That is, a full agonist like carbachol apparently produces much more Ca^{2+} than is necessary for maximal activation of the phosphodiesterase. This is revealed as large shifts to the right of the phosphodiesterase activation curves for carbachol after receptor inactivation by propylbenzilylcholine mustard, after agonist-induced desensitization, or after treatment of cells with a phorbol ester.

Although mAChR-mediated activation of phosphodiesterase has only been studied in detail in dog thyroid slices and 1321N1 cells, this regulatory response is not restricted to these tissues. For example, Butcher and coworkers (Butcher, 1978; Barber et al., 1980) have presented kinetic data indicating that activation of mAChR on WI-38 fibroblasts results in an increase in cyclic AMP degradation. This conclusion was supported by the subsequent work of Nemecek and Honeyman (1982), who reported that the effect of carbachol on cyclic AMP levels in WI-38 fibroblasts was attenuated by the phosphodiesterase inhibitor, isobutylmethylxanthine. As intimated above, this receptor-regulated activation of a Ca^{2+}-calmodulin-sensitive phosphodiesterase also has been observed with other receptors that stimulate phosphoinositide hydrolysis and Ca^{2+} mobilization. For example, activation of α_1 adrenergic receptors in rat ventricular myocytes results in a decrease in cyclic AMP accumulation (Buxton and Brunton, 1985). The facts that phosphodiesterase inhibitors block the effect of α_1 receptor stimulation and that the rate of disappearance of cyclic AMP is enhanced by α_1 receptor agonists led to the conclusion that these receptors activate a cyclic AMP phosphodiesterase in the heart cells. The demonstration that α_1 receptor stimulation also increases inositol phosphate formation in the myocytes is consistent with the idea that the activation of phosphodiesterase occurs secondarily to mobilization of Ca^{2+} (Brown et al., 1985a).

The observation of mAChR-stimulated phosphodiesterase activity in a variety of tissues, together with the observation that activation of α-1 adrenergic receptors can produce a similar event, suggests that the model depicted in Fig. 4 summarizes a response to be expected of mAChR. The same model would potentially apply for any phospholipase C-linked receptor that increases phosphoinositide hydrolysis and cytoplasmic Ca^{2+} in cells possessing sufficient Ca^{2+}-calmodulin-regulated phosphodiesterase activity. Thus, activation of mAChR results in the breakdown of phosphatidylinositol 4,5-bisphosphate to Ins(1,4,5)P_3 and diacylglycerol. Elevation of cytoplasmic Ca^{2+} levels occurs as a consequence of the increased Ins(1,4,5)P_3 levels, and an activation of a Ca^{2+}-calmodulin-regulated phosphodiesterase ensues. Therefore, the target cell responses that proceed secondarily to the mobilization of Ca^{2+} and activation of protein kinase C do so in the presence of reduced levels of cyclic AMP. In essence, the Ca^{2+}-mediated activation of phosphodiesterase provides a means whereby a hormone acting through the phosphoinositide signaling pathway can at the same time "turn down" the activity of another important second-messenger-mediated pathway. Thus, in this important way, the cyclic AMP-lowering effects of mAChR by the two mechanisms discussed in detail in this review should not be considered equivalent. On the contrary, the mechanism involving inhibition of adenylate cyclase can be considered more direct and, in effect, less complicated, in that it occurs in a situation where the decrement in cyclic AMP levels may be the primary hormonally regulated event experienced by the target cell. In contrast, the lowering of cyclic AMP levels that occurs upon elevation of cytoplasmic Ca^{2+} levels probably is always experienced under conditions in which various other Ca^{2+}- and protein kinase C-activated processes occur.

Although the evidence in support of occurrence of a mAChR-stimulated cascade of events culminating in activation of a Ca^{2+}-calmoduin-regulated phosphodiesterase is strong, this may not be the only means of regulation of phosphodiesterase. For example, Hartzell and Fischmeister (1986) have proposed on the basis of relatively indirect data that mAChR-mediated effects on the slow inward Ca^{2+} current in frog ventricular cells results from

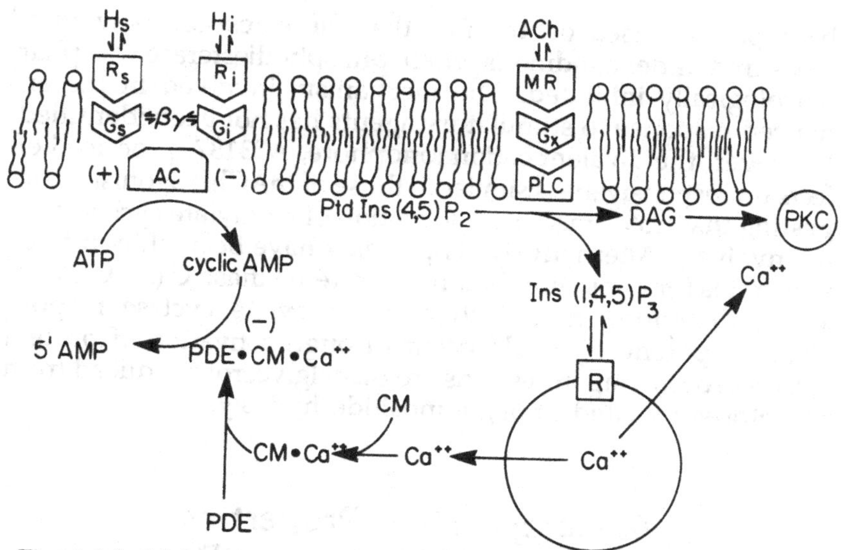

Fig. 4. Model for mAChR-stimulated activation of phosphodiesterase. Acetylcholine(ACh)-stimulated muscarinic receptors (MR) activate phospholipase C (PLC) through a yet-to-be-identified guanine nucleotide regulatory protein (G_x). As a result, PtdIns(4,5)P_2 is hydrolyzed to two second messengers: diacylglycerol (DAG), which activates protein kinase C (PKC), and Ins(1,4,5)P_3, which interacts with Ca^{2+} storage sites to release Ca^{2+}. It is proposed that the elevated cytoplasmic Ca^{2+} results in increased binding of Ca^{2+} to calmodulin (CM), which in turn results in activation of a CM-regulated cyclic AMP phosphodiesterase. Cyclic AMP is then hydrolyzed to 5′ AMP.

a cyclic GMP-stimulated phosphodiesterase that lowers intracellular cyclic AMP levels. Since elevation of cyclic GMP levels is a common response to mAChR activation, the extent of occurrence and properties of this proposed mechanism need to be examined in greater detail.

One complicating factor in measuring mAChR-mediated effects on cyclic AMP levels in target tissues in which the mAChR couples to phospholipase C is the potential modulation of adenylate cyclase responsiveness by one or more of the messengers produced as a consequence of phosphoinositide hydrolysis. For example, activation of mAChR in 1321N1 astrocytoma cells in the presence of a phosphodiesterase inhibitor greatly increases the cyclic AMP response to isoproterenol, PGE_1, and forskolin (Meeker and Harden, 1982; 1983). Kinetic analyses indicated that an alteration in cyclic AMP synthesis occurred, which also would be expected based on the fact that these responses were all measured under conditions where phosphodiesterase was pharmacologically inhibited. The mechanism involved in the enhanced response to stimulatory agents has not been identified. However, the divalent cation ionophore, A23187, produced effects on responsiveness similar to those of mAChR agonists, suggesting that Ca^{2+}-mediated activation of adenylate cyclase may be involved. Alternatively, G-proteins have been shown to be substrates for phosphorylation by protein kinase C (*see* Gilman, 1987). Therefore, modification of adenylate cyclase responsiveness potentially could occur through activation of protein kinase C occurring in response to diacylglycerol produced from receptor-stimulated phosphoinositide hydrolysis.

4. Comparative Properties of the Two Second-Messenger Response Systems Involved in mAChR-Mediated Lowering of Cyclic AMP Levels

Approaches for distinguishing between the two mechanisms of regulation of cyclic AMP levels by mAChR in intact cells have been implied from much of the discussion that has preceded. Thus, mAChR-mediated attenuation of cyclic AMP accumulation involving an inhibition of adenylate cyclase should not be modified by pharmacological antagonists of Ca^{2+}-calmodulin-stimulated phosphodiesterase. Likewise, pertussis toxin, which blocks G_i-mediated events and, thus, should block mAChR-mediated inhibition of adenylate cyclase, should have no effect on mAChR-mediated mobilization of intracellular Ca^{2+}. Other less straightforward differences in the two responses should be noted. For example, Brown and Brown (1984) have stressed that differences in maximal intrinsic activity of agonists occur for the two mAChR-activated mechanisms. Carbachol is a full agonist for both stimulation of phosphoinositide hydrolysis and inhibition of adenylate cyclase, whereas oxotremorine is a full agonist as an inhibitor of adenylate cyclase but a very poor activator of the phosphoinositide response. Since the activation of phosphodiesterase involves a multistep process expressing much "receptor reserve" in regard to activation of the enzyme by Ca^{2+} (at least in 1321N1 cells; Harden et al., 1986a), differences in apparent efficacy of oxotremorine vs carbachol might not be seen when activation of phosphodiesterase is measured. On the other hand, inactivation of mAChR as described by Harden et al. (1986a) would cause shifts to the right of the phosphodiesterase activation curve for carbachol and other full agonists, whereas the predominant effect observed on the oxotremorine activation curve would be a loss of maximal effect. As discussed in Chapter 7 of this volume, stimulation of phosphoinositide hydrolysis by mAChR usually shows little receptor reserve during inactivation of receptors by alkylation or down regulation, whereas a considerable reserve is encountered in the mAChR/adenylate cyclase mechanism.

As is considered in several venues in this volume, much interest has been directed toward establishing whether putative mAChR subtypes identified on the basis of their relative affinities for pirenzepine can be associated with one or the other of the two major biochemical responses activated by mAChR agonists. The overall view of the reported data is that they cannot. This is in spite of the fact that Gil and Wolfe (1985) published convincing data illustrating that the inhibition of adenylate cyclase by mAChR agonists in rat brain was blocked by pirenzepine at concentrations that consistently and markedly differed from those necessary for antagonism of the phosphoinositide response. Studies with other tissues have appeared, however, that either indicate that the apparent inhibitory affinity of pirenzepine is not different between the two kinds of responses (Brown et al., 1985b), or that pirenzepine antagonized the mAChR-mediated phosphoinositide response with differing affinities depending on the brain region (Lazareno et al., 1985; Fisher, 1986, and *see* Chapter 7).

Evans et al. (1984) have reported that pirenzepine shows little difference in affinity between the mAChR-mediating activation of phosphodiesterase in 1321N1 cells and the mAChR-mediating inhibition of adenylate cyclase in NG108-15 neuroblastoma × glioma cells. Furthermore, no remarkable difference in affinity between the two responses has been observed with a series of new putative subtype-selective antagonists (May and Harden, 1988).

Since no clear pharmacological delineation of receptor subtypes mediating the two second-messenger responses has been made, perhaps the same receptor protein mediates both second-messenger responses. However, this would not appear to be the case on several accounts. First, from a physiological point of view, this would not be an evolutionarily wise situation. The phosphoinositide/Ca^{2+} signaling response mediates a specific set of responses that differ from those mediated by cyclic AMP. Also, although it could be argued that a receptor-mediated decrease in cyclic AMP certainly provides a second-messenger response opposite to that for hormone receptors that stimulate adenylate cyclase, the events eventuating from the decrement in cyclic AMP would be in no way redundant with those occurring as a consequence of phosphoinositide hydrolysis. Furthermore, most mammalian cells contain the plasma membrane components necessary for both inhibition of adenylate cyclase and activation of phospholipase C. Thus, if a single mAChR type is capable of functionally coupling to both second-messenger response systems, both mechanisms would be activated upon interaction of agonists with mAChR in most cells. This does not appear to be the case, although this is a difficult problem to assess in mammalian tissues. Cloned cell lines have been used to address this question most clearly. For example, 1321N1 human astrocytoma cells (*see* Harden et al., 1986b) have been shown to express fully functional G_i as evidenced by:

1. The presence of a 41-kdalton substrate for pertussis toxin
2. The occurrence of guanine nucleotide-mediated inhibition of adenylate cyclase that is sensitive to inactivation by pertussis toxin (Evans et al., 1985b; Martin et al., 1985) and
3. The occurrence of pertussis toxin-sensitive adenosine receptor-mediated inhibition of adenylate cyclase in membranes and of isoproterenol-, forskolin, and PGE_1-stimulated accumulation of cyclic AMP in intact cells (Hughes and Harden, 1986).

Although these cells express a high concentration of otherwise "normal" mAChR, no mAChR-mediated inhibiton of adenylate cyclase occurs in membranes prepared from these cells (Meeker and Harden, 1982) and the mAChR-mediated inhibition of cyclic AMP accumulation in intact 1321N1 cells is completely blocked by phosphodiesterase inhibitors and is unaffected by pertussis toxin (Hughes et al., 1984). Similarly, pertussis toxin has no effect on mAChR-stimulated increases in inositol phosphate formation or Ca^{2+} mobilization in these cells (Masters et al., 1985). An opposite set of mAChR-related responses are observed in

NG108-15 cells. First, these cells express a functionally normal phosphoinositide/Ca^{2+} signaling system in that activation of bradykinin receptors results in marked increases in inositol phosphate formation and corresponding increases in cytoplasmic Ca^{2+} (Yano et al., 1985; Hepler et al., 1987). However, activation of mAChR on these cells affects neither phosphoinositide hydrolysis or Ca^{2+} levels (Hepler et al., 1987), but does markedly inhibit adenylate cyclase activity and cyclic AMP accumulation in a pertussis toxin-sensitive manner (Kurose et al., 1983; Hughes et al., 1984). Pertussis toxin also blocks GTP-sensitive high affinity binding of agonists to mAChR in membranes derived from NG108-15 cells (Kurose et al., 1983; Evans et al., 1985b). The most logical conclusion from these studies is that different mAChR exist on the two cells: one mAChR type is capable of coupling to G_i and adenylate cyclase, whereas the other mAChR type couples to the putative G-protein involved in activation of phospholipase C. Whether these mAChR types identified on the basis of their interaction with different G-proteins are truly selective in their interaction with the two second-messenger signaling systems is obviously not established. However, data published on the interaction of β-adrenergic receptors and rhodopsin with a series of G-proteins suggests that receptor-G-protein interaction is clearly selective, but is not necessarily absolute (Asano et al., 1984; Cerione et al., 1985).

The current view thus holds that at least two mAChR types exist that selectively activate one or the other of two second-messenger response systems. The response to a mAChR agonist experienced by any given target cell, therefore, would depend on the type of mAChR that is expressed by the cell. Whether a given target cell can express both mAChR types and, thus, both second-messenger responses has not been fully established. However, this would appear to be a likely occurrence, particularly in functionally complex target cells, e.g., neurons, myocardial cells, that receive input from multiple cholinergic neurons. Finally, molecular cloning studies indicate that at least four subtypes of mAChR exist (Bonner et al., 1987). Whether a single one of these receptors couples to the adenylate cyclase/cyclic AMP signaling system and another couples to phospholipase C is not yet known. However, it may turn out that several mAChR types couple to each of the second-messenger response systems with varying degrees of efficiency.

5. Conclusions

In summary, two mechanisms of mAChR-mediated attenuation of cyclic AMP accumulation occur. The most widely studied involves coupling of mAChR through G_i to inhibit adenylate cyclase. The second mechanism involves activation of a cyclic AMP phosphodiesterase and occurs secondary to a mAChR-stimulated rise in cytoplasmic Ca^{2+} levels. There is no doubt that the inhibition of adenylate cyclase represents a major signaling mechanism for mAChR. The mAChR-mediated attenuation of cyclic AMP accumulation that occurs as a consequence of activation of phosphodiesterase probably should be considered more as a "modulatory" mechanism in that its occurrence in most target cells would accompany all of the sequelae that follow from receptor-stimulated phosphoinositide hydrolysis. However, it is very possible that, because of the particular phenotype of certain target cells, the Ca^{2+}-activated increase in cyclic AMP degradation might represent the principal response to mAChR activation.

Progress in this area of research in the next several years will hopefully proceed with the following questions in mind. First, although it can be safely stated that mAChR inhibit adenylate cyclase and that this represents a major second-messenger response to mAChR activation, the mechanism whereby mAChR and other inhibitory receptors attenuate adenylate cyclase activity is still not fully understood. Thus, there is need to define unambiguously whether receptor activation results in inhibition of adenylate cyclase through the β or the α subunit of G_i. To date, the full significance of the mAChR-stimulated cyclic AMP phosphodiesterase mechanism has not been established. The extent to which this mechanism operates in tissues in which mAChR couple to phospholipase C is not clear. Furthermore, the physiological significance of the attenuation of cyclic AMP in the face of the other effects produced by phosphoinositide hydrolysis has not been established. Understanding of this mechanism at the level of the mAChR requires identification of the involved G-protein and its mechanism of receptor-stimulated activation of phospholipase C. It is reasonably clear that the two second-messenger responses discussed in this review emanate from at least two different mAChR types. However, the clear pharmacological and structural delineation of these receptors has not yet been made. With the substantial tools that rapidly are being introduced with the molecular biology of the mAChR, it is anticipated that, in the near future, a structurally defined receptor protein(s) can be unambiguously associated with each of the second-messenger responses. With the power then available with identified cloned mAChR, pharmacological agents can be identified that will permit the clear delineation of the two mAChR-mediated mechanisms under most conditions.

Acknowledgments

The author is indebted to Margaret Tapp, Amy Bunch, and Rachael Morgan for their assistance in preparation of the manuscript. Work from the author's laboratory was supported by USPHS grants 29536 and 38213.

References

Aktories, K. and Jakobs, K. H. (1981) Epinephrine inhibits adenylate cyclase and stimulates a GTPase in human platelet membranes via α-adrenoceptors. *FEBS Lett.* **130**, 235–238.

Aktories, K., Schultz, G., and Jakobs, K. H. (1982) Cholera toxin inhibits protaglandin E_1 but not adrenaline-stimulation of GTP hydrolysis in human platelet membranes. *FEBS Lett.* **146**, 65–68.

Aktories, K., Schultz, G., and Jakobs, K. H. (1983a) Islet activating protein impairs α_2-adrenoceptor mediated inhibitory regulation of human platelet adenylate cyclase. *Naunyn Schmiedebergs Arch. Pharmacol.* **324**, 196–200.

Aktories, K., Schultz, G., and Jakobs, K. H. (1983b) Somatostatin-induced stimulation of a high-affinity GTPase in membranes of S49 lymphoma cyc^- and $H21_a$ variants. *Mol. Pharmacol.* **24**, 183–188.

Asano, T., Katada, T., Gilman, A. G., and Ross, E. M. (1984) Activation of the inhibitory GTP-binding protein of adenylate cyclase, G_i, by β-adrenergic receptors in reconstituted phospholipid vesicles. *J. Biol. Chem.* **259**, 9351–9354.

Aurbach, G. D., Fedak, S. A., Woodward, C. J., Palmer, J. S., Hauser, D., and Troxler, F. (1974) β-Adrenergic receptor: Stereospecific interaction of iodinated β-blocking agent with a high affinity site. *Science* **186**, 1223–1224.

Barber, R., Ray, K. P., and Butcher, R. W. (1980) Turnover of adenosine 3′, 5′-monophosphate in WI-38 cultured fibroblasts. *Biochem.* **19**, 2560–2567.

Beavo, J. A., Hansen, R. S., Harrison, S. A., Hurwitz, R. L., Martins, T. J., and Mumby, M. C. (1982) Identification and properties of cyclic nucleotide phosphodiesterases. *Mol. Cell. Endocrinol.* **28**, 387–410.

Berridge, M. J. and Irvine, R. F. (1984) Inositol trisphosphate, a novel second messenger in signal transduction. **312**, 315–321.

Berrie, C. P., Birdsall, N. J. M., Burgen, A. S. V., and Hulme, E. C. (1979) Guanine nucleotides modulate muscarinic receptor binding in the heart. *Biochem. Biophys. Res. Commun.* **27**, 1000–1005.

Birnbaumer, L. (1987) Which G-protein subunits are the active mediators in signal transduction. *Trends Pharmacol. Sci.* **8**, 209–211.

Birnbaumer, L., Codina, J., Mattera, R., Cerione, R. A., Hildebrandt, J. D., Sunyer, T., Rojas, F. J., Caron, M. G., Lefkowitz, R. J., and Iyengar, R. (1985) Regulation of hormone receptors and adenylate cyclases by guanine nucleotide binding N-proteins. *Rec. Prog. Hormone Res.* **41**, 41–99.

Bokoch, G. M., Katada, T., Northup, J. K., Hewlett, E. L., and Gilman, A. G. (1983) Identification of the predominant substrate for ADP-ribosylation by islet activating protein. *J. Biol. Chem.* **258**, 2072–2075.

Bokoch, G. M., Katada, T., Northup, J. K., Ui, M., and Gilman, A. G. (1984) Purification and properties of the inhibitory guanine nucleotide-binding regulatory component of adenylate cyclase. *J. Biol. Chem.* **259**, 3560–3567.

Bonner, T. I., Buckley, N. J., Young, A. C., and Brann, M. R. (1987) Identification of a family of muscarinic acetylcholine receptor genes. *Science* **237**, 527–532.

Bourne, H. R., Coffino, P., and Tomkins, G. M. (1975) Selection of a variant lymphoma cell deficient in adenylate cyclase. *Science* **187**, 750–752.

Brown, B. S., Polson, J. G., and Krzanowski, J. J. (1979) Methacholine-induced attenuation of methylisobutylxanthine- and isoproterenol-elevated cyclic AMP levels in isolated rat atria. *Biochem. Pharmacol.* **28**, 948–951.

Brown, J. H. (1979) Cholinergic inhibition of catecholamine-stimulable cyclic AMP accumulation in murine atria. *J. Cyclic Nucleotide Res.* **5**, 423–433.

Brown, J. H. and Brown, S. L. (1984) Agonists differentiate muscarinic receptors that inhibit cyclic AMP formation from those that stimulate phosphoinositide metabolism. *J. Biol. Chem.* **259**, 2777–2781.

Brown, J. H., Buxton, I. L., and Brunton, L. L. (1985a) α_1-Adrenergic and muscarinic cholinergic stimulation of phosphoinositide hydrolysis in adult rat cardiomyocytes. *Circ. Res.* **57**, 532–537.

Brown, J. H., Goldstein, D., and Masters, S. B. (1985b) The putative M_1 muscarinic receptor does not regulate phosphoinositide hydrolysis. *Mol. Pharmacol.* **27**, 525–532.

Burns, D. L., Hewlett, E. L., Moss, J., and Vaughan, M. (1983) Pertussis toxin inhibits enkephalin stimulation of GTPase of NG108-15 cells. *J. Biol. Chem.* **258**, 1435–1438.

Butcher, R. W. (1978) Decreased cAMP levels in human diploid cells exposed to cholinergic stimuli. *J. Cyclic Nucleotide Res.* **4**, 411–421.

Buxton, I. L. and Brunton, L. L. (1985) Action of the cardiac α_1 receptor: activation of cyclic AMP degradation. *J. Biol. Chem.* **260**, 6733–6737.

Cassel, D., and Pfeuffer, T. (1978) Mechanism of cholera toxin action: Covalent modification of the guanyl nucleotide-binding protein of the adenylate cyclase system. *Proc. Natl. Acad. Sci. USA* **75**, 2669–2673.

Cassel, D. and Selinger, Z. (1977) Mechanism of adenylate cyclase activation by cholera toxin: Inhibition of GTP hydrolysis at the regulatory site. *Proc. Natl. Acad. Sci. USA* **74**, 3307–3311.

Cassel, D. and Selinger, Z. (1978) Mechanism of adenylate cyclase activation through the β-adrenergic receptor: Catecholamine-induced displacement of bound GDP by GTP. *Proc. Natl. Acad. Sci. USA* **75**, 4155–4159.

Cassel, D., Levkovitz, H., and Selinger, Z. (1977) The regulatory GTPase cycle of turkey erythrocyte adenylate cyclase. *J. Cyclic Nucleotide Res.* **3**, 373–406.

Cerione, R. A., Stainiszewski, C., Benovic, J. L., Lefkowitz, R. J., Caron, M. G., Gierschik, P., Somers, R., Spiegel, A. M., Codina, J., and Birnbaumer, L. (1985) Specificity of the functional interactions of the β-adrenergic receptor and rhodopsin with guanine nucleotide regulatory proteins reconstituted in phospholipid vesicles. *J. Biol. Chem.* **260**, 1493–1500.

Champion, S., Haye, B., and Jacquemin, C. (1974) Cholinergic control by endogenous prostaglandins of cAMP accumulation under TSH stimulation in the thyroid. *FEBS Lett.* **46**, 289.

Cochaux, P., Van Sande, J., and Dumont, J. E. (1985) Islet-activating protein discriminates between different inhibitors of thyroidal cyclic AMP systems. *FEBS Lett.* **179**, 303–306.

Codina, J., Hildebrandt, J. D., Iyengar, R., Birnbaumer, L., Sekura, R. D., and Manclark, C. R. (1983) Pertussis toxin substrate, the putative N_i component of adenylate cyclases, is an αβ heterodimer regulated by guanine nucleotide and magnesium. *Proc. Natl. Acad. Sci. USA* **80**, 4276–4280.

Cooper, D. M. F. (1982) Bimodal regulation of adenylate cyclase. *FEBS Lett.* **138**, 157–163.

Cooper, D. M. F., Schlegel, W., Lin, M. C., and Rodbell, M. (1979) The fat cell adenylate cyclase system. *J. Biol. Chem.* **254**, 8927–8930.

Decoster, C., Mockel, J., Van Sande, J., Unger, J., and Dumont, J. E. (1980) The role of calcium and guanosine 3′, 5′-monophosphate in the action of acetylcholine on thyroid metabolism. *Eur. J. Biochem.* **104**, 199–208.

Erneux, C., Van Sande, J., Dumont, J. E., and Boeynaems, J. M. (1977) Cyclic nucleotide hydrolysis in the thyroid gland. *Eur. J. Biochem.* **72**, 137–147.

Evans, T., Hepler, J. R., Masters, S. B., Brown, J. H., and Harden, T. K. (1985a) Guanine nucleotide regulation of agonist binding to muscarinic cholinergic receptors: relation to efficacy of agonists for stimulation of phosphoinositide breakdown and Ca^{2+} mobilization. *Biochem. J.* **131**, 751–757.

Evans, T., Martin, M. W., Hughes, A. R., and Harden, T. K. (1985b) Guanine nucleotide sensitive, high affinity binding of carbachol to muscarinic cholinergic receptors of 1321N1 astrocytoma cells is insensitive to pertussis toxin. *Mol. Pharmacol.* **27**, 32–37.

Evans, T., Smith, M. M., Tanner, L. I., and Harden, T. K. (1984) Muscarinic cholinergic receptors of two cell lines that regulate cyclic AMP metabolism by different molecular mechanisms. *Mol. Pharmacol.* **26**, 395–404.

Farfel, Z., Brothers, V. M., Brickman, A. S., Conte, F., Neer, R., and Bourne, H. R. (1981) Pseudohypoparathyroidism: inheritance of deficient receptor-cyclase coupling activity. *Proc. Natl. Acad. Sci. USA* **78**, 3098–3102.

Fisher, S. K. (1986) Inositol lipids and signal transduction at CNS muscarinic receptors. *Trends Pharmacol. Sci. (Suppl.)* **7**, 61–65.

Fleming, J. W., Strawbridge, R. A., and Watanabe, A. M. (1987) Muscarinic receptor regulation of cardiac adenylate cyclase activity. *J. Mol. Cell. Cardiol.* **19**, 47–61.

Gardner, R. M. and Allen, D. O. (1972) The relationship between cyclic nucleotide levels and glycogen phosphorylase activity in isolated rat hearts perfused with epinephrine and acetylcholine. *J. Pharmacol. Exp. Ther.* **202**, 346–353.

George, W. J., Wilkerson, R. D., and Kadowitz, P. J. (1973) Influence of acetylcholine on contractile force and cyclic nucleotide levels in the isolated perfused rat heart. *J. Pharmacol. Exp. Ther.* **184**, 228–235.

Gil, D. W. and Wolfe, B. B. (1985) Pirenzepine distinguishes between muscarinic receptor-mediated phosphoinositide breakdown and inhibition of adenylate cyclase. *J. Pharmacol. Exp. Ther.* **232**, 608–616.

Gill, D. M. and Meren, R. (1978) ADP-ribosylation of membrane proteins catalyzed by cholera toxin: Basis of the activation of adenylate cyclase. *Proc. Natl. Acad. Sci. USA* **75**, 3050–3054.

Gilman, A. G. (1984) Guanine nucleotide-binding regulatory proteins and dual control of adenylate cyclase. *J. Clin. Invest.* **73**, 1–4.

Gilman, A. G. (1987) G-proteins: transducers of receptor-generated signals. *Ann. Rev. Biochem.* **56**, 615–649.

Gross, R. A. and Clark, R. B. (1977). Regulation of adenosine 3′, 5′ monophosphate content in human astrocytoma cells by isoproterenol. *Mol. Pharmacol.* **13**, 242–250.

Haga, T., Ross, E. M., Anderson, H. J., and Gilman, A. G. (1977) Adenylate cyclase permanently uncoupled from hormone receptors in a novel variant of S49 mouse lymphoma cells. *Proc. Natl. Acad. Sci. USA* **74**, 2016–2020.

Harden, T. K., Heng, M. M., and Brown, J. H. (1986a) Receptor reserve in the calcium-dependent cyclic AMP response of astrocytoma cells to muscarinic receptor stimulation: demonstration by agonist-induced desensitization, receptor inactivation, and phorbol ester treatment. *Mol. Pharmacol.* **30**, 200–206.

Harden, T. K., Scheer, A. G., and Smith, M. M. (1982) Differential modification of the interaction of cardiac muscarinic cholinergic and β-adrenergic receptors with a guanine nucleotide binding site(s). *Mol. Pharmacol.* **21**, 570–580.

Harden, T. K., Tanner, L. I., Martin, M. W., Nakahata, N., Hughes, A. R., Hepler, J. R., Evans, T., Masters, S. B., and Brown, J. H. (1986b) Characteristics of two biochemical responses to stimulation of muscarinic cholinergic receptors. *Trends Pharmacol. Sci.* 7 **(supplement)**, 14–18.

Hartzell, H. C. and Fischmeister, R. (1986) Opposite effects of cyclic GMP and cyclic AMP on Ca^{++} current in single heart cells. *Nature* **323**, 273–275.

Harwood, J. P., Low, H., and Rodbell, M. (1973) Stimulating and inhibitory effects of guanyl nucleotides on fat cell adenylate cyclase. *J. Biol. Chem.* **248**, 6239–6245.

Hazeki, O. and Ui, M. (1981) Modification by islet-activating protein of receptor-mediated regulation of cyclic AMP accumulation in isolated rat heart cells. *J. Biol. Chem.* **256**, 2856–2862.

Heasley, L. E., Azari, J., and Brunton, L. L. (1985) Export of cyclic AMP from avian red cells: independence from major membrane transporters and specific inhibition by prostaglandin A_1. *Mol. Pharmacol.* **27**, 60–65.

Hepler, J. R., Hughes, A. R., and Harden, T. K. (1987) Evidence that muscarinic cholinergic receptors selectively interact with either the cyclic AMP or the inositol phosphate second messenger response systems. *Biochem. J.* **247**, 793–796.

Hildebrandt, J. D., Codina, J., and Birnbaumer, L. (1984) Interaction of the stimulatory and inhibitory regulatory proteins of the adenylate cyclase system with the catalytic component of cyc^- S49 mouse lymphoma cell membrane adenylate cyclase. *J. Biol. Chem.* **259**, 13178–13185.

Hildebrandt, J. D., Hanoune, J., and Birnbaumer, L. (1982) Guanine nucleotide inhibition of cyc^-S49 mouse lymphoma cell membrane adenylate cyclase. *J. Biol. Chem.* **257**, 14723–14725.

Hildebrandt, J. D., Sekura, R. D., Codina, J., Iyengar, R., Manclark, C. R., and Birnbaumer, L. (1983) Stimulation and inhibition of adenylate cyclase mediated by distinct regulatory proteins. *Nature* **302**, 706–709.

Hoffman, B. B. and Lefkowitz, R. J. (1980) Radioligand binding studies of adrenergic receptors: new insights into molecular and physiological regulation. *Ann. Rev. Pharmacol. Toxicol.* **20**, 580–608.

Hoffman, B. B., Mullikin-Kilpatrick, D., and Lefkowitz, R. J. (1980) Heterogeneity of radioligand binding to α-adrenergic receptors: analysis of guanine nucleotide regulation of agonist binding in relation to receptor subtypes. *J. Biol. Chem.* **255**, 4645–4652.

Hoffman, B. B., Yim, S., Tsai, B. S., and Lefkowitz, R. J. (1981) Preferential uncoupling by manganese of alpha-adrenergic receptor-mediated inhibition of adenylate cyclase in human platelets. *Biochem. Biophys. Res. Commun.* **100**, 724–731.

Hughes, A. R. and Harden, T. K. (1986) Adenosine and muscarinic cholinergic receptors attenuate cyclic AMP accumulation by different mechanisms in 1321N1 astrocytoma cells. *J. Pharmacol. Exp. Ther.* **237**, 173–178.

Hughes, A. R., Martin, M. W., and Harden, T. K. (1984) Pertussis toxin differentiates between two mechanisms of regulation of cyclic AMP accumulation by muscarinic cholinergic receptors. *Proc. Natl. Acad. Sci. USA* **81**, 5680–5684.

Iyengar, R. (1981) Hysteretic activation of adenylate cyclase II. Mg^{++} ion regulation of the activation of the regulatory component as analyzed by reconstitution. *J. Biol. Chem.* **256**, 11042–11050.

Iyengar, R. and Birnbaumer, L. (1982) Hormone receptor modulates the regulatory component of adenylate cyclase by reducing its requirement for Mg^{++} and enhancing its extent of activation by guanine nucleotides. *Proc. Natl. Acad. Sci. USA* **79**, 5179–5183.

Jakobs, K. H. (1979) Inhibition of adenylate cyclase by hormones and neurotransmitters. *Mol. Cell. Endocrinol.* **16**, 147–156.

Jakobs, K. H., Aktories, K., and Schultz, G. (1979) GTP-dependent inhibition of cardiac adenylate cyclase by muscarinic cholinergic agonists. *Naunyn-Schmiedebergs Arch. Pharmacol.* **310**, 113–119.

Jakobs, K. H., Aktories, K., and Schultz, G. (1983) A nucleotide regulatory site for somatostatin inhibition of adenylate cyclase in S49 lymphoma cells. *Nature* **303**, 177–178.

Jakobs, K. H., Aktories, K., and Schultz, G. (1984) Mechanisms and components involved in adenylate cyclase inhibition by hormones. *Adv. Cyclic Nucleotide Res.* **17**, 135–143.

Jakobs, K. H., Aktories, K., Minuth, M., and Schultz, G. (1985) Inhibition of adenylate cyclase. *Adv. Cyclic Nucleotide Res.* **19**, 137–150.

Jakobs, K. H., Saur, W., and Schultz, G. (1976) Reduction of adenylate cyclase activity in lysates of human platelets by the alpha-adrenergic component of epinephrine. *J. Cyclic Nucleotide Res.* **2**, 381–392.

Jakobs, K. H., Saur, W., and Schultz, G. (1978) Inhibition of platelet adenylate cyclase by epinephrine requires GTP. *FEBS Lett.* **85**, 167–170.

Jacobs, K. H., Lasch, P., Minuth, M., Aktories, K., and Schultz, G. (1982) Uncoupling of α-adrenergic-mediated inhibition of human platelet adenylate cyclase by *N*-ethylmaleimide. *J. Biol. Chem.* **257**, 2829–2833.

Johnson, G. L., Kaslow, H. R., Farfel, Z., and Bourne, H. R. (1980) Genetic analysis of hormone-sensitive adenylate cyclase. *Adv. Cyclic Nucleotide Res.* **13**, 1–38.

Katada, T. and Ui, M. (1980) Slow interaction of islet-activating protein with pancreatic islets during primary culture to cause reversal of α-adrenergic inhibition of insulin secretions. *J. Biol. Chem.* **255**, 9580–9588.

Katada, T. and Ui, M. (1981) Islet activating protein: a modifier of receptor-mediated regulation of rat islet adenylate cyclase. *J. Biol. Chem.* **256**, 8310–8317.

Katada, T. and Ui, M. (1982a) ADP ribosylation of the specific membrane protein of C6 cells by islet-activating protein associated with modification of adenylate cyclase activity. *J. Biol. Chem.* **257**, 7210–7216.

Katada, T. and Ui, M. (1982b) Direct modification of the membrane adenylate cyclase system by islet-activating protein due to ADP-ribosylation of a membrane protein. *Proc. Natl. Acad. Sci. USA* **79**, 3129–3133.

Katada, T., Bokoch, G. M., Northup, J. K., and Gilman, A. G. (1984a) The inhibitory guanine nucleotide-binding regulatory component of adenylate cyclase. Properties and function of the purified protein. *J. Biol. Chem.* **259**, 3567–3577.

Katada, T., Bokoch, G. M., Smigel, M. D., Ui, M., and Gilman, A. G. (1984b) The inhibitory guanine nucleotide-binding regulatory component of adenylate cyclase: subunit association and the inhibition of adenylate cyclase in S49 lymphoma cyc^- and wild type membranes. *J. Biol. Chem.* **259**, 3586–3595.

Katada, T., Northup, J. K., Bokoch, G. M., Ui, M., and Gilman, A. G. (1984c) The inhibitory guanine nucleotide-binding regulation component of adenylate cyclase. Subunit dissociation and guanine nucleotide-dependent hormonal inhibition. *J. Biol. Chem.* **259**, 3578–3585.

Kebabian, J. W., Steiner, A. L., and Greengard, P. (1975) Muscarinic cholinergic regulation of cyclic guanosine 3′, 5′-monophosphate in autonomic ganglia: possible role in synaptic transmission. *J. Pharmacol. Exp. Ther.* **193**, 474–488.

Koski, G. and Klee, W. A. (1981) Cyclic nucleotide-dependent protein kinases. *J. Biol. Chem.* **78**, 4185–4189.

Kuo, J. F. and Kuo, W. N. (1973) Regulation by β-adrenergic receptor and muscarinic cholinergic receptor activation of intracellular cyclic AMP and cyclic GMP levels in rat lung slices. *Biochem. Biophys. Res. Commun.* **55**, 660–665.

Kuo, J. F., Lee, T. P., Reyes, P. L., Walton, K. G., Donnelly, T. E., and Greengard, P. (1972) Cyclic nucleotide-dependent protein kinases. *J. Biol. Chem.* **247**, 16–22.

Kurose, H., Katada, T., Amano, T., and Ui, M. (1983) Specific uncoupling by islet-activating protein, pertussis toxin, of negative signal transduction via α-adrenergic, cholinergic, and opiate receptors in neuroblastoma × glioma hybrid cells. *J. Biol. Chem.* **258**, 4870–4875.

Lazareno, S., Kendall, D. A., and Nahorski, S. R. (1985) Pirenzepine indicates heterogeneity of muscarinic receptors linked to cerebral inositol phospholipid metabolism. *Neuropharmacol.* **24**, 593–595.

Lee, T. P., Kuo, J. F., and Greengard, P. (1972) Role of muscarinic cholinergic receptors in regulation of guanosine 3′:5′-cyclic monophosphate content in mammalian brain, heart muscle and intestinal smooth muscle. *Proc. Natl. Acad. Sci. USA* **69**, 3287–3291.

Lefkowitz, R. J., Mukherjee, C., Coverstone, M., and Caron, M. G. (1974) Stereospecific [^{3}H] (-)-alprenolol binding sites, β-adrenergic receptors and adenylate cyclase. *Biochem. Biophys. Res. Commun.* **60**, 706–709.

Levitzki, A., Atlas, D., and Steer, M. L. (1974) The binding characteristics and number of beta-adrenergic receptors on the turkey erythrocyte. *Proc. Natl. Acad. Sci. USA* **71**, 2773–2776.

Lichtshtein, D., Boone, G., and Blume, A. (1979) Muscarinic receptor regulation of NG108-15 adenylate cyclase: Requirement for Na^+ and GTP. *J. Cyclic Nucleotide Res.* **5**, 367–375.

Limbird, L. E. (1981) Activation and attenuation of adenylate cyclase. *Biochem. J.* **195**, 1–13.

Lohmann, S. M., Miech, R. P., and Butcher, F. R. (1977) Effects of isoproterenol, theophylline and carbachol on cyclic nucleotide levels and relaxation of bovine tracheal smooth muscle. *Biochim. Biophys. Acta* **499**, 238–250.

Londos, C., Cooper, D. M. F., Schlegel, W., and Rodbell, M. (1978) Adenosine analogs inhibit adipocyte adenylate cyclase by a GTP-dependent process: Basis for actions of adenosine and methylxanthines on cyclic AMP production and lipolysis. *Proc. Natl. Acad. Sci. USA* **75**, 5362–5366.

Martin, M. W., Evans, T., and Harden, T. K. (1985) Further evidence that muscarinic cholinergic receptors of 1321N1 astrocytoma cells couple to a guanine nucleotide regulatory protein that is not N_i. *Biochem. J.* **229**, 539–544.

Masters, S. B., Harden, T. K., and Brown, J. H. (1984) Relationships between phosphoinositide and calcium responses to muscarinic agonists in astrocytoma cells. *Mol. Pharmacol.* **26**, 149–155.

Masters, S. B., Martin, M. W., Harden, T. K., and Brown, J. H. (1985) Pertussis toxin does not inhibit muscarinic receptor-mediated phosphoinositide hydrolysis or calcium mobilization. *Biochem. J.* **227**, 933–937.

May, J. M. and Harden, T. K. (1988) Blockade of the muscarinic cholinergic receptor-stimulated inositol phosphate and cyclic AMP second messenger responses by antagonists of putative muscarinic receptor subtypes. Submitted for publication.

McDonough, P. M., Eubanks, J. H., and Brown, J. H. (1987) Desensitization and recovery of muscarinic and histaminergic calcium mobilization: evidence for a common hormone sensitive calcium store in astrocytoma cells *Biochem. J.* **249**, 135–141.

Meeker, R. B. and Harden, T. K. (1982) Muscarinic cholinergic receptor-mediated control of cyclic AMP metabolism: agonist-induced changes in nucleotide synthesis and degradation. *Mol. Pharmacol.* **23**, 384–392.

Meeker, R. B. and Harden, T. K. (1983) Muscarinic cholinergic receptor-mediated control of cyclic AMP metabolism: agonist-induced changes in nucleotide synthesis and degradation. *Mol. Pharmacol.* **23**, 261–266.

Miot, F. C., Erneux, C., Wells, J. N., and Aumont, J. E. (1984) The effects of alkylated xanthines on cyclic AMP accumulation in dog thyroid slices exposed to carbamylcholine. *Mol. Pharmacol.* **25**, 261–266.

Moss, J. and Vaughan, M. (1977) Mechanism of action of choleragen: evidence for ADP-ribosyltransferase activity with arginine as an acceptor. *J. Biol. Chem.* **252**, 2455–2462.

Motulsky, H. J., Hughes, R. J., Brickman, A. S., Farfel, Z., Bourne, H. R., and Insel, P. A. (1982) Platelets of pseudohypoparathyroid patients: Evidence that distinct receptor-cyclase coupling proteins mediate stimulation and inhibition of adenylate cyclase. *Proc. Natl. Acad. Sci. USA* **79**, 4193–4197.

Murad, F., Chi, Y.-M., Rall, T. W., and Sutherland, E. W. (1962) Adenyl cyclase. III. The effect of catecholamines and choline esters on the formation of adenosine 3′, 5′-phosphate by preparations of cardiac muscle and liver. *J. Biol. Chem.* **237**, 1231–1238.

Nakahata, N. and Harden, T. K. (1987) Regulation of inositol trisphosphate accumulation by muscarinic cholinergic and H_1-histamine receptors in human astrocytoma cells: differential induction of desensitization by agonists. *Biochem. J.* **241**, 337–344.

Nemecek, G. M. and Honeyman, T. W. (1982) The role of cyclic nucleotide phosphodiesterase in the inhibition of cyclic AMP accumulation by carbachol and phosphatidate. *J. Cyclic Nucleotide Res.* **8**, 395–408.

Northup, J. K., Smigel, M. D., and Gilman, A. G. (1982) The guanine nucleotide activating site of the regulatory component of adenylate cyclase: identification by ligand binding. *J. Biol. Chem.* **257**, 11416–11423.

Northup, J. K., Sternweis, P. C., Smigel, M. D., Schleifer, L. S., and Gilman, A. G. (1980) Purification of the regulatory component of adenylate cyclase. *Proc. Natl. Acad. Sci. USA* **77**, 6516–6520.

Pfeuffer, T. (1977) GTP-binding proteins in membranes and the control of adenylate cyclase activity. *J. Biol. Chem.* **252**, 7224–7234.

Pfeuffer, T. and Helmreich, E. J. M. (1975) Activation of pigeon erythrocyte membrane adenylate cyclase by guanyl nucleotide analogue and separation of a nucleotide binding protein. *J. Biol. Chem.* **250**, 867–876.

Robison, G. A., Butcher, R. W., and Sutherland, E. W. (1971) *Cyclic AMP.* Academic Press, New York.

Rodbell, M. (1975) On the mechanism of activation of fat cell adenylate cyclase by guanine nucleotides: an explanation for the biphasic inhibitory and stimulatory effects of the nucleotides and the role of hormones. *J. Biol. Chem.* **250**, 5826–5834.

Rodbell, M. (1980) The role of hormone receptors and GTP-regulatory proteins in membrane transductions. *Nature (London)* **284**, 17–22.

Rodbell, M., Birnbaumer, L., Pohl, S. L., and Krans, H. M. J. (1971) The glucagon-sensitive adenylate cyclase system in plasma membranes of rat liver. *J. Biol. Chem.* **246**, 1877–1882.

Rodbell, M., Lin, M. C., Salomon, Y., Londos, C., Harwood, J. P., Martin, B. R., Rendell, M., and Berman, M. (1975) Role of adenine and guanine nucleotides in the activity and response of adenylate cyclase systems to hormone: evidence of multisite transition states. *Advances Cyclic Nucleotide Res.* **5**, 3–29.

Rosenberger, L. B., Yamamura, H. I., and Roeske, W. R. (1980) Cardiac muscarinic cholinergic binding is regulated by Na^+ and guanyl nucleotides. *J. Biol. Chem.* **255**, 820–823.

Ross, E. M. and Gilman, A. G. (1977) Reconstitution of catecholamine-sensitive adenylate cyclase activity: Interaction of solubilized components with receptor-replete membranes. *Proc. Natl. Acad. Sci. USA* **74**, 3715–3719.

Ross, E. M. and Gilman, A. G. (1980) Biochemical properties of hormone-sensitive adenylate cyclase. *Annu. Rev. Biochem.* **49**, 533–564.

Ross, E. M., Howlett, A. C., Ferguson, K. M., and Gilman, A. G. (1978) Reconstitution of hormone-sensitive adenylate cyclase activity with resolved components of the enzyme. *J. Biol. Chem.* **253**, 6401–6412.

Smith, M. M. and Harden, T. K. (1984) Modification of inhibitory coupling of receptors to adenylate cyclase in NG 108-15 neuroblastoma × glioma cells by N-ethylmaleimide. *J. Pharmacol. Exp. Ther.* **228**, 425–433.

Smith, M. M. and Harden, T. K. (1985) The mechanism(s) of muscarinic cholinergic receptor-mediated attentuation of adenylate cyclase activity in rat heart membranes. *J. Cyclic Nucleotide Protein Phosphorylation Res.* **10**, 197–210.

Smith, S. K. and Limbird, L. E. (1981) Solubilization of human platelet α-adrenergic receptors: evidence that agonist occupancy of the receptor stabilizes receptor-effector interactions. *Proc. Natl. Acad. Sci. USA* **78**, 4026–4030.

Sternweis, P. C., Northup, J. K., Smigel, M. D., and Gilman, A. G. (1981) The regulatory component of adenylate cyclase. *J. Biol. Chem.* **256**, 11517–11526.

Strada, S. J., Martin, M. W., and Thompson, W. J. (1984) General properties of multiple molecular forms of cyclic nucleotide phosphodiesterase in the nervous system. *Adv. Cyclic Nucleotide and Protein Phosphorylation Res.* **16**, 13–30.

Stryer, L. and Bourne, H. R. (1986) G-proteins: a family of signal transducers. *Ann. Rev. Cell Biol.* **2**, 391–419.

Stryer, L., Hurley, J. B., and Fung, B. K.-K. (1981) First stage of amplification in the cyclic nucleotide cascade of vision. *Curr. Top. Membr. Transp.* **15**, 93–108.

Su, Y.-F., Johnson, G. L., Cubeddu, L. X., Leichtling, B. H., Ortmann, R., and Perkins, J. P. (1976) Regulation of adenosine 3′:5′ monophosphate content of human astrocytoma cells: mechanism of agonist-specific desensitization. *J. Cyclic Nucleotide Res.* **2**, 271–285.

Tanner, L. I., Harden, T. K., Wells, J. N., and Martin, M. W. (1986) Identification of the phosphodiesterase regulated by muscarinic cholinergic receptors of 1321N1 human astrocytoma cells. *Mol. Pharmacol.* **29**, 445–460.

Triner, L., Vulliemoz, Y., Verosky, M., and Nahas, G. G. (1972) Acetylcholine and the cyclic AMP system in smooth muscle. *Biochem. Biophys. Res. Commun.* **46**, 1866.

Ui, M. (1984) Islet activating protein, pertussis toxin: a probe for functions of the inhibitory guanine nucleotide regulatory components of adenylate cyclase. *Trends Pharmacol. Sci.* **5**, 277–279.

Van Sande, J., Decoster, C., and Dumont, J. E. (1979) Effects of carbamylcholine and ionophore A23187 on cyclic 3′, 5′-AMP and cyclic 3′, 5′-GMP accumulation in dog thyroid slices. *Mol. Cell. Endocrinol.* **14**, 45–57.

Van Sande, J., Erneux, C., and Dumont, J. E. (1977) Negative control of TSH action by iodide and acetylcholine: mechanism of action in intact thyroid cells. *J. Cyclic Nucleotide Res.* **3**, 335–345.

Watanabe, A. M., McConnaughey, M. M., Strawbridge, R. A., Fleming, J. W., Jones, L. R., and Besch, H. R. (1978) Muscarinic cholinergic receptor modulation of β-adrenergic receptor affinity for catecholamines. *J. Biol. Chem.* **253**, 4833–4836.

Wells, J. N. and Hardman, J. G. (1977) Cyclic nucleotide phosphodiesterase. *Adv. Cyclic Nucleotide Res.* **8**, 119–143.

Yamamura, H., Lad, P. M., and Rodbell, M. (1977) GTP stimulates and inhibits adenylate cyclase in fat cell membranes through distinct regulatory processes. *J. Biol. Chem.* **252**, 7964–7966.

Yano, K., Higashida, H., Hattori, H., and Nozawa, Y. (1985) Bradykinin-induced transient accumulation of inositol trisphosphate in neuron-like cell line NG108-15 cells. *FEBS Lett.* **181**, 403–406.

7

Muscarinic Cholinergic Receptor Regulation of Inositol Phospholipid Metabolism and Calcium Mobilization

Joan Heller Brown
and
Patrick M. McDonough

1. The Phosphoinositide Cycle

It is now well accepted that the breakdown of inositol phospholipids is a major mechanism for signal transduction in mammalian cells. The "PI" pathway is thought to generate second messengers responsible for effects of neurotransmitters, hormones, and growth factors on responses as diverse as muscle contraction, altered membrane potential, and cell growth. The history of the discovery that PI turnover is hormonally regulated, and the scientific trail to our present understanding of the importance of this pathway, is detailed in an excellent chapter by Michell (1986). It is noteworthy, in the context of this volume, that the earliest description of the PI response (Hokin and Hokin, 1953) concerned the observation that phosphatidylinositol metabolism was stimulated by acetylcholine, through a muscarinic cholinergic receptor.

Several excellent reviews describe our current understanding of how inositol phospholipid metabolism is regulated (Berridge,

1984; Hokin, 1985; Abdel-Latif, 1986; Berridge, 1987), and examine the evidence that $InsP_3$ is a second messenger in calcium mobilization (Berridge and Irvine, 1984; Berridge, 1986a) and that protein kinase C is an effector in this biological pathway (Nishizuka, 1986; Niedel and Blackshear, 1986). Only the key features of the inositol phospholipid pathway are described here, as schematized in Fig. 1. The inositol phospholipids shown on the left side of the figure are phosphatidylinositol (PI), the major inositol phospholipid in the cell, and its less abundant phosphorylated derivatives, the polyphosphoinositides (PIP and PIP_2). All of these lipids are possible substrates for an inositol phospholipid-specific phospholipase C, but under physiological conditions, agonists appear to increase specifically phospholipase C activity directed against PIP_2.

Phospholipase C releases inositol trisphosphate ($InsP_3$) from phosphatidylinositol 4,5-bisphosphate. The isomer of $InsP_3$ released from this lipid has phosphates in the 1, 4, and 5 positions. It is this 1,4,5 isomer of $InsP_3$ that has been shown to release calcium from intracellular sites and from isolated endoplasmic reticulum. There are two primary metabolic fates for Ins (1,4,5) P_3. It can be acted upon by a 5-phosphatase to yield Ins (1,4) P_2, a compound that is not very effective in releasing calcium, or it can be acted upon by a 3-kinase and thereby phosphorylated to inositol tetrakisphosphate ($InsP_4$), which is subsequently dephosphorylated to Ins (1,3,4) P_3 (Irvine et al., 1986). Inositol tetrakisphosphate and Ins (1,3,4) P_3 have some effects on calcium homeostasis (*see* Section 8), but their physiological roles have not yet been clearly elucidated. Inositol (1,3,4) trisphosphate is sequentially dephosphorylated to inositol bisphosphate, then to inositol monophosphate, and finally to inositol. The phosphomonoesterase that converts inositol 1-phosphate to inositol is inhibited by LiCl as originally described by Hallcher and Sherman (1980). Berridge et al. (1982) greatly simplified study of the PI response by describing an assay that used LiCl to allow the accumulation of inositol monophosphate and, thus, the detection of a signal resulting from inositol phospholipid hydrolysis in cells labeled with [^{3}H]inositol.

The other metabolite that serves as a second messenger for the PI pathway is diacylglycerol. The function of diacylglycerol as a second messenger was legitimized by independent discoveries in the laboratories of Nishizuka (Takai et al., 1979) and J. F. Kuo (Kuo et al., 1980) of a calcium- and phospholipid-

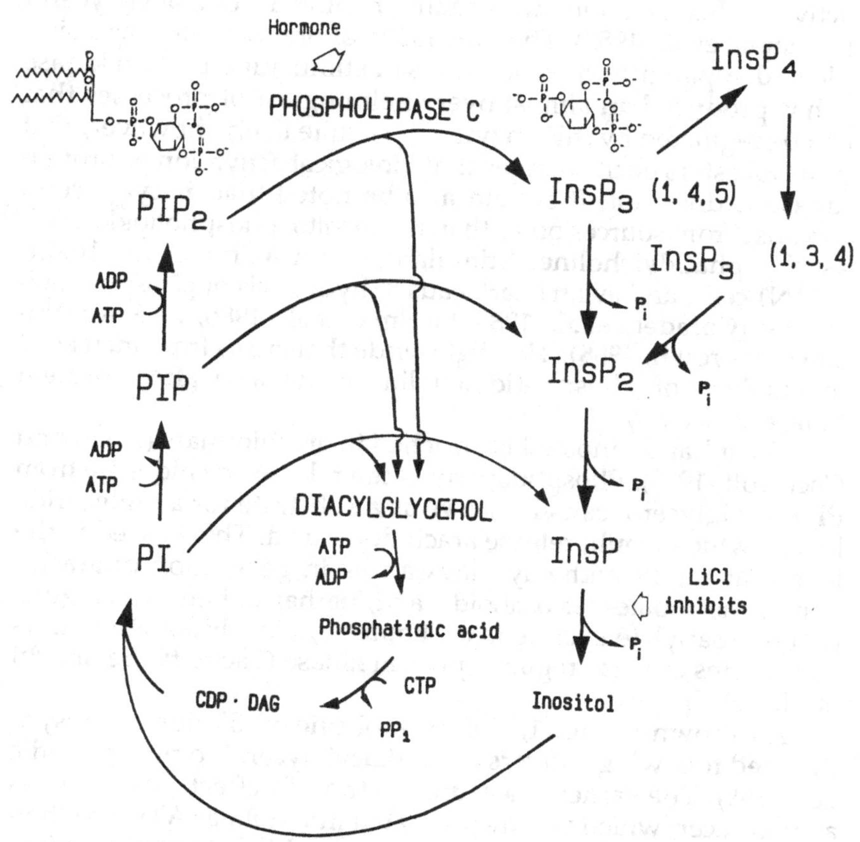

Fig. 1. The phosphoinositide cycle. Phosphatidylinositol 4,5 bisphosphate (PIP_2) is hydrolyzed by a phospholipase C to yield inositol 1,4,5 trisphosphate ($InsP_3$) and diacylglycerol. The $InsP_3$ is converted to other inositol phosphates by the action of kinases and phosphomonoesterases. Breakdown of InsP (inositol monophosphate) to inositol via inositol 1-phosphomonoesterase is inhibited by LiCl. Inositol reenters the cycle as PI, which is further phosphorylated to PIP_2.

dependent protein kinase. This kinase is activated by diacylglycerol, which lowers the calcium requirement for activity. The calcium- and phospholipid-dependent protein kinase(s), referred to as protein kinase C or C-kinase, have been cloned, and there now appears to be a family of C-kinase isozymes (Ono and Kikkawa, 1987). Protein kinase C is also the major cellular receptor for several tumor promoters, including the phorbol esters, which activate this enzyme in a manner similar to diacylglycerol (Castagna et al., 1982). The availability of phorbol esters as a simple and apparently specific means of stimulating protein kinase C has produced an almost overwhelming list of processes that can be regulated by this enzyme. It is quite likely, however, that phorbol esters elicit changes that biological activation of protein kinase C does not. It should also be noted that diacylglycerol can arise from sources other than the inositol phospholipids, such as phosphatidylcholine. Stimulation of mAChR in the brain, 1321N1 cells, and avian heart leads to hydrolysis of phosphatidylcholine (Coradetti et al., 1983; Lindmar et al., 1986, 1988; Martinson and Brown, 1988). The diglyceride that arises from increased metabolism of phosphatidylcholine could also affect protein kinase C activity.

Phosphatidylinositol is enriched in arachidonate (Allan and Cockcroft, 1983). Phospholipase A_2 can release arachidonate from PI; diacylglycerol can also serve as a substrate for a diglyceride lipase, which would release arachidonic acid. This has led to the idea that the PI pathway plays a role in generation of arachidonate metabolites (eicosanoids) and, perhaps, thereby in regulation of guanylate cyclase (*see* Chapter 8). Arachidonate and its metabolites can also regulate protein kinase C activity (Sekiguchi et al., 1987).

As shown in Fig. 1, the inositol phospholipids are resynthesized following conversion of diacylglycerol to phosphatidic acid (PA). The earliest measures of the "PI effect" used $^{32}P_i$ as a radiotracer, which was incorporated into cellular ATP and then into phosphatidic acid. Phosphatidic acid is activated by CTP, and a kinase catalyzes the transfer of inositol back onto the activated diglyceride, forming PI. Phosphorylation of PI to PIP and PIP_2 restores the levels of the polyphosphoinositides, which are dependent upon ATP for their maintenance.

2. Tissue Localization of Muscarinic Receptor-Stimulated Inositol Phospholipid Hydrolysis

The initial studies carried out by Hokin and Hokin demonstrated that muscarinic agonists increased ^{32}P incorporation into

PA and PI in a great number of tissues, including pancreas, brain, superior cervical ganglia, and avian salt gland (*see* Hokin, 1985; Hokin, 1987). Other pioneering workers in this field also examined changes in ^{32}P-labeled lipids in response to muscarinic agonists (Yagihara and Hawthorne, 1972; Schact and Agranoff, 1973; Jones and Michell, 1974; Jafferji and Michell, 1976; Abdel-Latif and Akhtar, 1976; Weiss and Putney, 1981). Phospholipid determinations are time-consuming, and the changes seen with muscarinic stimulation are generally small and often transient. In addition, the increase in ^{32}P incorporation into PA and PI measures a secondary event and not the primary receptor-mediated hydrolysis of the phospholipid. For these reasons, the information available from these early seminal studies is limited in comparison to what has emerged recently through analysis of the formation of the ^{3}H-labeled inositol phosphate products.

A partial list of tissues in which muscarinic receptor activation has been shown to be coupled to phosphoinositide hydrolysis is given in Table 1. Prominent among these are the effector sites innervated through the parasympathetic cholinergic division of the autonomic nervous system: glands, smooth muscle, and cardiac muscle. Salivary and secretory glands that have been studied in detail are the parotid (Jones and Michell, 1974; Oron et al., 1975; Weiss and Putney, 1981; Miller and Kowal, 1981; Downes and Wusteman, 1983; Aub and Putney, 1985; Irvine et al., 1984), pancreas (Calderon et al., 1979; Best and Malaisse, 1983; Rubin et al., 1984; Hokin-Neaverson and Sadghian, 1984; Streb et al., 1985; Trimble et al., 1987; Best et al., 1987), thyroid (Graff et al., 1987), and avian salt gland (Snider et al., 1986). Smooth muscles in which muscarinic agonists stimulate PI hydrolysis include those of the iris (Abdel-Latif and Akhtar, 1976; Akhtar and Abdel-Latif, 1984), trachea (Takuwa et al., 1986; Grandordy et al., 1986), gastrointestinal tract (Jafferji and Michell, 1976; Sekar and Roufagalis, 1984; Bielkiewicz-Vollraht et al., 1987; Gardner et al., 1988), myometrium (Marc et al., 1986), and vasculature (Sasaguri et al., 1986). Muscarinic-receptor stimulated phosphoinositide hydrolysis also occurs in the heart (Quist, 1982; Brown and Brown, 1983; Brown et al., 1985a; Brown and Jones, 1986; Poggioli et al., 1986; Ransnas et al., 1986; Woodcock et al., 1987b). We have previously commented (Brown and Masters, 1984) that it is paradoxical that this metabolic response would occur in the heart, because in this preparation the muscarinic response is "inhibitory" (acetylcholine decreases cardiac rate and contractility). However, recent evidence indicates that, under cer-

Table 1
Some Tissues in which Muscarinic Receptors Stimulate Inositol Phospholipid Hydrolysis

Parasympathetic effector sites	Nervous system	Continuous cell lines
Glands	Peripheral	PC12 (adrenal
Parotid	Sympathetic	medulla)
Pancreas	ganglia	1321N1
Thyroid	Adrenal medulla	(astrocytoma)
Avian salt		N1E115
gland		(neuroblastoma)
	Central	SK-N-SH
Smooth	Cortex	(neuroblastoma)
Muscle	Hippocampus	SH-SY5Y
Iris	Neostriatum	(neuroblastoma)
Trachea	Retina	NCB-20
Gastrointestinal		(neurohybridoma)
Myometrium		R1Nm5F (parotid
Vascular		acinar)
		FLOW 9000
		(pituitary)
Cardiac Muscle		C62B (glioma)
Atrium		
Ventricle		

tain conditions, acetylcholine can also increase contractility and depolarize cardiac muscle, an "excitatory" response that may be secondary to inositol phospholipid hydrolysis (Tajima et al., 1987) and associated with elevation of cytosolic Ca^{2+} (Korth et al., 1988).

The nervous system and tissues of neural origin also contain muscarinic receptors that regulate phosphoinositide metabolism. This includes sympathetic ganglia (Hokin, 1966; Lapetina et al., 1976; Bone et al., 1984; Bone and Michell, 1984; Horwitz et al., 1985, 1986), their analog, the adrenal medullary chromaffin cell (Mohd Adnan and Hawthorne, 1981; Forsberg et al., 1986; Malhotra et al., 1988), and the PC12 cell line derived from an adrenal chromaffin cell tumor (Vicentini et al., 1985a, 1986). Several laboratories have investigated the PI response in brain slices, nerve-ending preparations (synaptosomes or synaptoneurosomes), and brain cell cultures, and have used rat or guinea pig tissues to examine specific brain regions (e.g., cerebral cortex, hippocampus, and neostriatum) known to have cholinergic innervation (Hokin and Hokin, 1958; Schact and Agranoff, 1973; Fisher et al., 1980, 1981, 1983; Gonzales and Crews, 1984; Brown et al., 1984; Jacobson et al., 1985; Xu and Chuang, 1987; Audigier et al., 1988; White, 1988; Monsma et al., 1988). The question of whether the mAChR that regulate PI hydrolysis are presynaptic or postsynaptic was addressed by Fisher and his colleagues in studies in which they lesioned cholinergic input to the hippocampus (Fisher et al., 1980) or chemically destroyed the intrinsic cholinoceptive hippocampal neurons (Fisher et al., 1981). The conclusion from these studies is that the mAChR that regulate phosphoinositide turnover are derived from the cholinoceptive (i.e., postsynaptic) neurons. Brain slices and neonatal brain cell cultures contain glial as well as neuronal cells. The response to muscarinic agonists may be greater in the neuronal than in the glial elements of rat brain (Gonzales et al., 1985), but astrocyte-enriched rat brain cultures clearly show mAChR-stimulated inositol phosphate production (Pearce et al., 1985, 1986). Furthermore, the 1321N1 cell line, derived from a human astrocytoma, is very responsive to muscarinic agonists (Masters et al., 1984; Masters et al., 1985b; Nakahata and Harden, 1987; Ambler et al., 1987).

A variety of cell lines in continuous culture have muscarinic receptors that couple to PI hydrolysis. These include N1E-115, SK-N-SH and SH-SY5Y neuroblastoma cells (Cohen et al., 1983; Fisher and Snider, 1987; Serra et al., 1988; Mei et al., 1988), RINm5F acinar cells (Wolheim and Biden, 1986), NCB-20 neurohybridomas (Chuang, 1986), AtT-20 pituitary cells (Akiyama et al., 1986), C62B glioma cells (Brooks et al., 1987), and the PC-12 and 1321N1 cells mentioned above. There are also data suggesting that ionic responses to muscarinic agonists in *Xenopus* oocytes are mediated via the phosphoinositide system (Oron et al., 1985; Nomura et al., 1987), and retinal ganglion cells of *Xenopus* respond to ACh with an increase in PI metabolism (Anderson and Hollyfield, 1984).

It would appear from the extensive list cited above that PI hydrolysis is stimulated wherever muscarinic receptors are located. Michell et al. (1976) suggested such a conclusion in postulating that the phosphoinositide response may be intrinsic to the mechanism by which the muscarinic receptor operates. In light of recent evidence that there are at least five muscarinic receptor subtypes, it is gratifying to find some examples of tissues in which the mAChR that are present do not apper to regulate phosphoinositide turnover. These are the mAChR of the human erythrocyte (Sekar and Hokin, 1986) and that of the NG108 neuroblastoma cell; in NG108 cells, the mAChR couple to adenylate cyclase, but not to phospholipase C (Hepler et al., 1987; and *see* Chapter 6).

3. Relationship Between Agonist Occupancy and the Phosphoinositide Response

Michell, Jafferji, and Jones published a paper in 1976 in which they compared the EC_{50} values for mAChR-stimulated phosphoinositide turnover with the EC_{50} values obtained from radioligand binding studies. They noted that the dose–response curve for phospholipid labeling by ACh or carbachol closely followed the receptor occupancy curve, and suggested that the phosphoinositide turnover response might be intrinsic to receptor binding. These relationships have subsequently been studied in more detail, using [^{3}H]InsP formation as a measure of phospholipase C activation, and assessing agonist binding in membranes or whole cells.

It has routinely been noted that high concentrations of cholinergic agonists are required to maximally stimulate PI hydrolysis. This is shown in Table 2, which includes data from various tissues in which the carbachol dose–response curve for [^{3}H]InsP formation was determined. The values range from approximately 10 μM to 200μM, values similar to those obtained in the earlier phospholipid-labeling experiments. In describing these as high concentrations of carbachol, comparison is made with the concentrations required to elicit physiological responses such as contraction (Burgen and Spero, 1968; Grandordy et al., 1986), or with the concentration needed to affect other primary responses such as adenylate cyclase inhibition (Brown and Brown, 1984) or changes in K^+ conductance (Hunter and Nathanson, 1985).

The higher concentration requirement for the phosphoinositide response suggests that a high degree of receptor occupancy is needed to maximally activate phospholipase C (or the intermediate G-protein, *see below*). The relationship between

Table 2
EC_{50} Values for Carbachol-Stimulated Inositol Phosphate Formation in Various Tissues

Tissue	Carbachol EC_{50} (μM)*
Rat parotid (1;6)	3;10
Rat heart cell (2)	5
Cerebellar granule cell (3)	7
Guinea pig striatum (4)	8
Avian salt gland (5)	8
PC12 cells (7)	12
Guinea pig ventricle (8)	13
Guinea pig myometrium (9)	15
Chick heart cell (10)	20
SK-N-SH cells (4)	30
Tracheal smooth muscle (11)	38
1321N1 cells (12)	30
Rat brain (1;13)	43;47
NCB-20 cells (14)	50
Rat hippocampus (15)	50
Rat cerebral cortex (6;16;17)	68;100;120
Guinea pig hippocampus (4)	90
Rat medulla pons (17)	151
N1E115 cells (4)	163
AtT-20 cells (18)	170
Guinea pig cerebral cortex (4)	200

*EC_{50} values for carbachol are taken from the following references: (1) Gil and Wolfe, 1985; (2) Brown et al., 1985; (3) Xu and Chuang, 1987; (4) Fisher and Snider, 1987; (5) Snider et al., 1986; (6) Jacobson et al., 1985; (7) Vicentini et al., 1986; (8) Woodcock et al., 1987a; (9) Marc et al., 1986; (10) Brown and Brown, 1984; (11) Grandordy et al., 1986; (12) Masters et al., 1984; (13) Gonzalez and Crews, 1984; (14) Chuang, 1986; (15) Eva and Costa, 1986; (16) Jope et al., 1987; (17) Lazareno et al., 1985; (18) Akiyama et al., 1986.

receptor occupancy and response can be ascertained using experiments like those described by Furchgott and Bursztyn (1967), in which receptor number is decreased by irreversible alkylation and changes in the functional response are then assessed. We used propylbenzilylcholine mustard (PrBCM) to alkylate muscarinic receptors, and examined changes in the phosphoinositide and cyclic AMP responses in chick heart cells (Brown and Goldstein, 1986). Our data demonstrated that carbachol failed to give a maximal PI response when receptor number was reduced by alkylation and that the magnitude of the response decreased progressively as the receptor number decreased with further alkylation. Reductions in the maximal PI response to muscarinic agonists are also observed when the mAChR number is decreased by long-term agonist exposure (Masters et al., 1985b) or treatment with cholinesterase inhibitors (Costa et al., 1986). This is typical of systems in which there is little receptor reserve, i.e., in which there is a nearly one-to-one relationship between the number of receptors occupied and the size of the response that is seen. By comparison, when more than 95% of the mAChR of chick heart cells are alkylated, the dose–response curve for inhibition of adenylate cyclase shifts rightward, but the maximal response is not depressed; thus, there is far greater receptor reserve for adenylate cyclase inhibition than for PI hydrolysis (Brown and Goldstein, 1986).

Fisher and Snider (1987) recently carried out experiments comparing the PI response in slices from different brain regions and in two neuroblastoma cell lines. In guinea pig cortex and hippocampus and in N1E-115 cells, the EC_{50} for the PI response is high (Table 2), and receptor alkylation studies demonstrate little or no receptor reserve for this response. It is interesting to note, however, that in the neostriatum and in SK-N-SH cells, where the EC_{50} for carbachol is relatively low, partial receptor alkylation causes a rightward shift in the dose–response curve without decreasing the maximal response; only with greater alkylation does the maximal PI response decrease. Thus, there are differences in the degree of receptor reserve for the PI response among various tissues.

It should also be appreciated that the definition of receptor reserve depends upon which functional response in a possible sequence of responses is analyzed. In the 1321N1 astrocytoma cell, there is a little mAChR reserve for stimulation of PI hydrolysis; the same holds true for what appears to be the subsequent response, mobilization of intracellular calcium. However, there is significant "reserve" for the activation of cyclic AMP phosphodiesterase, which occurs in response to the increase in calcium (Harden et al., 1986; *see also* Chapter 6). Thus, it is possible to fully activate phosphodiesterase under conditions (or by agonists) that cause a limited PI and calcium response, a phenomenon that may be explained by amplification of the intracellular signal.

4. Differential Effects of Muscarinic Agonists on Phosphoinositide Hydrolysis

Several years ago, we demonstrated that oxotremorine was a full agonist for inhibiting adenylate cyclase in chick heart cells, whereas it was a weak partial agonist for stimulating phosphoinositide hydrolysis in this same preparation (Brown and Brown, 1984). At the same time, Fisher et al. (1983, 1984) compared a series of mAChR agonists and reported that they varied markedly in their ability to stimulate PI turnover in the brain. The efficacies of a series of agonists are compared to that of carbachol in Table 3, which summarizes data from several laboratories and different preparations. Carbachol, acetylcholine, and oxotremorine-M are similar in their ability to maximally stimulate inositol phospholipid hydrolysis, and methacholine and muscarine, where tested, are also nearly full agonists. A second group are partial agonists, and it is notable that there is much more variability in the response of various preparations to these agents. Bethanechol, for example, is 70–80% as effective as the full agonists in neostriatum and in SK-N-SH human neuroblastoma cells, but only 20–30% as effective in guinea pig cerebral cortex or N1E115 cells. Oxotremorine and arecoline are also partial agonists with greater efficacy in SK-N-SH cells and neostriatum.

The explanation for this variability may be that partial agonists need to occupy a greater number of receptors than a full agonist to give equivalent responses; the maximal response to a partial agonist is, therefore, most dependent on the absolute number of functional receptors. Indeed, the best responses to these partial agonists are seen in the systems in which some receptors reserve is demonstrable (Fisher and Snider, 1987). The agent McNA343 stimulates muscarinic receptors in ganglia and has been tested for its effects on PI hydrolysis because of its supposed selectivity for the M_1 receptor subtype (Goyal and Rattan, 1978). In systems in which the PI response to McNA343 has been examined, it is without agonist activity (Brown et al., 1985b; Gil and Wolfe, 1985; Fisher and Snider, 1987). However, it must bind to the receptor, because it blocks the PI response to other muscarinic agonists. The conclusion from these studies is that McNA343 is a very weak agonist, which is unlikely to effect

Table 3
Efficacies of Different Cholinergic Agonists on Muscarinic Receptor-Mediated Phosphoinositide Turnover*

	Rat cerebral cortex			Guinea pig Cortex	Guinea pig Striatum	1321N1	SK-N-SH	N1E115
	(1)	(2)	(3)	(4)	(5)	(6)	(7)	(8)
Carbamylcholine	100	100	100	100	100	100	(100)	(100)
Acetylcholine	140	119	100	135	ND	ND	ND	ND
Methacholine	96	ND	ND	83	ND	ND	86	ND
Muscarine	84	ND	ND	74	ND	ND	ND	ND
Bethanechol	59	ND	ND	19	68	49	77	27
Arecoline	42	ND	31	10	25	26	40	15
Pilocarpine	32	ND	14	9	6	2	18	9
Oxotremorine	23	12	12	12	40	11	22	8
Oxotremorine-M	ND	114	ND	ND	ND	ND	100	100

*Responses to the various agonists that were examined are expressed relative to the response to carbachol, which is taken as a 100% response, or in (7), as equal to Oxo-M. Data are from (1) Gonzales and Crews, 1984; (2) Jacobson et al., 1985; (3) Brown et al., 1984; (4) Fisher et al., 1984; (5) Fisher and Bartus, 1985; (6) Evans et al., 1985; (7) Fisher and Snider, 1987. ND indicates not determined.

responses through the PI pathway, although it may effect responses such as inhibition of adenylate cyclase where there is more efficient receptor–effector coupling.

5. Relationship of Muscarinic Receptor Subtypes to the Regulation of Phospholipase C

The evidence that there are muscarinic receptor subtypes is discussed in detail in Chapters 2, 3, and 4. Initially, the only evidence that there was more than a single type of mAChR was pharmacological, with subtypes defined by the occurrence of low- and high-affinity binding sites for pirenzepine (Hammer et al., 1980). High-affinity pirenzepine binding sites, termed M_1 receptors, were localized primarily in certain brain regions (cortex, striatum) and in ganglia, whereas mAChR in the heart and cerebellum had low affinity for pirenzepine and were termed M_2. The functional significance of these different receptors is still unknown. One hypothesis put forth was that the M_1 receptor subtype was coupled to phospholipase C, whereas the M_2 receptor was coupled to adenylate cyclase. This hypothesis has been tested in several systems, and the results are summarized in Table 4.

Studies carried out in our laboratory (Brown et al., 1985b) compared the effects of the antagonists atropine and pirenzepine on the dose–response relationships for carbachol-stimulated [^{3}H]InsP formation and for carbachol-induced inhibition of cyclic AMP production in intact chick heart cells. A range of antagonist concentrations was used, and Schild plots were generated to give K_i value for the antagonists. As expected, the nonselective antagonist atropine blocked both responses with nearly the same K_i (Table 4). Pirenzepine showed some discrimination, since the K_i values for the two responses differed fourfold. In contrast to what was initially proposed, however, pirenzepine had slightly higher affinity for blocking the cyclic AMP response than for inhibiting [^{3}H]InsP formation.

Gil and Wolfe (1985) carried out similar experiments at the same time. Initially they used the rat parotid to examine the PI response and the rat heart to examine adenylate cyclase inhibition. They also examined both responses in rat cortex. Gil and Wolfe demonstrated a 15–20-fold higher affinity of pirenzepine

Table 4
Selectivity of Pirenzepine vs Atropine for Phosphoinositide vs Cyclic AMP Responses

	Atropine			Pirenzepine			
	K_i (nM)			K_i (nM)			
	PI	cAMP	PI/cAMP selectivity	PI	cAMP	PI/cAMP selectivity	PZ/atropine relative selectivity
Brown et al., 1985b							
Chick heart cells	0.8	1.9	2	240	60	0.2	0.1
Gil and Wolfe, 1985							
Parotid vs. heart	1.0	6.0	6	124	2323	19	3.2
Brain	1.0	2.0	2	21	310	15	7.4
Woodcock et al., 1987a							
Atria	0.8	5.0	6	35	692	20	3.3
Ventricle	0.3	3.2	11	22	372	17	1.5
Akiyama et al., 1986							
AtT-20	0.3	0.4	1.3	7	280	40	31

for blocking the PI response (PI/cAMP selectivity; *see* Table 4). In the peripheral tissues, atropine also showed somewhat higher affinity (sixfold) for the PI response. Woodcock et al. (1987a) obtained similar results using guinea pig atria and ventricle. Like Gil and Wolfe, they found an approximately 20-fold higher affinity of pirenzepine for blocking the PI response. However they too found that atropine had unexpectedly higher affinity (6–11-fold) for blocking the PI response. Like Gil and Wolfe, Woodcock et al. examined the PI response in tissue slices and the cyclic AMP response by measuring inhibition of adenylate cyclase in broken cell homogenates; this may contribute in part to the "PI/cAMP selectivity" seen with atropine. If the selectivity seen with atropine is taken into account, the selectivity of pirenzepine for the PI response appears smaller, ranging between 1.5- and 7.4-fold.

Akiyama et al. (1986) studied AtT-20 cells using a protocol similar to that which we had applied in our work on the chick heart cell; both responses were measured in the intact cell and under similar conditions. In agreement with our findings, atropine blocked both responses with virtually the same K_i. In contrast to our findings with the chick cell, however, pirenzepine showed marked selectivity, inhibiting the PI response with 40-fold higher affinity than the cyclic AMP response. Thus, in these cells, it does appear that the M_1 receptor, as defined by pirenzepine affinity, regulates PI metabolism, whereas a receptor of lower affinity mediates inhibition of adenylate cyclase.

The data described above indicate that a receptor with high affinity for pirenzepine can regulate PI hydrolysis in some tissues, but that this functional response may be regulated through a receptor with relatively low affinity for pirenzepine in others. This conclusion is strengthened by the data summarized in Table 5, which show that the K_i for blockade of PI hydrolysis by pirenzepine ranges from 0.7 n*M*–240 n*M*. A value of 450 n*M* was obtained for the PI-linked mAChR in guinea-pig bladder, although in this tissue the K_i for atropine was also high (Noronhona-Blob et al., 1987). In rat striatum, the inhibition of PI-turnover by pirenzepine was best described by assuming that more than one mAChR subtype coupled to this response (Monsma et al., 1988). If pirenzepine affinity is used to distinguish M_1 and M_2 mAChR, one must conclude that either subtype can couple to inositol phosphate formation. Indeed, recent data from Ashkenazi et al. (1987) demonstrates that the M_2 receptor, ex-

Table 5
Inhibition Constants of Pirenzepine and Atropine for Blocking Phosphoinositide Hydrolysis*

	K_i (nM)		Ratio
	Atropine	Pirenzepine	PZ/Atr
Chick heart cell (1)	0.8	240	300
GP striatum (2)	0.5	160	320
Cerebellar granule cell (3)	0.5	120	240
SK-N-SH (4)	1.0	234	234
Parotid (5)	1.0	124	124
Salt gland (6)	0.5	90	180
GP ventricle (7)	0.3	20	66
NCB-20 (8)	0.5	25	50
N1E-115 (4)	0.3	15	51
Rat brain (5)	1.0	21	21
GP cortex (2)	0.5	12	24
Rat cortex (10)	0.8	16	20
PC12 (11)	1.5	16	11
AtT-20 (9)	0.3	0.7	2.3

*Data are from (1) Brown et al., 1985b; (2) Fisher and Bartus, 1985; (3) Xu and Chuang, 1987; (4) Fisher and Snider, 1987; (5) Gil and Wolfe, 1985; (6) Snider et al., 1986; (7) Woodcock et al., 1987a; (8) Chuang, 1986; (9) Akiyama et al., 1986; (10) Audigier et al., 1988; (11) Vicentini et al., 1986.

pressed in cells transfected with a vector containing porcine atrial M_2 mAChR cDNA, cannot only inhibit adenylate cyclase, but also activate inositol phospholipid hydrolysis if enough receptors are expressed. On the other hand, higher concentrations of agonist are required to elicit a response through the expressed M_2 (vs M_1) receptor, and the M_2 receptor-mediated ionositol phosphate response is smaller and less sustained (Peralta et al., 1988). Recent evidence indicates that the M_1 and M_2 receptors couple to phospholipase through different G-proteins (Ashkenazi et al., 1989).

6. Evidence that a GTP-Binding Protein Transduces Muscarinic Receptor Binding into Activation of Phospholipase C

The mechanism by which GTP-binding proteins couple receptors to adenylate cyclase is described in Chapter 6. It now appears that G-proteins play a role not only in coupling mAChR to adenylate cyclase, but also in coupling mAChR to K^+ and other ion channels (*see* Chapters 9 and 10) and to phospholipase C (Joseph, 1985; Litosch and Fain, 1986; Cockcroft, 1987). That muscarinic receptors couple to phospholipase C through a GTP-binding protein was first suggested by the observation that guanine nucleotides had effects on the binding of agonists to mAChR in tissues where the mAChR did not regulate adenylate cyclase. The 1321N1 astrocytoma cell is such an example. Harden and colleagues demonstrated that guanine nucleotides decreased the affinity of muscarinic agonists for the mAChR in these cells (Evans et al., 1985). The effect of GTP on agonist affinity was not lost following treatment of 1321N1 cells with pertussis toxin, an intervention known to block the effects of GTP on mAChR that couple to adenylate cyclase through G_i (Hughes et al., 1984). The effect of guanine nucleotides on binding of carbachol to mAChR in rat pancreatic islet cells is also insensitive to pertussis toxin (Dunlop and Larkins, 1986).

In 1321N1 cells, there is a good correlation between the ability of an agonist to induce high-affinity GTP-sensitive agonist binding and its ability to stimulate PI hydrolysis and calcium mobilization (Evans et al., 1985). Thus, carbachol and methacholine are highly effective at inducing (or binding to) a high-affinity state of the mAChR (*see* Chapter 2) and are full agonists for generating inositol phosphate formation. In contrast, agonists like oxotremorine and pilocarpine not only show low efficacy for increasing PI hydrolysis (as described in section 4 and *see* Chapter 5), but also fail to induce or bind to a significant fraction of mAChR with high affinity. The correlation between the ability of an agonist to cause PI hydrolysis and to induce a high-affinity state (presumably the state in which the receptor is coupled to a GTP-binding protein) suggests that interaction of the mAChR with the GTP-binding protein is necessary for the agonist–receptor complex to stimulate phospholipase C.

Additional evidence suggesting that a GTP-binding protein transduces responses in the PI pathway came from studies of Gomperts (1983), demonstrating that a secretory response could be elicited by guanine nucleotides introduced into permeabilized mast cells. At about the same time, Haslam and Davidson (1984) demonstrated that GTPγS increased the formation of diacylglycerol in the permeabilized platelet. Both observations suggested effects of GTP on a G-protein regulating inositol phospholipid hydrolysis, although they could also result from effects of GTP on the release process or on breakdown of other phospholipids. Direct evidence that guanine nucleotides affect inositol phospholipid hydrolysis came from the work of Cockcroft and Gomperts (1985), in which GTPγS was shown to decrease inositol phospholipid levels in neutrophil membranes. Evidence was also obtained by Litosch et al. (1985), who demonstrated that guanine nucleotides stimulated inositol phosphate formation in membranes from the blowfly salivary gland. Litosch et al. (1985) also made the important observation that the response to agonist stimulation depended upon the presence of guanine nucleotides, indicating that the agonist receptor coupled to phospholipase C through the G-protein.

The mAChR has recently been shown to couple to phospholipase C via a G-protein in several systems. Using electrically permeabilized pancreatic acinar cells, Merritt et al. (1986) showed that GTPγS and other guanine nucleotides stimulated $InsP_3$ accumulation and potentiated the effect of carbachol. We have shown similar effects of guanine nucleotides on basal and carbachol-stimulated $InsP_3$ accumulation in saponin-permeabilized chick heart cells (Jones et al., 1988) and adult rat cardiomyocytes (Jones and Brown, 1988).

Washed membrane preparations from the brain show GTP-sensitive inositol phosphate production, although these preparations fail to respond synergistically to mAChR agonists and GTP (Gonzales and Crews, 1985; Jope et al., 1987). In pancreatic islet cell membranes, carbachol stimulates the breakdown of the polyphosphoinositides most effectively when GTP is present (Dunlop and Larkins, 1986). In membranes from porcine coronary artery, the breakdown of the exogenous PIP_2 is also stimulated by acetylcholine to a greater extent if GTPγs is present (Sasaguri et al., 1986). Finally, studies carried out by Hepler and Harden (1986) and in our laboratory (Orellana et al., 1987) using washed membranes from 1321N1 astrocytoma cells show that carbachol has little effect on inositol phosphate formation when used alone, but potentiates the effect of guanine nucleotides on inositol phospholipid hydrolysis. This is what one would expect if the effect of agonist binding to the mAChR is obligatorily transduced through a G-protein to effect changes in phospholipase C.

The G-protein that regulates phospholipase C cannot yet be described in molecular terms, although some of its properties are known. Pertussis toxin does not block coupling of the mAChR to phospholipase C in chick heart cells (Masters et al., 1985a), 1321N1 astrocytoma cells (Masters et al., 1985a; Hepler and Harden, 1986), pancreas (Merritt et al., 1986; Dunlop and Larkins, 1986), guinea pig coronary artery smooth muscle cells (Sasaguri et al., 1986), PC12 cells (Vicentini et al., 1986), pituitary cells (Lo and Hughes, 1987), cerebellar granule cells (Xu and Chiang, 1987), or brain slices (Kelly et al., 1985). Several of these studies showed that pertussis toxin did block mAChR-mediated inhibition of adenylate cyclase or ribosylate protein(s) of approximately 41 kdaltons in the same cells (Masters et al., 1985a; Kelly et al., 1985; Hepler and Harden, 1986; Sasaguri et al., 1986; Lo and Hughes, 1987). There is one published report of blockade of mAChR-mediated inositol phosphate production in SH-SY5Y cells (Mei et al., 1988); however, inhibition was incomplete and most significant with a high concentration (10 μg/mL) of pertussis toxin. The G-protein that endogenously couples the mAChR to phospholipase C does not, therefore, appear to be functionally blocked by pertussis toxin.

Although the G-protein that endogenously couples the mAChR to phospholipase C has not been defined, the PI response can be "reconstituted" using various G-proteins. The most striking example of this is the development of a mAChR-mediated phosphoinositide response in normally unresponsive 3T3 fibroblasts that have been transformed with mutated *ras* protein (Chiarugi et al., 1985). One interpretation of these data and those published in a later paper by the same authors (Chiarugi et al., 1986) is that the *ras* protein is the G-protein coupling the mAChR to phospholipase C. The same conclusion was suggested by another group examining phosphoinositide responses to growth factors in cells transfected with genes encoding *ras* proteins (Wakelam et al., 1986). Although these data demonstrate that the *ras* protein is capable of transducing mAChR occupancy into a phosphoinositide response, they do not prove that it normally serves this function. Furthermore, in the *ras*-transformed 3T3 cells (Chiarugi et al., 1985), the effect of carbachol on $InsP_3$ formation was inhibited by pertussis toxin. Since *ras* is not a substrate for pertussis toxin, the G-protein mediating the PI response in these cells may actually be G_i or G_o, and the presence of *ras* may somehow cause these proteins to couple more effectively to phospholipase C. In CHO cells transfected with M_2 muscarinic receptor clones, PI hydrolysis is also activated through a pertussis toxin-sensitive protein (Ashkenazi et al., 1987,

1989). In this preparation, the pertussis toxin sensitivity is different for mAChR effects on adenylate cyclase and phospholipase C, suggesting that these cells may have two forms of G_i that selectively regulate adenylate cyclase or phospholipase C. Thus, a pertussis toxin-sensitive G_i, G_o, or other G-protein is capable of coupling mAChR to phospholipase C, although these pertussis toxin-sensitive proteins are apparently not the G-proteins normally activated by mAChR that regulate PI hydrolysis.

7. Regulation of Muscarinic Receptor-Mediated Phosphoinositide Hydrolysis by Protein Kinase C

Phosphoinositide metabolism is altered in cells treated with phorbol esters, suggesting that this pathway is regulated by protein kinase C. The most frequent finding is that agonist-stimulated increases in inositol phosphate metabolism are inhibited by phorbol ester treatment. This was initially shown in brain slices in which muscarinic as well as alpha-adrenergic agonists were used to stimulate PI hydrolysis (Labarca et al., 1984). We subsequently showed that mAChR-mediated inositol phosphate formation was inhibited by phorbol esters in 1321N1 cells (Orellana et al., 1985), and similar data have been obtained for mAChR-mediated inositol phosphate formation in PC12 cells (Vicentini et al., 1985b), SH-SY5Y cells (Serra et al., 1986), NCB-20 cells (Chuang, 1986), cerebellar granule cells (Yu and Chuang, 1987), and C62B cells (Brooks et al., 1987). Although phorbol ester treatment could affect the binding properties of the muscarinic receptor, as shown for N1E-115 neuroblastoma cells (Liles et al., 1986), this does not appear to be the primary mechanism underlying inhibition of the PI response in 1321N1 cells because mAChR binding properties are not altered (Orellana et al., 1985). Additionally, blockade of the PI response by phorbol esters is not confined to responses activated through the mAChR, but includes blockade of responses mediated through histamine, bradykinin, and alpha-adrenergic receptors (Labarca et al., 1984; Orellana et al., 1985, 1987).

We proposed that the site of phorbol ester action was distal to the mAChR, possibly at the level of the G-protein (Fig. 2). To test this proposal, we used membranes isolated from 1321N1

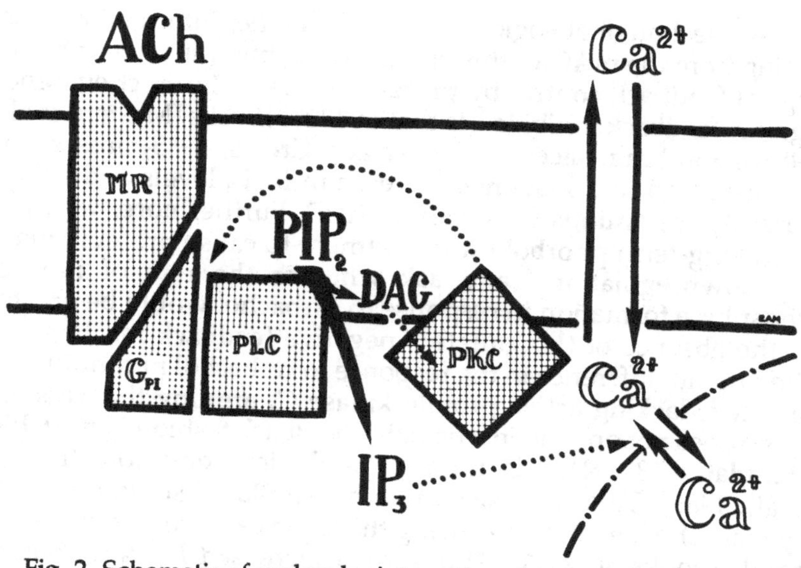

Fig. 2. Schematic of molecular interactions in muscarinic cholinergic receptor-stimulated phosphoinositide hydrolysis and second messenger formation. Abbreviations are: Acetylcholine (ACh); Muscarinic cholinergic receptor (MR); G-protein involved in PI response (G_{PI}); PI-specific phospholipase C (PLC); Inositol trisphosphate ($InsP_3$); Protein kinase C (PKC).

cells, and activated the phospholipase with exogenous guanine nucleotide. The ability of GTPγS to increase inositol phosphate accumulation was markedly attenuated in membranes from cells treated with phorbol ester (Orellana et al., 1987). The inhibition could also be mimicked by adding purified protein kinase C to the isolated membranes (Orellana et al., 1987), and was lost if protein kinase C was "downregulated" by long-term phorbol ester treatment (Brown et al., 1988). These data indicate that inhibition by phorbol ester was mediated through activation of protein kinase C and suggest that the G-protein might be phosphorylated. Experiments by Smith et al. (1987) using membranes from polymorphonuclear leukocytes also suggested that the G-protein rather than the receptor is the target of phorbol ester action. One cannot rule out, however, that protein kinase C might phosphorylate phospholipase C, thereby decreasing its ability to interact with the G-protein. Indeed, recent evidence has demonstrated that phospholipase C is a substrate for protein kinase C and that its activity is affected by phosphorylation (Bennett and Crooke, 1987).

A question of biological interest is whether information processing from the mAChR through the PI pathway is under physiological feedback control by protein kinase C. Indirect evidence against feedback inhibition is that in 1321N1 cells (Masters et al., 1985b) and brain slices (Gonzales and Crews, 1985) muscarinic agonists lead to a sustained increase in $InsP_3$ levels and an apparently continuous formation of InsP. Furthermore, we have used long-term phorbol ester treatment to cause protein kinase C "downregulation" and ask whether the rate of inositol phosphate formation in response to carbachol would be greater in the absence of this putative negative feedback mechanism. We find no difference in the response to carbachol in control cells and cells lacking active protein kinase C, arguing that this enzyme does not operate in a negative feedback fashion in mAChR-stimulated 1321N1 cells under physiological conditions (Brown et al., 1988). This is consistent with a similar observation made by Blackshear et al. (1987) using the same cell line. On the other hand, Brooks et al. (1987) observed increased $InsP_3$ formation in C62B cells treated with the protein kinase C inhibitor sphingosine. They interpreted this as the result of removing the tonic inhibition by protein kinase C.

8. Muscarinic Receptor Stimulation of Calcium Mobilization

Muscarinic agonists cause smooth muscle contraction (Somlyo and Somlyo, 1970; Ticku and Triggle, 1976; Siegel et al., 1984; Howe et al., 1986) and glandular secretion (Schulz and Stolze, 1980; Malaisse, 1986). Because these processes are initiated through elevation of cytoplasmic calcium (Rubin, 1982; Somlyo, 1985; Kamm and Stull, 1985; Rasmussen et al., 1987), it is supposed that muscarinic agonists stimulate these responses by increasing calcium availability.

Direct evidence that muscarinic stimulation alters calcium homeostasis was initially obtained through measurement of $^{45}Ca^{2+}$ fluxes in the pancreas (Case and Clausen, 1973; Stolze and Schulz, 1980), parotid gland (Miller and Nelson, 1977; Haddas et al., 1979), smooth muscle (Ticku and Triggle, 1976; Bitar et al., 1986), and astrocytoma cells (Masters et al., 1984). In preparations previously equilibrated with $^{45}Ca^{2+}$, muscarinic agents induce a marked increase in $^{45}Ca^{2+}$ efflux. This has been widely interpreted as evidence for agonist-induced mobilization of calcium from intracellular calcium depots into the cytoplasm, the mobilization elevating the cytoplasmic calcium concentration and increasing calcium export from the cell. In experiments in which $^{45}Ca^{2+}$ is added to unlabeled cells, muscarinic agents can also increase the influx of ^{45}Ca (Ticku and Triggle, 1976; Masters et al., 1984), suggesting the presence of receptor-operated calcium channels. Further evidence for a muscarinic-induced calcium influx is that the contractile or secretory response to muscarinic agonists may depend, in part, on the presence of extracellular calcium (Ticku and Triggle, 1976; Muallem and Sachs, 1984; Foraeus et al., 1987). Often, a muscarinic receptor-induced $^{45}Ca^{2+}$ efflux and influx can both be demonstrated for the same cell type (Poggioli and Putney, 1982; Masters et al., 1984).

The development of the fluorescent intracellular calcium indicators, quin-2 and fura-2 (Grynkiewicz et al., 1985), has made it possible to measure directly the cytoplasmic calcium concentration. Use of these compounds has yielded detailed records of muscarinic effects on cytoplasmic calcium concentrations in a number of preparations (Table 6). In smooth muscle, parotid gland, gastric gland, pancreas, and RINm5F and 1321N1 cells, the addition of carbachol or acetylcholine initiates a nearly instantaneous increase in cytoplasmic calcium concentration from resting levels of 100–200 n*M* to stimulated calcium concentrations of 300–1500 n*M*. A typical example of the pattern of change in cytoplasmic calcium is shown in Fig. 3. The initial increase in cytoplasmic calcium is transient, reaching a maximum within 5–20 s and then declining back to near basal levels in a few minutes. The increase generally occurs to nearly the same extent in the absence of extracellular calcium, confirming that it results from mobilization of calcium from intracellular stores. In many studies, a smaller sustained elevation in cytoplasmic calcium has also been observed to follow the initial peak response and remain as long as the agonist (carbachol or acetylcholine) is present or until an antagonist is added (*see* Fig. 3). The sustained component of the muscarinic calcium response, in contrast to the initial peak, depends upon extracellular calcium and may correspond to the muscarinic-induced calcium influx monitored with $^{45}Ca^{2+}$.

Some systems deviate from the typical mAChR-mediated calcium response pattern described above. Bovine adrenal medullary and chromaffin cells respond to muscarinic stimula-

Table 6
Calcium Responses to Muscarinic Stimulation Monitored with Quin-2 or Fura-2*

	Cytoplasmic calcium (n*M*)	
	Basal	Peak
Smooth muscle (1–4)	175	960
Pancreas (5–12)	100–200	300–2000
Parotid (13–15)	200	750
Submandibular gland (16)	77	280
Gastric gland (17–19)	100–140	500–750
Thyroid (20,21)	70	110
Adrenal medulla (22,23)	150	200
Adrenal chromaffin cell (24,25)	100–150	150–210
Avian salt gland (26)	200	760
1321N1 cell (27,28)	177–250	230–1500
RINm5F cell (29)	180	300
PC12 cell (30–33)	100–200	200–300
T_{84} cell (34)	120	160
HSG-PA cell (35)	120	230
IMR32 cell (36)	90	170
GH_3 cell (37)	100	60†

*Data are from: (1) Bitar et al., 1986; (2) Williams et al., 1987; (3) Reynolds and Dubyak, 1986; (4) Ishii et al., 1987; (5) Pandol et al., 1985; (6) Krims and Pandol, 1988; (7) Muallem et al., 1988a; (8) Pandol et al., 1987; (9) Ochs et al., 1985; (10) Trimble et al., 1987; (11) Chien and Warren, 1986; (12) Bruzzone et al., 1986; (13) Merritt and Rink, 1987a; (14) Merritt and Rink, 1987b; (15) Takemura, 1985; (16) Morris et al., 1987; (17) Negulescu and Machen, 1988a; (18) Muallem et al., 1986; (19) Chew, 1986; (20) Raspe et al., 1986; (21) Rani et al., 1985; (22) Misbahuddin et al., 1985; (23) Cheek and Burgoyne, 1985; (24) Kao and Schneider, 1986; (25) Kao and Schneider, 1985; (26) Snider et al., 1986; (27) Noronha-Blob et al., 1987; (28) McDonough et al., 1988; (29) Wollheim and Biden, 1986; (30) Pozzan et al., 1986; (31) Vicentini et al., 1986; (32) Reynolds and Dubyak, 1986; (33) Vicentini et al., 1985a,b; (34) Dharmsathaphorn and Pandol, 1986; (35) He et al., 1988; (36) Sher et al., 1988; (37) Schlegel et al., 1985.

†Cytoplasmic calcium was diminished by muscarinic stimulation in GH_3 cells.

tion by releasing calcium from intracellular stores, but only a small increase in cytoplasmic calcium (50–60 n*M* above basal levels) is seen. Relatively small muscarinic calcium responses are also observed with T_{84} epithelial cells and thyroid cells. In the avian salt gland, mAChR stimulation elicits a large increase in cytoplasmic calcium, which depends entirely upon extracellular calcium, whereas in GH_3 pituitary cells, muscarinic stimulation actually diminishes the cytoplasmic calcium concentration. In car-

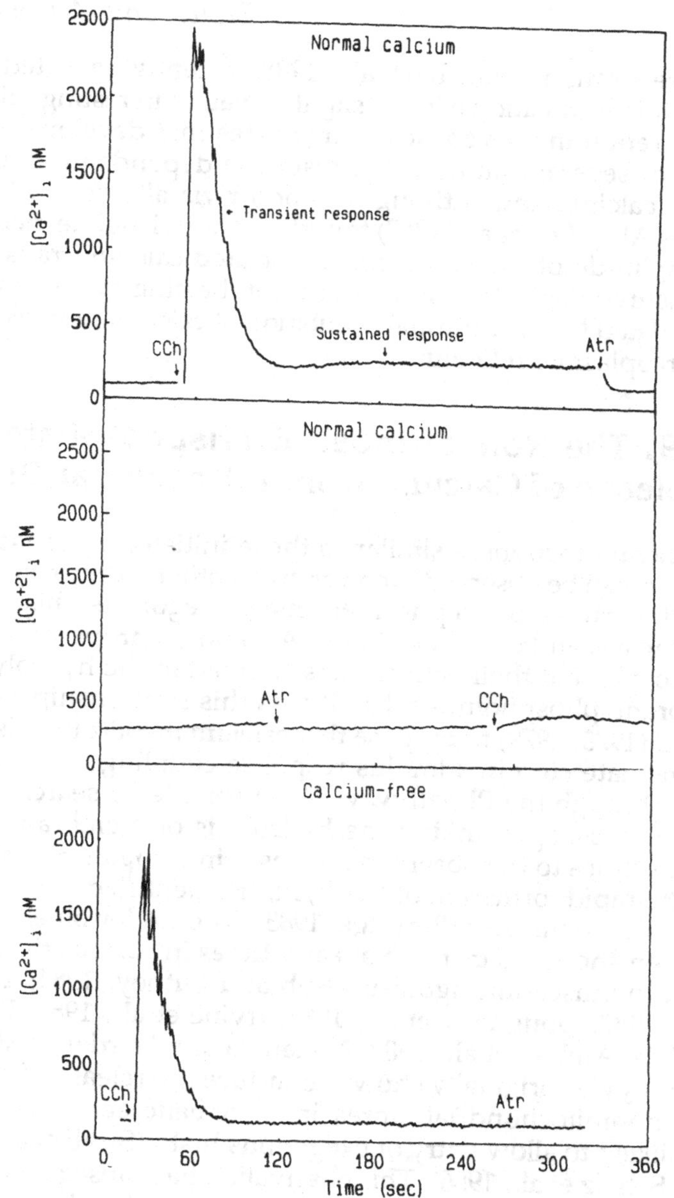

Fig. 3. Muscarinic elevation of cytoplasmic calcium concentration in 1321N1 astrocytoma cells. Fura-2 fluorescence (340 nm excitation, 510 nm emission) was monitored from suspensions of 1321N1 cells and calcium concentrations were calculated according to McDonough et al. (1988). Drug additions were carbachol (CCh, 300 *μM*) and atropine (Atr, .5 *μM*). The "calcium-free" buffer contained 1.8 m*M* $CaCl_2$ plus 6 m*M* EGTA.

diac preparations, Korth et al. (1988) recently reported that muscarinic stimulation elicits a slight elevation in resting calcium concentration in rat ventricular myocytes that develops over a period of several minutes; this response depends upon extracellular calcium and is thought to occur via altered Na:Ca exchange. Also, Lee et al. (1987) found that acetylcholine increases the amplitude of the contractile-associated calcium transients, as measured via indo-1 fluorescence, in beating rat hearts; this is postulated to occur through inhibition of calcium reuptake into the sarcoplasmic reticulum.

9. The Role of Inositol Trisphosphate in Release of Calcium from Intracellular Stores

Calcium responses similar to those initiated by muscarinic receptors can be observed for a variety of other receptor systems, including those for alpha-1 adrenergic agonists, histamine, angiotensin, and cholecystokinin. A common property of these receptors is that their activation is coupled to the hydrolysis of membrane phosphoinositides. It was this relationship that led Michell (1975, 1979) to propose that calcium mobilization is a key intermediate step in stimulus-response coupling for receptors acting through the PI pathway. This prompted a search for the second messenger linking the hydrolysis of membrane phosphoinositides to the observed increases in cytoplasmic calcium.

The rapid formation of Ins (1,4,5) P_3 identified it as a prime candidate for this role (Berridge, 1983). The 1,4,5-isomer of $InsP_3$ has been shown to increase at early times in various tissues exposed to muscarinic agonists (Aub and Putney, 1984; Trimble et al., 1987; Doughney et al., 1987; Irvine et al., 1985; Batty et al., 1985; Ambler et al., 1987; Nakahata and Harden, 1987). Ins (1,4,5) P_3 was originally shown to induce the release of calcium from nonmitochondrial stores in pancreatic acinar cells permeabilized to allow entry of exogenous $InsP_3$ (Streb et al., 1983, 1984; Schulz et al., 1986). This observation has subsequently been repeated on numerous cell preparations including hepatocytes, smooth muscle, and platelets (reviewed by Berridge, 1986b). Other metabolites from the PI cycle either do not release calcium in these preparations or are not as potent as the 1,4,5-isomer of $InsP_3$. In the heart, the 1,4,5-isomer of $InsP_3$ is formed in response to muscarinic stimulation (Renard and Poggioli, 1987; Woodcock et al., 1987b; Jones et al., 1988), but it is controversial whether or not $InsP_3$ induces calcium release (Hirata et al., 1984; Movsesian et al., 1985; Nosek et al., 1986; Benevolensky et al., 1987).

Through differential centrifugation of pancreatic cell homogenates, Streb et al. (1984) further determined that the cell fraction that contains the $InsP_3$-releasable store was also enriched with RNA and NADPH cytochrome c reductase, markers for the endoplasmic reticulum. In contrast, there was no correlation between mitochondrial markers and sensitivity to $InsP_3$. Thus, the endoplasmic reticulum is presumed to be the site of the agonist-sensitive calcium pool. Perhaps the most convincing demonstration of the link between the PI cycle and calcium mobilization was the experiments in which Streb et al. (1985) used permeabilized pancreatic acinar cells to compare $InsP_3$ formation and calcium release in response to carbachol. They showed that the formation of $InsP_3$ and release of calcium had a similar time course. Furthermore, inhibition of $InsP_3$ formation with neomycin suppressed calcium release, whereas inhibition of $InsP_3$ breakdown with 2,3-bisphosphoglycerate or by reducing the Mg^{2+} concentration resulted in greater calcium release.

Although considerable evidence has accrued suggesting that Ins (1,4,5) P_3 can act as a second messenger for calcium release from intracellular stores, there is far less known regarding the mechanism for agonist-induced calcium influx. Concepts that have been suggested include the idea that phosphatidic acid resulting from the PI cycle might act as a calcium ionophore, that calcium influx is actually a result of inhibition of calcium extrusion by the plasma membrane, or that calcium entry may occur nonspecifically via monovalent cation channels (reviewed by Putney, 1986a,b). Recently, Irvine and colleagues (1986, 1988) suggested that calcium entry into sea urchin eggs may be activated through the combined effects of $InsP_4$ and $InsP_3$. Consistent with this hypothesis, $InsP_4$ binding sites on the plasma membrane have recently been demonstrated (Bradford and Irvine, 1987). Other recent evidence suggests that Ins (1,4,5) P_3 can directly activate calcium entry under patch-clamp conditions (Kuno and Gardner, 1987). Both $InsP_4$ and the 1,3,4-isomer of $InsP_3$ are formed in response to muscarinic receptor stimulation in parotid (Irvine et al., 1985; Hawkins et al., 1986; Merritt and Rink, 1987b), brain (Batty et al., 1985), ileal smooth muscle (Bielkiewicz-Vollraht et al., 1987), and 1321N1 cells (Nakahata and Harden, 1987; Ambler et al., 1987). The possibility that an isomer of $InsP_3$ and/or $InsP_4$ regulates muscarinic receptor-stimulated calcium influx awaits confirmation.

10. Depletion and Refilling of Calcium Stores

As mentioned above, the calcium responses to muscarinic stimulation are similar to those observed with other agonists that act via the hydrolysis of membrane phosphoinositides. Thus, muscarinic receptors belong to a more general class of "calcium-mobilizing" receptor systems that operate via a common pathway. This relationship is confirmed by the study of systems in which two or more PI-linked receptor types are present. Examples include pancreatic and gastric gland cells (muscarinic and cholecystokin receptors; Schulz and Stolze, 1980; Pandol et al., 1985; Chew, 1986; Negulescu and Machen, 1988a; Negulescu and Machen, 1988b) parotid glands (muscarinic, alpha-1 adrenergic, and substance P receptors; Putney, 1977, Aub et al., 1982), and astrocytoma cells (muscarinic and histaminergic receptors; McDonough et al., 1988). In each of these systems, the calcium response to one PI-linked agent is quite similar to the responses to the other(s). Furthermore, in these studies, treatment of the tissues or cells with one agonist inhibited subsequent calcium responses to the second agonist. Detailed studies of the mechanism for this "heterologous" desensitization suggests that the initial agonist treatment inhibits the response to the second agent by depleting the agonist-sensitive calcium pool, which is shared by the two receptor systems.

In the parotid gland, gastric gland, pancreas, and 1321N1 astrocytoma cells, the presence of two calcium-releasing receptor types has been used to characterize the refilling of the agonist-releasable calcium pool. In each system, cells were exposed to carbachol followed by either substance P (parotid gland—Aub et al., 1982), cholecystokinin (pancreas—Muallem et al., 1988a,b), or histamine (astrocytoma cells—McDonough et al., 1988; gastric glands—Negulescu and Machen, 1988a). The second hormone did not elicit a calcium response unless atropine was added to block the actions of carbachol. In each case, the intervention of the muscarinic antagonist atropine resulted in a "recovery" of the response to the second calcium-mobilizing agonist, and this recovery was shown to depend upon the presence of extracellular calcium. The half-times for recovery, which are taken to be a measure of the times for refilling the agonist-sensitive calcium pools, ranged from 1 to 5 min. For pancreatic cells, $^{45}C^{2+}$ influx has been demonstrated to increase transiently after the termination of agonist exposure (Schulz and Stoltze, 1980; Muallem et al., 1988b), as if to replenish the agonist-releaseable store. La^{3+} will block the refilling process. Organic calcium channel antagonists, however, are only effective at a relatively high concentration (e.g., 1 m*M* verapamil) (Pandol et al., 1987; Negulescu and Machen, 1988b). This result suggests that, if a specific receptor-operated calcium channel is involved, it is distinct from the voltage-dependent calcium channels. At present, the exact route by which calcium enters the agonist-releaseable pool is in question; Putney (1986a) argues that the calcium may enter the agonist-sensitive calcium pool via a pathway independent of the cytoplasm, whereas other evidence has been obtained for participation of the cytoplasm in refilling of calcium stores (Negulescu and Machen, 1988b).

11. Role of Calcium Elevation in Muscarinic Responses

The biochemical and physiological processes initiated by muscarinic elevation of cytoplasmic calcium have been elucidated for several systems. For smooth muscle, the initial, transient elevation of cytoplasmic calcium correlates well with the phosphorylation of myosin light chain (Kamm and Stull, 1985). Furthermore, myosin light chain kinase can be activated by calmodulin complexed with calcium. Thus, muscarinic stimulation of contraction is thought to occur via the calcium-dependent phosphorylation of myosin light chain; the phosphorylated myosin then forms crossbridges with actin, and contraction is initiated. This model is not completely satisfactory, however, because the contractile force often continues to increase after the cytoplasmic calcium concentration and the level of phosphorylated myosin have maximized and begun to decline. One possible explanation for this is that contraction is sustained because of the formation of "latch-bridges." These have been proposed to result from activation of protein kinase C by calcium and diacylglycerol (Rasmussen et al., 1987).

A similar dual pathway has been proposed for the stimulation of amylase secretion in the pancreas and parotid gland (Pandol et al., 1985; Putney et al., 1984). Muscarinic stimulation results in an initial, rapid release of amylase thought to be induced by the calcium transient, followed by a slower sustained release of amylase. Since phorbol esters also initiate a sustained release of amylase from these cells, it has been proposed that DAG formation and the resulting protein kinase C activation is responsible for the slow, sustained release of amylase that follows the calcium transient. On the other hand, it has alternatively been proposed that protein kinase C may play an inhibitory role in amylase secretion (Pandol and Schoeffield, 1986).

A variety of additional physiological responses stimulated through the mAChR also appear to be secondary to calcium elevation. These include the activation of potassium (Marty, 1987) and chloride (Dascal et al., 1984; Fukuda et al., 1987) channels, the activation of cyclic AMP phosphodiesterase (Meeker and Harden, 1982; discussed in detail in Chapter 6), and the activation of phospholipase A_2 (DeGeorge et al., 1986), which may in turn lead to formation of arachidonate metabolites, production of cyclic GMP (*see* Chapter 8), and production of endothelium-derived relaxant factor (EDRF; Furchgott, 1984).

12. Conclusions and Future Directions

The data presented in this chapter demonstrate that, in most cells, mAChR activation can lead to the breakdown of inositol phospholipids and mobilization of calcium. The efficacy of weak agonists for stimulating inositol phosphate formation varies depending on the preparation, however, and there are some cells in which PI hydrolysis does not occur at all. What determines whether agonist stimulation of mAChR will cause PI hydrolysis in a given tissue? It is likely that the molecular structure of the mAChR will prove to be a very important determinant of this function. Although the simple division of mAChR into high and low affinity for pirenzepine fails to define the receptor that couples to phospholipase C, one must now recognize the existence of at least five mAChR subtypes with varying affinities for pirenzepine. Some subtypes may not be capable of coupling to phospholipase C, others may do so poorly, and one may be best suited to serve this function. Recent data comparing cells transfected with cloned DNA for each of four receptor subtypes suggest that this is indeed the case (Peralta et al., 1988). Why would receptors of differing structure vary in their ability to activate phospholipase C? This may be explained by small differences in mAChR structure that determine the efficacy with which an activated muscarinic receptor interacts with the GTP-binding protein G_{PI}.

Although "G_{PI}" has not yet been identified, there is little doubt that a G_{PI} exists, i.e., that there is a G-protein made to regulate the PI-specific phospholipase C. On the other hand, just as more than one mAChR may interact with G_{PI}, relationships between G-proteins and phospholipase C may not be exclusive. Thus, G-proteins other than G_{PI}, particularly if overexpressed, may also be able to recognize and activate phospholipase C, albeit with less efficiency. The future holds the challenge of discovering the structures of G_{PI} and the mAChR that regulates it, as well as elucidating the molecular events involved in interactions between the mAChR and G_{PI} and between G_{PI} and phospholipase C.

There is still much to be learned concerning the role of the PI pathway as a mediator of muscarinic responses. As with other second-messenger pathways, it will be a continuous challenge to prove that mAChR-mediated effects on inositol phospholipid hydrolysis and calcium mobilization are necessary and causal links to more distal cellular responses.

Acknowledgment

Research in the author's laboratory was supported by Grants GM 36927 and HL 28143 from the National Institutes of Health. We are very grateful to David Goldstein for his superb editorial assistance with this chapter and in preparing the entire volume.

References

Abdel-Latif, A. A. (1986) Calcium-mobilizing receptors, polyphosphoinositides, and the generation of second messengers. *Pharmacol. Rev.* **38**, 228–272.

Abdel-Latif, A. A. and Akhtar, R. A. (1976) Acetylcholine causes an increase in the hydrolysis of triphosphoinositide pre-labeled with [^{32}P]phosphate or [^{3}H]myo-inositol and a corresponding increase in the labeling of phosphatidylinositol and phosphatidic acid in rabbit iris muscle. *Biochem. Soc. Trans.* **4**, 317–321.

Akhtar, R. A. and Abdel-Latif, A. A. (1984) Carbachol causes rapid phosphodiesteratic cleavage of phosphatidylinositol 4,5-bisphosphate and accumulation of inositol phosphates in rabbit iris smooth muscle; prazosin inhibits noradrenalin- and ionophore A23187-stimulated accumulation of inositol phosphates. *Biochem. J.* **224**, 291–300.

Akiyama, K., Vickroy, T. W., Watson, M., Roeske, W. R., Reisine, T. D., Smith, T. L., and Yamamura, H. I. (1986) Muscarinic cholinergic ligand binding to intact mouse pituitary tumor cells (AtT-20/D16-16) coupling with two biochemical effectors: adenylate cyclase and phosphatidylinositol turnover. *J. Pharmacol. Exp. Ther.* **236**, 653–661.

Allan, D. and Cockcroft, S. (1983) The fatty acid composition of 1,2-diacylglycerol and polyphosphoinositides from human erythrocyte membranes. *Biochem. J.* **213**, 555–557.

Ambler, S. K., Solski, P. A., Brown, J. H., and Taylor, P. (1987) Receptor-mediated inositol phosphate formation in relation to calcium mobilization: A comparison of two cell lines. *Mol. Pharmacol.* **32**, 376–383.

Anderson, R. E. and Hollyfield, J. G. (1984) Inositol incorporation into phosphoinositides in retinal horizontal cells of Xenopus laevis: Enhancement by acetylcholine, inhibition by glycine. *J. Cell. Bio.* **99**, 686–691.

Ashkenazi, A., Peralta, E. G., Winslow, J. W., Ramachandran, J., and Capon, D. J. (1989) Functionally distinct G proteins selectively couple different receptors to PI hydrolysis in the same cell. *Cell* **56**, 487–493.

Askhenazi, A., Winslow, J. W., Peralta, E. G., Peterson, G. L., Schimerlik, M. I., Capon, D. J., and Ramachandran, J. (1987) An M2 muscarinic receptor subtype coupled to both adenylyl cyclase and phosphoinositide turnover. *Science* **238**, 672–675.

Aub, D. L. and Putney, J. W. (1984) Metabolism of inositol phosphates in parotid cells: Implications for the pathway of the phosphoinositide effect and for the possible messenger role of inositol trisphosphate. *Life Sci.* **34**, 1347–1355.

Aub, D. L. and Putney, J. W. (1985) Properties of receptor-controlled inositol trisphosphate formation in parotid acinar cells. *Biochem. J.* **225**, 263–266.

Aub, D. L., McKinney, J. S., and Putney, J. W. (1982) Nature of the receptor-regulated calcium pool in the rat parotid gland. *J. Physiol.* **331**, 557–565.

Audigier, S. M. P., Wang, J. K. T., and Greengard, P. (1988) Membrane depolarization and carbamoylcholine stimulate phosphatidylinositol turnover in intact nerve terminals. *Proc. Natl. Acad. Sci. USA* **85**, 2859–2863.

Batty, I. R., Nahorski, S. R., and Irvine, R. F. (1985) Rapid formation of inositol 1,3,4,5-tetrakisphosphate following muscarinic receptor stimulation of rat cerebral cortical slices. *Biochem. J.* **232**, 211–215.

Benevolensky, D. S., Menshikova, E. V., Watras, J., Levitsky, D. O., and Ritov, V. B. (1987) Characterization of Ca^{2+} release from the sarcoplasmic reticulum of myocardium and vascular smooth muscle. *Biomed. Biochim. Acta.* **8/9**, s393–s398.

Bennett, C. F. and Crooke, S. T. (1987) Purification and characterization of a phosphoinositide-specific phospholipase C from guinea pig uterus: phosphorylation by protein kinase C in vivo. *J. Biol. Chem.* **262**, 13789–13797.

Berridge, M. J. (1983) Rapid accumulation of inositol trisphosphate reveals that agonists hydrolyse polyphosphoinositides instead of phosphatidylinositol. *Biochem. J.* **212**, 849–858.

Berridge, M. J. (1984) Inositol trisphosphate and diacylglycerol as second messengers. *Biochem. J.* **220**, 345–360.

Berridge, M. J. (1986a) Inositol phosphates as second messengers, in *Phosphoinositides and Receptor Mechanisms* (Putney, J. W., Jr., ed.) Alan R. Liss, New York, 25–45.

Berridge, M. J. (1986b) Regulation of ion channels by inositol trisphosphate and diacylglycerol. *J. Exp. Biol.* **124**, 323–335.

Berridge, M. J. (1987) Inositol trisphosphate and diacylglycerol: two interacting second messengers. *Ann. Rev. Biochem.* **56**, 159–193.

Berridge, M. J. and Irvine, R. F. (1984) Inositol trisphosphate, a novel second messenger in cellular signal transduction. *Nature* **312**, 315–321.

Berridge, M. J., Downes, C. P., and Hanley, M. R. (1982) Lithium amplifies agonist-dependent phosphatidylinositol responses in brain and salivary glands. *Biochem.* **206**, 587–595.

Best, L. and Malaisse, W. J. (1983) Stimulation of phosphoinositide breakdown in rat pancreatic islets by glucose and carbamylcholine. *Biochem. Biophys. Res. Commun.* **116**, 9–16.

Best, L., Tomlinson, S., Hawkins, P. T., and Downes, C. P. (1987) Production of inositol trisphosphates and inositol tetrakisphosphate in stimulated pancreatic islets. *Biochim. Biophys. Acta.* **927**, 112–116.

Bielkiewicz-Vollraht, B., Carpenter, J. R., Schulz, R., and Cook, A. (1987) Early production of 1,4,5-inositol trisphosphate and 1,3,4,5-inositol tetrakisphosphate by histamine and carbachol in ileal smooth muscle. *Mol. Pharmacol.* **31**, 513–522.

Bitar, K. N., Bradford, P., Putney, J. W., and Makhlouf, G. M. (1986) Cytosolic calcium during contraction of isolated mammalian gastric muscle cells. *Science* **232**, 1143–1145.

Blackshear, P. J., Stumpo, D. J., Huang, J-K., Nemenoff, R. A., and Spach, D. H. (1987) Protein kinase C-dependent and -independent pathways of proto-oncogene induction in human astrocytoma cells. *J. Biol. Chem.* **262**, 7774–7781.

Bone, E. A. and Michell, R. H. (1984) Accumulation of inositol phosphates in sympathetic ganglia: effects of depolarization and of amine and peptide neurotransmitters. *Biochem. J.* **227**, 263–269.

Bone, E. A., Fretten, P., Palmer, S., Kirk, C. J., and Michell, R. H. (1984) Rapid accumulation of inositol phosphates in isolated rat superior cervical sympathetic ganglia exposed to V1-vasopressin and muscarinic cholinergic stimuli. *Biochem. J.* **221**, 803–811.

Bradford, P. G. and Irvine, R. F. (1987) Specific binding sites for [^{3}H]inositol (1,3,4,5) tetrakisphosphate on membranes of HL-60 cells. *Biochem. Biophys. Res. Commun.* **149**, 680–685.

Brooks, R. C., Morell, P., DeGeorge, J. J., McCarthy, K. D., and Lapetina, E. G. (1987) Differential effects of phorbol ester and diacylglycerols on inositol phosphate formation in C62B glioma cells. *Biochem. Biophys. Res. Commun.* **148**, 701–708.

Brown, E., Kendall, D. A., and Nahorski, S. R. (1984) Inositol phospholipid hydrolysis in rat cerebral cortical slices: I. Receptor characterization. *J. Neurochem.* **42**, 1379–1387.

Brown, J. H. and Brown, S. L. (1984) Agonists differentiate muscarinic receptors that inhibit cyclic AMP formation from those that stimulate phosphoinositide metabolism. *J. Biol. Chem.* **259**, 3777–3781.

Brown, J. H., Buxton, I. L., and Brunton, L. L. (1985a) α_1-Adrenergic and muscarinic cholinergic stimulation of phosphoinositide hydrolysis in adult rat cardiomyocytes. *Circ. Res.* **57**, 532–537.

Brown, J. H., Goldstein, D., and Masters, S. B. (1985b) The putative M1 muscarinic receptor does not regulate phosphoinositide hydrolysis: Studies with pirenzepine and McN-A343 in chick heart and astrocytoma cells. *Mol. Pharmacol.* **27**, 525–531.

Brown, J. H. and Goldstein, D. (1986) Differences in muscarinic receptor reserve for inhibition of adenylate cyclase and stimulation of phosphoinositide hydrolysis in chick heart cells. *Mol. Pharmacol.* **30**, 566–570.

Brown, J. H. and Jones, L. G. (1986) Phosphoinositide metabolism in the heart, in *Phosphoinositides and Receptor Mechanisms* (Putney, J. W., ed.) Alan R. Liss, New York, 245–270.

Brown, J. H. and Masters, S. B. (1984) Does phosphoinositide hydrolysis mediate 'inhibitory' as well as 'excitatory' muscarinic responses. *Trends Pharmacol. Sci.* **5**, 417–419.

Brown, J. H., Orellana, S. A., Buss, J. E., and Quilliam, L. A. (1988) Regulation of phosphoinositide hydrolysis by GTP-binding proteins, phorbol esters and botulinum toxin type D, in *Neuroreceptors and Signal Transduction* (Kito, S., ed.) Plenum Press, N.Y.

Brown, S. L. and Brown, J. H. (1983) Muscarinic stimulation of phosphatidylinositol metabolism in atria. *Mol. Pharmcol.* **24**, 351–356.

Bruzzone, R., Pozzan, T., and Wolheim, C. B. (1986) Caerulein and carbamylcholine stimulate pancreatic amylase release at resting cytosolic free Ca^{2+}. *Biochem. J.* **235**, 139–143.

Burgen, A. S. V. and Spero, L. (1968) The action of acetylcholine and other drugs on the efflux of potassium and rubidium from smooth muscle of the guinea-pig intestine. *Br. J. Pharmacol.* **34**, 99–115.

Calderon, P., Furnelle, J., and Christophe, J. (1979) In vitro lipid metabolism in the rat pancreas: III. Effects of carbamylcholine and pancreozymin on the turnover of phosphatidylinositols, 1,2-diacylglycerols and phosphatidylcholines. *Biochim. Biophys. Acta* **574**, 404–413.

Case, R. M. and Clausen, T. (1973) The relationship between calcium exchange and enzymatic secretion in the isolated rat pancreas. *J. Physiol.* **235**, 75–102.

Castagna, M., Takai, Y., Kaibuchi, K., Sano, K., Kiddawa, U., and Nishizuka, Y. (1982) Direct activation of calcium-activated, phospholipid-dependent protein kinase by tumor promoting phorbol esters. *J. Biol. Chem.* **257**, 7847–7851.

Cheek, T. R. and Burgoyne, R. D. (1985) Effect of activation of muscarinic receptors on intracellular free calcium and secretion in bovine adrenal chromaffin cells. *Biochim. Biophys. Acta* **846**, 167–173.

Chew, C. S. (1986) Cholecystokinin, carbachol, gastrin, histamine and forskolin increase $[Ca^{2+}]i$ in gastric cells. *Am. J. Physiol.* **250**, G814–823.

Chiarugi, V., Porciatti, F., Pasquali, F., and Bruni, P. (1985) Transformation of Balb/3T3 cells with EJ/T24/H-ras oncogene inhibits adenylate cyclase response to β-adrenergic agonist while increasing muscarinic receptor dependent hydrolysis of inositol lipids. *Biochem. Biophys. Res. Commun.* **132**, 900–907.

Chiarugi, V. P., Pasquali, F., Vannucchi, S., and Ruggiero, M. (1986) Point-mutated p21 ras couples a muscarinic receptor to calcium channels and polyphosphoinositide hydrolysis. *Biochem. Biophys. Res. Commun.* **141**, 591–599.

Chien, J. L. and Warren, J. R. (1986) Muscarinic receptor coupling to intracellular calcium release in rat pancreatic acinar carcinoma. *Cancer Res.* **46**, 5706–5714.

Chuang, D-M. (1986) Carbachol-induced accumulation of inositol-1-phosphate in neurohybridoma NCB-20 cells: Effects of lithium and phorbol esters. *Biochem. Biophys. Res. Commun.* **136**, 622–629.

Cockcroft, S. (1987) Polyphosphoinositide phosphodiesterase: regulation by a novel guanine nucleotide binding protein, Gp. *Trends Biochem. Sci.* **12**, 75–78.

Cockcroft, S. and Gomperts, B. D. (1985) Role of a guanine nucleotide binding protein in the activation of polyphosphoinositide phosphodiesterase. *Nature* **314**, 534–536.

Cohen, N. M., Schmidt, D. M., McGlennen, R. C., and Klein, W. L. (1983) Receptor-mediated increases in phosphatidylinositol turnover in neuron-like cell lines. *J. Neurochem.* **40**, 547–554.

Corradetti, R., Lindmar, R., and Loffelholz, K. (1983) Mobilization of cellular choline by stimulation of muscarinic receptors in isolated chicken heart and rat cortex in vivo. *J. Pharmacol. Exp. Ther.* **226**, 826–832.

Costa, L. G., Kaylor, G., and Murphy, S. D. (1986) Carbachol- and norepinephrine-stimulated phosphoinositide metabolism in rat brain: Effect of chronic cholinesterase inhibition. *J. Pharmacol. Exp. Ther.* **239**, 32–37.

Dascal, N., Landau, E. M., and Lass, Y. (1984) Xenopus oocyte resting potential, muscarinic responses and the role of calcium and guanosine 3′,5′-cyclic monophosphate. *J. Physiol.* **352**, 551–574.

DeGeorge, J. J., Morell, P., McCarthy, K. D., and Lapetina, E. G. (1986) Cholinergic stimulation of arachidonic acid and phosphatidic acid metabolism in C62B glioma cells. *J. Biol. Chem.* **261**, 3428–3433.

Dharmsathaphorn, K. and Pandol, S. J. (1986) Mechanism of chloride secretion induced by carbachol in a colonic epithelial cell line. *J. Clin. Invest.* **77**, 348–354.

Doughney, C., Brown, G. R., McPherson, M. A., and Dormer, R. L. (1987) Rapid formation of inositol 1,4,5-trisphosphate in rat pancreatic acini stimulated by carbamylcholine. *Biochim. Biophys. Acta* **928**, 341–348.

Downes, C. P. and Wusteman, M. M. (1983) Breakdown of polyphosphoinositides and not phosphatidylinositol accounts for muscarinic agonist-stimulated inositol phospholipid metabolism in rat parotid glands. *Biochem. J.* **216**, 633–640.

Dunlop, M. E. and Larkins, R. G. (1986) Muscarinic agonist and guanine nucleotide activation of polyphosphoinositide phosphodiesterase in isolated islet-cell membranes. *Biochem. J.* **204**, 731–737.

Eva, C. and Costa, E. (1986) Potassium ion facilitation of phosphoinositide turnover activation by muscarinic receptor agonists in rat brain. *J. Neurochem.* **44**, 540–543.

Evans, T., Hepler, J. R., Masters, S. B., Brown, J. H., and Harden, T. K. (1985) Guanine nucleotide regulation of agonist binding to muscarinic cholinergic receptors: Relation to efficacy of agonists for stimulation of phosphoinositide breakdown and Ca^{2+} mobilization. *Biochem. J.* **232**, 751–757.

Fisher, S. K. and Bartus, R. T. (1985) Regional differences in the coupling of muscarinic receptors to inositol phospholipid hydrolysis in guinea pig brain. *J. Neurochem.* **45**, 1085–1095.

Fisher, S. K. and Snider, R. M. (1987) Differential receptor occupancy requirements for muscarinic cholinergic stimulation of inositol lipid hydrolysis in brain and in neuroblastomas. *Mol. Pharmacol.* **32**, 81–90.

Fisher, S. K., Boast, C. A., and Agranoff, B. W. (1980) The muscarinic stimulation of phospholipid labeling in hippocampus is independent of its cholinergic input. *Brain Res.* **189**, 284–288.

Fisher, S. K., Figueiredo, J. C., and Bartus, R. T. (1984) Differential stimulation of inositol phospholipid turnover in brain by analogs

of oxotremorine. *J. Neurochem.* **43**, 1171–1179.
Fisher, S. K., Frey, K. A., and Agranoff, B. W. (1981) Loss of muscarinic receptors and of stimulated phospholipid labeling in ibotenate-treated hippocampus. *J. Neurosci.* **1**, 1407–1413.
Fisher, S. K., Klinger, P. D., and Agranoff, B. W. (1983) Muscarinic agonist binding and phospholipid turnover in brain. *J. Biol. Chem.* **258**, 7358–7363.
Foraeus, M., Anderson, K.-E., Batra, S., Morgan, E., and Sjogren, C. (1987) Effects of calcium, calcium channel blockers and BAY K8644 on contractions induced by muscarinic receptor stimulation of isolated bladder muscle from rabbit and man. *J. Urology* **137**, 798–803.
Forsberg, E. J., Rojas, E., and Pollard, H. B. (1986) Muscarinic receptor enhancement of nicotine-induced catecholamine secretion may be mediated by phosphoinositide metabolism in bovine adrenal chromaffin cells. *J. Biol. Chem.* **261**, 4915–4920.
Fukuda, K., Kubo, T., Akiba, I., Maeda, A., Mishina, M., and Numa, S. (1987) Molecular distinction between muscarinic acetylcholine receptor subtypes. *Nature* **327**, 623–625.
Furchgott, R. F. (1984) The role of endothelium in the responses of vascular smooth muscle to drugs. *Ann. Rev. Pharmacol. Toxicol.* **24**, 175–197.
Furchgott, R. F. and Bursztyn, P. (1967) Comparison of dissociation constants and of relative efficacies of selected agonists acting on parasympathetic receptors. *Ann. N.Y. Acad. Sci.* **144**, 882–899.
Gardner, A. L., Choo, L. K., and Mitchelson, F. (1988) Comparison of the effects of some muscarinic agonists on smooth muscle function and phosphatidylinositol turnover in the guinea-pig taenia caeci. *Br. J. Pharmacol.* **94**, 199–211.
Gil, D. W. and Wolfe, B. B. (1985) Pirenzepine distinguishes between muscarinic receptor-stimulated phosphoinositide breakdown and inhibition of adenylate cyclase. *J. Pharmacol. Exp. Ther.* **232**, 608–616.
Gomperts, B. D. (1983) Involvement of guanine-nucleotide binding protein in the gating of Ca^{2+} by receptors. *Nature* **306**, 64–66.
Gonzales, R. A. and Crews, F. T. (1984) Characterization of the cholinergic stimulation of phosphoinositide hydrolysis in rat brain slices. *J. Neurosci.* **4**, 3120–3127.
Gonzales, R. A. and Crews, F. T. (1985) Cholinergic- and adrenergic-stimulated inositide hydrolysis in brain: Interaction, regional distribution, and coupling mechanisms. *J. Neurochem.* **45**, 1076–1084.
Gonzales, R. A., Feldstein, J. B., Crews, F. T., and Raizada, M. K. (1985) Receptor-mediated inositide hydrolysis is a neuronal response: comparison of primary neuronal and glial cultures. *Brain Res.* **345**, 350–355.
Goyal, R. K. and Rattan, S. (1978) Neurohumoral, hormonal, and drug receptors for the lower esophageal sphincter. *Gastroenterology* **74**, 598–619.

Graff, I., Mockel, J., Laurent, E., Erneux, C., and Dumont, J. E. (1987) Carbachol and sodium fluoride, but not TSH, stimulate the generation of inositol phosphates in the dog thyroid. *FEBS Lett.* **210**, 204–210.

Grandordy, B. M., Cuss, F. M., Sampson, A. S., Palmer, J. B., and Barnes, P. J. (1986) Phosphatidylinositol response to cholinergic agonists in airway smooth muscle: Relationship to contraction and muscarinic receptor occupancy. *J. Pharmacol. Exp. Ther.* **238**, 273–279.

Grynkiewicz, G., Poenie, M., and Tsien, R. Y. (1985) A new generation of Ca^{2+} indicators with greatly improved fluorescence properties. *J. Biol. Chem.* **260**, 3440–3450.

Haddas, R. A., Landis, C. A., and Putney, J. W. (1979) Relationship between calcium release and potassium release in rat parotid gland. *J. Physiol.* **291**, 457–465.

Hallcher, L. M. and Sherman, W. R. (1980) The effects of lithium ion and other agents on the activity of myo-inositol-1-phosphatase from bovine brain. *J. Biol. Chem.* **255**, 10896–10901.

Hammer, R., Berrie, C. P., Birdsall, N. J. M., Burgen, A. S. V., and Hulme, E. C. (1980) Pirenzepine distinguishes between different subclasses of muscarinic receptors. *Nature* **283**, 90–92.

Harden, T. K., Heng, M. M., and Brown, J. H. (1986) Receptor reserve in the calcium-dependent cyclic AMP response of astrocytoma cells to muscarinic receptor stimulation: Demonstration by agonist-induced desensitization, receptor inactivation, and phorbol ester treatment. *Mol. Pharmacol.* **30**, 200–206.

Haslam, R. J. and Davidson, M. M. L. (1984) Receptor-induced diacylglycerol formation in permeabilized platelets: possible role for a GTP-binding protein. *J. Receptor Res.* **4**, 605–629.

Hawkins, P. T., Stephens, L., and Downes, C. P. (1986) Rapid formation of inositol 1,3,4,5-tetrakisphosphate and inositol 1,3,4-trisphosphate in rat parotid glands may both result indirectly from receptor-stimulated release of inositol 1,4,5-trisphosphate from phosphatidylinositol 4,5-bisphosphate. *Biochem. J.* **238**, 507–516.

He, X., Wu, X., and Baum, B. J. (1988) Protein kinase C differentially inhibits muscarinic receptor operated Ca^{2+} release and entry in human salivary cells. *Biochem. Biophys. Res. Commun.* **152**, 1062–1069.

Hepler, J. R. and Harden, T. K. (1986) Guanine nucleotide-dependent pertussis toxin-insensitive stimulation of inositol phosphate formation by carbachol in a membrane preparation from human astrocytoma cells. *Biochem. J.* **239**, 141–146.

Hepler, J. R., Hughes, A. R., and Harden, T. K. (1987) Evidence that muscarinic cholinergic receptors selectively interact with either cyclic AMP or the inositol phosphate second messenger response systems. *Biochem. J.* **247**, 793–796.

Hirata, M., Suematsu, E., Hashimoto, T., Hamachi, T., and Koga, T. (1984) Release of Ca^{2+} from a non-mitochondrial store site in peritoneal macrophages treated with saponin by inositol 1,4,5-trisphosphate. *Biochem. J.* **223**, 229–236.

Hokin, L. E. (1966) Effects of acetylcholine on the incorporation of ^{32}P into various phospholipids in slices of normal and denervated superior cervical ganglia of the cat. *J. Neurochem.* **13**, 179–184.

Hokin, L. E. (1985) Receptors and phosphoinositide-generated 2nd messengers. *Ann. Rev. Biochem.* **55**, 205–235.

Hokin, L. E. (1987) The road to the phosphoinositide-generated second messengers. *Trends Pharmacol. Sci.* **8**, 53–56.

Hokin, L. E. and Hokin, M. R. (1958) Acetylcholine and the exchange of phosphate in phosphatidic acid in brain microsomes. *J. Biol. Chem.* **233**, 822–826.

Hokin, M. R. and Hokin, L. E. (1953) Enzyme secretion and the incorporation of P^{32} into phospholipids of pancreas slices. *J. Biol. Chem.* **203**, 967–977.

Hokin-Neaverson, M. and Sadghian, K. (1984) Lithium-induced accumulation of inositol 1-phosphate during cholecystokinin octapeptide- and acetylcholine-stimulated phosphatidylinositol breakdown in dispersed mouse pancreas acinar cells. *J. Biol. Chem.* **259**, 4346–4352.

Horwitz, J., Anderson, C. H., and Perlman, R. L. (1986) Comparison of the effects of muscarine and vasopressin on inositol phospholipid metabolism in the superior cervical ganglion of the rat. *J. Pharmacol. Exp. Ther.* **237**, 312–317.

Horwitz, J., Tsymbalov, S., and Perlman, R. L. (1985) Muscarine stimulates the hydrolysis of inositol-containing phospholipids in the superior cervical ganglion. *J. Pharmacol. Exp. Ther.* **233**, 235–241.

Howe, P. H., Akhtar, R. A., Naderi, S., and Abdel-Latif, A. A. (1986) Correlative studies on the effect of carbachol on myo-inositol trisphosphate accumulation, myosin light chain phosphorylation and contraction in sphincter smooth muscle of rabbit iris. *J. Pharmacol. Exp. Ther.* **239**, 574–583.

Hughes, A. R., Martin, M. W., and Harden, T. K. (1984) Pertussis toxin differentiates between two mechanisms of attenuation of cyclic AMP accumulation by muscarinic cholinergic receptors. *Proc. Natl. Acad. Sci. USA* **81**, 5680–5684.

Hunter, D. D. and Nathanson, N. M. (1985) Assay of muscarinic acetylcholine receptor function in cultured cardiac cells by stimulation of $^{86}Rb^+$ efflux. *Anal. Biochem.* **149**, 392–398.

Irvine, R. F. and Moor, R. M. (1986) Micro-injection of inositol 1,3,4,5-tetrakisphosphate activates sea urchin eggs by a mechanism dependent on external Ca^{2+}. *Biochem. J.* **240**, 917–920.

Irvine, R. F., Moor, R. M., Pollock, W. K., Smith, P. M., and Wreggett, K. A. (1988) Inositol phosphates: proliferation, metabolism and function. *Phil. Trans. R. Soc. Lond. B* **320**, 281–298.

Irvine, R. F., Anggard, E. E., Letcher, A. J., and Downes, C. P. (1985)

Metabolism of inositol 1,4,5-trisphosphate and inositol 1,3,4-trisphosphate in rat parotid glands. *Biochem. J.* **229**, 237–243.

Irvine, R. F., Letcher, A. J., Heslop, J. P., and Berridge, M. J. (1986) The inositol tris/tetrakisphosphate pathway-demonstration of Ins(1,4,5)P_3 3-kinase activity in animal tissues. *Nature* **320**, 631–634.

Irvine, R. F., Letcher, A. J., Lander, D. J., and Downes, C. P. (1984) Inositol trisphosphates in carbachol-stimulated rat parotid glands. *Biochem. J.* **223**, 237–243.

Ishii, K., Sasakawa, N., and Kato, R. (1987) Simultaneous recording of tension and fluorescence ratio of the fura-2-loaded longitudinal muscle of guinea-pig ileum: responses to muscarinic agents. *Eur. J. Pharmacol.* **135**, 455–456.

Jacobson, M. D., Wusteman, M., and Downes, C. P. (1985) Muscarinic receptors and hydrolysis of inositol phospholipids in rat cerebral cortex and parotid gland. *J. Neurochem.* **44**, 465–472.

Jafferji, S. S. and Michell, R. H. (1976) Muscarinic cholinergic stimulation of phosphatidylinositol turnover in the longitudinal smooth muscle of guinea-pig ileum. *Biochem. J.* **154**, 653–657.

Jones, L. G. and Brown, J. H. (1988) Guanine nucleotide-regulated inositol polyphosphate production in adult rat cardiomyocytes, in *Biology of Isolated Adult Cardiac Myocytes* (Clark, W. A., Decker, R. S., and Bork, T. K., eds.) Elsevier, New York, 257–260.

Jones, L. G., Goldstein, D., and Brown, J. H. (1988) Guanine nucleotide-dependent inositol trisphosphate formation in chick heart cells. *Circ. Res.* **62**, 299–305.

Jones, L. M. and Michell, R. H. (1974) Breakdown of phosphatidylinositol provoked by muscarinic cholinergic stimulation of rat parotid-gland fragments. *Biochem. J.* **142**, 583–590.

Jope, R. S., Casebolt, T. L., and Johnson, G. V. W. (1987) Modulation of carbachol-stimulated inositol phospholipid hydrolysis in rat cerebral cortex. *Neurochem. Res.* **12**, 693–700.

Joseph, S. K. (1985) Receptor-stimulated phosphoinositide metabolism: a role for GTP-binding proteins. *Trends Biochem. Sci.* **10**, 297–298.

Kamm, K. E. and Stull, J. T. (1985) The function of myosin and myosin light chain kinase phosphorylation in smooth muscle. *Ann. Rev. Pharmacol. Toxicol.* **25**, 593–620.

Kao, L-S. and Schneider, A. S. (1985) Muscarinic receptors on bovine chromaffin cells mediate a rise in cytosolic calcium that is independent of extracellular calcium. *J. Biol. Chem.* **260**, 2019–2022.

Kao, L-S. and Schneider, A. S. (1986) Calcium mobilization and catecholamine secretion in adrenal chromaffin cells. *J. Biol. Chem.* **261**, 4881–4888.

Kelly, E., Rooney, T. A., and Nahorski, S. R. (1985) Pertussis toxin separates two muscarinic receptor-effector mechanisms in the striatum. *Eur. J. Pharmacol.* **119**, 129–130.

Korth, M., Sharma, V. K., and Sheu, S-S. (1988) Stimulation of muscarinic receptors raises free intracellular Ca^{2+} concentration in rat ventricular myocytes. *Circ. Res.* **62**, 1080–1087.

Krims, P. E. and Pandol, S. J. (1988) Free cytosolic calcium and secretagogue-stimulated initial pancreatic exocrine secretion. *Pancreas* **3**, 383–390.

Kuno, M. and Gardner, P. (1987) Ion channels activated by inositol 1,4,5-trisphosphate in plasma membrane of human T-lymphocytes. *Nature* **326**, 301–304.

Kuo, J. F., Andersson, R. G. G., Wise, B. C., Mackerlova, L., Salomonsson, I., Brackett, N. L., Katoh, N., Shoji, M., and Wrenn, R. W. (1980) Calcium-dependent protein kinase: Widespread occurrence in various tissues and phyla of the animal kingdom and comparison of effects of phospholipid, calmodulin, and trifluoperazine. *Proc. Natl. Acad. Sci. USA* **77**, 7039–7043.

Labarca, R., Janowsky, A., Patel, J., and Paul, S. M. (1984) Phorbol esters inhibit agonist-induced [^{3}H] inositol-1-phosphate accumulation in rat hippocampal slices. *Biochem. Biophys. Res. Commun.* **123**, 703–709.

Lapetina, E. G., Brown, W. E., and Michell, R. H. (1976) Muscarinic cholinergic stimulation of phosphatidylinositol turnover in isolated rat superior cervical sympathetic ganglia. *J. Neurochem.* **26**, 649–651.

Lazareno, S., Kendall, D. A., and Nahorski, S. R. (1985) Pirenzepine indicates heterogeneity of muscarinic receptors linked to cerebral inositol phospholipid metabolism. *Neuropharmacology* **244**, 282–285.

Lee, H-C., Smith, N., Mohabir, R., and Clusin, W. T. (1987) Cytosolic calcium transients from the beating mammalian heart. *Proc. Natl. Acad. Sci. USA* **84**, 7793–7797.

Liles, W. C., Hunter, D. D., Meier, K. E., and Nathanson, N. M. (1986) Activation of protein kinase C induces rapid internalization and subsequent degradation of muscarinic acetylcholine receptors in neuroblastoma cells. *J. Biol. Chem.* **261**, 5307–5313.

Lindmar, R., Loffelholz, K., and Sandmann, J. (1986) Characterization of choline efflux from the perfused heart at rest and after muscarinic receptor activation. *Naunyn-Schmiedeberg's Arch. Pharmacol.* **332**, 224–229.

Lindmar, R., Loffelholz, K., and Sandmann, J. (1988) The mechanism of muscarinic hydrolysis of choline phospholipids in the heart. *Biochem. Pharmacol.* **37**, 4689–4695.

Litosch, I. and Fain, J. N. (1986) Regulation of phosphoinositide breakdown by guanine nucleotides. *Life Sci.* **39**, 187–194.

Litosch, I., Wallis, C., and Fain, J. N. (1985) 5-Hydroxytryptamine stimulated inositol phosphate production in a cell-free system from blowfly salivary glands. Evidence for a role of GTP in coupling receptor activation to phosphoinositide breakdown. *J. Biol. Chem.* **260**, 5464–5471.

Lo, W. W. Y. and Hughes, J. (1987) Pertussis toxin distinguishes between muscarinic receptor-mediated inhibition of adenylate cyclase and stimulation of phosphoinositide hydrolysis in Flow 9000 cells. *FEBS Lett.* **220**, 155–158.

Malaisse, W. J. (1986) Stimulus-secretion coupling in the pancreatic B-cell: The cholinergic pathway for insulin release. *Diabetes/Metabolism Reviews* **2**, 243–259.

Malhotra, R. K., Wakade, T. D., and Wakade, A. R. (1988) Vasoactive intestinal polypeptide and muscarine mobilize intracellular Ca^{2+} through breakdown of phosphoinositides to induce catecholamine secretion. Role of IP_3 in exocytosis. *J. Biol. Chem.* **263**, 2123–2126.

Marc, S., Leiber, D., and Harbon, S. (1986) Carbachol and oxytocin stimulate the generation of inositol phosphates in the guinea pig myometrium. *FEBS Lett.* **201**, 9–14.

Martinson, E. A. and Brown, J. H. (1989) Muscarinic receptors activate hydrolysis of phosphatidylcholine by a dual mechanism, in preparation.

Marty, A. (1987) Control of ionic currents and fluid secretion by muscarinic agonists in exocrine glands. *Trends Neurosci.* **10**, 373–377.

Masters, S. B., Harden, T. K., and Brown, J. H. (1984) Relationships between phosphoinositide and calcium responses to muscarinic agonists in astrocytoma cells. *Mol. Pharmacol.* **26**, 149–155.

Masters, S. B., Martin, M. W., Harden, T. K., and Brown, J. H. (1985a) Pertussis toxin does not inhibit muscarinic-receptor-mediated phosphoinositide hydrolysis or calcium mobilization. *Biochem. J.* **227**, 933–937.

Masters, S. B., Quinn, M. T., and Brown, J. H. (1985b) Agonist-induced desensitization of muscarinic receptor-mediated calcium efflux without concomitant desensitization of phosphoinositide hydrolysis. *Mol. Pharmacol.* **27**, 325–332.

McDonough, P. M., Eubanks, J. H., and Brown, J. H. (1988) Desensitization and recovery of muscarinic and histaminergic calcium mobilization: Evidence for a common hormone sensitive calcium store in 1321N1 astrocytoma cells. *Biochem. J.* **249**, 135–141.

Meeker, R. B. and Harden, T. K. (1982) Muscarinic cholinergic receptor-mediated activation of phosphodiesterase. *Mol. Pharmacol.* **22**, 310–319.

Mei, L., Yamamura, H. I., and Roeske, W. R. (1988) Muscarinic receptor-mediated hydrolysis of phosphatidylinositols in human neuroblastoma (SH-SY5Y) cells is sensitive to pertussis toxin. *Brain Res.* **447**, 360–363.

Merritt, J. E. and Rink, T. J. (1987a) Rapid increases in cytosolic free calcium in response to muscarinic stimulation of rat parotid acinar cells. *J. Biol. Chem.* **262**, 4958–4960.

Merritt, J. E. and Rink, T. J. (1987b) The effects of substance P and carbachol on inositol tris- and tetrakisphosphate formation and cytosolic free calcium in rat parotid acinar cells. *J. Biol. Chem.* **262**, 14912–14916.

Merritt, J. E., Taylor, C. W., Rubin, R. P., and Putney, J. W. (1986) Evidence suggesting that a novel guanine nucleotide regulatory protein couples receptors to phospholipase C in exocrine pancreas. *Biochem. J.* **236**, 337–343.

Michell, R. H. (1975) Inositol phospholipids and cell surface receptor function. *Biochim. Biophys. Acta* **415**, 81–147.

Michell, R. H. (1979) Inositol phospholipids in membrane function. *Trends Biochem. Sci.* **4**, 128–131.

Michell, R. H. (1986) Inositol lipids and their role in receptor function: History and general principles, in *Phosphoinositides and Receptor Mechanisms* (Putney, J. W., ed.) Alan R. Liss, Inc., New York, 1–24.

Michell, R. H., Jafferji, S. S., and Jones, L. M. (1976) Receptor occupancy dose–response curve suggests that phosphatidylinositol breakdown may be intrinsic to the mechanism of the muscarinic cholinergic receptor. *FEBS Lett.* **69**, 1–5.

Miller, B. E. and Nelson, D. L. (1977) Calcium fluxes in isolated acinar cells from rat parotid. *J. Biol. Chem.* **252**, 3629–3636.

Miller, J. C. and Kowal, C. N. (1981) The relationship between the incorporation of ^{32}P into phosphatidic acid and phosphatidylinositol in rat parotid acinar cells. *Biochem. Biophys. Res. Commun.* **102**, 999–1007.

Misbahuddin, M., Isosaki, M., Houchi, H., and Oka, M. (1985) Muscarinic receptor-mediated increase in cytoplasmic free Ca^{2+} in isolated bovine adrenal medullary cells. *FEBS Lett.* **190**, 25–28.

Mohd Adnan, N. A. and Hawthorne, J. N. (1981) Phosphatidylinositol labelling in response to activation of muscarinic receptors in bovine adrenal medulla. *J. Neurochem.* **36**, 1858–1860.

Monsma, F. J., Abood, L. G., and Hoss, W. (1988) Inhibition of phosphoinositide turnover by selective muscarinic antagonists in the rat striatum. *Biochem. Pharmacol.* **37**, 2437–2443.

Morris, A. P., Fuller, C. M., and Gallacher, D. V. (1987) Cholinergic receptors regulate a voltage-insensitive but Na^+-dependent calcium influx pathway in salivary acinar cells. *FEBS Lett.* **211**, 195–199.

Movsesian, M. A., Thomas, A. P., Selak, M., and Williamson, J. R. (1985) Inositol trisphosphate does not release Ca^{2+} from permeabilized cardiac myocytes and sarcoplasmic reticulum. *FEBS Lett.* **185**, 328–332.

Muallem, S. and Sachs, G. (1984) Changes in cytosolic free Ca^{2+} in isolated parietal cells: differential effects of secretagogues. *Biochim. Biophys. Acta* **805**, 181–185.

Muallem, S., Fimmel, C. J., Pandol, S. J., and Sachs, G. (1986) Regulation of free cytosolic Ca^{2+} in the peptic and parietal cells of the rabbit gastric gland. *J. Biol. Chem.* **261**, 2660–2667.

Muallem, S., Schoeffield, M. S., Fimmel, C. J., Krims, P. E., and Pandol, S. J. (1988a) The agonist-sensitive calcium pool in the pancreatic acinar cell. I. Permeability properties at rest and during stimulation. *Am. J. Physiol.* **255**, G221–G228.

Muallem, S., Schoeffield, M. S., Fimmel, C. J., and Pandol, S. J. (1988b) The agonist-sensitive calcium pool in the pancreatic acinar cell. Characterization of reloading. *Am. J. Physiol.* **255**, G229-G235.

Nakahata, N. and Harden, T. K. (1987) Regulation of inositol trisphosphate formation by muscarinic cholinergic and H_1-histamine receptors on human astrocytoma cells: Differential induction of desensitization by agonists. *Biochem. J.* **241**, 337–344.

Negulescu, P. A. and Machen, T. E. (1988a) Intracellular Ca^{2+} regulation during secretagogue stimulation of the parietal cell. *Am. J. Physiol.* **254C130**, C130–C140.

Negulescu, P. A. and Machen, T. E. (1988b) Release and reloading of intracellular Ca^{2+} stores following cholinergic stimulation of the parietal cell. *Am. J. Physiol.* **254**, C498–C504.

Niedel, J. E. and Blackshear, P. J. (1986) Protein kinase C, in *Phosphoinositides and Receptor Mechanisms* (Putney, J. W., ed.) Alan R. Liss, Inc., New York, 47–88.

Nishizuka, Y. (1986) Studies and perspectives of protein kinase C. *Science* **233**, 305–312.

Nomura, Y., Kaneko, S., Kato, K-I., Yamagashi, S-I., and Sugiyama, H. (1987) Inositol phosphate formation and chloride current responses induced by acetylcholine and serotonin through GTP-binding proteins in Xenopus oocyte after injection of rat brain messenger RNA. *Mol. Brain Res.* **2**, 113–123.

Noronha-Blob, L., Lowe, V. C., Hanson, R. C., and U'Prichard, D. C. (1987) Heterogeneity of muscarinic receptors coupled to phosphoinositide breakdown in guinea pig brain and peripheral tissues. *Life Sci.* **41**, 967–975.

Noronha-Blob, L., Richard, C., and U'Prichard, D. C. (1987) Calcium mobilization by muscarinic receptors in human astrocytoma cells: measurements with quin-2. *Biochem. Biophys. Res. Commun.* **14**, 182–188.

Nosek, T. M., Williams, M. F., Zeigler, S. T., and Godt, R. E. (1986) Inositol trisphosphate enhances calcium release in skinned cardiac and skeletal muscle. *Am. J. Physiol.* **250**, C807–C811.

Ochs, D. L., Korenbrot, J. I., and Williams, J. A. (1985) Relation between free cytosolic calcium and amylase release by pancreatic acini. *Am. J. Physiol.* **249**, G389–G398.

Ono, Y. and Kikkawa, U. (1987) Do multiple species of protein kinase C transduce different signals? *Trends Biochem. Sci.* **12**, 421–423.

Orellana, S., Solski, P. A., and Brown, J. H. (1987) Guanosine 5′-0-(thiotriphosphate)-dependent inositol trisphosphate formation in membranes is inhibited by phorbol ester and protein kinase C. *J. Biol. Chem.* **262**, 1638–1643.

Orellana, S. A., Solski, P. A., and Brown, J. H. (1985) Phorbol ester inhibits phosphoinositide hydrolysis and calcium mobilization in cultured astrocytoma cells. *J. Biol. Chem.* **260**, 5236–5239.

Oron, Y., Dascal, N., Nadler, E., and Lupu, M. (1985) Inositol 1,4,5-trisphosphate mimics muscarinic response in Xenopus oocytes. *Nature* **313**, 141–143.

Oron, Y., Lowe, M., and Selinger, Z. (1975) Incorporation of inorganic [^{32}P]phosphate into rat parotid phosphatidylinositol. *Mol. Pharmacol.* **11**, 79–86.

Pandol, S. J. and Schoeffield, M. S. (1986) 1,2-Diacylglycerol, protein kinase C, and pancreatic enzyme secretion. *J. Biol. Chem.* **261**, 4438–4444.

Pandol, S. J., Schoeffield, M. S., Fimmel, C. J., and Muallem, S. (1987) The agonist-sensitive calcium pool in the pancreatic acinar cell. III. Activation of a plasma membrane Ca^{2+} channel. *J. Biol. Chem.* **262**, 16963–16968.

Pandol, S. J., Schoeffield, M. S., Sachs, G., and Muallem, S. (1985) Role of free cytosolic calcium in secretagogue-stimulated amylase release from dispersed acini from guinea pig pancreas. *J. Biol. Chem.* **260**, 10081–10086.

Pearce, B., Cambray-Deakin, M., Morrow, C., Grimble, J., and Murphy, S. (1985) Activation of muscarinic and of α_1-adrenergic receptors on astrocytes results in the accumulation of inositol phosphates. *J. Neurochem.* **45**, 1534–1540.

Pearce, B., Morrow, C., and Murphy, S. (1986) Receptor-mediated inositol phospholipid hydrolysis in astrocytes. *Eur. J. Pharmacol.* **121**, 231–243.

Peralta, E. G., Ashkenazi, A., Winslow, J. W., Ramachandran, J., and Capon, D. J. (1988) Differential regulation of PI hydrolysis and adenylyl cyclase by muscarinic receptor subtypes. *Nature* **334**, 434–437.

Poggioli, J. and Putney, J. W. (1982) Net calcium fluxes in rat parotid acinar cells: evidence for a hormone-sensitive calcium pool in or near the plasma membrane. *Pflugers Arch.* **392**, 239–243.

Poggioli, J., Sulpice, J. C., and Vassort, G. (1986) Inositol phosphate production following α_1-adrenergic, muscarinic or electrical stimulation in isolated rat heart. *FEBS Lett.* **206**, 292–298.

Pozzan, T., Divirgilio, F., Vicentini, L. M., and Meldolesi, J. (1986) Activation of muscarinic receptors in PC12 cells. *Biochem. J.* **234**, 547–553.

Putney, J. W. (1977) Muscarinic, alpha$_1$-adrenergic and peptide receptors regulate the same calcium influx sites in the parotid gland. *J. Physiol.* **268**, 139–149.

Putney, J. W. (1986a) A model for receptor-regulated calcium entry. *Cell Calcium* **7**, 1–12.

Putney, J. W. (1986b) Identification of cellular activation mechanisms associated with salivary secretion. *Ann. Rev. Physiol.* **48**, 75–88.

Putney, J. W., McKinney, J. S., Aub, D. L., and Leslie, B. A. (1984) Phorbol ester induced protein secretion in rat parotid gland. *Mol.*

Pharmacol. **26**, 261–266.

Quist, E. E. (1982) Evidence for a carbachol stimulated phosphatidylinositol effect in heart. *Biochem. Pharmacol.* **31**, 3130–3133.

Rani, C. S., Boyd, A. E., and Field, J. B. (1985) Effects of acetylcholine, TSH and other stimulators on intracellular calcium concentration in dog thyroid cells. *Biochem. Biophys. Res. Commun.* **131**, 1041–1047.

Ransnas, L., Gjorstrup, P., Hjalmarson, A., Sjogren, C-G., and Jacobsson, B. (1986) Muscarinic receptors in mammalian myocardium: effects of atrial and ventricular receptors on phosphatidylinositol metabolism and adenylate cyclase. *J. Mol. Cell. Cardiol.* **18**, 807–814.

Rasmussen, H., Takuwa, Y., and Park, S. (1987) Protein kinase C in the regulation of smooth muscle contraction. *FASEB J.* **1**, 177–185.

Raspe, R., Roger, P. P., and Dumont, J. E. (1986) Carbamylcholine, TRH, $PGF_{2\alpha}$ and fluoride enhance free intracellular Ca^{2+} and Ca^{2+} translocation in dog thyroid cells. *Biochem. Biophys. Res. Commun.* **141**, 569–577.

Renard, D. and Poggioli, J. (1987) Does the inositol tris/tetrakisphosphate pathway exist in rat heart. *FEBS Lett.* **217**, 117–123.

Reynolds, E. E. and Dubyak, G. R. (1986) Agonist-induced calcium transients in cultured smooth muscle cells: measurements with fura-2 loaded monolayers. *Biochem. Biophys. Res. Commun.* **136**, 927–934.

Rubin, R. P. (1982) *Calcium and Cellular Secretion*. Plenum, New York, 1–276.

Rubin, R. P., Godfrey, P. P., Chapman, D. A., and Putney, J. W. (1984) Secretagogue-induced formation of inositol phosphates in rat exocrine pancreas. *Biochem. J.* **219**, 655–659.

Sasaguri, T., Hirata, M., Itoh, T., Koga, T., and Kuriyama, H. (1986) Guanine nucleotide binding protein involved in muscarinic responses in the pig coronary artery is insensitive to islet-activating protein. *Biochem. J.* **239**, 567–574.

Schact, J. and Agranoff, B. W. (1973) Acetylcholine stimulates hydrolysis of ^{32}P-labeled phosphatidic acid in guinea pig synaptosomes. *Biochem. Biophys. Res. Commun.* **50**, 934–941.

Schlegel, W., Wuarin, F., Zbaren, C., and Zahnd, G. R. (1985) Lowering of cytosolic free Ca^{2+} by carbachol, a muscarinic cholinergic agonist, in clonal pituitary (GH3 cells). *Endocrinology* **117**, 976–981.

Schulz, I. and Stolze, H. H. (1980) The exocrine pancreas: the role of secretagogues, cyclic nucleotides, and calcium in enzyme secretion. *Ann. Rev. Physiol.* **42**, 127–156.

Schulz, I., Streb, H., Bayerdorffer, E., and Imamura, K. (1986) Intracellular messengers in stimulus-secretion coupling of pancreatic acinar cells. *J. Cardiovasc. Pharmacol.* **8 (Suppl.)**, 591–596.

Sekar, M. C. and Hokin, L. E. (1986) Phosphoinositide metabolism and cGMP levels are not coupled to the muscarinic-cholinergic receptor in human erythrocyte. *Life Sci.* **39**, 1257–1262.

Sekar, M. C. and Roufagalis, B. D. (1984) Muscarinic-receptor stimulation enhances polyphosphoinositide breakdown in guinea-pig ileum smooth muscle. *Biochem. J.* **223**, 527–531.

Sekiguchi, K., Tsukuda, M., Ogita, K., Kikkawa, U., and Nishizuka, Y. (1987) Three distinct forms of rat brain protein kinase C: Differential response to unsaturated fatty acids. *Biochem. Biophys. Res. Commun.* **145**, 797–802.

Serra, M., Lin, M., Roeske, W. R., Lui, G. K., Watson, M., and Yamamura, H. I. (1988) The intact human neuroblastoma cell (SH-SY5Y) exhibits high-affinity [^{3}H]pirenzepine binding associated with hydrolysis of phosphatidylinositols. *J. Neurochem.* **50**, 1513–1521.

Serra, M., Smith, T. L., and Yamamura, H. I. (1986) Phorbol esters alter muscarinic receptor binding and inhibit phosphoinositide breakdown in human neuroblastoma (SH-SY5Y) cells. *Biochem. Biophys. Res. Commun.* **140**, 160–166.

Sher, E., Gotti, C., Pandiella, A., Madeddu, L., and Clementi, F. (1988) Intracellular calcium homeostasis in a human neuroblastoma cell line: Modulation by depolarization, cholinergic receptors, and α-latrotoxin. *J. Neurochem.* **50**, 1708–1713.

Siegel, H., Jim, K., Bolger, G. T., Gengo, P., and Triggle, D. J. (1984) Specific and non-specific desensitization of guinea-pig ileal smooth muscle. *J. Auton. Pharmacol.* **4**, 109–126.

Smith, C. D., Uhing, R. J., and Snyderman, R. (1987) Nucleotide regulatory protein-mediated activation of phospholipase C in human polymorphonuclear leukocytes is disrupted by phorbol esters. *J. Biol. Chem.* **262**, 6121–6127.

Snider, R. M., Roland, R. M., Lowy, R. J., Agranoff, B. W., and Ernst, S. A. (1986) Muscarinic receptor-stimulated Ca^{2+} signaling and inositol lipid metabolism in avian salt gland cells. *Biochim. Biophys. Acta* **889**, 216–224.

Somlyo, A. P. (1985) Excitation-contraction coupling and the ultrastructure of smooth muscle. *Circ. Res.* **57**, 497–507.

Somlyo, A. P. and Somlyo, A. V. (1970) Vascular smooth muscle. II. Pharmacology of normal and hypertensive vessels. *Pharmacol. Rev.* **22**, 249–353.

Stolze, H. and Schulz, I. (1980) Effect of atropine, ouabain, antimycin A, and A23187 on "trigger-Ca^{2+} pool" in exocrine pancreas. *Am. J. Physiol.* **238**, G338–G348.

Streb, H., Bayerdorffer, E., Haase, W., Irvine, R. F., and Schulz, I. (1984) Effect of inositol 1,4,5-trisphosphate on isolated subcellular fractions of rat pancreas. *J. Mem. Biol.* **81**, 241–253.

Streb, H., Heslop, J. P., Irvine, R. F., Schulz, I., and Berridge, M. J. (1985) Relationship between secretagogue-induced Ca^{2+} release and inositol polyphosphate production in permeabilized pancreatic acinar cells. *J. Biol. Chem.* **260**, 7309–7315.

Streb, H., Irvine, R. F., Berridge, M. J., and Schulz, I. (1983) Release of Ca^{2+} from a nonmitochondrial intracellular store in pancreatic acinar cells by inositol-1,4,5 trisphosphate. *Nature* **306**, 67–69.

Tajima, T., Tsuji, Y., Brown, J. H., and Pappano, A. J. (1987) Pertussis-toxin insensitive phosphoinositide hydrolysis, membrane depolarization, and positive inotropic effect of carbachol in chick atria. *Circ. Res.* **61**, 436–445.

Takai, Y., Kishimoto, A., Iwasa, Y., Kawahara, Y., Mori, T., and Nishizuka, Y. (1979) Calcium-dependent activation of a multifunctional protein kinase by membrane phospholipids. *J. Biol. Chem.* **254**, 3692–3695.

Takemura, H. (1985) Changes in cytosolic free calcium concentration in isolated rat parotid cells by cholinergic and beta-adrenergic agonists. *Biochem. Biophys. Res. Commun.* **131**, 1048–1055.

Takuwa, Y., Takuwa, N., and Rasmussen, H. (1986) Carbachol induces a rapid and sustained hydrolysis of polyphosphoinositide in bovine tracheal smooth muscle: Measurements of the mass of polyphosphoinositides,1,2-diacylglycerol, and phosphatidic acid. *J. Biol. Chem.* **261**, 14670–14675.

Ticku, M. K. and Triggle, D. J. (1976) Calcium and the muscarinic receptor. *Gen. Pharmacol.* **7**, 133–140.

Trimble, E. R., Bruzzone, R., Meehan, C. J., and Biden, T. J. (1987) Rapid increases in inositol 1,4,5-trisphosphate,inositol 1,3,4,5-tetrakisphosphate and cytosolic free Ca^{2+} in agonist-stimulated pancreatic acini of the rat. *Biochem. J.* **242**, 289–292.

Vicentini, L. M., Ambrosini, A., Di Virgilio, F., Meldolesi, J., and Pozzan, T. (1986) Activation of muscarinic receptors in PC12 cells: Relation between cytosolic Ca^{2+} rise and phosphoinositide hydrolysis. *Biochem. J.* **234**, 555–562.

Vicentini, L. M., Ambrosini, A., Di Virgilio, F., Pozzan, T., and Meldolesi, J. (1985a) Muscarinic receptor-induced phosphoinositide hydrolysis at resting cytosolic Ca^{2+} concentration in PC12 cells. *J. Cell. Biol.* **100**, 1330–1333.

Vicentini, L. M., Di Virgilio, F., Ambrosini, A., Pozzan, T., and Meldolesi, J. (1985b) Tumor promoter phorbol 12-myristate 13-acetate inhibits phosphoinositide hydrolysis and cytosolic Ca^{2+} rise induced by the activation of muscarinic receptors in PC12 cells. *Biochem. Biophys. Res. Commun.* **127**, 310–317.

Wakelam, M. J. O., Davies, S. A., Houslay, M. D., McKay, I., Marshall, C. J., and Hall, A. (1986) Normal p21 N-ras couples bombesin and other growth factor receptors to inositol phosphate production. *Nature* **323**, 173–176.

Weiss, S. J. and Putney, J. W. (1981) The relationship of phosphatidylinositol turnover to receptors and calcium-ion channels in rat parotid acinar cells. *Biochem. J.* **194**, 463–468.

White, H. L. (1988) Effects of acetylcholine and other agents on ^{32}P-prelabeled phosphoinositides and phosphatidate in crude synaptosomal preparations. *J. Neurosci. Res.* **20**, 122–128.

Williams, D. A., Becker, P. L., and Fay, F. S. (1987) Regional changes in calcium underlying contraction of single smooth muscle cells. *Science* **235**, 1644–1648.

Wolheim, C. B. and Biden, T. J. (1986) Second messenger function of inositol 1,4,5-trisphosphate: Early changes in inositol phosphates, cytosolic Ca^{2+}, and insulin release in carbamylcholine-stimulated RINm5F cells. *J. Biol. Chem.* **261**, 8314–8319.

Woodcock, E. A., Leung, E., and McLeod, J. K. (1987a) A comparison of muscarinic acetylcholine receptors coupled to phosphatidylinositol turnover and to adenylate cyclase in guinea-pig atria and ventricles. *Eur. J. Pharmacol.* **133**, 283–289.

Woodcock, E. A., White, L. B. S., Smith, A. I., and McLeod, J. K. (1987b) Stimulation of phosphatidylinositol mechanism in the isolated, perfused rat heart. *Circ. Res.* **61**, 625–631.

Xu, J. and Chuang, D. (1987) Muscarinic acetylcholine receptor-mediated phosphoinositide turnover in cultured cerebellar granule cells: desensitization by receptor agonists. *J. Pharmacol. Exp. Ther.* **242**, 238–244.

Yagihara, Y. and Hawthorne, J. N. (1972) Effects of acetylcholine on the incorporation of [^{32}P]orthophosphate in vitro into the phospholipids of nerve-ending particles from guinea pig brain. *J. Neurochem.* **19**, 355–367.

8

Muscarinic Receptor Regulation of Cyclic GMP and Eicosanoid Production

Michael McKinney
and
Elliott Richelson

1. Physiology

1.1. Muscarinic Receptor-Mediated Cyclic GMP Formation in Cells

1.1.1. Overview

For many of the cells or tissues in which muscarinic receptors are located, activation of these receptors results in elevated tissue levels of cyclic guanosine 3′,5′-monophosphate (cyclic GMP). Cyclic GMP and the enzyme that synthesizes this cyclic nucleotide are widely distributed (Goldberg and Haddox, 1977), and the muscarinic receptor is one of several receptors that mediates increased intracellular levels of cyclic GMP. Of the two forms of guanylate cyclase that exist, soluble and particulate, the muscarinic receptor is thought to activate the soluble form. This activation has generally been considered an indirect effect, involving a second messenger, since muscarinic receptors cannot activate guanylate cyclase in broken cell preparations. Calcium or a metabolite of arachidonic acid are the two most likely candidates for this second messenger.

1.1.2. Occurrence of Muscarinic Receptor-Mediated Cyclic GMP Formation

Cyclic GMP formation in response to acetylcholine was first reported in the heart (George et al., 1970). Responses in other tissues with muscarinic receptors were soon described (Lee et al., 1972; Ferrendelli et al., 1970), and the dependence on extracellular calcium ions was established early (Schultz et al., 1973). Cyclic GMP elevation in response to acetylcholine or muscarinic analogs has since been reported in many tissues; a complete list appears in Goldberg and Haddox (1977). In general, all the major target organs of parasympathetic cholinergic fibers contain muscarinic receptors that have been shown to mediate an elevation of cyclic GMP. In addition, muscarinic receptors in brain and sympathetic ganglia also mediate cyclic GMP formation. Cyclic GMP elevation in response to muscarinic receptor activation also occurs in certain neuron-like tumor cells, for example the murine neuroblastoma clone N1E-115 (Matsuzawa and Nirenberg, 1975).

1.2. Function of Cyclic GMP in Various Systems

1.2.1. Overview

Physiological roles for cyclic GMP are less well-defined than those for cyclic adenosine 3′,5′-monophosphate (cyclic AMP). The types of data that suggest specific physiological functions for cyclic GMP are:

1. the wide distribution of cyclic GMP and the enzymes that synthesize and degrade it
2. several receptors appear to mediate cyclic GMP formation in various tissues (often in association with a physiological response) and
3. the presence of kinases that are activated by cyclic GMP and receptors that elevate cyclic GMP.

A consideration of such data leads one to the conclusion that there are specific cellular functions for this cyclic nucleotide, even though the detailed mechanisms of receptor–effector coupling and exact physiological consequences of the subsequent alterations in the intracellular levels of cyclic GMP remain obscure. For example, the data indicate that cyclic GMP is involved in regulating cationic conductance in the retina and in altering neuronal membrane resistance in the cerebral cortex. In the latter case, the data strongly suggest that muscarinic receptor-mediated increases in neuronal excitability involves cyclic GMP. This cyclic nucleotide is probably involved in the muscarinic receptor-mediated endothelium-dependent relaxation of some vascular tissues, in the muscarinic receptor-mediated contraction in certain smooth muscle tissues, and in certain muscarinic receptor-mediated effects on cardiac contractility. Some of the evidence supporting these conclusions is summarized below.

1.2.2. Effects in Nervous Tissue

By immunocytochemical techniques, cyclic GMP has been localized in brain cells (Chan-Palay and Palay, 1979; Ariano and Matus, 1981; Nakane et al., 1983) and in neurons of the sympathetic ganglion (Kebabian et al., 1975), where its level appears to be modulated by synaptic transmission or by neurotransmitter activation (Weight et al., 1974; Kebabian et al., 1975; Wamsley et al., 1979). (However, for a conflicting report, *see* Frey and McIsaac, 1981.) In the striatum, the subcellular distribution of cyclic GMP appears to correlate with the distribution of cyclic GMP-dependent protein kinase, cyclic GMP phosphodiesterase, and guanylate cyclase (Ariano, 1983; Nakane et al., 1983). Iontophoresis of cyclic GMP onto cortical pyramidal cells leads to neuronal activation in a way similar to muscarinic activation (Stone et al., 1975). Intracellular injection of cyclic GMP into cortical pyramids increases membrane resistance in a way identical to that seen with the activation of muscarinic receptors by administration of an agonist extracellularly (Woody et al., 1978; Swartz and Woody, 1979; Woody et al., 1986a). Intracellular injection of antibodies to cyclic GMP blocks the effect of extracellular application of muscarinic agonists (Swartz and Woody, 1984). Finally, intracellular injection of cyclic GMP-dependent protein kinase causes an increase in neuronal membrane resistance in the cortex (Woody et al., 1986b). Therefore, there is circumstantial evidence that cyclic GMP formation is involved in increases in muscarinic receptor-mediated neuronal excitation in the cortex. It has long been known that muscarinic receptors can mediate an increase in cyclic GMP levels in brain tissue (Ferrendelli et al., 1970; Lee et al., 1972; Palmer and Duszynski, 1975; Dinnendahl and Stock, 1975; Opmeer et al., 1976; Hanley and Iversen, 1977; Black et al., 1979; Lenox et al., 1980). Neuroblastoma cells also possess muscarinic receptors that mediate cyclic GMP formation (Matsuzawa and Nirenberg, 1975) and a cyclic GMP-dependent hyperpolarization (Wastek et al., 1980; reviewed in Richelson and El-Fakahany, 1981 and in McKinney and Richelson, 1984; also *see below*).

Cyclic GMP mediates an effect on rod plasma membrane cation conductance without an apparent stimulation of cyclic GMP-dependent protein phosphorylation (Fesenko et al., 1985; Yau and Nakatani, 1985; Haynes et al., 1986). In these studies involving patches of the plasma membrane of rod outer segments, cyclic GMP was shown to activate a conductance directly. The hyperpolarization of the rods that occurs in response to light is the result of reduction in a cationic conductance induced by the hydrolysis of cyclic GMP by a light-activated phosphodiesterase, which lowers intracellular cyclic GMP levels, leading to the closure of the channels. Cones appear to have the same mechanism (Cobbs et al., 1985). The action of cyclic GMP is observable at the single-channel level (Haynes et al., 1986).

1.2.3. Cyclic GMP-Dependent Protein Phosphorylation

Cyclic GMP-dependent kinases have been described in several tissues (Hofmann and Sold, 1972; Kuo, 1974). Reports of muscarinic receptor-mediated or cyclic GMP-dependent phosphorylation are evidence in favor of an involvement of this receptor–effector system in cellular function. Cyclic GMP-dependent protein phosphorylation has been described in various smooth muscles (Casnellie and Greengard, 1974; Rapoport et al., 1982), and a cerebellar protein that is a substrate for a cyclic GMP-dependent kinase has been described (Schlichter et al., 1978; Aswad and Greengard, 1981a,b). This latter cyclic GMP-dependent kinase appears to be developmentally regulated in association with synaptogenesis (Levitt et al., 1984). Muscarinic receptor-mediated phosphorylation of proteins has been described in tracheal smooth muscle (Silver and Stull, 1983), vascular smooth muscle (Rapoport et al., 1983), and sympathetic ganglia (Cahill and Perlman, 1986). Muscarinic receptors modulate the beta-adrenergic receptor stimulation of protein phosphorylation in the heart (Iwasa and Hosey, 1983). These findings suggest that tissue responses to the muscarinic receptor, e.g., contraction or relaxation, may be mediated by protein phosphorylation regulated by the intracellular level of cyclic GMP.

1.2.4. Effects in Smooth Muscle Tissue

In some vascular smooth muscles, the relaxation elicited by the muscarinic receptor and by the NO-containing vasodilators is thought by many workers to be linked to elevation of cyclic GMP (*see* for example, Ignarro et al., 1984b; Rapoport and Murad, 1983). The muscarinic receptors involved are localized on the vascular endothelium, whereas the chemical vasodilators act directly on the vascular muscle where the cyclic GMP elevation actually occurs (Furchgott, 1984). An endothelium-dependent factor that is generated by muscarinic and other receptors appears to be a transient lipid or oxygen radical (*see below*); the formation of this factor is closely linked with receptor-mediated relaxation of the muscle (Furchgott, 1984). The muscarinic receptor-mediated endothelium-dependent relaxation is clearly associated with elevation of smooth muscle cyclic GMP levels (Rapoport and Murad, 1983; Ignarro et al., 1984a). It has been directly demonstrated that soluble guanylate cyclase can be activated in vitro by this endothelium-dependent factor, generated in response to receptors for acetylcholine, bradykinin, and vasoactive intestinal polypeptide (Ignarro et al., 1986b; Forstermann et al., 1986). Agents that block guanylate cyclase, such as methylene blue (Gruetter et al., 1981), or interfere with or destroy the endothelium-derived factor, such as arachidonic acid metabolic inhibitors (Furchgott, 1984) or hemoglobin (Murad et al., 1978; Martin et al., 1984), will block cyclic GMP formation and vasorelaxation (Furchgott and Zawadzki, 1980; Gruetter et al., 1981; Singer and Peach, 1983; Martin et al., 1984; Ignarro et al., 1984b, 1986; Gruetter and Lemke, 1986; also *see* other references given in Furchgott, 1984). Since vascular protein phosphorylation is associated with either endothelium-dependent muscarinic receptor activation (Rapoport et al., 1983) or with the direct action of chemical vasodilators or 8-bromo cyclic GMP (Rapoport et al., 1982), it is probable that the formation of cyclic GMP and cyclic GMP-dependent phosphorylation of specific proteins are involved in physiological responses to the muscarinic receptor in this tissue.

In tracheal smooth muscle, where the muscarinic receptor causes cyclic GMP elevation and muscular contraction (Katsuki and Murad, 1976), and where cyclic GMP addition induces protein phosphorylation (Hogaboom et al., 1982), the muscarinic receptor mediates a calcium-dependent phosphorylation of myosin light chain (Silver and Stull, 1983; Park and Rasmussen, 1986). Agents like sodium nitroprusside relax tracheal muscle and elevate cyclic GMP (Katsuki and Murad, 1976), so it is improbable that the elevation of cyclic GMP by the muscarinic receptor is involved in contraction *per se*. The patterns of tracheal protein phosphorylation elicited by phorbol ester treatment and by the muscarinic receptor appear similar, suggesting that the receptor activates protein kinase C (Park and Rasmussen, 1986). However, with the use of a rapid assay technique, it has been shown in tracheal muscle that the muscarinic receptor also increases the activity of a cyclic GMP-dependent kinase (Fiscus et al., 1984). The cyclic GMP formed in response to the muscarinic receptor appears to be involved in cellular mechanisms in this tissue unrelated to contraction.

1.2.5. Effects in Cardiac Tissue

In the heart, cyclic GMP levels seem to correlate with the negative inotropic effects of acetylcholine (George et al., 1970). In the ventricle, the muscarinic receptor mediates parasympathetic negative inotropic effects by its interaction with beta-adrenergic receptor–effector system; it is only when the sympathetic system is active and the beta-adrenergic receptor is stimulated that pronounced muscarinic effects in the ventricle are observable. β-receptor-mediated activation of a cyclic AMP-dependent protein kinase system in the ventricle appears to be the mechanism whereby the sympathetic system controls calcium fluxes. Several studies have shown that the ventricular muscarinic receptor reduces beta-adrenergic receptor-mediated cyclic AMP elevation, in association with cyclic GMP elevation (Endoh, 1979; Inui et al., 1982). Studies attempting to clarify the possible mechanism by which cyclic GMP acts to affect contractile force have been conflicting (for review, *see* Linden and Brooker, 1979). However, it has recently been shown that cyclic GMP is involved in regulating isoproterenol-stimulated calcium fluxes in the ventricle (Fischmeister and Hartzell, 1986; Hartzell and Fischmeister, 1986). These authors suggest that cyclic GMP, generated in response to muscarinic receptors, activates a phosphodiesterase that degrades cyclic AMP, leading to decreased calcium currents. Muscarinic activation inhibits the beta-adrenergic receptor stimulation of protein phosphorylation in the ventricle (Lindemann and Watanabe, 1985; Iwasa and Hosey, 1983), which is thought to mediate the beta-adrenergic receptor effects on cardiac calcium fluxes. Thus, cyclic GMP elevation and its subsequent alteration of the level of cyclic AMP could be one explanation for how the muscarinic receptor interacts with the beta-adrenergic receptor in the ventricle. Using LY83583 to lower intracellular cyclic GMP concentrations (Schmidt et al., 1985),

MacLeod and Diamond (1986) demonstrated that, in the presence of forskolin (to elevate cyclic AMP levels), the negative inotropic response to ventricular muscarinic receptors was completely blocked when cyclic GMP elevation was prevented. However, the physiology of the cardiac muscarinic receptor-cyclic GMP system appears to be even more complex than this, because LY83583 also blocks muscarinic receptor-mediated cyclic GMP formation in the atrium, but does so without blocking either the direct muscarinic receptor-mediated inotropic effect (Diamond and Chu, 1985), or the indirect inotropic effect mediated in the presence of elevated cyclic AMP (MacLeod and Diamond, 1986). Thus, it is only in the ventricle that there appears to be an interaction between cyclic AMP and cyclic GMP generating systems. However, in the ventricle, the mechanism of negative inotropism involves more than a cyclic GMP-mediated cyclic AMP reduction, because the muscarinic receptor, although reducing the chronotropic effects of forskolin in the ventricle, does not reduce ventricular forskolin-stimulated cyclic AMP levels (MacLeod and Diamond, 1986; Lindemann and Watanabe, 1985).

2. Muscarinic Receptor Subtype Mediating Cyclic GMP Formation

2.1. N1E-115 Neuroblastoma Cells

2.1.1. Muscarinic Receptor Subtypes in N1E-115

Both muscarinic and nicotinic cholinergic receptors are present in this murine neuroblastoma clone. Muscarinic and certain other receptors mediate cyclic GMP formation in these cells, and several studies of muscarinic receptor–effector coupling have employed this clone as a model system. In the last several years, binding studies have shown that both M_1 and M_2 subtypes of the muscarinic receptor are present. These subtypes appear to be functionally coupled to cyclic nucleotide levels. With the use of pirenzepine, which has selective antagonistic effects at subtypes of muscarinic receptors, we showed that the M_1 subtype mediates cyclic GMP formation, whereas the M_2 subtype mediates inhibition of hormone-induced cyclic AMP formation (McKinney et al., 1985). The two binding constants determined for this "atypical" muscarinic antagonist by pharmacological methods (Schild analysis) were in close agreement with those determined in biochemical assays (competition with [^{3}H]quinuclidinyl benzilate or [^{3}H]*N*-methyl scopolamine).

2.1.2. Agonist-Receptor Conformations

In a study of the activity of a series of muscarinic agonists in responses to the M_1 and M_2 receptors of N1E-115 cells, it was shown that most agonists were full or good partial agonists at the M_2-cyclic AMP systems, whereas only "acetylcholine-like" agonists were able to induce significantly the formation of cyclic GMP (McKinney et al., 1985). Without knowledge of the effects of pirenzepine indicating that these two responses are mediated by separate receptor–effector complexes, one would be tempted to invoke the "spare receptor" hypothesis to explain these differences in efficacies and potencies. However, binding studies in N1E-115 cells indicate that both low affinity and high affinity binding sites for full agonists exist with characteristics similar to those in cerebral cortex. Those agonists that are unable to mediate much of an increase in cyclic GMP levels (like oxotremorine, arecoline, and pilocarpine) appear to bind to a single class of sites, insofar as the technique could reveal. Detailed studies of their properties indicated that these drugs do bind to the M_1 receptor, but do so without inducing a low affinity conformation. Thus, the tertiary amine agonists appear unable to lead to cyclic GMP elevation by virtue of their inability to induce a low affinity agonist-receptor conformation. Full agonists (acetylcholine, methacholine, and carbachol) bind to two populations of receptors, and mediate both M_1 and M_2 responses. Thus, it was inferred that a high affinity conformation is involved in the inhibition of cyclic AMP levels, whereas a low affinity conformation is involved in the elevation of cyclic GMP. This was confirmed in a study where the binding constants for carbachol were determined by the method of partial receptor inactivation (McKinney and Richelson, 1986b). This study, although confirming a previous finding that the low affinity agonist-receptor conformation mediates cyclic GMP with a constant of about 400 μM (El-Fakahany and Richelson, 1981), also showed that the inhibition of cyclic AMP levels are mediated by an agonist-receptor conformation of the "high affinity" type (about 10 μM). These parameters are in reasonable agreement with binding constants determined directly by radioligand binding assays. Furthermore, the data indicate that few spare receptors exist in either receptor–effector system.

2.2. Correlation of Cyclic GMP and Phosphoinositide Metabolism

2.2.1. N1E-115 and Cortical Muscarinic Receptors

A comparison of N1E-115 and cortical muscarinic receptors reveals remarkable similarities. Both high and low affinity binding sites for agonists exist in approximately the same ratios and with similar binding constants in both tissues. In addition, pirenzepine binds to two classes of sites (M_1 and M_2) in both tissues (McKinney et al., 1985; Luthin and Wolfe, 1984).

2.2.2. Correlation with Phosphoinositide Metabolism

The cortical M_1 receptor stimulates phosphoinositide metabolism with a pharmacologic profile essentially identical to that of the N1E-115 M_1-cyclic GMP system: the categorization of full and partial agonists described in the section above for N1E-115 cells holds for the cortical M_1 receptor (Fisher et al., 1983). In the cortical system, the induction of a low affinity agonist-receptor conformation appears to be required (Fisher et al., 1983, 1984), as it does for cyclic GMP formation in N1E-115 cells (McKinney et al., 1985; McKinney and Richelson, 1986b). The N1E-115 M_1 receptor mediates phosphoinositide metabolism with an agonist efficacy profile quite similar to that of cyclic GMP formation (Snider and Agranoff, 1985). Further, several other N1E-115 receptors also coordinately mediate cyclic GMP and phosphoinositide metabolism with similar relative efficacies (Richelson and Pfenning, unpublished data), suggesting that the two biochemical responses are linked. Phorbol ester treatment blocks both these responses in N1E-115 cells, further suggesting some commonality (Kanba et al., 1986). The muscarinic receptor has been shown to elevate cyclic GMP levels in both the cortex (Lee et al., 1972) and striatum (Hanley and Iversen, 1978), tissues where muscarinic-phosphoinositide responses also have been described. A link between the two biochemical responses is also suggested by the finding that phosphatidic acid, a metabolite of the phosphoinositide cycle and a calcium ionophore, elicits cyclic GMP formation and calcium influx in N1E-115 cells (Ohsako and Deguchi, 1981). A recent study, however, concluded that phosphoinositide and cyclic GMP metabolism were not linked in the brain (Kendall, 1986). Further, cyclic GMP responses to the muscarinic receptor rapidly desensitize (Richelson, 1978; DeCoster et al., 1984), whereas the phosphoinositide response does not appear to, indicating that the effector systems may not be linked. Unfortunately, pirenzepine's binding profile or its differential blocking effects at biochemical responses are not definitive in establishing links between muscarinic receptor subtypes and biochemical responses. For example, although pirenzepine binds to a similar proportion of M_1 and M_2 sites in the cortex and striatum (Luthin and Wolfe, 1984), it blocks the phosphoinositide responses differentially in these two tissues (Fisher and Bartus, 1985). Conversely, if an M_1 response is defined by high affinity pirenzepine blockade, then both M_1 and M_2 receptors can mediate phosphoinositide responses (*see* Chapter 7).

3. Possible Mechanisms of Activation of Guanylate Cyclase

3.1. Calcium Channels

3.1.1. Requirement for Divalent Cations in Guanylate Cyclase Activation

Guanylate cyclase is an ubiquitous enzyme that can exist as a soluble or particulate type (Kimura and Murad, 1974; Goldberg and Haddox, 1977). Muscarinic stimulation does not elevate cyclic GMP in homogenates, which seems to indicate that the receptor is not coupled directly to the particulate guanylate cyclase. Since the cyclic GMP response mediated by muscarinic receptors requires the presence of extracellular calcium ions (Schultz et al., 1973), which often act as second messengers, it appears that muscarinic receptors stimulate the soluble form of guanylate cyclase in intact cells. However, calcium ions do not stimulate guanylate cyclase directly (e.g., Limbird and Lefkowitz, 1975), even though divalent cations are known to be important for guanylate cyclase activity (Chrisman et al., 1975; Goldberg and Haddox, 1977). Either manganese or magnesium ions will activate the soluble enzyme, although only manganese ions activate the particulate enzyme. In vitro, though calcium ions in the range of physiological cytosolic concentrations do not activate the soluble enzyme, they will enhance the activation by manganese ions. Calcium ions inhibit the particulate enzyme (Chrisman et al., 1975; Goldberg and Haddox, 1977). This and other differences between the soluble and particulate guanylate cyclase indicate that there are separate physiological functions for these two enzymes. Recently, it was shown that the particulate enzyme and a receptor for atrial natriuretic factor copurify as the same macromolecule (Kuno et al., 1986).

3.1.2. N1E-115 Calcium Channels

It has been generally assumed that muscarinic receptors are coupled in some way to calcium channels. Several studies have been directed towards characterizing the calcium channels or effectors to which muscarinic receptors seem to be coupled. Some of these studies have involved the calcium requirements of the muscarinic receptor/guanylate cyclase system of N1E-115 neuroblastoma cells. About 85% of the guanylate cyclase in these cells exists in the soluble form (Bartfai et al., 1978). In broken N1E-115 cell preparations, the soluble form of guanylate cyclase is not activated by calcium ions except at millimolar concentrations, although it possesses an allosteric site at which divalent cations bind and it will metabolize Ca-GTP (Bartfai et al., 1978). Depolarizing the cells to cause calcium ion influx elevates cyclic GMP levels in a way additive to that elevation caused by activation of the muscarinic receptor (Study et al., 1978). The cyclic GMP response mediated by the voltage-sensitive mechanism is more sensitive to blockade by cobalt than is that response elicited by the muscarinic receptor. The effects of manganese ions on cyclic GMP formation and receptor desensitization in N1E-115 cells suggest that receptors that stimulate cyclic GMP do so by activating calcium channels (El-Fakahany and Richelson, 1980b). The lanthanides affect muscarinic receptor–effector coupling in N1E-115 cells in such a way as to suggest the existence of calcium channels (El-Fakahany and Richelson, 1980a,b). Calcium channel antagonists have effects on muscarinic receptor-mediated cyclic GMP formation (El-Fakahany and Richelson, 1983). Calcium influx is not measurable with the photoprotein aequorin (Snider et al., 1984) that luminesces upon binding calcium ions, but use of quin 2, a probe that changes its fluorescence properties upon binding these ions, made possible the demonstration of a small cytosolic calcium increase in response to the muscarinic receptor (Ohsako and Deguchi, 1984). We recently confirmed these results in preliminary studies with single cells and another fluorescent probe, fura 2 (Kanba et al., unpublished observations). It is possible that $InsP_3$, formed in response to muscarinic activation of the phosphoinoside metabolic cycle, releases the calcium ions from cytoplasmic stores. Paradoxically, the calcium ions required for cyclic GMP formation appear to originate extracellularly. Though calcium ion is unquestionably required for cyclic GMP stimulation in N1E-115 cells and in many tissues, and various studies indicate that the N1E-115 muscarinic receptor activates calcium channels, it seems unlikely that a sufficiently high cytosolic calcium concentration results from this activation to activate directly guanylate cyclase to the degree observed (from 10–100-fold in N1E-115 cells). This suggests that additional second-messenger processes may be involved, such as the formation of eicosanoids (Snider et al., 1984).

3.2. Oxygen and Lipids

3.2.1. Heme in the Guanylate Cyclase

Agents from which nitric oxide can be generated activate both the soluble and particulate forms of guanylate cyclase; studies in various tissues have shown that these agents interact with a heme moiety and sulfhydryl groups of the enzyme to regulate the activity of the enzyme. Purification of soluble guanylate cyclase often causes the loss of responsiveness to the NO-containing vasodilators, and responsiveness is restored with heme or certain porphyrins (e.g., Craven and DeRubertis, 1978; Ohlstein et al., 1982; Ignarro et al., 1984a; *see also* Ignarro et al., 1986a). Purified soluble enzyme from lung retains its sensitivity to the vasodilators because heme is retained during the purification (Gerzer et al., 1981a, 1982). Many vasodilators appear to interact directly with the heme to activate guanylate cyclase, whereas certain ones (e.g., phenylhydrazine and sodium azide) require the additional presence of catalase (Gerzer et al., 1981b; Ignarro et al., 1984c) or thiols (Ignarro and Gruetter, 1980) to activate guanylate cyclase. For retention of enzyme activity, it is thought that the heme iron must be maintained in the ferrous state (Craven and DeRubertis, 1978); agents like methylene blue might therefore block formation of cyclic GMP in tissues by directly oxidizing the heme iron of the guanylate cyclase. Protoporphyrin IX, a demetalated heme precursor, can bind to the apoenzyme and activate it in the absence of NO (Wolin et al., 1982). Thus, the mechanism of guanylate cyclase activation by vasodilators appears to involve the formation of NO or S-nitrosothiol, which interacts directly with the iron of the guanylate cyclase heme (Craven and DeRubertis, 1978) to cause it to be structurally similar to protoporphyrin IX (Ignarro et al., 1984a). Studies of the particulate guanylate cyclase indicate differences with regard to regulation by porphyrins (Waldman et al., 1984). The heme group may be important in the regulation of the enzyme in vivo, for example, by lipids, oxygen, and their radicals.

3.2.2. Oxidation Conditions and Guanylate Cyclase Activation

The enzyme's activity in various tissues is increased by a variety of oxidizing or oxidizable agents, e.g., hydrogen peroxide (White et al., 1976), hydroxyl radicals (Murad et al., 1977; Mittal and Murad, 1977), fatty acids (Glass et al., 1977; Gerzer et al., 1983), fatty acid hydroperoxides (Hidaka and Asano, 1977; Goldberg et al., 1977), and certain other chemical redox perturbants (Goldberg et al., 1977; Deguchi, 1977; Murad et al., 1977; Arnold et al., 1977). The enzyme in some preparations will spontaneously activate by an oxidation process (White et al., 1976, 1982; Goldberg et al., 1977; Yoshikawa and Kuriyama, 1980). It would appear from these data that the cyclase is activated in vitro by some type of oxidation-reduction reaction, probably involving the heme group and perhaps also involving sulfhydryl groups (Goldberg et al., 1977; Braughler, 1982). It is thus conceivable that receptors initiate a similar mechanism for elevating cyclic GMP in vivo. It has been shown that incubating smooth muscle (human umbilical cord) in oxygen induces cyclic GMP formation, even in the presence of mitochondrial inhibitors, whereas purging the tissue with nitrogen prevents agonist-induced cyclic GMP formation (Clyman et al., 1975). Lipoxygenase when added to some preparations can activate guanylate cyclase (White et al., 1982; Mittal and Murad, 1978). A catalase-like enzyme has been implicated in guanylate cyclase activation by some agents (Mittal et al., 1977; Miki et al., 1976). Superoxide dismutase can participate in the generation of oxygen-free radicals to activate guanylate cyclase (Mittal and Murad, 1977; Yoshikawa and Kuriyama, 1980). Thus, several intracellular or membrane-bound enzymes that interact with oxygen might be involved in modulating soluble guanylate cyclase.

3.2.3. Activation of Guanylate Cyclase by Lipids

Guanylate cyclase of the spleen (Goldberg et al., 1977), liver (Braughler et al., 1979), adrenal cortex (Struck and Glossmann, 1978), platelets (Hidaka and Asano, 1977; Glass et al., 1977; Gerzer et al., 1983), kidney (Craven and DeRubertis, 1982), mammary gland (Rillema, 1978), and guinea-pig myometrium (Leiber and Harbon, 1982) is activated by unsaturated fatty acids or fatty acid hydroperoxides. These studies suggest that receptors in these tissues activate guanylate cyclase by stimulating the release and metabolism of arachidonic acid or other unsaturated fatty acids. This release could be effected by a phospholipase A_2, which is calcium-activated in many tissues, or by diacylglycerol lipase acting on a product of the phosphoinositide cycle. Thus, an involvement of arachidonate metabolism in guanylate cyclase-activating mechanisms would share several features with other biochemical responses to the muscarinic receptor. Gerzer et al. (1983) showed that calcium ions mediate a release of unsaturated fatty acids from platelet membranes to activate platelet guanylate cyclase. Similarly, calcium ions mediate arachidonate release in the kidney to elevate cyclic GMP levels (Craven and DeRubertis, 1982). It appears therefore that, in various tissues, unsaturated fatty acid released from membrane phospholipids may either stimulate guanylate cyclase directly or be metabolized before interacting with the enzyme. This is inferred from the fact that fatty acid hydroperoxides stimulate the guanylate cyclase and that inhibitors of fatty acid hydroperoxide or endoperoxide formation often block the formation of cyclic GMP (Spies et al., 1980; Snider et al., 1984; Craven and DeRubertis, 1980; Briggs and DeRubertis, 1979; Leiber and Harbon, 1982; Rapoport and Murad, 1983; Kendall, 1986). However, lipoxygenase inhibitors can inhibit guanylate cyclase directly (Clark and Linden, 1986; Gerzer et al., 1986); therefore, it is not yet clear whether a lipoxygenase intervenes to produce a metabolite of arachidonate to activate guanylate cyclase. It is possible that unsaturated lipids could be oxidized by the guanylate cyclase itself, since it contains heme, which could catalyze the oxidation of eicosanoids (Tappel, 1955; Radmark et al., 1984). However, Gerzer et al. (1986) could find no evidence that guanylate cyclase mediated an oxidation of arachidonate, and the activation of the enzyme in vitro by this lipid did not require the presence of oxygen.

4. Muscarinic Receptor-Mediated Arachidonate Metabolism

4.1. Blockade of Muscarinic Responses by Arachidonic Metabolic Inhibitors

4.1.1. Smooth Muscle Tissue

The muscarinic receptor of vascular endothelial cells stimulates the release of "endothelium-dependent relaxation factor" (EDRF); because the effect of EDRF is blocked by lipoxygenase inhibitors, this factor may be an arachidonic acid metabolite (Furchgott and Zawadzki, 1980; Singer and Peach, 1983; Forstermann and Neufang, 1984). However, recently workers have proposed that nitric oxide may be somehow generated naturally in response to receptor activation (Palmer et al., 1987; Ignarro et al., 1988). The muscarinic receptor of guinea-pig myometrium mediates a calcium-dependent cyclic GMP resonse that is blocked by eicosatetraynoic acid (ETYA) (Leiber and Harbon, 1982). In the rat ductus deferens (Spies et al., 1980), phospholipase A_2 or lipoxygenase inhibitors block the muscarinic receptor-mediated cyclic GMP formation.

4.1.2. N1E-115 Neuroblastoma Cells

A lipoxygenase appears to be involved in the muscarinic receptor-mediated elevation of cyclic GMP in murine neuroblastoma cells (Snider et al., 1984). This conclusion was based on the blockade of this response by ETYA and nordihydroguaiaretic acid (NDGA), another lipoxygenase inhibitor. Indomethacin, a cyclooxygenase inhibitor, was without effect on the cyclic GMP response. Further studies with a more extensive series of metabolic inhibitors substantiated this view (McKinney and Richelson, 1986a), but direct evidence of a lipoxygenase activity in these cells has not been obtained. Additionally, the mechanism of cyclic GMP stimulation may be more complex than activation of lipoxygenase. Exogenous arachidonic acid would be expected to stimulate cyclic GMP in N1E-115 cells; to the contrary, it mediates an inhibition that is increased in potency by oxidation (McKinney and Richelson, 1986a). This led to the finding that specific arachidonate enzymatic products, like hydroxyeicosatetraenoic acids (HETEs), were also inhibitors. 5-HETE, 12-HETE, and 15-HETE are equipotent with ETYA in the blockade of muscarinic receptor-mediated cyclic GMP in these cells (McKinney, 1987). Arachidonate inhibitors like indomethacin, ETYA, and 15-HETE elevate free arachidonate in N1E-115 cells, consistent with blockade of metabolic pathways. Additionally, the HETEs appear to be esterified in N1E-115 phospholipids, and they block the esterification of arachidonate. A redistribution of arachidonate pools occurs when cells are incubated with a HETE. Thus, arachidonate metabolic inhibitors may influence receptor function by mechanisms other than lipoxygenase blockade, e.g., by altering membrane fluidity or by changing the characteristics of lipid substrates or activators of phospholipases or protein kinase C. As pointed out above, lipoxygenase inhibitors can directly inhibit guanylate cyclase. Esterification of 12-HETE has been reported in another neural tissue, the retina (Birkle and Bazan, 1984). With the multiple actions of inhibitors that are possible, one cannot at the present draw any firm conclusions regarding the involvement of a lipoxygenase pathway in receptor-mediated cyclic GMP formation in N1E-115 cells. Many further complexities are possible, since there are several points where "crosstalk" or "feedback" in the biochemical pathways could exist. For example, protein kinase C can be activated by arachidonic acid in neutrophils (McPhail et al., 1984); rat brain guanylate cyclase is phosphorylated and activated by protein kinase C (Zwiller et al., 1985); HETEs can regulate the activity of platelet phospholipase A_2 (Chang et al., 1985); diacylglycerol can activate mobilization of arachidonic acid in macrophages (Emilsson et al., 1986); and protein kinase C is implicated in the desensitization of the N1E-115 muscarinic receptor (Liles et al., 1986).

4.2. Release of Eicosanoids by the Muscarinic Receptor

4.2.1. N1E-115 Cells

There is good evidence that calcium-dependent release of unsaturated fatty acids from the plasma membrane is a key event in guanylate cyclase activation (Gerzer et al., 1983; Briggs and DeRubertis, 1979; Craven and DeRubertis, 1980). As phospholipids are the primary depot for arachidonic acid, it is believed that receptors must activate an enzyme that releases arachidonic acid. However, evidence for the muscarinic receptor-mediated elevation in cellular levels of unesterified arachidonic acid is rare. In N1E-115 cells, agents that elevate cyclic GMP levels (carbachol, histamine, ionophore X537A, and melittin) also elevate free arachidonate levels (Snider et al., 1984). The apparent release by receptor activation is only a minor fraction over basal release, but it is blocked by receptor antagonists, as well as by quinacrine, which is thought to block the activity of phospholipase A_2. Further studies of arachidonic acid metabolism in N1E-115 cells indicate that this lipid cycles rapidly into and out of membrane phospholipids (McKinney, 1987). A rapid reesterification process might explain why it is difficult to elicit a robust rise in free arachidonate levels with receptor activation. Though arachidonic acid appears to be oxidatively metabolized in these cells, only very low levels of polar metabolites are found. Hydroxyeicosatetraenoic acids are also rapidly esterified into N1E-115 phospholipids (McKinney, 1987); this may provide an explanation for why it has not been possible to demonstrate the muscarinic receptor-stimulated lipoxygenation of arachidonic acid.

4.2.2. C62B Glioma and Astroglioma Cells

The muscarinic receptor mediates a transient release of arachidonic acid in C62B glioma cells (DeGeorge et al., 1986a). Carbachol is a full agonist, and oxotremorine and pilocarpine are ineffective. Phosphatidic acid levels appear also to rise transiently, indicating the activation of a phospholipase C. The arachidonic acid released in response to muscarinic receptor activation in C62B cells is further metabolized to an unidentified lipoxygenase product. Very small amounts of prostaglandins are formed in response to the muscarinic receptor, but prolonged ionophore stimulation provokes the release and metabolism of arachidonate to several prostaglandins, as well as to 5-HETE and 12-HETE. Lapetina and coworkers showed also that muscarinic receptors in primary astroglial cultures stimulate the release of arachidonic acid (DeGeorge et al., 1986b). Other receptors in these cultures also stimulate release of arachidonate, but ionophore treatment was necessary to demonstrate the concurrent metabolism of arachidonate to prostaglandins and an unidentified metabolite. As glial cultures can synthesize cyclic GMP (Bloch-Tardy et al., 1980), and since muscarinic M_1 (mediating phosphoinositide metabolism) and M_2 receptors (mediating cyclic AMP reduction) have been shown to exist in astrocytes (Murphy et al., 1986), this system may permit molecular studies of a putative relationship between muscarinic receptors, arachidonic acid metabolism, and guanylate cyclase activation.

4.2.3. Other Tissues

The muscarinic receptor mediates a calcium-dependent, indomethacin-sensitive release of prostacyclin from the endothelial layer of rabbit aorta (Boeynaems and Galand, 1983), indicating the probable stimulation of arachidonic acid release by the muscarinic receptor in this smooth muscle tissue. Though prostacyclin is a vasodilator, it cannot be EDRF, since the vasodilation mediated by the latter substance is insensitive to cyclooxygenase blockade. Muscarinic receptors in the rabbit iris mediate phosphoinositide metabolism, and arachidonic acid and PGE_2 release (Yousufzai and Abdel-Latif, 1984). Muscarinic receptors also stimulate the release of arachidonic acid and PGE_2 in the Torpedo electric organ (Pinchasi et al., 1984). The muscarinic receptor mediates the release of prostaglandins PGE_2, PGI_2, and $PGF_{2\alpha}$ in the rat urinary bladder (Jeremy et al., 1986). In the latter system, in which parasympathetic activity elicits a muscarinic receptor-mediated contraction, it was shown with the use of pirenzepine that the M_2 subtype of the muscarinic receptor was involved in the release of the eicosanoids, and interestingly, the tertiary amine agonist arecoline was equal to carbachol and methacholine in intrinsic activity. The effects of a metal ion chelator indicated that calcium ions are involved in the receptor-mediated stimulation of eicosanoid metabolism.

4.3. Lipoxygenation in Neural Tissue and Relevance to Disease

4.3.1. Evidence for Lipoxygenase Activity in Neural Tissue

Metabolism of arachidonic acid via the cyclooxygenase pathway in neural tissue has been described and implicated in modulation of neurotransmission (Bergstrom et al., 1973; Brody and Kadowitz, 1974; Hayaishi, 1983; Hemker and Aiken, 1980; Hillier and Templeton, 1980). If an arachidonic acid metabolic cascade is involved in receptor-mediated cyclic GMP formation in N1E-115 cells (Snider et al., 1984; McKinney and Richelson, 1984), this might indicate that such a process is involved in the brain. Kendall (1986) showed that depolarization-induced cyclic GMP formation in the brain is blocked by lipoxygenase inhibitors.

Lapetina and coworkers found evidence that astrocytes possessed a lipoxygenase pathway (DeGeorge et al., 1986b). Brain slices that were made free of blood by perfusion of the brain before slice preparation synthesize 5-HETE and 12-HETE (Adesuyi et al., 1985). There is evidence that leukotrienes, which are important products of the lipoxygenase pathway, are formed in brain tissue (Lingren et al., 1984; Moskowitz et al., 1984). Thus, there may be fundamental roles for lipoxygenase products, as well as cyclooxygenase products in neurotransmission.

4.3.2. Potential Neuropathologic Implications

Receptor mechanisms involving the oxidation of unsaturated fatty acids would likely be particularly perturbed in brain states, such as ischemia or edema, which are associated with the elevation and rapid peroxidation of unesterified fatty acids (Lunt and Rowe, 1968; Bazan, 1970). Studies of N1E-115 cells indicate that most polar metabolites of arachidonic acid inhibit muscarinic receptor-mediated cyclic GMP (McKinney and Richelson, 1986a). Increased levels of lipid peroxidation products have been reported in Alzheimer's disease (Chia et al., 1984) and in Down's syndrome (Brooksbank and Balazs, 1983). Pathological processes that give rise to these products could have detrimental effects by a variety of mechanisms at multiple points in receptor–effector systems (Halliwell and Gutteridge, 1984).

Acknowledgments

The writing of this manuscript and the work cited from the authors' laboratories was sponsored in part by Mayo Foundation and U.S.P.H.S. grants NS21319 and MH27692.

References

Adesuyi, S. A., Cockrell, C. S., Gamache, D. A., and Ellis, E. F. (1985) Lipoxygenase metabolism of arachidonic acid in brain. *J. Neurochem.* **45**, 770–776.

Ariano, M. A. (1983) Distribution of components of the guanosine 3′,5′-phosphate system in rat caudate-putamen. *Neuroscience* **10**, 707–723.

Ariano, M. A. and Matus, A. I. (1981) Ultrastructural localization of cyclic GMP and cyclic AMP in rat striatum. *J. Cell. Bio.* **91**, 287–292.

Arnold, W. P., Mittal, C. K., Katsuki, S., and Murad, F. (1977) Nitric oxide activates guanylate cyclase and increases guanosine 3′:5′-cyclic monophosphate levels in various tissue preparations. *Proc. Natl. Acad. Sci. USA* **74**, 3203–3207.

Aswad, D. W. and Greengard, P. (1981a) A specific substrate from rabbit cerebellum for guanosine 3′:5′-monophosphate-dependent protein kinase. I. Purification and characterization. *J. Biol. Chem.* **256**, 3487–3493.

Aswad, D. W. and Greengard, P. (1981b) A specific substrate from rabbit cerebellum for guanosine 3′:5′-monophosphate-dependent protein kinase. II. Kinetic studies on its phosphorylation by guanosine 3′:5′-monophosphate-dependent and adenosine 3′:5′-monophosphate-dependent protein kinases. *J. Biol. Chem.* **256**, 3494–3500.

Bartfai, T., Breakefield, X. O., and Greengard, P. (1978) Regulation of synthesis of guanosine 3′:5′-cyclic monophosphate in neuroblastoma cells. *Biochem. J.* **176**, 119–127.

Bazan, N. G. (1970) Effects of ischemia and electroconvulsive shock on free fatty acid pool in the brain. *Biochem. Biophys. Acta* **218**, 1–10.

Bergstrom, S., Farnebo, L. O., and Fuxe, K. (1973) Effect of prostaglandin E_2 on central and peripheral catecholamine neurons. *Eur. J. Pharmacol.* **21**, 362–368.

Birkle, D. L. and Bazan, N. G. (1984) Effect of K^+ depolarization on the synthesis of prostaglandins and hydroxyeicosatetra(5,8,11,14)-enoic acids (HETE) in the rat retina. Evidence for esterification of 12-HETE in lipids. *Biochim. Biophys. Acta* **795**, 564–573.

Black, A. C., Sandquist, D., West, J. R., Wamsley, J. K., and Williams, T. H. (1979) Muscarinic cholinergic stimulation increases cyclic GMP levels in rat hippocampus. *J. Neurochem.* **33**, 1165–1168.

Bloch-Tardy, M., Fages, C., and Gonnard, P. (1980) Cyclic guanosine monophosphate in primary cultures of glial cells. *J. Neurochem.* **35**, 612–615.

Boeynaems, J. M. and Galand, N. (1983) Cholinergic stimulation of vascular prostaglandin synthesis. *Prostaglandins* **26**, 531–544.

Braughler, J. M. (1982) Soluble guanylate cyclase activation by nitric oxide and its reversal. *Biochem. Pharmacol.* **32**, 811–818.

Braughler, J. M., Mittal, C. K., and Murad, F. (1979) Purification of soluble guanylate cyclase from rat liver. *Proc. Natl. Acad. Sci. USA* **76**, 219–222.

Briggs, R. G. and DeRubertis, F. R. (1979) Calcium-dependent modulation of guanosine 3′,5′-monophosphate in renal cortex. Possible relationship to calcium-dependent release of fatty acid. *Biochem. Pharmacol.* **29**, 717–722.

Brody, M. J. and Kadowitz, P. J. (1974) Prostaglandins as modulators of the autonomic nervous system. *Fed. Proc.* **33**, 48–60.

Brooksbank, B. W. L. and Balazs, R. (1983) Superoxide dismutase and lipoperoxidation in Down's syndrome fetal brain. *Lancet* **1**, 881–882.

Cahill, A. L. and Perlman, R. L. (1986) Nicotinic and muscarinic agonists, phorbol esters, and agents which raise cyclic AMP levels phosphorylate distinct groups of proteins in the superior cervical ganglion. *Neurochem. Res.* **11**, 327–338.

Casnellie, J. E. and Greengard, P. (1974) Guanosine 3′:5′-cyclic monophosphate-dependent phosphorylation of endogenous substrate proteins in membranes of mammalian smooth muscle. *Proc. Natl. Acad. Sci. USA* **71**, 1891–1895.

Chan-Palay, V. and Palay, S. L. (1979) Immunocytochemical localization of cyclic GMP: light and electron microscope evidence for involvement of neuroglia. *Proc. Natl. Acad. Sci. USA* **76**, 1485–1488.

Chang, J., Blazek, E., Kreft, A. F., and Lewis, A. J. (1985) Inhibition of platelet and neutrophil phospholipase A2 by hydroxyeicosatetraenoic acids (HETEs). A novel pharmacological mechanism for regulating free fatty acid release. *Biochem. Pharmacol.* **34**, 1571–1575.

Chia, L. S., Thompson, J. E., and Moscarello, M. A. (1984) X-ray diffraction evidence for myelin disorder in brain from humans with Alzheimer's disease. *Biochim. Biophys. Acta* **775**, 308–312.

Chrisman, T. D., Garbers, D. L., Parks, M. A., and Hardman, J. G. (1975) Characterization of particulate and soluble guanylate cyclases from rat lung. *J. Biol. Chem.* **250**, 374–381.

Clark, D. L. and Linden, J. (1986) Modulation of guanylate cyclase by lipoxygenase inhibitors. *Hypertension* **8**, 947–950.

Clyman, R. I., Blacksin, A. S., Manganiello, V. C., and Vaughan, M. (1975) Oxygen and cyclic nucleotides in human umbilical artery. *Proc. Natl. Acad. Sci. USA* **72**, 3883–3887.

Cobbs, W. H., Barkdoll, A. E. III, and Pugh, E. N., Jr. (1985) Cyclic GMP increases photocurrent and light sensitivity of retinal cones. *Nature* **317**, 64–66.

Craven, P. A. and DeRubertis, F. R. (1978) Restoration of the responsiveness of purified guanylate cyclase to nitrosoguanidine, nitric oxide, and related activators by heme and hemeproteins. *J. Biol. Chem.* **253**, 8433–8443.

Craven, P. A. and DeRubertis, F. R. (1980) Calcium and O_2-dependent control of inner medullary cGMP: possible role for Ca^{+2}-dependent arachidonate release and prostaglandin synthesis in expression of the action of osmolality on renal inner medullary guanosine 3′,5′-monophosphate. *Metab.* **29**, 842–853.

Craven, P. A. and DeRubertis, F. R. (1982) Relationship between calcium stimulation of cyclic GMP and lipid peroxidation in the rat kidney: Evidence for involvement of calmodulin and separate pathways of peroxidation in cortex versus inner medulla. *Metab.* **31**, 103–116.

DeCoster, C., Moreau, C., and Dumont, J. E. (1984) Desensitization of carbamylcholine-mediated cyclic GMP accumulation in dog thyroid slices. *Biochim. Biophys. Acta* **798**, 235–239.

DeGeorge, J. J., Morell, P., McCarthy, K. D., and Lapetina, E. G. (1986a) Cholinergic stimulation of arachidonic acid and phosphatidic acid metabolism in C62B glioma cells. *J. Biol. Chem.* **261**, 3428–3433.

DeGeorge, J. J., Morell, P., McCarthy, K. D., and Lapetina, E. G., (1986b) Adrenergic and cholinergic stimulation of arachidonate and phosphatidate metabolism in cultured astroglial cells. *Neurochem. Res.* **11**, 1061–1071.

Deguchi, T. (1977) Activation of guanylate cyclase in cerebral cortex of rat by hydroxylamine. *J. Biol. Chem.* **252**, 596–601.

Diamond, J. and Chu, E. B. (1985) A novel cyclic GMP-lowering agent, LY83583, blocks carbachol-induced cyclic GMP elevation in rabbit atrial strips without blocking the negative inotropic effects of carbachol. *Can. J. Physiol. Pharmacol.* **63**, 908–911.

Dinnendahl, V. and Stock, K. (1975) Effects of arecoline and cholinesterase inhibitors on cyclic guanosine 3′,5′-monophosphate and adenosine 3′,5′-monophosphate in mouse brain. *Naunyn-Schmied. Arch. Pharmacol.* **290**, 297–306.

El-Fakahany, E. and Richelson, E. (1980a) Effects of lanthanides on muscarinic acetylcholine receptor function. *Mol. Pharmacol.* **19**, 282–290.

El-Fakahany, E. and Richelson, E. (1980b) Involvement of calcium channels in short-term desensitization of muscarinic receptor-mediated cyclic GMP formation in mouse neuroblastoma cells. *Proc. Natl. Acad. Sci. USA* **77**, 6897–6901.

El-Fakahany, E. and Richelson, E. (1981) Phenoxybenzamine and dibenamine interactions with calcium channel effectors of the muscarinic receptor. *Mol. Pharmacol.* **20**, 519–525.

El-Fakahany, E. and Richelson, E. (1983) Effect of some calcium antagonists on muscarinic receptor-mediated cyclic GMP formation. *J. Neurochem.* **40**, 705–710.

Emilsson, A., Wijkander, J., and Sundler, R. (1986) Diacylglycerol induces deacylation of phosphatidylinositol and mobilization of arachidonic acid in mouse macrophages. Comparison with induction with phorbol ester. *Biochem. J.* **239**, 685–690.

Endoh, M. (1979) Correlation of cyclic AMP and cyclic GMP levels with changes in contractile force of dog ventricular myocardium during cholinergic antagonism of positive inotropic actions of histamine, glucagon, theophylline and papaverine. *Jpn. J. Pharmacol.* **29**, 855–864.

Ferrendelli, J. A., Steiner, A. L., McDougal, J. R., and Kipnis, D. M. (1970) The effect of oxotremorine and atropine on cGMP and cAMP levels in mouse cerebral cortex and cerebellum. *Biochem. Biophys. Res. Comm.* **41**, 1061–1067.

Fesenko, E. E., Kolesnikow, S. S., and Lyubarsky, A. L. (1985) Induction by cyclic GMP of cationic conductance in plasma membrane of retinal rod outer segment. *Nature* **313**, 310–313.

Fischmeister, R. and Hartzell, C. (1986) Mechanism of action of acetylcholine on calcium current in single cells from frog ventricle. *J. Physiol. Lond.* **376**, 183–202.

Fiscus, R. R., Torphy, T. J., and Mayer, S. E. (1984) Cyclic GMP-dependent protein kinase activation in canine tracheal smooth muscle by methacholine and sodium nitroprusside. *Biochim. Biophys. Acta* **805**, 382–392.

Fisher, S. K. and Bartus, R. T. (1985) Regional differences in the coupling of muscarinic receptors to inositol phospholipid hydrolysis in guinea pig brain. *J. Neurochem.* **45**, 1085–1095.

Fisher, S. K., Figueiredo, J. C., and Bartus, R. T. (1984) Differential stimulation of inositol phospholipid turnover in brain by analogs of oxotremorine. *J. Neurochem.* **43**, 1171–1179.

Fisher, S. F., Klinger, P. D., and Agranoff, B. W. (1983) Muscarinic agonist binding and phospholipid turnover in brain. *J. Biol. Chem.* **258**, 7358–7363.

Forstermann, U. and Neufang, B. (1984) The endothelium-dependent vasodilator effect of acetylcholine: Characterization of the endothelial relaxing factor with inhibitors of arachidonic acid metabolism. *Eur. J. Pharmacol.* **103**, 65–70.

Forstermann, U., Mulsch, A., Bohme, E., and Busse, R. (1986) Stimulation of soluble guanylate cyclase by an acetylcholine-induced endothelium-derived factor from rabbit and canine arteries. *Circ. Res.* **58**, 531–538.

Frey, E. A. and McIsaac, R. J. (1981) A comparison of cyclic guanosine 3′:5′-monophosphate and muscarinic excitatory responses in the superior cervical ganglion of the rat. *J. Pharmacol. Exp. Ther.* **218**, 115–121.

Furchgott, R. F. (1984) The role of endothelium in the responses of vascular smooth muscle to drugs. *Ann. Rev. Pharmacol. Toxicol.* **24**, 175–197.

Furchgott, R. F. and Zawadzki, J. V. (1980) The obligatory role of endothelial cells in the relaxation of arterial smooth muscle by acetylcholine. *Nature* **288**, 373–376.

George, W. J., Polson, J. B., O'Toole, A. G., and Goldberg, N. D. (1970) Elevation of guanosine 3′,5′-cyclic phosphate in rat heart after perfusion with acetylcholine. *Proc. Natl. Acad. Sci. USA* **66**, 398–403.

Gerzer, R., Bohme, E., Hofmann, F., and Schultz, G. (1981a) Soluble guanylate cyclase purified from bovine lung contains heme and copper. *FEBS Lett.* **132**, 71–74.

Gerzer, R., Hofmann, F., and Schultz, G. (1981b) Purification of a soluble, sodium-nitroprusside-stimulated guanylate cyclase from bovine lung. *Eur. J. Biochem.* **116**, 479–486.

Gerzer, R., Brash, A. R., and Hardman, J. G. (1986) Activation of soluble guanylate cyclase by arachidonic acid and 15-lipoxygenase products. *Biochim. Biophys. Acta* **886**, 383–389.

Gerzer, R., Hamet, P., Ross, A. H., Lawson, J. A., and Hardman, J. G. (1983) Calcium-induced release from platelet membranes of fatty acids that modulate soluble guanylate cyclase. *J. Pharmacol. Exp. Ther.* **226**, 180–186.

Gerzer, R., Radany, E. W., and Garbers, D. L. (1982) The separation of the heme and apoheme forms of soluble guanylate cyclase. *Biochem. Biophys. Res. Comm.* **108**, 678–686.

Glass, D. B., Frey, W. II, Carr, D. W., and Goldberg, N. D. (1977) Stimulation of human platelet guanylate cyclase by fatty acids. *J. Biol. Chem.* **252**, 1279–1285.

Goldberg, N. D. and Haddox, M. K. (1977) Cyclic GMP metabolism and involvement in biological regulation. *Ann. Rev. Biochem.* **46**, 823–896.

Goldberg, N. D., Graff, G., Haddox, M. K., Stephenson, J. H., Glass, D. B., and Moser, M. E. (1977) Redox modulation of splenic cell soluble guanylate cyclase activity: activation by hydrophilic and hydrophobic oxidants represented by ascorbic and dehydroascorbic acids, fatty acid hydroperoxides, and prostaglandin endoperoxides. *Adv. Cycl. Nuc. Res.* **9**, 101–130.

Gruetter, C. A. and Lemke, S. M. (1986) Comparison of endothelium-dependent relaxation in bovine intrapulmonary artery and vein by acetylcholine and A23187. *J. Pharmacol. Exp. Ther.* **238**, 1055–1062.

Gruetter, C. A., Kadowitz, P. J., and Ignarro, L. J. (1981) Methylene blue inhibits coronary arterial relaxation and guanylate cyclase activation by nitroglycerin, sodium nitrite, and amyl nitrite. *Can. J. Physiol. Pharmacol.* **59**, 150–156.

Halliwell, B. and Gutteridge, J. M. C. (1984) Oxygen toxicity, oxygen radicals, transition metals and disease. *Biochem. J.* **219**, 1–14.

Hanley, M. R. and Iversen, L. L. (1978) Muscarinic cholinergic receptors in rat corpus striatum and regulation of guanosine cyclic 3′,5′-monophosphate. *Mol. Pharmacol.* **14**, 246–255.

Hartzell, H. C. and Fischmeister, R. (1986) Opposite effects of cyclic GMP and cyclic AMP on Ca^{2+} current in single heart cells. *Nature* **323**, 273–275.

Hayaishi, O. (1983) Prostaglandin D_2: A neuromodulator. *Adv. Prost. Thromb. Leuk. res.* **12**, 333–337.

Haynes, L. W., Kay, A. R., and Yau, K-W. (1986) Single cyclic GMP-activated channel activity in excised patches of rod outer segment membrane. *Nature* **321**, 66–70.

Hemker, D. P. and Aiken, J. W. (1980) Modulation of autonomic neurotransmission by PGD_2: Comparison with effects of other prostaglandins in anesthetized cats. *Prostaglandins* **20**, 321–332.

Hidaka, H. and Asano, T. (1977) Stimulation of human platelet guanylate cyclase by unsaturated fatty acid peroxides. *Proc. Natl. Acad. Sci. USA* **74**, 3657–3661.

Hillier, K. and Templeton, W. W., Jr. (1980) Regulation of noradrenaline overflow in rat cortex by prostaglandin E_2. *J. Pharmacol.* **70**, 469–473.

Hofmann, F. and Sold, G. (1972) A protein kinase activity from rat cerebellum stimulated by guanosine-3′:5′-monophosphate. *Biochem. Biophys. Res. Comm.* **49**, 1100–1107.

Hogaboom, G. K., Emler, C. A., Butcher, F. R., and Fedan, J. S. (1982) Concerted phosphorylation of endogenous tracheal smooth muscle membrane proteins by Ca^{2+}:calmodulin-, cyclic GMP-, and cyclic AMP-dependent protein kinases. *FEBS Lett.* **139**, 309–312.

Ignarro, L. J. and Gruetter, C. A. (1980) Requirement of thiols for activation of coronary arterial guanylate cyclase by glyceryl trinitrate and sodium nitrite. *Biochim. Biophys. Acta* **631**, 221–231.

Ignarro, L. J., Adams, J. B., Horwitz, P. M., and Wood, K. S. (1986a) Activation of soluble guanylate cyclase by NO-hemoproteins involves NO-heme exchange. Comparison of heme-containing and heme-deficient enzyme forms. *J. Biol. Chem.* **261**, 4997–5002.

Ignarro, L. J., Harbison, R. G., Wood, K. S., and Kadowitz, P. J. (1986b) Activation of purified soluble guanylate cyclase by endothelium-derived relaxing factor from intrapulmonary artery and vein: Stimulation by acetylcholine, bradykinin and arachidonic acid. *J. Pharmacol. Exp. Ther.* **237**, 893–900.

Ignarro, L. J., Ballot, B., and Wood, K. S. (1984a) Regulation of soluble guanylate cyclase activity by porphyrins and metalloporphyrins. *J. Biol. Chem.* **259**, 6201–6207.

Ignarro, L. J., Buga, G. M., Byrns, R. E., Wood, K. S., and Chandhuri, G. (1988) Endothelium-derived relaxing factor and nitric oxide possess identical pharmacological properties as relaxants of bovine arterial and venous smooth muscle. *J. Pharmacol. Exp. Ther.* **246**, 218–226.

Ignarro, L. J., Burke, T. M., Wood, K. S., Wolin, M. S., and Kadowitz, P. J. (1984b) Association between cyclic GMP accumulation and acetylcholine-elicited relaxation of bovine intrapulmonary artery. *J. Pharmacol. Exp. Ther.* **228**, 682–690.

Ignarro, L. J., Wood, K. S., Ballot, B., and Wolin, M. S. (1984c) Guanylate cyclase from bovine lung. Evidence that enzyme activation by phenylhydrazine is mediated by iron-phenyl hemoprotein complexes. *J. Biol. Chem.* **259**, 5923–5931.

Inui, J., Brodde, O-E., and Shumann, H. J. (1982) Influence of acetylcholine on the positive inotropic effect evoked by α- and β-adrenoceptor stimulation in the rabbit heart. *Naunyn-Schmied. Arch. Pharmacol.* **320**, 152–159.

Iwasa, Y. and Hosey, M. M. (1983) Cholinergic antagonism of beta-adrenergic stimulation of cardiac membrane protein phosphorylation in situ. *J. Biol. Chem.* **258**, 4571–4575.

Jeremy, J. Y., Mikhailidis, D. P., and Dandona, P. (1986) Prostanoid synthesis by the rat urinary bladder: evidence for stimulation through muscarine-linked calcium channels. *Naunyn-Schmied. Arch. Pharmacol.* **334**, 463–467.

Kanba, S., Kanba, K. S., and Richelson, E. (1986) The protein kinase

C activator, 12-O-tetradecanoylphorbol-13-acetate (TPA) inhibits muscarinic (M_1) receptor-mediated inositol phosphate release and cyclic GMP formation in murine neuroblastoma cells (clone N1E-115). *Eur. J. Pharmacol.* **125**, 155–156.

Katsuki, S. and Murad, F. (1976) Regulation of adenosine cyclic 3′,5′-monophosphate and guanosine cyclic 3′,5′-monophosphate levels and contractility in bovine tracheal smooth muscle. *Mol. Pharmacol.* **13**, 330–341.

Kebabian, J. W., Steiner, A. L., and Greengard, P. (1975) Muscarinic cholinergic regulation of cyclic guanosine 3′,5′-monophosphate in autonomic ganglia: possible role in synaptic transmission. *J. Pharmacol. Exp. Ther.* **193**, 474–488.

Kendall, D. A. (1986) Cyclic GMP and inositol phosphate accumulation do not share common origins in rat brain slices. *J. Neurochem.* **47**, 1483–1489.

Kimura, H. and Murad, F. (1974) Evidence for two different forms of guanylate cyclase in rat heart. *J. Biol. Chem.* **249**, 6910–6916.

Kuno, T., Andresen, J. W., Kamisaki, Y., Waldman, S. A., Chang, L. Y., Saheki, S., Leitman, D. C., Nakane, M., and Murad, F. (1986) Co-purification of atrial natriuretic factor receptor and particulate guanylate cyclase from rat lung. *J. Biol. Chem.* **261**, 5817–5823.

Kuo, J. F. (1974) Guanosine 3′,5′-monophosphate-dependent protein kinases in mammalian tissues. *Proc. Natl. Acad. Sci. USA* **71**, 4037–4041.

Lee, T-P., Kuo, J. F., and Greengard, P. (1972) Role of muscarinic cholinergic receptors in regulation of guanosine 3′:5′-cyclic monophosphate content in mammalian brain, heart muscle, and intestinal smooth muscle. *Proc. Natl. Acad. Sci. USA* **69**, 3287–3291.

Leiber, D. and Harbon, S. (1982) The relationship between the carbachol stimulatory effect on cyclic GMP content and activation by fatty acid hydroperoxides of a soluble guanylate cyclase in the guinea pig myometrium. *Mol. Pharmacol.* **21**, 654–663.

Lenox, R. H., Kant, G. J., and Meyerhoff, J. L. (1980) Regional sensitivity of cyclic AMP and cyclic GMP in rat brain to central cholinergic stimulation. *Life Sci.* **26**, 2201–2209.

Levitt, P., Rakic, P., De Camilli, P., and Greengard, P. (1984) Emergence of cyclic guanosine 3′:5′-monophosphate-dependent protein kinase immunoreactivity in developing rhesus monkey cerebellum: Correlative immunocytochemical and electron microscopic analysis. *J. Neurosci.* **4**, 2553–2564.

Liles, W. C., Hunter, D. D., Meier, K. E., and Nathanson, N. M. (1986) Activation of protein kinase C induces rapid internalization and subsequent degradation of muscarinic acetylcholine receptors in neuroblastoma cells. *J. Biol. Chem.* **261**, 5307–5313.

Limbird, L. E. and Lefkowitz, R. J. (1975) Myocardial guanylate cyclase: properties of the enzyme and effects of cholinergic agonists in vitro. *Biochim. Biophys. Acta* **377**, 186–196.

Lindemann, J. P. and Watanabe, A. M. (1985) Muscarinic cholinergic inhibition of beta-adrenergic stimulation of phospholamban phosphorylation and Ca^{2+} transport in guinea pig ventricles. *J. Biol. Chem.* **260**, 13122–13129.

Linden, J. and Brooker, G. (1979) The questionable role of cyclic guanosine 3′:5′ monophosphate in heart. *Biochem. Pharmacol.* **28**, 3351–3360.

Lingren, J. A., Hulting, A., Dahlen, S.-E., Hokfelt, T., Werner, S., and Samuelsson, B. (1984) Evidence for the occurrence of leukotrienes (LT) in the central nervous system and a neuroendocrine role of LT. *Soc. Neurosci. Abstr.* **10**, 1129.

Lunt, G. G. and Rowe, C. E. (1968) The production of unesterified fatty acid in brain. *Biochim. Biophys. Acta* **152**, 681–693.

Luthin, G. R. and Wolfe, B. B. (1984) Comparison of [^{3}H]pirenzepine and [^{3}H]quinuclidinyl benzilate binding to muscarinic cholinergic receptors in rat brain. *J. Pharmacol. Exp. Ther.* **228**, 648–655.

MacLeod, K. M. and Diamond, J. (1986) Effects of the cyclic GMP lowering agent LY83583 on the interaction of carbachol with forskolin in rabbit isolated cardiac preparations. *J. Pharmacol. Exp. Ther.* **238**, 313–318.

Martin, W., Villani, G. M., Jothianandan, D., and Furchgott, R. F. (1984) Selective blockade of endothelium-dependent and glyceryl trinitrate-induced relaxation by hemoglobin and by methylene blue in the rabbit aorta. *J. Pharmacol. Exp. Ther.* **232**, 708–716.

Matsuzawa, H. and Nirenberg, M. (1975) Receptor-mediated shifts in cGMP and cAMP levels in neuroblastoma cells. *Proc. Natl. Acad. Sci. USA* **72**, 3472–3476.

McKinney, M. (1987) Blockade of receptor-mediated cyclic GMP formation by hydroxyeicosatetraenoic acid. *J. Neurochem.* **49**, 331–341.

McKinney, M. and Richelson, E. (1984) The coupling of the neuronal muscarinic receptor to responses. *Ann. Rev. Pharmacol. Toxicol.* **24**, 121–146.

McKinney, M. and Richelson, E. (1986a) Blockade of N1E-115 murine neuroblastoma muscarinic receptor function by agents that affect the metabolism of arachidonic acid. *Biochem. Pharmacol.* **35**, 2389–2397.

McKinney, M. and Richelson, E. (1986b) Muscarinic responses and binding in a murine neuroblastoma clone (N1E-115): Cyclic GMP formation is mediated by a low affinity agonist-receptor conformation and cyclic AMP reduction is mediated by a high affinity agonist-receptor conformation. *Mol. Pharmacol.* **30**, 207–211.

McKinney, M., Stenstrom, S., and Richelson, E. (1985) Muscarinic responses and binding in a murine neuroblastoma clone (N1E-115): Mediation of separate responses by high affinity and low affinity agonist-receptor conformations. *Mol. Pharmacol.* **27**, 223–235.

McPhail, L. C., Clayton, C. C., and Snyderman, R. (1984) A potential second messenger role for unsaturated fatty acids: activation of Ca^{2+}-dependent protein kinase. *Science* **224**, 622–625.

Miki, N., Nagano, M., and Kuriyama, K. (1976) Catalase activates cerebral guanylate cyclase in the presence of sodium azide. *Biochem. Biophys. Res. Comm.* **72**, 952–959.
Mittal, C. K. and Murad, F. (1977) Activation of guanylate cyclase by superoxide dismutase and hydroxyl radical: a physiological regulator of guanosine 3',5'-monophosphate formation. *Procl. Natl. Acad. Sci. USA* **74**, 4360–4364.
Mittal, C. K. and Murad, F. (1987) Effects of phospholipase A2 and lipoxygenase on rat cerebral cortex guanylate cyclase activity. *Fed. Proc.* **37**, 1689.
Mittal, C. K., Kimura, H., and Murad, F. (1977) Purification and properties of a protein required for sodium azide activation of guanylate cyclase. *J. Biol. Chem.* **252**, 4384–4390.
Moskowitz, M. A., Kiwak, K. J., Hekimian, K., and Levine, L. (1984) Synthesis of compounds with properties of leukotrienes C4 and D4 in gerbil brains after ischemia and reperfusion. *Science* **224**, 886–889.
Murad, F., Mittal, C. K., Arnold, W. P., Katsuki, S., and Kimura, H. (1977) Guanylate cyclase: activation by azide, nitro compounds, nitric oxide, and hydroxyl radical and inhibition by hemoglobin and myoglobin. *Adv. Cyc. Nuc. Res.* **9**, 145–158.
Murphy, S., Pearce, B., and Morrow, C. (1986) Astrocytes have both M_1 and M_2 muscarinic receptor subtypes. *Brain Res.* **364**, 177–180.
Nakane, M., Ichikawa, M., and Deguchi, T. (1983) Light and electron microscopic demonstration of guanylate cyclase in rat brain. *Brain Res.* **273**, 9–15.
Ohlstein, E. H., Wood, K. S., and Ignarro, L. J. (1982) Purification and properties of heme-deficient hepatic soluble guanylate cyclase: effects of heme and other factors on enzyme activation by NO, NO-heme, and protoporphyrin IX. *Arch. Biochem. Biophys.* **218**, 187–198.
Ohsako, S. and Deguchi, T. (1981) Stimulation by phosphatidic acid of calcium influx and cyclic GMP synthesis in neuroblastoma cells. *J. Biol. Chem.* **256**, 10945–10948.
Ohsako, S. and Deguchi, T. (1984) Receptor-mediated regulation of calcium mobilization and cyclic GMP synthesis in neuroblastoma cells. *Biochem. Biophys. Res. Comm.* **122**, 333–339.
Opmeer, F. A., Gumulka, S. W., Dinnendahl, V., and Schonhofer, P. S. (1976) Effects of stimulatory and depressant drugs on cyclic guanosine 3',5'-monophosphate and adenosine 3',5'-monophosphate levels in mouse brain. *Naunyn-Schmied. Arch. Pharmacol.* **292**, 259–265.
Palmer, G. C. and Duszynski, C. R. (1975) Regional cyclic GMP content in incubated tissue slices of rat brain. *Eur. J. Pharmacol.* **32**, 375–379.
Palmer, R. M. J., Ferrige, A. G., and Moncada, S. (1987) Nitric oxide release accounts for the biological activity of endothelium-derived relaxing factor. *Nature* **327**, 524–526.

Park, S. and Rasmussen, H. (1986) Carbachol-induced protein phosphorylation changes in bovine tracheal smooth muscle. *J. Biol. Chem.* **261**, 15734–15739.

Pinchasi, I., Burstein, M., and Michaelson, D. M. (1984) Metabolism of arachidonic acid and prostaglandins in the Torpedo electric organ: Modulation by the presynaptic muscarinic acetylcholine receptor. *Neurosci.* **13**, 1359–1364.

Radmark, O., Shimizu, T., Fitzpatrick, F., and Samuelsson, B. (1984) Hemoprotein catalysis of leukotriene formation. *Biochim. Biophys. Acta* **792**, 324–329.

Rapoport, R. M. and Murad, F. (1983) Agonist-induced endothelium-dependent relaxation in rat thoracic aorta may be mediated through cGMP. *Cir. Res.* **52**, 352–357.

Rapoport, R. M., Draznin, M. B., and Murad, F. (1982) Sodium nitroprusside-induced protein phosphorylation in intact rat aorta is mimicked by 8-bromo cyclic GMP. *Proc. Natl. Acad. Sci. USA* **79**, 6470–6474.

Rapoport, R. M., Draznin, M. B., and Murad, F. (1983) Endothelium-dependent relaxation in rat aorta may be mediated through cyclic GMP-dependent protein phosphorylation. *Nature* **306**, 174–176.

Richelson, E. (1978) Desensitisation of muscarinic receptor-mediated cyclic GMP formation by cultured nerve cells. *Nature* **272**, 366–368.

Richelson, E. and El-Fakahany, E. (1981) The molecular basis of neurotransmission at the muscarinic receptor. *Biochem. Pharmacol.* **30**, 2887–2891.

Rillema, J. A. (1978) Activation of guanylate cyclase by arachidonic acid in mammary gland homogenates from mice. *Prostaglandins* **15**, 857–865.

Schlichter, D. J., Casnelli, J. E., and Greengard, P. (1978) An endogenous substrate for cGMP-dependent protein kinase in mammalian cerebellum. *Nature* **273**, 61–72.

Schmidt, M. J., Sawyer, B. D., Truex, L. L., Marshall, W. S., and Fleisch, J. H. (1985) LY83583: An agent that lowers intracellular levels of cyclic guanosine 3′,5′.monophosphate, *J. Pharmacol. Exp. Ther.* **232**, 764–769.

Schultz, G., Hardman, J. G., Schultz, K., Baird, C. E., and Sutherland, E. W. (1973) The importance of calcium ions for the regulation of guanosine 3′:5′-cyclic monophosphate levels. *Proc. Natl. Acad. Sci. USA* **70**, 3889–3893.

Silver, P. J. and Stull, J. T. (1983) Phosphorylation of myosin light chain and phosphorylase in tracheal smooth muscle in response to KCl and carbachol. *Mol Pharmacol.* **25**, 267–274.

Singer, H. A. and Peach, M. J. (1983) Endothelium-dependent relaxation of rabbit aorta. II. Inhibition of relaxation stimulated by methacholine and A23187 with antagonists of arachidonic acid metabolism. *J. Pharmacol. Exp. Ther.* **226**, 796–801.

Snider, R. M. and Agranoff, B. W. (1985) Neurotransmitter-stimulated Ca^{2+} mobilization and inositol phospholipid metabolism in neuroblastoma cells. *Soc. Neurosci. Abstr.* **11**, 96.

Snider, R. M., McKinney, M., Forray, C., and Richelson, E. (1984) Neurotransmitter receptors mediate cyclic GMP formation by involvement of arachidonic acid and lipoxygenase. *Proc. Natl. Acad. Sci. USA* **81**, 3905–3909.

Spies, C., Schultz, K-D., and Schultz, G. (1980) Inhibitory effects of mepacrine and eicosatetraynoic acid on cyclic GMP elevations caused by calcium and hormonal factors in rat ductus deferens. *Naunyn-Schmied. Arch. Pharmacol.* **311**, 71–77.

Stone, T. W., Taylor, D. A., and Bloom, F. E. (1975) Cyclic AMP and cyclic GMP may mediate opposite neuronal responses in the rat cerebral cortex. *Science* **187**, 845–847.

Struck, C-J. and Glossmann, H. (1987) Soluble bovine adrenal cortex guanylate cyclase: Effect of sodium nitroprusside, nitrosamines, and hydrophobic ligands on activity, substrate specificity and cation requirement. *Naunyn-Schmiedeberg's Arch. Pharmacol.* **304**, 51–61.

Study, R. E., Breakefield, X. O., Bartfai, T., and Greengard, P. (1978) Voltage-sensitive calcium channels regulate guanosine 3′,5′-cyclic monophosphate levels in neuroblastoma cells. *Proc. Natl. Acad. Sci. USA* **75**, 6295–6299.

Swartz, B. E. and Woody, C. D. (1979) Correlated effects of acetylcholine and cyclic guanosine monophosphate on membrane properties of mammalian neocortical neurons. *J. Neurobiol.* **10**, 465–488.

Swartz, B. E. and Woody, C. D. (1984) Effects of intracellular antibodies to cGMP on responses of cortical neurons of awake cats to extracellular application of muscarinic agonists. *Exp. Neurol.* **86**, 388–404.

Tappel, A. L. (1955) Unsaturated lipid oxidation catalyzed by hematin compounds. *J. Biol. Chem.* **217**, 721–733.

Waldman, S. A., Sinacore, M. S., Lewicki, J. A., Chang, L. Y., and Murad, F. (1984) Selective activation of particulate guanylate cyclase by a specific class of porphyrins. *J. Biol. Chem.* **259**, 4038–4042.

Wamsley, J. K., West, J. R., Blach, A. C., and Williams, T. H. (1979) Muscarinic cholinergic and preganglionic physiological stimulation increase cyclic GMP levels in guinea pig superior cervical ganglion *J. Neurochem.* **32**, 1033–1035.

Wastek, G. J., Lopez, J. R., and Richelson, E. (1980) Demonstration of a muscarinic receptor-mediated cyclic GMP-dependent hyperpolarization of the membrane potential of mouse neuroblastoma cells using [^{3}H]tetraphenylphosphonium. *Mol. Pharmacol.* **19**, 15–20.

Weight, F. F., Petzold, G., and Greengard, P. (1974) Guanosine 3′,5′-monophosphate in sympathetic ganglia: increase associated with synaptic transmission. *Science* **186**, 942–944.

White, A. A., Crawford, K. M., Patt, C. S., and Lad, P. J. (1976) Activation of soluble guanylate cyclase from rat lung by incubation or by hydrogen peroxide. *J. Biol. Chem.* **251**, 7304–7312.

White, A. A., Karr, D. B., and Patt, C. S. (1982) Role of lipoxygenase in the O_2-dependent activation of soluble guanylate cyclase from rat lung. *Biochem. J.* **204**, 383–392.

Wolin, M. S., Wood, K. S., and Ignarro, L. J. (1982) Guanylate cyclase from bovine lung. A kinetic analysis of the regulation of the purified soluble enzyme by protoporphyrin IX, heme, and nitrosyl-heme. *J. Biol. Chem.* **257**, 13312–13320.

Woody, C., Gruen, E., Sakai, H., Sakai, M., and Swartz, B. (1986a) Responses of morphologically identified cortical neurons to intracellularly injected cyclic GMP. *Exp. Neurol.* **91**, 580–595.

Woody, C. D., Bartfai, T., Gruen, E., and Nairn, A. C. (1986b) Intracellular injection of cGMP-dependent protein kinase results in increased input resistance in neurons of the mammalian motor cortex. *Brain Res.* **386**, 379–385.

Woody, C. D., Swartz, B. E., and Gruen, E. (1978) Effects of acetylcholine and cyclic GMP on input resistance of cortical neurons in awake cats. *Brain Res.* **158**, 373–395.

Yau, K-W. and Nakatani, K. (1985) Light-suppressible, cyclic GMP-sensitive conductance in the plasma membrane of a truncated rod outer segment. *Nature* **317**, 252–255.

Yoshikawa, K. and Kuriyama, K. (1980) Superoxide dismutase catalyzes activation of synaptosomal soluble guanylate cyclase from rat brain. *Biochem. Biophys. Res. Comm.* **95**, 529–534.

Yousufzai, S. Y. K. and Abdel-Latif, A. A. (1984) The effects of alfa$_1$-adrenergic and muscarinic cholinergic stimulation on prostaglandin release by rabbit iris. *Prostaglandins* **28**, 399–415.

Zwiller, J., Revel, M-O., and Malviya, A. N. (1985) Protein kinase C catalyzes phosphorylation of guanylate cyclase in vitro. *J. Biol. Chem.* **260**, 1350–1353.

9

Muscarinic Cholinergic Receptor Regulation of Ion Channels

R. Alan North

1. Introduction

The action of acetylcholine at muscarinic receptors on excitable tissues has profoundly influenced the development of ideas on chemical transmission. Much of the early progress was the result of the availability and long historical use of belladonna alkaloids. In experiments never published, carried out in 1906, Dixon showed that atropine blocked the inhibitory effect on the exposed frog heart of an extract partially purified from the heart of a dog under vagal inhibition (Dale, 1934). In 1921, Loewi of Graz in Austria reported his experiments on the effects of *vagusstoff*, providing convincing evidence for chemical transmission. However, even in those long ago days, it was clear to some that the muscarinic actions of esters of choline were not uniformly inhibitory; Sir Henry Dale was "struck by the remarkable fidelity with which [acetylcholine] reproduced the various effects of parasympathetic nerves, inhibitor on some organs and augmentor on others" (Dale, 1934).

Whether or not the action of acetylcholine on an excitable tissue is inhibitory or excitatory eventually depends on the constellation of membrane ion channels that are opened and closed. There may be additional direct actions on the contractile machinery in the case of muscle, or on the secretory apparatus in the case of glands or nerves, but in nerve, gland, and muscle those functions of the tissues that are embodied by changes in membrane potential are brought about ultimately by the opening and closing of membrane ion channels. This chapter deals with the mechanisms by which acetylcholine, by binding to muscarinic receptors, brings about the opening and/or closing of membrane ion channels.

Two such mechanisms might be distinguished for transmitters and hormones in general. In the first, the agonist binds to some extracellular domain of the channel molecule itself. Acetylcholine acts at nicotinic receptors in this way, and increases the probability that the channel molecule will adopt the open, conducting, conformation. Such a mechanism is also used by GABA acting at $GABA_A$ receptors and by glycine, although in both these cases, the channels allow chloride ions rather than cations to pass; it provides for control of the open/closed state of the channel on the ms time scale. The second mechanism involves the activated receptor interacting with a G-protein: the G-protein dissociates into its subunits, which can then influence other molecules within the cell. The other molecule might be an ion channel or an enzyme; the altered activity of the enzyme changes the level of a product, which diffuses within the cell and opens or closes one or more ion channels. The time course of action of substances utilizing the second mechanism typically ranges from tens of ms to a few s when the G-protein interacts directly with the ion channel, but the time course of the opening or closing of ion channels as a result of the receptor activation might range from tens of ms to min or h when a freely diffusible intracellular second messenger is involved. Indeed, it is an arbitrary line that excludes the role of second messengers that alter transcription, and ultimately affect the synthesis and/or expression of the ion channels in the membrane.

There is no evidence that muscarinic agonists bind directly to ion channels and bring about their effects by the first mechanism; this seems to be the prerogative of receptors for acetylcholine belonging to the nicotinic family. The clearest evidence for the first kind of action involving G-proteins is provided by the actions on potassium channels in the heart; in other cases, it is not yet known whether the G-protein subunits directly affect ion channels and evidence for the involvement of a further diffusible second messenger relies in many cases on mimicry of the action of muscarine by the putative messenger molecule.

2. Potassium Conductance Increase

Most excitable tissues maintain a resting potential that is not the same as the equilibrium potential for any of the ions separated by their membrane. In other words, ions are continuously passing through membrane channels down their electrochemical gradients, which are maintained in the long term by the expenditure of metabolic energy. Increasing the conductance of the membrane to one species of ion alone (i.e., opening channels that are selectively permeable to the ion) will cause the membrane potential to be weighted more by the equilibrium potential for that ion. In the case of potassium, a few ions will leave the cell, and a hyperpolarization will ensue. The meaning of few depends primarily on cell size (and, hence, the membrane capacitance); in mammalian cells, the quantity of potassium ions that leaves the cell during a hyperpolarization of up to several mV sustained for a few seconds causes a negligible change in the intracellular potassium concentration. Experimentally, a potassium conductance increase is shown by setting the membrane potential arbitrarily to any required level. This can be done simply by passing electric current across it. When the membrane potential is set to the potassium equilibrium potential, an increase in conductance (i.e., an opening of potassium channels) will result in no movement of potassium ions, since the forces on the ions to enter or leave the cell are equal. The potassium equilibrium potential is determined by the concentrations of ions on both sides of the membrane, according to the Nernst equation; thus, systematic changes in these potassium concentrations enable the experimenter to demonstrate unequivocally that a potassium conductance increase is involved.

2.1. Heart

It has been known since 1845 (Weber and Weber, 1845) that electrical stimulation of the exposed vagus nerve slows the heart; the first suggestion that the mobilization of potassium ions was involved was that of Howell (Howell, 1906; Howell and Duke, 1908), who measured their concentration in the blood following vagal activity. A hyperpolarization of tortoise atrium results from stimulation of the vagus nerve: this was recorded by Gaskell in

1887 as an increase in the demarcation potential between healthy and crushed tissue. The hyperpolarization was blocked by atropine. Burgen and Terroux (1953) and Trautwein and Dudel (1958) used electrical measurements to show that the application of acetylcholine and its congeners resulted in a membrane hyperpolarization and an increase in potassium conductance; del Castillo and Katz (1955) found a hyperpolarization when they stimulated the vagus nerve in the frog, and Pott (1979) carried out a similar experiment. Much of the work of the subsequent 25 years was carried out with sucrose gap and/or microelectrode recording, and has been reviewed by Brown (1982) and by Löffelholz and Pappano (1985). These studies provided information on the time course of action of acetylcholine, which, despite subsequent controversies relating to the interpretation of experiments regarding diffusion barriers in a syncytial tissue, suggested two new things: first, that a significant delay was intrinsic to the muscarinic action that was lacking from the nicotinic action of acetylcholine, and second, that the potassium conductance increased by acetylcholine exhibited inward (anomalous) rectification, and was preferentially to be found in atrial and nodal tissue.

Rapid advances in knowledge followed the application of patch-clamp methods to dissociated cardiac muscle cells. Activity of the individual potassium channels opened by acetylcholine in the rabbit atrium was recorded by Sakmann et al. (1983) and by Soejima and Noma (1984). These experiments indicated that the channels involved had a conductance of about 50 pS (measured with approximately equal potassium concentrations on either side of the membrane); the channels opened by acetylcholine carried inward current more readily than outward current, but could be distinguished from other acetylcholine-insensitive potassium channels that also showed strong inward rectification (I_{K1}). The most important aspect of these studies was the finding that the acetylcholine opened the potassium channel only when it was applied within the recording electrode, the tip of which enclosed a patch of membrane a few square μm in size. This indicates that the potassium channels were not opened as the result of diffusion of a second-messenger molecule throughout the cytoplasm.

The whole cell recording configuration has also allowed improved description of the properties of the potassium conductance. It was thought by some that the conductance opened by acetylcholine was the same as the inward rectifier conductance (I_{K1}) already present in cardiac cells. Recordings of whole cell current, particularly for the frog atrium, support the finding from the single-channel recordings that the potassium current increased by acetylcholine is distinct from I_{K1}; the acetylcholine activated channel opens more slowly and passes more current in the outward direction (*see* Simmons and Hartzell, 1988; and references therein).

Sorota et al. (1985), Breitweiser and Szabo (1985), and Pfaffinger et al. (1985) next showed that hydrolysis of GTP by a pertussis-toxin-sensitive G-protein was necessary for coupling between muscarinic receptor and potassium channel, and that hydrolysis of GTP was essential for the termination of the action of acetylcholine. Intracellular cyclic AMP, which was known to increase calcium currents in the cells (Tsien, 1973), was found not to be involved in the muscarinic increase in potassium conductance. It is now known that the α subunit of a G-protein purified from erythrocytes (G_K, probably identical with the $G_{i\alpha 3}$ subunit; *see* Gilman, 1987) can directly activate the potassium channels when it is applied to the inner surface of guinea-pig atrial cells at concentrations of 1 pM or less (Yatani et al., 1987a; Codina et al., 1987). Potassium channels in isolated patches of membrane from the chick heart were opened by the $\beta\gamma$ subunits from cow brain; comparable concentrations of α subunit had no effect (Logothetis et al. 1987).

The same individual potassium channels that are opened by muscarinic agonists can also be opened by adenosine (Kurachi et al., 1986a,b,c; 1987a); the two agonists activate distinct cell surface receptors (muscarinic M_2 and adenosine A_1), but open the same population of potassium channels. Experiments in which desensitization to the effects of the two agonists has been compared can be interpreted to suggest that the same pool of G-protein is shared by agonists at the two different receptors (Kurachi et al., 1987b).

Taken together, these experiments provide a reasonably complete picture of the action of muscarinic agonists to increase the potassium conductance of cardiac pacemaking tissue. When acetylcholine is bound to the cardiac muscarinic receptor, an associated G-protein (G_K or a G_i) alters its conformation in such a way as to promote the dissociation of GDP (and its replacement with GTP) from the α subunit. The binding of GTP is believed to result in dissociation of the G-protein into α and $\beta\gamma$ subunits. The liberated α subunit, with bound GTP, combines with the channel protein. The probability that the channel is open, which is otherwise very low, increases markedly; potassium ions leave the cell, and hyperpolarization follows. The GTPase activity of the α subunit results in hydrolysis of the bound GTP; as this occurs, the α subunit, freed from the channel, can reassociate with $\beta\gamma$ into the oligomer, and the open channel probability falls to the basal level. It is generally assumed that this sequence of molecular steps also ensues when action potentials in the vagus nerve are used to release acetylcholine. This has been shown in the case of the frog (del Castillo and Katz, 1955), tortoise (Gaskell, 1887), and guinea pig (Pott, 1979). However, transmission through the cardiac ganglion and other distinct actions of acetylcholine (*see below*), as well as the possibility that other compounds are released at the same time might each complicate the simple interpretation offered here.

2.2. Neurons

The first hint that acetylcholine might inhibit the firing of neurons by increasing the membrane potassium conductance was provided by Tauc and Gerschenfeld (1962); this was studied in detail by Kehoe (1972a,b,c) and Ascher (1972), who noted that the potassium conductance showed inward rectification. Those experiments were carried out on pleural ganglion of *Aplysia*; the medial pleural neurons showed hyperpolarizations in response to exogenous acetylcholine and synaptic potentials representing the release of endogenous acetylcholine. The increase in potassium conductance is also produced by dopamine and histamine (Ascher, 1972; Gruol and Weinreich, 1979); Ascher and Chesnoy-Marchais (1982) found that the three agonists act on a common set of potassium channels and that there was cross-desensitization among them. They also showed that changes in intracellular calcium ion concentration were not likely to be involved in the increase in potassium conductance. Sasaki and Sato (1987) subsequently demonstrated that a pertussis-toxin-sensitive G-protein couples the three receptors to the potassium channel, thus extending the similarity with the observations made for the potassium conductance in cardiac tissue.

A direct action of acetylcholine to increase the potassium conductance of vertebrate nerve cells was demonstrated first by Hartzell et al. (1977) for the parasympathetic ganglion cells in the atrial septum of the mudpuppy, and later by Dodd and Horn (1983) for the C cells of the bullfrog sympathetic ganglia. In both these cases, one is compelled by the evidence that the muscarinic receptors also bind the acetylcholine released by action potentials in presynaptic fibres (vagus or sympathetic preganglionics) and, thus, bring about the slow inhibitory postsynaptic potential.

Mammalian neurons of the parasympathetic (cat bladder ganglia: Cole and Shinnick-Gallagher, 1984; Gallagher et al., 1982) and sympathetic (rabbit superior cervical ganglion: Ashe and Yarosh, 1984; Yarosh and Ashe, 1987) ganglia are also hyperpolarized by muscarine, and this appears to underlie the slow inhibitory postsynaptic potential in these cells. In the case of the rabbit sympathetic ganglion, available evidence indicates that the hyperpolarization results from activation of M_2 (or gallamine-sensitive, pirenzepine-insensitive) receptors; the depolarization of the same cells by muscarine is the result of activation of M_1 receptors (Yavari and Weight, 1987).

Some neurons of the mammalian central nervous system are also inhibited by muscarinic agonists, but bear in mind that the most common response is excitation (*see* Krnjevic, 1974). This has been shown to result directly from an increase in membrane potassium conductance (rat parabrachial neurons: Egan and North, 1986; guinea-pig thalamic reticular neurons: McCormick and Prince, 1986; guinea-pig and cat lateral and medial geniculate neurons; McCormick and Prince, 1987). The receptor type involved is the cardiac M_2 receptor, at least in the rat nucleus parabrachialis; this is indicated by dissociation equilibrium constants (K_D) for pirenzepine (Egan and North, 1986) and AFDX-116 (Christie and North, 1988). Agonists at $GABA_B$ and at μ opioid receptors increase the same conductance as do agonists at muscarinic receptors, and this conductance shows inward rectification (Christie and North, 1988). Thus, in all respects so far examined, the mechanism by which muscarinic agonists increase the potassium conductance of nerve cells appears similar to that found in the heart. However, it must be stressed that definitive experiments on the coupling between muscarinic receptor and potassium channel have not been carried out with isolated membrane patches from neurons, and the participation of a diffusible cytoplasmic messenger molecule cannot be excluded.

2.3. Glands

Muscarinic agonists increase the potassium conductance in gland-like tissue by a mechanism not involving diffusible cytoplasmic second messengers. The muscarinic receptor is coupled to the potassium channel through an intervening G-protein, and there is no requirement for any diffusible second messenger. This situation has been established for a cell line from rat anterior pituitary that secretes growth hormone (Yatani et al. 1987b). The same 55 pS channels are opened both by acetylcholine and by somatostatin. There are few details of this action available, and the receptor type involved is not known.

An increase in potassium conductance can also be produced by muscarinic agonists that stimulate the hydrolysis of phosphatidylinositol, because the resultant inositol trisphosphate ($InsP_3$) that is formed will release calcium; increases in intracellular calcium result in opening of potassium channels of several kinds (it also causes several other channels to open; *see below*). This mechanism of action seems to underlie the increase in potassium conductance caused by muscarinic agonists in pancreatic acinar cells of the pig, guinea pig, and human (in the rat and mouse, the predominant channel opened is a calcium-dependent chloride channel; *see below*). A similar increase in potassium conductance is evoked by muscarinic agonists in lacrimal gland of the mouse (Suzuki and Petersen, 1985) and in salivary gland (Gallacher and Morris, 1986). The particular properties of the potassium conductance affected vary from cell to cell. For example, bombesin, CCK-gastrin peptides, and acetylcholine hyperpolarize guinea-pig acinar cells by activating a low conductance (about 30 pS), TEA-insensitive channel; but in the pig exocrine pancreas (Maruyama et al., 1983), rat and mouse salivary glands (Maruyama et al., 1983), and lacrimal glands (Trautmann and Marty, 1984; Findlay, 1984), the channel has a high conductance (200–250 pS) and strong sensitivity to TEA and membrane potential. The large unit conductance channels dominate the resting permeability of these cells, each cell having about 100 channels. The overall pattern of secretion in response to a muscarinic agonists depends not only on the opening of these potassium channels in the basolateral membrane of the cell, but also on other effects of the rise of intracellular calcium concentration (e.g., closure of gap junctions between acini, increase in cation and/or chloride conductances: *see* Petersen and Findlay, 1987; Marty, 1987 for recent reviews).

2.4. Oocytes

The oocytes of *Xenopus laevis* have a native muscarinic receptor. One of the consequences of applying muscarinic agonists to the denuded oocyte is a slow outward potassium current (Dascal et al., 1984).

3. Potassium Conductance Decrease

3.1. Neurons

A role for the muscarinic action of acetylcholine in the mediation of excitatory synaptic transmission has been known for some time (Eccles and Libet, 1961), and there is now good evidence for this form of synaptic transmission in sympathetic ganglia (Nishi and Koketsu, 1968; Kuba and Koketsu, 1978; Libet, 1979), the enteric nervous system (North and Tokimasa, 1982; North et al., 1985), and the hippocampus (Cole and Nicoll, 1983). Most of the neurons on which the action of muscarinic agonists has been tested exhibit several different potassium conductances; some are affected and some are not.

3.1.1. Background Potassium Conductance

This is also sometimes known as the resting or leak conductance; it is one that exhibits no voltage-dependence other than that expected from the difference in potassium ion concentrations across the membrane (constant field rectification) and that is not sensitive to changes in intracellular calcium concentration. This is the potassium conductance that contributes predominantly to the resting potential of the neuron. In most neurons at the "resting" potential, persistent inward currents carried by sodium, calcium, and/or chloride ions are balanced by equal and opposite outward currents across the cell membrane. Any decrease in the conductance of the cell to potassium ions therefore causes the cell to depolarize in the face of the persistent inward currents. Both the decrease in conductance and the inward current can be measured in voltage clamp experiments. When membrane voltage is recorded, the depolarization is usually associated with an increase in the input resistance of the cell; if the current pulses passed across the membrane to measure input resistance are carried by movement of potassium ions, then the closure of these channels by the action of muscarine (or any other transmitter) will result in a greater potential change for a given current. A simple equivalent circuit shows that, in these circumstances, the depolarizing response to a muscarinic agonist that results from a reduction in the *background* conductance will reverse polarity at the potassium equilibrium potential. The situation is more complicated if the potassium conductance involved is itself sensitive to membrane potential (*see below*); obviously, muscarine cannot close potassium channels that have already closed at the membrane potential under test. The failure of the muscarinic depolarization to reverse polarity when the membrane is strongly hyperpolarized can also result from a combination of actions—for example, a reduction in background conductance coupled with an increase in cation conductance (*see* Kuba and Koketsu, 1978).

The first effort to demonstrate a potassium conductance decrease was made by Weight and Votava (1970) for bullfrog sympathetic neurons and was proposed for mammalian cortical neurons by Krnjevic et al. (1971). Reduction in the background conductance has since been shown in some sympathetic neurons of the frog (*see* Kuba and Koketsu, 1978), enteric neurons of guinea pig (Morita et al., 1982; North et al., 1985; Galligan et al., 1988), hippocampal pyramidal cells of the rat (Dodd et al., 1981; Bernardo and Prince, 1982a,b; Madison et al., 1987), rat striatal cells (Muller and Misgeld, 1986) and nucleus accumbens neurons (Uchimura and North, unpublished observations). The reduction in the background conductance leads to a depolarization of the neuron at all membrane potentials less negative than E_K; the depolarization itself is excitatory, and the decrease in cell conductance tends to increase the amplitude and prolong the duration of excitatory postsynaptic potentials generated by conductance increases (e.g., glutamate or nicotinic acetylcholine actions). A slow cholinergic excitatory postsynaptic potential is mediated by acetylcholine released from presynaptic nerves in autonomic neurons (Eccles and Libet, 1961; Nishi and Koketsu, 1968), enteric neurons (North and Tokimasa, 1982; North et al., 1985), and hippocampal pyramidal cells (Cole and Nicoll, 1982).

The receptor type involved in the reduction in the background potassium conductance seems to belong to the M_1 class. The K_D of pirenzepine measured by Schild analysis in the case of the enteric neurons (North et al., 1985) was about 10 n*M*. The dose–response curves for the agonists McN A343, oxotremorine, bethanechol, or muscarine were the same as those when presynaptic inhibition of acetylcholine was measured in the same preparation: the presynaptic inhibition is the result of activation of an M_2 type of receptor (pirenzepine K_D about 200 n*M*). Pirenzepine has been used in several other experiments in a less quantitative manner; these indicate that, in the hippocampus (Muller and Misgeld, 1986; and Dutar and Nicoll, 1988), the depolarization also results from M_1 receptor activation. In several other studies, neuronal depolarization (ionic mechanism not determined) has been shown to be associated with the M_1 receptor (rat superior cervical ganglion: Brown et al., 1980; Gardier et al., 1978; Newberry et al., 1985).

There is little evidence available that addresses the question of coupling between this background potassium conductance and the muscarinic receptor. The group of Nicoll (Malenka et al., 1987) showed that this action of acetylcholine on hippocampal pyramidal cells could be well mimicked by phorbol esters. On the other hand, in guinea-pig enteric neurons, the potassium conductance decrease can also be produced by forskolin or cyclic AMP (Palmer et al., 1987).

3.1.2 Calcium-Dependent Potassium Conductance

One class of calcium-dependent potassium conductance is also reduced by muscarinic agonists. These experiments have been carried out with bullfrog ganglion cells (Koketsu et al., 1982; Tokimasa, 1984), in guinea-pig myenteric neurons (North and Tokimasa, 1983; Galligan et al., 1988), guinea-pig sympathetic neurons (Cassell and McLachlan, 1987), rat hippocampal pyramidal cells (Müller and Misgeld, 1986; Madison et al., 1987), and rat olfactory cortex neurons (Constanti and Sim, 1987). In these experiments, the effect of muscarinic agonists is observed on the afterhyperpolarization that follows the action potential, which results from calcium entering the cell and increasing the potassium conductance during the subsequent several hundred ms to several s. A difficulty with interpretation this always presents is whether the action of the muscarinic agonist is to reduce the entry of calcium, or directly to inhibit the calcium-activated potassium conductance. Two kinds of experiments have demonstrated the latter action. First, acetylcholine applied to the surface of a myenteric neuron *after* the calcium has entered the neuron (during the action potential) is still effective to reduce the afterhyperpolarization (or aftercurrent recorded in voltage clamp) (North and Takimasa, 1983; Galligan et al., 1988). Second, the depression of the afterhyperpolarization by muscarine is mainly a reduction in duration; in contrast, the action of cobalt, which is presumed to act by blocking calcium entry, is on the peak amplitude of the afterhyperpolarization (Tokimasa, 1984; North and Tokimasa, 1983).

Several neurons have a persistent calcium current entering the cell at the resting potential, and this current continuously activates the potassium conductance of the cell. The contribution of such a persistent calcium-dependent potassium current can often be judged by observing the effect on membrane potential and conductance of blocking the inward calcium current with, for example, cobalt or calcium ion removal. The reduction by muscarine in the calcium-dependent potassium current also depolarizes the neuron directly if this current is contributing to the maintenance of the resting potential of the cell (North and Tokimasa, 1987; Galligan et al., 1988).

The muscarinic receptor type involved in the suppression of the calcium-dependent potassium conductance is also M_1 in the case of the myenteric neurons (pirenzepine K_D from Schild analysis was 10 n*M*; Galligan et al. 1988) and the hippocampal pyramidal cells (insensitive to gallamine and relatively insensitive to pirenzepine; Dutar and Nicoll, 1988). In the striatal neurons and olfactory cortex of the rat, the receptor type may be M_2 (Constanti and Sim, 1987; Muller and Misgeld, 1986), although in those tissues, it is less clear whether the reduction in the afterhyperpolarization results from a reduction in calcium entry (e.g., Misgeld et al., 1986).

The functional consequences of a reduction in the afterhyperpolarization that follows the action potential can be very marked, since the calcium-dependent potassium current is one of the currents that limits the frequency of action potential firing. Thus, synaptic potentials resulting from the action of acetylcholine at M_1 receptors to reduce this potassium conductance are associated with high-frequency action potential discharge (*see* North and Tokimasa, 1982; Cole and Nicoll, 1983; Morita and North, 1985).

3.1.3. M Potassium Current

This is a time- and voltage-sensitive potassium current observed in some vertebrate cells. The M-conductance is slightly operative at the resting potential of the cell, but as the neuron is depolarized, the potassium channels open. The rate of opening is rather slow for small depolarizations (say, to −40 or −30 mV), but may be faster for the large depolarization subserving the action potential or perhaps a very strong excitatory postsynaptic potential. The opening of potassium channels with depolarization tends to limit the depolarization itself, and the current is sometimes described as a "braking" current, since it limits the frequency of firing of action potentials; muscarine reduces this influence and thus allows the cell to fire more rapidly. This leads to the same functional consequence as reduction in the calcium-dependent potassium current referred to above; the main difference is that M-current inhibition can only occur when the channels are open (by depolarization from about −60 mV), whereas inhibition of the calcium-dependent potassium current will occur at all membrane potentials.

Inhibition of the M current by muscarinic agonists has been described in several different neurons. In some cases, one or both of the other two potassium currents mentioned above are also inhibited. For example, in bullfrog ganglion cells, muscarine reduces both M current and calcium-dependent current and, in some cells, the background potassium current. In hippocampal pyramidal cells, muscarinic agonists reduce the M current, but 10-fold higher concentrations of agonist are required than those needed to depress the background and the calcium-dependent potassium current.

Bullfrog sympathetic cells (Adams and Brown, 1982; Adams et al., 1982a,b), rat sympathetic ganglion cells (Constanti and Brown, 1981), phasic-type inferior mesenteric or coeliac neurons of the guinea pig (Cassell and McLachlan, 1987), hippocampal pyramidal cells (Halliwell and Adams, 1982; Madison et al., 1987), olfactory cortex neurons (Constanti and Galvan, 1983), and cultured mouse spinal cord neurons (Nowak and Macdonald, 1983) have been shown to express an M conductance that is reduced by muscarinic agonists. In some other cells, recordings of membrane voltage have shown that acetylcholine causes a depolarization that fails to reverse when the membrane is hyperpolarized; this is the behavior expected from a decrease in voltage-dependent conductance such as the M conductance (or a calcium-dependent potassium conductance in which the voltage sensitivity results from the voltage sensitivity of the persistent calcium current; *see* Galligan et al., 1988). Such an action of muscarine has been described in about half of the neurons of the cat and guinea-pig medial and lateral geniculate nuclei (McCormick and Prince, 1987), and guinea-pig cerebral cortex (McCormick and

Prince, 1986). Unfortunately, voltage recordings of this kind cannot distinguish effectively between suppression of the M current and a combination of simple reduction in background potassium conductance coupled with an increase in cation conductance.

Only in one experiment has an effort been made to demonstrate directly the kind of muscarinic receptor involved in the inhibition of the M current, although the view is strangely prevalent that the M_1 receptor is involved. A recent paper by Dutar and Nicoll (1988) presents evidence that gallamine can block the inhibition of the M potassium current by carbachol (as well as the presynaptic inhibitory action), while sparing the reduction by carbachol of the background current and the calcium-dependent potassium current. Experiments with pirenzepine were not conclusive in the absence of dose–response curves; they nonetheless indicated that the inhibition of the M current was less sensitive to blockade by pirenzepine than the resting depolarization (background potassium conductance) or the inhibition of the afterhyperpolarization.

The coupling between the muscarinic receptor and the M potassium channels has been investigated. A G-protein is involved in the sympathetic ganglion cells; however, neither in sympathetic ganglion cells nor in hippocampus does this protein seem to be a substrate for pertussis-toxin (Pfaffinger, 1987; Hones, 1987; Lopez et al., 1987; Dutar and Nicoll, 1988). Dutar and Nicoll (1988) suggest that $InsP_3$ (but not altered levels of intracellular calcium) may be involved in the reduction in the M current; evidence that diacylglycerol and activation of C kinase are involved is weak, although, as mentioned above, phorbol esters mimic the other actions of muscarinic agonists in the hippocampal cells (Malenka et al., 1987). Whatever the details of the coupling mechanism, it seems likely to involve a diffusible cytoplasmic messenger and to be shared by several other membrane receptors. The increase in M conductance occurs after a considerable delay when the agonist is applied, and among the other receptors that close the same channels are those for substance P, bradykinin, and luteinizing hormone-releasing hormone (*see* Brown, 1983).

3.1.4. Other Potassium Conductances

In tonic neurons of the guinea-pig inferior mesenteric ganglion (Cassell and McLachlan, 1987) and cultured rat hippocampal cells (Nakajima et al., 1986), acetylcholine reduces a transient potassium current (I_A) that is activated when the membrane is depolarized from a relatively hyperpolarized level.

3.2. Smooth Muscle

Acetylcholine and substance P both reduce the same potassium conductance in freshly dissociated smooth muscle cells of the toad stomach (Sims et al., 1986, 1988). The properties of the potassium conductance involved clearly define it as the M conductance (Sims et al., 1985); it has the same time and voltage dependence as the M conductance in various neurons. The lack of effectiveness of even quite high concentrations of pirenzepine is taken to indicate that the receptor involved is more likely to be M_2 than M_1.

4. Calcium Channels

Calcium currents can be measured directly only after suppression of other currents that are activated by the same voltage change. Removal of sodium ions, or addition of tetrodotoxin, is often used; in many experiments, barium rather than calcium is used as the charge carrier because it helps to reduce outward potassium currents. Potassium currents are usually eliminated by substituting cesium or tetraethylammonium for potassium. Clearly, these manipulations should be taken into account in the interpretation of experiments in which agonist actions on whole-cell calcium currents are measured. Equally important to bear in mind is that the usual protocol to evoke the calcium current (a depolarization lasting tens or hundreds of ms) is a very poor mimic of the depolarization that occurs under more natural circumstances during an action potential: one significant difference is that the amount of calcium (or barium) that enters the cell under these experimental conditions is likely much greater than that associated with an action potential. This excess of calcium (or barium) might significantly alter the coupling through an intracellular second messenger between muscarinic receptors and the calcium channels.

4.1. Heart

The regulation of the heartbeat by acetylcholine depends not only on the changes in potassium conductance; other ionic currents are also affected, including calcium currents. In pacemaking tissue (sinoatrial node, atrioventricular node, and Purkinje fibres), acetylcholine opens potassium channels (*see above*). In ventricular muscle, acetylcholine affects primarily calcium currents, and in atrial muscle both calcium currents and potassium currents are affected (Giles and Noble, 1976; Iijima et al., 1985; *see* review by Loffelholz and Pappano, 1985).

It is well known that agonists at β-adrenergic receptors increase calcium currents in cardiac tissue by increasing the level of cyclic AMP; a cyclic AMP-dependent protein kinase then phosphorylates the calcium channel or a closely associated molecule. Thus, acetylcholine might act by reducing the formation of cyclic AMP (it is known to inhibit adenylate cyclase in heart and other tissues), by reducing the effectiveness of cyclic AMP, or by increasing the hydrolysis of cyclic AMP (Hescheler et al., 1986; Rardon and Pappano, 1986; Fischmeister and Hartzell, 1986; Kameyama et al., 1985). In dissociated cells of the frog (Fischmeister and Hartzell, 1986) or guinea-pig (Hescheler et al., 1986; Rardon and Pappano, 1986) ventricle, the decrease in calcium current produced by acetylcholine can be accounted for by a decrease in the intracellular levels of cyclic AMP. Hescheler et al. (1986) concluded that the main action of muscarinic agonists was to inhibit the formation of cyclic AMP, through the activation of pertussis-toxin-sensitive G_i negatively coupled to adenylate cyclase. Rardon and Pappano (1986) concluded that carbachol acted at a site beyond the formation of cyclic AMP. Fischmeister and Hartzell (1986, 1987) carried out experiments in which the interior of the cardiac myocyte was perfused with various concentrations of cyclic AMP and found acetylcholine to be ineffective under these conditions. That is to say, acetylcholine most probably acted by inhibiting adenylate cyclase. On the other hand, it was found that increased levels of intracellular cyclic GMP reduced the effectiveness of isoprenaline or cyclic AMP to increase the calcium current, strongly suggesting that the level of cyclic AMP (and hence the calcium current) was reduced because cyclic GMP increased the activity of a cyclic GMP-dependent phosphodiesterase responsible for the breakdown of the cyclic AMP (Fischmeister and Hartzell, 1986, 1987). Acetylcholine had been known for some time to increase the levels of cyclic GMP within cardiac cells, so this provides an additional mechanism whereby calcium currents might be reduced, and one consistent with the findings of Rardon and Pappano (1986). The possibility that calcium currents are reduced as a result of protein kinase C activation must also be considered in light of the evidence that cardiac M_2 receptors can also stimulate phophatidylinositol hydrolysis, when expressed in Chinese hamster ovary cells (Ashkenazi et al., 1987). On the other hand, coupling of M_2 receptors to phosphatidylinositol is rather weak, with small responses from high agonist concentrations. Such an action might be expected to be resistant to pertussis toxin (*see* Masters et al., 1985).

4.2. Neurons

Muscarinic agonists inhibit the release of numerous neurotransmitters; an M_2 receptor type has been implicated in most sites investigated (Fuder et al., 1982; Kilbinger et al., 1984; James and Cubeddu, 1986; Meyer and Otero, 1985; North et al., 1985; Dutar and Nicoll, 1988). It is generally thought that an important mechanism that could contribute to presynaptic inhibition of transmitter release is a reduction in calcium currents at the nerve terminal.

Reduction of calcium currents in the cell bodies of neurons has been described in several tissues (myenteric neurons: North and Tokimasa, 1983; rat sympathetic ganglion cells: Belluzi et al., 1985; Wanke et al., 1987; rat hippocampal pyramidal cells in culture: Gahwiler and Brown, 1987; rat striatum: Misgeld et al., 1986). The most complete study is that of Wanke et al. (1987) in the rat sympathetic neurons. In this case, the kind of calcium conductance involved has properties similar to the N type of dorsal root ganglion cells (Nowycky et al., 1985). The reduction in conductance by muscarine was quite selective for this transient conductance, little effect being observed on the sustained inward calcium current evoked by a 80-ms depolarizing command. Muscarinic inhibition was blocked by treatment of the cells with pertussis toxin (150 ng/mL for 3 h). There is little direct evidence available on the question of which muscarinic receptor type is involved in the reduction of calcium conductance; the finding that pirenzepine at 120 n*M* completely blocked the effect of a supramaximal concentration of acetylcholine was interpreted by Wanke et al. (1987) to implicate the M_1 receptor.

It is not known whether the coupling between muscarinic receptor and calcium channel involves a diffusible cytoplasmic messenger or is more similar to the coupling between the M_2

receptor and the potassium channel in the heart. Evidence against the involvement of either A kinase or C kinase was provided by Wanke et al. (1987).

4.3. Smooth Muscle

The opposite action of acetylcholine on a calcium current in toad stomach smooth muscle has recently been reported (Clapp et al., 1987a). In these experiments, a transient calcium current was evoked by voltage depolarizing steps from −78 mV. As long as the current was not close to maximal (steps to about 0 mV), acetylcholine increased its amplitude; there is no information available regarding the receptor type involved in this action. Diacylglycerol analogs that activate C kinase have similar effect, suggesting that the diacylglycerol may be an endogenous second messenger in the mediation of the muscarinic increase of calcium conductance (Clapp et al., 1987b).

5. Inward Currents from Increased Cation Conductance

5.1. Heart

It has been stressed above that any changes in the levels of diffusible intracellular second messengers might be expected to have effects on several different ion channels. In cardiac tissue, there is a conductance that is activated when the membrane is hyperpolarized, the "pacemaker current" (I_f) (DiFrancesco, 1985). Acetylcholine inhibits the activation of this current. Because the net current is inward (mixture of Na^+ and K^+ ions), it contributes to slow depolarization during diastole; inhibition of the current therefore slows the heart. Recent work by DiFrancesco and Tromba (1987) indicates that acetylcholine inhibits the current by a pertussis-toxin-sensitive mechanism. The effect on the current is opposite to that exhibited by catecholamines (Tsien, 1973), but the acetylcholine is effective whether or not the current is first increased by β-adrenergic receptor agonists. Acetylcholine was not effective, however, to reduce the current increased by forskolin, implying that it may exert its effect by reducing cyclic AMP levels within the cell.

There is a second kind of inward current that has recently been described in cardiac tissue. This can be seen only *after* treatment of the tissue with pertussis toxin (Pappano et al., 1988); application of carbachol to isolated guinea-pig ventricular myocytes caused a small inward current, almost voltage-independent but reduced by removal of sodium ions. It is suggested that another consequence of muscarinic receptor activation, but one not mediated by a pertussis-toxin-sensitive G-protein might be an increase in the sodium conductance of the cell. Of interest in this respect is the finding by Cohen-Armon and Sokolovsky (1986) and Cohen-Armon et al. (1988) that a pertussis-insensitive G-protein is involved in the action of muscarinic agonists in rat atria to increase the uptake of radioactive sodium ions.

5.2. Neurons

Several neurons also express a cation current activated by hyperpolarization from the resting level that is similar in many respects to I_f. This has been termed I_H and I_Q (Mayer and Westbrook, 1983). It is not known whether the current is affected by muscarinic agonists. However, it is clear from the early experiments of Kuba and Koketsu (1976, 1978) that one component of the muscarinic slow excitatory postsynaptic potential results from activation of a membrane current having a reversal potential less negative than rest, and blocked by removal of sodium ions from the solution.

An excitatory effect of muscarine on single neurons of rat locus ceruleus results from activation of cardiac-type M_2 receptors (Egan, 1985; Egan and North, 1986; Christie and North, 1988). This excitation also appears to involve a small inward voltage-independent current that requires external sodium (Egan, 1985); this action is also insensitive to pertussis toxin (Christie and North, unpublished). The finding of Cohen-Armon et al. (1988) may be relevant to this effect; synaptoneurosomes from rat brain stem (which contains a relatively high density of M_2 receptors) were found to accumulate radioactive sodium more readily in the presence of a muscarinic agonist. However, the sodium accumulation (which was blocked by tetrodotoxin) could be, of course, secondary to a depolarization by muscarine having a primary ionic mechanism that does not involve sodium channels.

5.3. Smooth Muscle

There is a wealth of experimental evidence that indicates that muscarinic agonists depolarize mammalian smooth muscles by increasing their permeability to sodium and/or calcium ions (Bolton, 1979; Bulbring et al., 1970; Bulbring and Shuba, 1976). Interpretation of the early work was complicated by the difficulties of voltage-clamping syncytial tissue; more recently, dispersed mammalian smooth muscle cells have been voltage clamped using the patch-clamp method (Benham et al., 1985). One of the actions of acetylcholine on cells from rabbit jejunum was an increase in the membrane conductance to cations; the reversal potential was in the region of −30 to 0 mV, implying a mixture of cations as the likely charge carriers. This action of acetylcholine presumably underlies the atropine-sensitive excitatory junction potential resulting from activity in the parasympathetic postganglionic fibers.

5.4. Gland Cells

It was mentioned above that the muscarinic agonists cause phosphatidylinositol hydrolysis in gland cells; one result of this is an increase in the intracellular calcium concentration and an opening of a class of potassium channels. Other channels are also opened by the rise in calcium concentration, particularly channels selective for chloride (*see below*) and for cations. The cation channel has a unit conductance of about 30 pS, is activated when intracellular calcium concentration reaches about 1 μM, and has been found in a wide variety of exocrine gland cells (Maruyama and Petersen, 1982; Petersen and Findlay, 1987; Marty, 1987). The physiological role of the calcium-sensitive cation channels activated by muscarinic agonists is not clear. Also, in chick chromaffin cells, muscarine causes an inward current that reverses at about 0 mV (Knight and Maconochie, 1987).

5.5. Oocytes

A current carried by a combination of sodium and potassium ions results from application of acetylcholine to *Xenopus* oocytes that have been injected previously with the mRNA, which encodes the synthesis of the cardiac M_2 muscarinic receptor (Fukuda et al., 1987). Such oocytes show an oscillatory chloride current (*see below*) that can be abolished by intracellular injection of a calcium chelator. The current that remains reverses polarity at about 10 mV in the usual sodium and potassium concentrations; it had smooth rather than oscillatory time course. This effect of expressed cardiac M_2 receptor is of particular interest in view of the pertussis-toxin-insensitive actions of muscarinic agonists in heart and brain stem mentioned above.

6. Chloride Channels

6.1. Neurons

There are calcium-activated chloride conductances in several neurons that could well be a target for muscarinic agonists acting to increase the intracellular calcium concentration. However, the only direct evidence that muscarinic agonists affect a chloride conductance in nerve cells seems to be the work of Brown and Selyanko (1985); in this case a chloride conductance is decreased rather than increased by acetylcholine.

6.2. Gland Cells

The calcium-dependent chloride current has been observed in the exocrine pancreas and the lacrimal gland of mouse and rat (Findlay and Petersen, 1985). The underlying channels have a very low conductance (1–2 pS), and are progressively activated as the calcium concentration inside the cell rises from 100 n*M* to about 1 μ*M*. The rise in calcium concentration following muscarinic receptor activation (O'Doherty et al., 1983) is thought to activate first the large unit conductance potassium channels (*see above*) and then, as the concentration rises further, the small unit conductance chloride channels. However, the total currents carried by potassium and chloride in the lacrimal gland are about the same, implying that there are about 100 times more chloride channels per cell than potassium channels.

6.3. Oocytes

Receptors that couple to phosphatidylinositol hydrolysis can be expressed in *Xenopus* oocytes, including the muscarinic M_1 receptor (Kubo et al., 1986). Indeed, the oocytes have a native muscarinic receptor that couples through this mechanism (Kusano et al., 1982: Dascal et al., 1984). The rise in intracellular calcium causes a large increase in the chloride conductance of the oocyte, and this results in an inward current under the usual recording conditions.

7. Conclusions

There are several distinct muscarinic receptors (Bonner et al., 1987). Can anything be said about the likely consequences to ion channels of activating one or another of these types? In a general sense, probably not. It was mentioned already that activation of receptors in Chinese hamster ovary cells that had been stably transfected so as to express pig atrial M_2 receptors can both stimulate phosphatidylinositol metabolism and inhibit cyclic AMP accumulation (Ashkenazi et al., 1987); thus, the potential is there for this receptor to open and close a variety of ion channels that might be sensitive to cyclic AMP, calcium, or diacylglycerol. Yet, the native receptor in the pig atrium presumably also couples to an inwardly rectifying potassium conductance that is independent of these messengers. In another case, an M_2 receptor defined pharmacologically by its affinity for pirenzepine and AFDX-116 is present in rat parabrachial neurons and in the closely adjacent locus ceruleus neurons; muscarine inhibits the parabrachial cells and excites the locus ceruleus neurons (Egan and North, 1985, 1986; Christie and North, 1988).

In a sense, it is disappointing that one can predict so little about the consequences to any excitable tissue of activating the muscarinic receptors upon its surface membrane. However, certain patterns begin to emerge: the same mechanisms are used in quite disparate tissues.

At the beginning of the chapter, it was noted that muscarinic receptors couple to ion channels through intermediate molecules. One class of coupling has been identified in which only a single intermediate molecule seems likely to be involved; in this case a G-protein intervenes between a cardiac muscarinic receptor and particular kind of inwardly rectifying potassium channel. Although best studied in the heart, it has also been demonstrated in clonal gland cells, and it appears that coupling of a similar kind will occur also in some mammalian neurons that are inhibited by acetylcholine. There is much ignorance about this form of coupling: we do not yet know whether other effectors are accessed through the same G-protein, we do not know under what circumstances the cell might elect to couple to one effector or another, and we do not know whether the result of such a coupling to another effector might feed back to alter the increase in potassium conductance. The potassium channel involved can be opened by several other receptors [adenosine in the heart

(Kurachi et al., 1986b), somatostatin in GH_3 cells (Yatani et al., 1987b), $GABA_B$ and μ opioid in parabrachial neurons (Christie and North, 1988)]. Indeed, the kind of potassium channel used here by the muscarinic receptor seems similar to the channels opened in various tissues by agonists at a number of other receptors (5-HT_1, D_2, δ opioid, α-2 adrenergic; *see* North et al. 1987; Surprenant and North, 1988).

It seems likely that this kind of coupling between receptor and channel is fundamentally important in a physiological sense; in the case of muscarinic receptors, it provides the principal substrate for the observation that activity in the vagus nerve slows the rate of beating of the heart (Weber and Weger, 1845). The activation by naturally released transmitter of muscarinic receptors or other receptors that couple to potassium channels in this way results in inhibitory postsynaptic potentials in several different kinds of neuron, from *Aplysia* to mammalian brain. These inhibitory postsynaptic potentials all have similar time courses—1–2 s in duration—which is consistent with the underlying molecular mechanism.

It is possible that muscarinic (and other) receptors might affect the activity of channels for ions other than potassium by a similar process. Biochemical experiments suggest an interaction between voltage-dependent sodium channels and pertussis-toxin-insensitive G-protein; whether or not calcium conductances can be affected directly by activated G-protein subunits is currently being investigated.

The second kind of coupling that has been distinguished is even less unique because many more intracellular molecules may be involved. When activation of muscarinic receptors leads to increases and/or decreases in the level of a diffusible cytoplasmic messenger, who can predict which channels will open and which will close? It is known that rises in intracellular calcium can lead to an increase in the probability of opening several different channels even in the same cell (potassium, cation, chloride); it is also known that the same messenger (for example, cyclic AMP) can open potassium channels in some cells and close them in others.

In those cases so far tested, the main second messengers implicated in transduction from liganded muscarinic receptor to open (or closed) channel are calcium ions liberated from intracellular storage sites by $InsP_3$ (and here the evidence is best advanced in glandular tissues) and diacylglycerol acting in concert with the calcium to activate protein kinase C (suggested in some neurones). How calcium ions open and close channels is largely unknown; we do not know whether channel proteins themselves are substrates for C kinase. In hippocampal pyramidal neurons, some but not all of the actions of muscarine are mimicked by phorbol esters, but that certainly does not prove that diacylglycerol is a necessary intermediate.

Clearly these aspects of signal transduction are not restricted to muscarinic receptors. In gland cells, a number of different agonists, in addition to acetylcholine acting at M_4 receptors, result in the opening of the same channels (for example, substance P, cholecystokinin); in smooth muscle, both acetylcholine and substance P reduce the conductance of the same potassium channels. The number of receptors is limited (perhaps five or six), the number of coupling mechanisms likely to be involved now seems to be only a handful, and only a few ion channels appear to provide significant targets. Yet, the way in which these elements are put together by various cell types will need to be worked out tissue by tissue, and circumstance by circumstance.

Acknowledgments

Work in the author's laboratory is supported by U.S. Department of Health and Human Services grants DA03160, DA03161, MH40416, and DK32979.

References

Adams, P. R. and Brown, D. A. (1982) Synaptic inhibition of the M-current: Slow excitatory post-synaptic potential mechanism in bullfrog sympathetic neurones. *J. Physiol. (Lond.)* **332**, 263–272.

Adams, P. R., Brown, D. A., and Constanti, A. (1982a) M-currents and other potassium-currents in bullfrog sympathetic neurones. *J. Physiol. (Lond.)* **330**, 537–572.

Adams, P. R., Brown, D. A., and Constanti, A. (1982b) Pharmacological inhibition of the M-current. *J. Physiol. (Lond.)* **332**, 223–262.

Ascher, P. (1972) Inhibitory and excitatory effects of dopamine on *Aplysia* neurones. *J. Physiol. (Lond.)* **225**, 173–209.

Ascher. P. and Chesnoy-Marchais, D. (1982) Interactions between three slow potassium responses controlled by three distinct receptors in *Aplysia* neurones. *J. Physiol. (Lond.)* **324**, 67–92.

Ashe, J. H., and Yarosh, C. A. (1984) Differential and selective antagonism of the slow-inhibitory postsynaptic potential and slow-excitatory postsynaptic potential by gallamine and pirenzepine in the superior cervical ganglion of the rabbit. *Neuropharmacology* **23**, 1321–1329.

Ashkenazi, A., Winslow, J. W., Peralta, E. G., Peterson, G. L., Schimerlik, M. I., Capon, D. J., and Ramachandran, J. (1987) An M_2 muscarinic receptor subtype coupled to both adenylyl cyclase and phosphoinositide turnover. *Science* **328**, 672–675.

Belluzzi, O., Sacchi, O., and Wanke, E. (1985) Identification of delayed potassium and calcium currents in the rat sympathetic neurone under voltage clamp. *J. Physiol. (Lond.)* **358**, 109–129.

Benham, C. D., Bolton, T. B., and Lang, R. J. (1985) Acetylcholine activates an inward current in single mammalian smooth muscle cells. *Nature* **316**, 345–347.

Bernardo, L. S. and Prince, D. A. (1982a) Cholinergic excitation of mammalian hippocampal pyramidal cells. *Brain Res.* **249**, 315–333.

Bernardo, L. S. and Prince, D. A. (1982b) Ionic mechanisms of cholinergic excitation in mammalian hippocampal pyramidal cells. *Brain Res.* **249**, 333–344.

Bolton, T. B. (1979) Mechanisms of action of transmitters and other substances on smooth muscle. *Physiol. Rev.* **59**, 606–718.

Bonner, T. I., Buckley, N. J., Young, A. C. and Brann, M. R. (1987) Identification of a family of muscarinic acetylcholine receptor genes. *Science* **237**, 527–532.

Breitwieser, G. E. and Szabo, G. (1985) Uncoupling of cardiac muscarinic and β-adrenergic receptors from ion channels by a guanine nucleotide analogues. *Nature* **317**, 538–540.

Brown, D. A. (1983) Slow cholinergic excitation—a mechanism for increased neuronal excitability. *Trends Neurosci.* **6**, 302–307.

Brown, D. A. and Selyanko, A. A. (1985) Two components of muscarine-sensitive membrane current in rat sympathetic neurones. *J. Physiol.* **358**, 335–363.

Brown, D. A., Forward, A., and Marsh, S. (1980) Antagonist discrimination between ganglionic and ileal muscarinic receptors. *Br. J. Pharmacol.* **71**, 362–364.

Brown, H. F. (1982) Electrophysiology of the sinoatrial node. *Physiol. Rev.* **62**, 505–530.

Bülbring, E. and Shuba, M. F. (1976) *Physiology of Smooth Muscle* (Raven, New York).

Bülbring, E., Brading, A. F., Jones, A. W., and Tomita, T. (1970) *Smooth Muscle* (William and Wilkins, London).

Burgen, A. S. V. and Terroux, K. G. (1953) On the negative inotropic effect in the cat's auricle. *J. Physiol. (Lond.)* **120**, 449–464.

Cassell, J. F. and McLachlan, M. (1987) Muscarinic receptors block five different potassium conductances in guinea-pig sympathetic neurones. *Br. J. Pharmacol.* **91**, 259–261.

Christie, M. J. and North, R. A. (1988) Agonists at μ opioid, M_2 muscarine and $GABA_B$ receptors increase the same potassium conductance in rat lateral parabrachial neurones. *Br. J. Pharmacol.*, **95**, 896–902.

Clapp, L. H., Vivaudou, M. B., Walsh, J. V., and Singer, J. J. (1987a) Acetylcholine increases voltage-activated Ca^{2+} current in freshly dissociated smooth muscle cells. *Proc. Natl. Acad. Sci. USA* **84**, 2092–2096.

Clapp, L. H., Singer, J. J., Vivaudou, M. B., and Walsh, J. V. (1987b) A diacylglycerol analogue enhances voltage-activated Ca^{2+} current in freshly dissociated smooth muscle cells from the stomach of the toad. *J. Physiol. (Lond.)* **394**, 44P.

Codina, J., Yatani, A., Grenet, D., Brown, A. M., and Birnbaumer, L. (1987) The α subunit of the GTP binding protein G_K opens atrial potassium channels. *Science* **236**, 442–445.

Cohen-Armon, M. and Sokolovsky, M. (1986) Interactions between the muscarinic receptors, sodium, channels, and guanine nucleotide-binding protein(s) in rat atria. *J. Biol. Chem.* **261**, 12498–12505.

Cohen-Armon, M., Garty, H., and Sokolovsky, M. (1988) G-protein mediates voltage regulation of agonist binding to muscarinic receptors: Effects on receptor Na^+ channel interaction. *Biochemistry* **27**, 368–374.

Cole, A. E. and Nicoll, R. A. (1983) Acetylcholine mediates a slow synaptic potential in hippocampal pyramidal cells. *Science* **221**, 1299–1301.

Cole, A. E. and Shinnick-Gallagher, P. (1984) Muscarinic inhibitory transmission in mammalian sympathetic ganglia mediated by increased potassium conductance. *Nature* **307**, 270–271.

Constanti, A. and Brown, D. A. (1981) M-currents in voltage-clamped mammalian sympathetic neurones. *Neurosci. Lett.* **24**, 289–294.

Constanti, A. and Galvan, M. (1983) M-current in voltage-clamped olfactory cortex neurones. *Neurosci. Lett.* **39**, 65–70.

Constanti, A. and Sim, J. A. (1987) Muscarinic receptors mediating suppression of the M-current in guinea-pig olfactory cortex neurones may be of the M_2-subtype. *Br. J. Pharmacol.* **90**, 3–5.

Dale, H. (1934) Chemical transmission of the effects of nerve impulses. *Br. Med. J.* **12**, 835–841.

Dascal, N., Landau, E. M., and Lass, Y. (1984) Xenopus oocyte resting potential, muscarinic responses and the role of calcium and guanosine 3′,5′-cyclic monophosphate. *J. Physiol. (Lond.)* **352**, 551–574.

del Castillo, J. and Katz, B. (1955) Production of membrane potential changes in the frog's heart by inhibitory nerve impulses. *Nature* **175**, 1035.

DiFrancesco, D. (1985) The cardiac hyperpolarizing-activated current, i_f. Origins and developments. *Prog. Biophys. Mol. Biol.* **6**, 163–183.

DiFrancesco, D. and Tromba, C. (1987) Acetylcholine inhibits activation of the cardiac hyperpolarizing-activated current, i_f. *Pflugers Arch.* **410**, 139–142.

Dodd, J. and Horn, J. P. (1983) Muscarinic inhibition of sympathetic C neurones in the bullfrog. *J. Physiol. (Lond.)* **325**, 145–159.

Dodd, J., Dingledine, R., and Kelly, J. S. (1981) The excitatory action of acetylcholine on hippocampal neurones of the guinea pig and rat maintained in vitro. *Brain Res.* **207**, 109–127.

Dutar, P. and Nicoll. R. A. (1988) Classification of muscarinic responses in hippocampus in terms of receptor subtypes and second messenger systems. *J. Neurosci.*, **8**, 4214–4224.

Eccles, R. M. and Libet, B. (1961) Origin and blockade of the synaptic responses of curarized sympathetic ganglia. *J. Physiol. (Lond.)* **157**, 484–503.

Egan, T. M. (1985) The action of acetylcholine on neurons of the rat locus coeruleus. Ph.D. thesis. M.I.T., Cambridge, Massachusetts, U.S.A.

Egan, T. M. and North, R. A. (1985) Acetylcholine acts on M_2-muscarinic receptors to excite rat locus coeruleus neurones. *Br. J. Pharmacol.* **85**, 733–735.

Egan, T. M. and North, R. A. (1986) Acetylcholine hyperpolarizes central neurones by acting on an M_2 muscarinic receptor. *Nature* **319**, 405–407.

Findlay, I. (1984) A patch clamp study of potassium channels and whole-cell currents in acinar cells of the mouse lacrimal gland. *J. Physiol. (Lond.)* **350**, 179–195.

Findlay, I. and Petersen, O. H. (1985) Acetylcholine stimulates a Ca^{2+}-dependent Cl-conductance in mouse lacrimal acinar cells. *Pflugers Arch.* **403**, 328–330.

Fischmeister, R. and Hartzell, H. C. (1987) Cyclic guanosine 3′,5′-monophosphate regulates the calcium current in single cells from frog ventricle. *J. Physiol. (Lond.)* **387**, 453–472.

Fischmeister, R. and Hartzell, H. C. (1986) Mechanism of action of acetylcholine on calcium current in single cells from frog ventricle. *J. Physiol. (Lond.)* **376**, 183–202.

Fuder, H., Rink, D., and Muscholl, E. (1982) Sympathetic nerve stimulation on the perfused rat heart. Affinities of n-methylatropine and pirenzepine at pre- and postsynaptic muscarinic receptors. *Naunyn-Schmiedebergs Arch. Pharmacol.* **318**, 210–219.

Fukuda, K., Kubo, T., Akiba, I., Maeda, A., Mishina, M., and Numa, S. (1987) Molecular distinction between muscarinic acetylcholine receptor subtypes. *Nature* **327**, 623–625.

Gahwiler, B. H. and Brown, D. A. (1987) Muscarine affects calcium-currents in rat hippocampal pyramidal cells in vitro. *Neurosci. Lett.* **76**, 301–306.

Gallacher, D. V. and Morris, A. P. (1986) A patch-clamp study of potassium currents in resting and acetylcholine-stimulated mouse submandibular acinar cells. *J. Physiol. (Lond.)* **373**, 379–396.

Gallagher, J. P., Griffith, W. H., and Shinnick-Gallagher, P. (1982) Cholinergic transmission in cat parasympathetic ganglia. *J. Physiol. (Lond.)* **332**, 473–486.

Galligan, J. J., North, R. A., and Tokimasa, T. (1989) Muscarinic agonists and potassium currents in guinea pig myenteric neurones. *Br. J. Pharmacol.*, **96**, 193–203.

Gardier, R. W., Tsevdos, E. J., and Jackson, D. B. (1978) The effect of pancuronium and gallamine on muscarinic tranmission in the superior cervical ganglion. *J. Pharmac. Exp. Ther.* **204**, 46–53.

Gaskell, W. H. (1887). On the action of muscarine upon the heart, and on the electrical changes in the non-beating cardiac muscle brought about by stimulation of the inhibitory and augmentor nerves. *J. Physiol. (Lond.)* **8**, 404–415.

Giles, W. and Noble, S. (1976) Changes in membrane currents in bullfrog atrium produced by acetylcholine. *J. Physiol. (Lond.)* **261**, 103–123.

Gilman, A. G. (1987) Receptor regulated G-proteins. *Trends Pharmacol. Sci.* **9**, 460–463.

Gruol, D. L. and Weinreich, D. (1979) Cooperative interactions of histamine and competitive antagonism by cimetidine at neuronal histamine receptors in the marine mollusc, *Aplysia californica. Neuropharmacology* **18**, 415–421.

Halliwell, J. V. and Adams, P. R. (1982) Voltage-clamp analysis of muscarinic excitation in hippocampal neurons. *Brain Res.* **250**, 71–92.

Hartzell, H. C., Kuffler, S. W., Stickgold, R., and Yoshikami, D. (1977) Synaptic excitation and inhibition resulting from direct action of acetylcholine on two types of chemoreceptors on individual amphibian parasympathetic neurons. *J. Physiol. (Lond.)* **271**, 817–846.

Hescheler, J., Kameyama, M., and Trautwein, W. (1986) On the mechanism of muscarinic inhibition of the cardiac Ca current. *Pflugers Arch.* **407**, 182–189.

Howel, W. H. (1906) Vagus inhibition of the heart in its relation to the inorganic salts of the blood. *Am. J. Physiol.* **15**, 280.

Howell, W. H. and Duke, W. W. (1908) The effect of vagus inhibition on the output of potassium from the heart. *Am. J. Physiol.* **21**, 51.

Iijima, T., Irisawa, H., and Kameyama, M. (1985) Membrane currents and their modification by acetylcholine in isolated single atrial cells of the guinea-pig. *J. Physiol. (Lond.)* **359**, 485–501.

James, M. K. and Cubeddu, L. X. (1986) Pharmacologic characterization and functional role of muscarinic autoreceptors in the rabbit striatum. *J. Pharmacol. Exp. Ther.* **204**, 206–215.

Jones, S. W. (1987) GTP-γS inhibits the M-current of dissociated bullfrog sympathetic neurons. *Soc. Neurosci. Abstr.* **13**, 533.

Kameyama, M., Hoffmann, F., and Trautwein, W. (1985) On the

mechanism of beta-adrenergic regulation of the Ca channel in the guinea pig heart. *Pflugers Arch.* **405**, 285–293.

Kehoe, J. (1972a) Ionic mechanisms of two-component cholinergic inhibition in *Aplysia* neurones. *J. Physiol. (Lond.)* **225**, 85–114.

Kehoe, J. (1972b) Three acetylcholine receptors in *Aplysia* neurones. *J. Physiol. (Lond.)* **225**, 115–146.

Kehoe, J. (1972c) The physiological role of three acetylcholine receptors in synaptic transmission in *Aplysia*. *J. Physiol. (Lond.)* **225**, 147–172.

Kilbinger, H., Halim, S., Lambrecht, G., Weiler, W., and Wessler, I. (1984) Comparison of affinities of muscarinic antagonists to pre- and postjunctional receptors in guinea-pig ileum. *Eur. J. Pharmacol.* **103**, 313–320.

Koketsu, K., Akasu, T., and Miyagawa, M. (1982) Identification of g_k systems activated by Ca^{2+}. *Brain Res.* **243**, 369–372.

Krnjevic, K. (1974) Chemical nature of synaptic transmission in vertebrates. *Physiol. Rev.* **54**, 418–540.

Krnjevic, K., Pumain, R., and Renaud, L. (1971) The mechanism of excitation by acetylcholine in the cerebral cortex. *J. Physiol. (Lond.)* **215**, 247–267.

Kuba, K. and Koketsu, K. (1976) Analysis of slow excitatory postsynaptic potential in bullfrog sympathetic ganglion cells. *Jpn. J. Physiol.* **26**, 647–664.

Kuba, K. and Koketsu, K. (1978) Synaptic events in sympathetic ganglia. *Prog. Neurobiol.* **11**, 77–169.

Kubo, T., Fukuda, K., Mikami, A., Maeda, A., Takahashi, H., Mishina, M., Haga, T., Haga, K., Ichiyama, A., Kangawa, K., Kojima, M., Matsuo, H., Hirose, T., and Numa, S. (1986) Cloning, sequencing and expression of complementary DNA encoding the muscarinic acetylcholine receptor. *Nature* **323**, 411–416.

Kurachi, Y., Nakajima, T., and Sugimoto, T. (1986a) On the mechanism of activation of muscarinic K^+ channels by adenosine in isolated atrial cells: involvement of GTP-binding proteins. *Pflugers Arch.* **407**, 264–274.

Kurachi, Y., Nakajima, T., and Sugimoto, T. (1986b) Acetylcholine activation of K^+ channels in cell-free membrane of atrial cells. *Am. J. Physiol.* **251**, H681–H684.

Kurachi, Y., Nakajima, T., and Sugimoto, T. (1986c) Role of intracellular Mg^{2+} in the activation of muscarinic K^+ channel in cardiac atrial cell membrane. *Pflugers Arch.* **407**. 572–574.

Kurachi, Y., Nakajima, T., and Sugimoto, T. (1987a) Quinidine inhibition of the muscarine receptor-activated K^+ channel current in atrial cells of guinea pig. *Naunyn-Schmiedeberg's Arch. Pharmacol.* **335**, 216–218.

Kurachi, Y., Nakajima, T., and Sugimoto, T. (1987b) Short-term desensitization of muscarinic K^+ channel current in isolated atrial myocytes and possible role of GTP-binding proteins. *Pflugers Arch.* **410**, 227–233.

Kusano, K., Miledi, R., and Stinnakre, J. (1982) Cholinergic and catecholaminergic receptors in the *Xenopus* oocyte membrane. *J. Physiol. (Lond.)* **328**, 143–170.

Libet, B. (1979) Slow postsynaptic actions in ganglionic function, in *Integrative Functions of the Autonomic Nervous System* (McC. Brooks, C., Koizumi, K., and Sato, A., eds.) Elsevier, Amsterdam, pp. 197–222.

Loewi, O. (1921) Uber humorale Ubertragbarkeit der Herznervenwirkung. *Pflugers Arch.* **189**, 239–242.

Löffelholz, H. and Pappano, A. J. (1985) The parasympathetic neuroeffector junction of the heart. *Pharmacol. Rev.* **37**, 1–24.

Logothetis, D. E., Yoshihisa, K., Galper, J., Neer, E. J., and Clapham, D. E. (1987) The $\beta\gamma$ subunits of GTP-binding proteins activate the muscarinic K^+ channel in heart. *Nature* **325**, 321–326.

Lopez, H. S., Brown, D., and Adams, P. R. (1987) Possible involvement of GTP-binding proteins in coupling of muscarinic receptors to M-current in bullfrog ganglion cells. *Soc. Neurosci. Abstr.* **13**, 532.

Madison, D. V., Lancaster, B., and Nicoll, R. A. (1987) Voltage clamp analysis of cholinergic action in hippocampus. *J. Neurosci.* **7**, 733–741.

Malenka, R. C., Madison, D. V., Andrade, R., and Nicoll, R. A. (1987) Phorbol esters mimic some cholinergic actions in hippocampal pyramidal neurons. *J. Neurosci.* **6**, 475–480.

Marty, A. (1987) Control of ionic currents and fluid secretion by muscarinic agonists in exocrine glands. *Trends Neurosci.* **10**, 373–377.

Maruyama, Y. and Petersen, O. H. (1982) Cholecystokinin activation of single channel currents is mediated by internal messenger in pancreatic acinar cells. *Nature* **300**, 61–63.

Maruyama, Y., Gallacher, D. V., and Petersen, O. H. (1983) Voltage and Ca^{2+}-activated K^+ channel in basolateral acinar cell membranes of mammalian salivary gland. *Nature* **302**, 827–829.

Maruyama, Y., Petersen, O. H., Flanagan, P., and Pearson, G. T. (1983) Quantification of Ca^{2+}-activated K^+ channels under hormonal control in pig pancreas acinar cells. *Nature* **305**, 228–232.

Masters, S. B., Martin, M. W., Harden, T. K., and Brown, J. H. (1985) Pertussis toxin does not inhibit muscarinic-receptor-mediated phosphoinositide hydrolysis or calcium mobilization. *Biochem. J.* **227**, 933–937.

Mayer, M. L. and Westbrook, G. L. (1983) A voltage-clamp analysis of inward (anomalous) rectification in mouse spinal sensory ganglion neurons. *J. Physiol. (Lond.)* **340**, 19–45.

McCormick, D. A. and Prince, D. A. (1986) Acetylcholine induces burst firing in thalamic reticular neurons by activating a potassium conductance. *Nature* **319**, 402–405.

McCormick, D. A. and Prince, D. A. (1987) Actions of acetylcholine in the guinea-pig and cat medial and lateral geniculate nuclei, in vitro. *J. Physiol. (Lond.)* **392**, 147–165.

Meyer, E. M. and Otero, D. H. (1985) Pharmacological and ionic characterizations of muscarinic receptors modulating [^{3}H] acetylcholine release from rat cortical synaptosomes. *J. Neurosci.* **5**, 1202–1207.

Misgeld, U., Calabresi, P., and Dodt, H. U. (1986) Muscarinic modulation of calcium dependent plateau potentials in rat neostriatal neurons. *Pflugers Arch.* **407**, 482–487.

Morita, K. and North, R. A. (1985) Significance of slow synaptic potentials for transmission of excitation in guinea-pig myenteric plexus. *Neuroscience* **14**, 95–101.

Morita, K., North, R. A., and Tokimasa, T. (1982) Muscarinic agonists inactivate potassium conductance of guinea-pig myenteric neurones. *J. Physiol. (Lond.)* **333**, 125–139.

Muller, W. and Misgeld, U. (1986) Slow cholinergic excitation of guinea pig hippocampal neurons is mediated by two muscarinic receptor subtypes. *Neurosci. Lett.* **67**, 107–112.

Nakajima, Y., Nakajima, S., Leonard, R. J., and Yamaguchi, K. (1986) Acetylcholine raises excitability by inhibiting the fast transient potassium current in cultured hippocampal neurons. *Proc. Natl. Acad. Sci. USA* **83**, 3022–3026.

Newberry, N. R., Priestley, T., and Woodruff, G. N. (1985) Pharmacological distinction between two muscarinic responses on the isolated superior cervical ganglion of the rat. *Eur. J. Pharmacol.* **116**, 191–192.

Nishi, S. and Koketsu, K. (1968) Early and late afterdischarges of amphibian sympathetic ganglion cells. *J. Neurophysiol.* **31**, 109–121.

North, R. A. and Tokimasa, T. (1987) Persistent calcium-sensitive potassium current and the resting properties of guinea-pig myenteric neurons. *J. Physiol. (Lond.)* **386**, 333–354.

North, R. A. and Tokimasa, T. (1983) Depression of calcium-dependent potassium conductance by muscarinic agonists. *J. Physiol. (Lond.)* **342**, 253–266.

North, R. A. and Tokimasa, T. (1982) Muscarinic synaptic potentials in guinea-pig myenteric plexus neurones. *J. Physiol. (Lond.)* **258**, 17–32.

North, R. A., Slack, B. E., and Surprenant, A. (1985) Muscarinic m_1 and m_2 receptors mediate depolarization and presynaptic inhibition in guinea-pig enteric nervous system. *J. Physiol. (Lond.)* **368**, 435–452.

North, R. A., Williams, J. T., Surprenant, A., and Christie, M. J. (1987) μ and δ receptors belong to a family of receptors that are coupled to potassium channels. *Proc. Natl. Acad. Sci. USA* **84**, 5487–5491.

Nowak, L. M. and MacDonald, R. L. (1983) Muscarine-sensitive voltage-dependent potassium current in cultured murine spinal cord neurons. *Neurosci. Lett.* **35**, 85–91.

Nowycky, M. C., Fox, A. P., and Tsien, R. W. (1985) Three types of neuronal calcium channel with different calcium agonist sensitivity. *Nature* **316**, 440–442.

O'Doherty, J., Stark, R. J., Crane, S. J., and Brugge, K. L. (1983) Changes in cytosolic calcium during cholinergic and adrenergic stimulation of the parotid salivary gland. *Pflugers Arch.* **398**, 241–246.

Palmer, J. M., Wood, J. D., and Zafirov, D. H. (1987) Purinergic inhibition in the small intestinal myenteric plexus of the guinea-pig. *J. Physiol. (Lond.)* **387**, 357–369.

Pappano, A. J., Matsumoto, K., Tajima, T., Agnarsson, U., and Webb, W. (1988) Pertussis toxin-insensitive mechanism for carbachol-induced depolarization and positive inotropic effect in heart muscle. *Trends Pharmacol. Sci. Suppl.* 35–39.

Petersen, O. H. and Findlay, I. (1987) Electrophysiology of the pancreas. *Physiol. Rev.* **6**, 1054–1116.

Pfaffinger, P. J., Martin, J. M., Hunter, D. D., Nathanson, N. M., and Hille, B. (1985) GTP-binding proteins couple cardiac muscarinic receptors to a K channel. *Nature* **317**, 536–538.

Pfaffinger, P. J. (1987) Control of M-current in dialysed sympathetic ganglion neurons. *Soc. Neurosci. Abstr.* **13**, 152.

Pott, L. (1979) On the time course of the acetylcholine-induced hyperpolarization in quiescent guinea-pig atria. *Pflugers Arch.* **380**, 71–77.

Rardon, D. P. and Pappano, A. J. (1986) Carbachol inhibits electrophysiological effects of cyclic AMP in ventricular myocytes. *Am. J. Physiol. (Lond.)* **251**, H601–H611.

Sakmann, B., Noma, A., and Trautwein, W. (1983) Acetylcholine activation of single muscarinic K^+ channels in isolated pacemaker cells of the mammalian heart. *Nature* **303**, 250–253.

Sasaki, K. and Sato, M. (1987) A single GTP-binding protein regulates K channels coupled with dopamine, histamine and acetylcholine receptors. *Nature* **325**, 259–262.

Simmons, M. A. and Hartzell, H. C. (1988) A quantitative analysis of the acetylcholine-activated potassium current in single cells from frog atrium. *Pflügers* Arch. **409**, 454–461.

Sims, S. M., Singer, J. J., and Walsh, J. V. (1985) Cholinergic agonists suppress a potassium current in freshly dissociated smooth muscle cells of the toad. *J. Physiol. (Lond.)* **367**, 503–529.

Sims, S. M., Walsh, J. V., and Singer, J. J. (1986) Substance P and acetylcholine both suppress the same K^+ current in dissociated smooth muscle cells. *Am. J. Physiol.* **251**, C580-C587.

Soejima, M. and Noma, A. (1984) Mode of regulation of the ACh-sensitive K-channel by the muscarinic receptor in rabbit atrial cells. *Pflügers* Arch. **400**, 424–431.

Sorota, S., Tsuji, Y., Tajima, T., and Pappano, A. (1985) Pertussis toxin treatment blocks hyperpolarization by muscarinic agonists in chick atrium. *Circ. Res.* **57**, 748–758.

Surprenant, A. and North, R. A. (1988) Mechanism of synaptic inhibition by noradrenaline acting at α_2 adrenoceptors. *Proc. Roy. Soc. Lond. B.* **234**, 85–114.

Suzuki, K. and Petersen, O. H. (1985) The effects of Na and Cl removal and of loop diuretics on acetylcholine-evoked membrane potential change in mouse lacrimal acinar cells. *Q. J. Exp. Physiol.* **70**, 437–445.

Tauc, L. and Gerschenfeld, H. M. (1962) A cholinergic mechanism of inhibitory synaptic transmission in a molluscan nervous system. *J. Neurophysiol.* **25**, 236–262.

Tokimasa, T. (1984) Muscarinic agonists depress calcium-dependent g_K in bullfrog sympathetic neurons. *J. Auton. Nerv. Syst.* **10**, 107–116.

Trautman, A. and Marty, A. (1984) Activation of Ca-dependent K channels by carbamoylcholine in rat lacrimal glands. *Proc. Natl. Acad. Sci. USA* **81**, 611–615.

Trautwein, W. and Dudel, J. (1958) Zum Mechanisms der Membranwirkung des Acetylcholin an der Herzmuskelfaser. *Pflugers Arch.* **266**, 324–334.

Tsien, R. W. (1973) Adrenaline-like effects of intracellular iontophoresis of cyclic AMP in cardiac Purkinje fibres. *Nature* **245**, 120–122.

Wanke, E., Ferroni, A., Malgaroli, A., Ambrosini, A., Pozzan, T., and Meldolesi, J. (1987) Activation of muscarinic receptor selectively inhibits a rapidly inactivated Ca^{2+} current in rat sympathetic neurons. *Proc. Natl. Acad. Sci. USA* **84**, 4313–4317.

Weber, E. F. W. and Weber, E. H. (1845) Experimenta, quibus probatur nervos vagos rotatione machinae galvano-magneticae irratatos motum cordis retardaro. *Ann. Univ. Med., Milano, 3 ser.* **20**, 227–233.

Weight, F. F. and Votava, J. (1970) Slow synaptic excitation in sympathetic ganglion cells: evidence for synaptic inactivation of potassium conductance. *Science* **170**, 775–758.

Yarosh, C. A. and Ashe, J. H. (1987) Antagonist discrimination of two muscarinic responses elicited by applied agonists and orthodromic stimuli in superior cervical ganglion of rabbit. *Neuropharmacology* **26**, 1549–1560.

Yatani, A., Codina, J., Brown, A. M., and Birnbaumer, L. (1987a) Direct activation of mammalian atrial muscarinic potassium channels by GTP regulatory protein G_K. *Science* **23**, 207–210.

Yatani, A., Codina, J., Sekura, R. D., Birnbaumer, L., and Brown, A. M. (1987b) Reconstitution of somatostatin and muscarinic receptor mediated stimulation of K channels by isolated G_K protein in clonal rat anterior pituitary cell membranes. *Mol. Endocrinol.* **1**, 283–288.

Yavari, P. and Weight, F. F. (1987) Antagonists discriminate muscarinic excitation and inhibition in sympathetic ganglion. *Brain Res.* **400**, 133–138.

SECTION 4

Regulation of Muscarinic Receptors

10

Allosteric Interactions of Muscarinic Receptors and Their Regulation by Other Membrane Proteins

Yoav I. Henis, Yoel Kloog, and Mordechai Sokolovsky

The complex nature of ligand binding to muscarinic acetylcholine (ACh) receptors in various tissues and the battery of compounds with antimuscarinic activity that affect muscarinic binding sites in a manner that does not fit simple competition (reviewed by Sokolovsky, 1984) suggest the possible existence of cooperative and allosteric interactions in the muscarinic system. In this review, we will discuss the current evidence and the experimental approaches (up to early 1987) employed to investigate such interactions in the muscarinic receptor system.

1. Allostery and Cooperativity: Definitions and Experimental Tools

1.1. Definitions

An allosteric effector (Monod et al., 1963) is a molecule that interacts with a specific allosteric site—a site topographically distinct from the substrate binding site (on an enzyme), or from the neurotransmitter or hormone binding site (in the case of a receptor). When site–site interactions occur between nonidentical sites, the interacting sites are heterologous, and the interactions are defined as heterotropic. When the interacting sites are identical (homologous), the interactions are defined as homotropic (Levitzki, 1978). Homotropic interactions give rise to the phenomenon of cooperativity in ligand binding—either positive (a progressive increase in the affinity as the degree of saturation by the ligand increases) or negative (reduced affinity at higher saturation degrees). Thus, cooperative interactions are part of the larger family of allosteric interactions.

1.2. Experimental Tools

1.2.1. Identification of Cooperative Interactions

Since cooperativity is defined according to receptor-ligand affinity as a function of saturation by the ligand, there are diagnostic plots of binding data that can help in the identification of cooperative interactions. The most popular plots are the Hill plot and the Scatchard plot. On the Hill plot, noncooperative binding yields a Hill coefficient (n_H; the slope of the Hill plot at mid-saturation) of 1.0, whereas in positive cooperativity, $n_H > 1.0$, and in negative cooperativity $n_H < 1.0$ (Dahlquist, 1978). On the Scatchard plot, noncooperative binding yields a straight line, positively cooperative binding—a concave downward curve, and negatively cooperativing binding—a concave upward curve (Levitzki, 1978). Although the identification of positive cooperativity by such a diagnostic binding plot is acceptable evidence for cooperative interactions, a diagnostic binding plot by itself does *not* prove negative cooperativity. The existence of a heterogeneous population of noninteracting binding sites will produce a binding pattern similar to that of negative cooperativity, and the diagnostic plots derived from direct ligand binding isotherms cannot distinguish between site heterogeneity and negatively cooperative interactions. A conclusive demonstration of negative cooperativity can be achieved by the following approaches:

1. Equilibrium competition studies. In this approach (Henis and Levitzki, 1979; Henis and Sokolovsky, 1983; detailed under section 2.1.), the effects of a competing ligand (whose binding pattern is known) on the binding pattern of the primary labeled ligand are examined. In the case of heterogeneous, noninteracting sites, one can calculate quantitatively the effects of competition on the

binding constants of the primary ligand. Significant deviations from the predicted effects demonstrate the existence of site–site interactions and their involvement in generating the negatively cooperative binding pattern.

2. Effects on dissociation kinetics. Following the binding of subsaturating levels of a labeled ligand, dissociation can be induced either by a large dilution with buffer or by dilution in the presence of a saturating concentration of the same unlabeled ligand. The inclusion of the unlabeled ligand is expected to affect the dissociation rate of the labeled ligand (elevate it in the case of negative cooperativity, and reduce it in the case of positive cooperativity) only if site–site interactions exist, since in this case, the affinity to the ligand (and therefore the k_{off} value) is altered as a function of the saturation by the ligand.
3. Use of covalent affinity labels. This approach is based on the classic "reordered alkylation" studies introduced by Koshland to investigate the mechanism of half-of-the-sites reactivity of oligomeric enzymes toward certain alkylating agents (reviewed by Levitzki and Koshland, 1976). In a modification of this procedure, a ligand that reveals a "negatively cooperative" binding pattern is tested for its ability to protect the receptors from the alkylating affinity label. The protecting ligand is then washed away, and its binding to the unalkylated sites (which should be only the high-affinity sites of this ligand in the case of noninteracting heterogeneous binding sites) is measured again. Alternatively, it is possible to alkylate the receptors with an affinity label that reacts similarly with all the receptor species (i.e., does not discriminate between sites with different affinities toward the "negatively cooperative" ligand), and study the effects of different degrees of alkylation on the binding pattern of the "negatively cooperative" ligand. In the case of noninteracting heterogeneous sites, this binding pattern is not expected to change with the degree of alkylation (Birdsall et al., 1978).

1.2.2. Identification of Allosteric Interactions

By analogy with the identification of cooperative interactions, the first indications for allosteric interactions are usually obtained from binding studies, through the use of diagnostic plots. Thus, when the compound suspected as an allosteric affector is capable of displacing the receptor-bound ligand, the displacement data can be plotted as a Hill plot; as in the case of cooperative interactions, n_H values significantly different from unity suggest the possibility of allosteric interactions. Similarly, a nonlinear Schild plot [log(affinity shift — 1) vs log(competitor)] may be the result of allosteric interactions. However, as in the case of homotropic negative cooperativity, caution has to be taken in cases where $n_H < 1$ or Schild plots demonstrating diminished "competition" at higher competitor concentrations are encountered; in such cases, the above phenomena may be the result of direct competition by the "allosteric" compound on the ligand binding sites, if the latter sites are heterogeneous with respect to binding of the "allosteric" substance. Thus, a conclusive demonstration of allosteric effects in such cases requires other approaches, as in the case of cooperative interactions. The most common approach is to demonstrate effects of the allosteric compound on the dissociation kinetics of a receptor-bound ligand (Waelbroeck et al., 1984b; Dunlap and Brown, 1983; Stockton et al., 1983; Kloog and Sokolovsky, 1985a). Another approach (e.g., Dunlap and Brown, 1983) is to test whether or not the allosteric compound is capable of inhibiting the modification of the receptor binding sites by a covalent affinity label; an allosteric affector is expected to be an inefficient inhibitor.

It is important to note that, even if these tests suggest the existence of allosteric or cooperative interactions, one has to be cautious if these effects are observed only at high concentrations of the affectors, especially when the latter have a hydrophobic character. Under such conditions, the affectors may alter the general properties of the membrane, resulting in nonspecific effects that are similar in appearance to allosteric phenomena (*see* section 3.3. with respect to the effects of verapamil, secoverine, or lidocaine).

Another common statement that has to be dealt with carefully is that B_{max} (the maximal concentration of bound ligand) has to be reduced in the case of allosteric (noncompetitive) inhibition (e.g., Burgermeister et al., 1978). Although it is true that, in many cases, B_{max} shows an apparent reduction in the presence of an allosteric inhibitor (presumably because of the decreased affinity for the radioligand, which owing to technical limitation, does not permit the binding to be measured at high ligand concentrations), the number of receptors cannot be truly decreased unless an irreversible reaction takes place. The situation differs from that encountered in steady-state kinetics, where noncompetitive inhibition occurs because of the formation of an inactive ESI (enzyme-substrate-inhibitor) ternary complex, which leads to a reduction in V_{max}; in the case of the measurement of substrate binding, the ESI complex is still active in binding S, and therefore, the formation of a ternary complex does *not* lead to a reduction in the amount of substrate bound, but only in the amount of substrate converted to product.

2. Homotropic Site–Site Interactions among Muscarinic Receptors

The binding of classical muscarinic antagonists to membrane preparations from brain and heart yields binding isotherms that fit a single dissociation constant (Snyder et al., 1975; Kloog and Sokolovsky, 1978; Hulme et al., 1978; Ehlert et al., 1981). On the other hand, the binding of agonists to these preparations does not obey a simple mass-action law, and yields n_H values below unity (Birdsall et al., 1978; Kloog et al., 1979; reviewed in Sokolovsky, 1984 and in Chapter 2, this volume). This phenomenon was interpreted as indicating the existence of two or three distinct populations of muscarinic binding sites (i.e., site heterogeneity) that are recognized differently by agonists, but not by antagonists (for review, *see* Sokolovsky, 1984). However, as discussed in section 1.2.1., an alternative explanation is the existence of negatively cooperative interactions between the muscarinic binding sites, which are more pronounced in the case of agonist binding. Binding isotherms and the diagnostic plots derived from them are not sufficient to distinguish between these two possibilities (see section 1.2.1.). In order to overcome this difficulty, attempts were made to demonstrate the existence of site–site interactions between muscarinic binding sites, using some of the approaches described in section 1.2.1.

2.1. Identification of Cooperative Interactions between Muscarinic Binding Sites in Rat Adenohypophysis Using Competition Studies

Equilibrium competition experiments can directly test the existence of cooperative interactions among ligand-occupied binding sites (*see* section 1.2.1.). The mechanisms of site heterogeneity (without site–site interactions) and negative cooperativity predict different effects of competition on the binding of the primary ligand (Henis and Levitzki, 1979; Henis et al., 1982, Henis and Sokolovsky, 1983). For site heterogenity without interactions, the presence of an unlabeled competing ligand will affect n_H and the dissociation (or association) constants of the primary labeled ligand to an extent that can be calculated quantitatively from the dissociation constants extracted from the direct binding curves of each ligand (primary or competitor) separately. In contrast, the effects of the competitor on the binding parameters of the primary ligand in negatively cooperative systems cannot be predicted from the separate binding curves, because of the involvement of interactions between sites occupied with different ligands (primary and competitor). Therefore, deviations of the binding parameters obtained in competition experiments from the expected values calculated assuming site heterogeneity indicate the involvement of cooperative interactions in the binding mechanism.

This method requires an accurate determination of the complete binding curves of the primary ligand, the competing ligand, and the primary ligand in the presence of a constant concentration of the competing ligand. Although radiolabeled muscarinic agonists are available, their affinities are low relative to antagonists and only their binding to the high-affinity sites can be measured (Gurwitz et al., 1984a; Ehlert et al., 1980; Birdsall et al., 1978; Waelbroek et al., 1984b). Binding studies employing labeled antagonists do not suffer such limitations. However, there are no indications for negative cooperativity or site heterogeneity in the binding of classical muscarinic antagonists to brain or heart preparations. In contrast, in rat adenohypophysis, curvilinear Scatchard plots and n_H values below unity were observed for the binding of several muscarinic antagonists (Avissar et al., 1981; Mukherjee et al., 1980). Thus, rat adenohypophysis homogenates provided a convenient system, wherein the existence of cooperative interactions among muscarinic receptors could be explored by competition experiments employing labeled and unlabeled antagonists (Henis et al., 1982; Henis and Sokolovsky, 1983). The inhibition of the binding of the primary ligand by the competing ligand was well above that expected assuming heterogeneous noninteracting sites. The deviations from the expectations of the site heterogeneity model were above the experimental error for all the ligands tested, and reached an order of magnitude for the effect of (−)-*N*-methylscopolamine [(−)NMS] on *N*-[^{3}H]methyl-4-piperidyl benzilate [^{3}H]4NMPB) binding. Such deviations cannot be generated by antagonist-mediated isomerization of noninteracting sites, and require the participation of cooperative interactions between antagonist-occupied binding sites (Henis et al., 1982). Moreover, a detailed mathematical analysis (Henis and Sokolovsky, 1983) demonstrated that, when the separate binding of the primary and competing ligands shows a similar extent of cooperativity (as was the case with the binding of muscarinic antagonists to rat adenohypophysis homogenates), deviations from the expectations of site heterogeneity in competition experiments can occur only if the primary and competing ligands induce *different* conformational alterations in the receptor upon binding. It therefore follows that different muscarinic antagonists induce different conformational transitions (isomerizations) in the receptor; the propagation of such transitions to neighboring sites produces cooperative effects, as seen in the competition experiments.

Interestingly, the deviations seen upon competition between 4NMPB and (−)NMS were more pronounced than for competition between 4NMPB and (−)-3-quinuclidinyl benzilate [(−)-3QNB]. This indicates that a larger difference exists between the transitions induced by 4NMPB and (−)NMS than between those induced by 4NMPB and (−)3QNB. Indeed, the latter two ligands are more closely related to each other (both are hydrophobic nonquaternary benzilate derivatives). Other studies have also shown differences between (−)3QNB and (−)NMS in their ability to induce isomerization (Galper et al., 1982) and in their binding to muscarinic receptors (Lee and El-Fakahany, 1985; Brown and Goldstein, 1986, and references therein).

Unlike the situation encountered in the adenohypophysis, the binding of muscarinic antagonists to homogenates of various brain regions yields linear Scatchard plots (Sokolovsky, 1984), and competition experiments analogous to those described above do not reveal significant deviations from the expectations of pure competition (Henis and Sokolovsky, 1983). The negative nature of this result does not exclude the possibility that site–site interactions do occur among brain muscarinic receptors, but they could not be detected by the experimental procedures employed (competition between antagonists).

2.2. Studies on Site–Site Interactions in Heart and Brain Muscarinic Receptors

2.2.1. Cooperative Interactions in Cardiac Muscarinic Receptors

Cardiac muscarinic receptors interact with guanine nucleotide binding proteins (G_i or G_o); the interaction of guanine nucleotides with these proteins modulates muscarinic agonist binding, usually by interconversion of high affinity agonist binding sites (R_H) to the low affinity form (R_L) (Berrie et al., 1979; Sokolovsky et al., 1980; Wei and Sulakhe, 1980; Ehlert et al., 1981; Harden et al., 1982; Burgisser et al., 1982; Waelbroek et al., 1982). These interactions are extremely sensitive to the experimental conditions, and it has been shown that, under specific conditions (in the absence of Na^+ ions, and in the presence of millimolar Mg^{2+} concentrations and EDTA), guanine nucleotides convert muscarinic antagonist binding from low to a higher affinity state in cardiac preparations from several sources (Hulme et al., 1981; Burgisser et al., 1982; Hosey, 1982). This phenomenon is masked in the presence of sodium ions (above 20 m*M*), which have an effect similar to that of guanine nucleotides on antagonist binding (Hosey, 1982). This sensitivity was used by Mattera et al. (1985) to generate conditions under which cooperative interactions between cardiac muscarinic binding sites can be detected. In experiments performed on sarcolemmal membranes prepared from canine heart in the absence of sodium ions and in the presence of Mg^{2+} and EDTA, these authors demonstrated that the binding of [^{3}H](−)3QNB, which fitted a simple binding isotherm (n_H = 1.0), became *positively cooperative* (n_H = 1.4) in the presence of guanyl-5′-yl imidodiphosphate [Gpp(NH)p]. Similar findings under analogous conditions in the presence of Gpp(NH)p were reported by Boyer et al. (1986). Since positive cooperativity can be generated only by site–site interactions and *not* by site heterogeneity, these findings demonstrate that, at least under the specific experimental conditions employed in this study, there are cooperative interactions among antagonist-occupied cardiac muscarinic receptors. Of course, in such a case, a cooperative model should also fit the data for agonist-antagonist competition curves. Indeed, although a site-heterogeneity model with three noninteracting sites also fitted the data, a model of a cooperative dimer (the simplest system where cooperative interactions can occur), probably in equilibrium with a noncooperative dimeric state, provided excellent fits to all the binding curves [either in the presence or absence of Gpp(NH)p] (Mattera et al., 1985).

2.2.2. Studies on Brain Muscarinic Receptors

As pointed out earlier (section 2.1.), ligand competition studies failed to demonstrate site–site interactions among muscarinic receptors in homogenates of rat brain regions. The possible existence of such interactions in homogenates of rat brain cortex was investigated by alkylation experiments employing propylbenzilylcholine mustard (PrBCM), a muscarinic antagonist that irreversibly binds to the receptor through an alkylation reaction (Birdsall et al., 1978). In the case of site heterogeneity (without cooperative interaction), the pattern of agonist binding (i.e., the deviation of the Scatchard plot from linearity) is not expected to change as a function of the percentage of alkylated binding sites. The results of these binding studies (employing carbachol as the ligand) were interpreted as not significantly deviating from the site heterogeneity model (Birdsall et al., 1978); however, in the absence of complete carbachol binding curves at each alkylation degree, it is not possible to definitely rule out site–site interactions. Another experiment with PrBCM reported in the same study employed carbachol at concentrations that saturate only the high affinity sites to protect from alkylation by PrBCM. After the alkylation, the bound agonist was removed by washing, and its binding to the unalkylated sites was remeasured. The binding was found to be noncooperative and of higher affinity, as expected in the case of site heterogeneity, where the protected sites are the high affinity sites. However, since the alkylating ligand (PrBCM) by itself does not show any cooperativity and binds to all sites with equal affinity, this behavior would be expected also in the case of negative cooperativity in agonist binding. In the latter case, the agonist (but not the alkylating antagonist) induced negative cooperativity upon binding; thus, after its dissociation, the protected binding sites return to their original conformation, and again bind the agonist at high affinity. The induction of low affinity sites would not occur, because of the blockade of the other site by the alkylating agent.

In spite of the lack of conclusive proof for site–site interactions among brain muscarinic receptors, the existence of such interactions cannot be completely ruled out at this stage. This notion is emphasized by binding (Hedlund et al., 1982) and kinetic (Jarv et al., 1980) studies suggesting that two ligands may bind simultaneously to the muscarinic receptors in rat cerebral cortex, thus indicating the possibility of cooperative site–site interactions.

In summary, the demonstration of site–site interactions among muscarinic receptors in the adenohypophysis and heart suggests that they are organized in these tissues as oligomers containing two or more binding sites. However, this does not exclude site heterogeneity, which could exist along with site-site interactions (Henis and Sokolovsky, 1983). The inability to demonstrate site–site interactions among, for example, cerebral cortex muscarinic receptors could reflect differences between muscarinic receptor subtypes as defined by pirenzepine binding —mostly M_1 in the cortex, vs M_2 and possibly glandular M_2 in the heart and adenohypophysis, respectively (Hammer et al., 1980; Hammer and Giachetti, 1984). Alternatively, the membrane lipid composition (Moscona-Amir et al., 1986) or tissue-specific interactions with additional membrane components may influence the interactions between the muscarinic binding sites.

3. Allosteric Interactions of Muscarinic Receptors

3.1. Interactions of Gallamine with Muscarinic Receptors

The neuromuscular blocking agent gallamine, which has nicotinic blocking properties, has long been shown to have antimuscarinic properties that appear to be selective for the heart (Riker and Wescoe, 1951; Rathburn and Hamilton, 1970; Clark and Mitchelson, 1976). Pharmacologic studies yielded contradictory results on whether it acts as a competitive or noncompetitive antagonist (Rathburn and Hamilton, 1970; Clark and Mitchelson, 1976). This prompted several studies on the mechanisms of gallamine-muscarinic receptor interaction. Depending on the preparation and the experimental conditions, some of the findings demonstrate allosteric interactions, whereas others suggest direct competition on heterogeneous site populations (Table 1).

Table 1
Interactions of Gallamine with Muscarinic Binding Sites

Experiment	Conditions	Ligand	Allosteric interactions[a]	Reference
Deviations from simple one-site competition	Rat heart or cerebral cortex; high ionic strength	[³H](–)NMS	+(>10 μM gal) –(<10 μM gal)	Stockton et al., 1983
	Rat atria; high ionic strength	[³H](–)3QNB	+(>30 μM gal) –(<30 μM gal)	Dunlap and Brown, 1983
	Rat cerebral cortex or cerebellum; low ionic strength	[³H](–)3QNB	–(<1 μM gal)	
		[³H]Pir	–(< 1 μM gal)	Burke, 1986
	Rat forebrain; low ionic strength	[³H](–)3QNB	—	Ellis and Lenox[b], 1985a Ellis and Hoss, 1982
Effect of gallamine on muscarinic antagonist dissociation rate	Rat heart or cerebral cortex; high ionic strength	[³H](–)NMS	+(3×10^{-4} M gal)	Stockton et al., 1983
	Rat atria; high ionic strength	[³H](–)3QNB	+(5×10^{-4} M gal)	Dunlap and Brown, 1983
	Rat atria; high ionic strength	[³H](–)3QNB	+(1 mM gal)	Nedoma et al., 1986
	Rat forebrain; low ionic strength	[³H](–)3QNB	–(100 μM gal)	Ellis and Lenox, 1985a
		[³H](–)NMS	+(100 μM gal)	
Gallamine protection from PBCM	Rat atria, high ionic strength	[³H](–)3QNB	+(no protection at 300 μM gal)	Dunlap and Brown, 1983
	Rat cerebral cortex or cerebellum, low ionic strength	[³H](–)3QNB	–(good protection)	Burke[b], 1986
		[³H]Pir	–(at 1 μM gal)	

[a]Gal designates gallamine, and [³H]Pir designates [³H]pirenzepine.
[b]These studies also suggested heterogeneity of [³H](–3QNB sites toward gallamine.

3.1.1. Indications for Allosteric Interactions

Indications for allosteric interactions between gallamine and muscarinic binding sites were provided in experiments measuring the displacement of [^{3}H](–)NMS by gallamine in homogenates of rat heart or cerebral cortex (Stockton et al., 1983). These experiments, conducted at normal ionic strength, yielded nonlinear Schild plots, which indicated that the inhibition of [^{3}H]-(–)NMS binding by gallamine is weakened at high gallamine concentrations. This phenomenon by itself could also stem from competition of gallamine at different affinities for subpopulations of muscarinic binding sites. However, high gallamine concentrations were able to inhibit the dissociation rate of [^{3}H](–)NMS, an effect that is clear evidence for allosteric interactions (Table 1).

Allosteric interactions were also demonstrated at high ionic strength in rat atria homogenates, employing [^{3}H](–)3QNB as a ligand (Dunlap and Brown, 1983). At concentrations up to 30 μM, gallamine inhibited [^{3}H](–)3QNB binding competitively, presumably by binding to the muscarinic binding site. However, at higher gallamine concentrations, [^{3}H](–)3QNB binding yielded curvilinear Scatchard plots, indicative either of allosteric interactions that weaken [^{3}H](–)3QNB binding or heterogeneity of the muscarinic binding sites toward gallamine. Further support for the existence of allosteric interactions was provided by the reduction in the dissociation rate of [^{3}H](–)3QNB by high gallamine concentrations (Table 1). Similar results were reported by Nedoma et al. (1986) for the effect of gallamine on the dissociation rate of [^{3}H](–)3QNB in rat atrial preparations, and by Ellis and Lenox (1985a) who demonstrated a reduction in the dissociation rate of [^{3}H](–)NMS in rat forebrain homogenates in the presence of gallamine (Table 1). In the latter experiments, which were performed in low ionic strength buffer, the allosteric effects of gallamine were clearly dependent on the muscarinic ligand employed, as gallamine did not alter the dissociation rate of [^{3}H](–)3QNB. A dependence of the allosteric interactions on the muscarinic ligand employed was evident also in the experiments of Stockton et al. (1983) (Table 1).

Dunlap and Brown (1983) also tested the ability of high gallamine concentrations (300 μM) to protect cardiac muscarinic receptors from alkylation by PrBCM. Only 1% of the sites were protected, an extremely low level considering the IC_{50} value of gallamine for competition with [^{3}H](–)3QNB (in the μM range). This was interpreted as an indication that at high concentrations gallamine binds to an allosteric site, reducing its own affinity to the muscarinic binding site (Dunlap and Brown, 1983). Interestingly, another study which employed homogenates of rat cerebral cortex and cerebellum (Burke, 1986) demonstrated that *low* gallamine concentrations can protect 65-85% of the [^{3}H]-(–)3QNB sites from alkylation by PrBCM. It is therefore possible that at low gallamine concentrations, where it competes for the muscarinic binding site but does not bind to the allosteric site, gallamine protects the [^{3}H](–)3QNB sites much more effectively than at higher concentrations, under which conditions it binds to the allosteric site and reduces its own affinity to the muscarinic ligand binding site. Indeed, at submicromolar concentrations, gallamine did not affect the dissociation rate of [^{3}H](–)3QNB or [^{3}H]pirenzepine, in accord with the notion that binding to the allosteric site occurs only at higher concentrations (Burke, 1986) (Table 1).

3.1.2. Indications for Competition

Experiments measuring the inhibition of [^{3}H](–)3QNB by gallamine in rat forebrain homogenates (Ellis and Hoss, 1982) were interpreted as demonstrating competition by gallamine for the muscarinic sites, which are divided to two subpopulations that bind gallamine with different affinities. However, allosteric interactions could not be ruled out, since the criteria employed to define the inhibition as competitive (no reduction in B_{max}, and a correlation between high affinity carbachol binding sites and low affinity sites for gallamine) do not exclude allosteric interactions (as discussed in section 1.2.2.). However, later studies (Ellis and Lenox, 1985a) demonstrated that, in this preparation, even high gallamine concentrations did not affect [^{3}H](–)3QNB dissociation rate, in accord with pure competitive inhibition (Table 1). On the other hand, the dissociation rate of [^{3}H](–)NMS was slowed down, suggesting that with this ligand allosteric interactions are more pronounced (Table 1). Interestingly, curvilinear Scatchard plots for [^{3}H](–)3QNB binding in the presence of gallamine were obtained in spite of the lack of measurable allosteric effects on this muscarinic ligand, suggesting that the [^{3}H](–)3QNB sites are heterogeneous toward gallamine binding. This phenomenon was evident especially under low ionic strength conditions. Site heterogeneity of [^{3}H](–)3QNB sites toward gallamine at low ionic strength was reported also by Burke (1986), who worked on rat cerebral cortex and rat cerebellar homogenates (Table 1). Under these conditions, the high affinity sites of gallamine appear to correlate, at least partially, with the low affinity sites for pirenzepine (M_2) (Burke, 1986). This was deduced from experiments in which pirenzepine was employed to protect its high affinity sites (M_1) from alkylation by PrBCM; the protected sites revealed *low* affinity for gallamine (Table 1). Alternatively, the site heterogeneity of [^{3}H](−)3QNB sites toward gallamine may reflect the heterogeneity of these sites toward quaternary muscarinic ligands, such as (−)NMS (Lee and El-Fakahany, 1985; Brown and Goldstein, 1986; Ellis and Lenox, 1985b).

Convincing evidence for competition of gallamine at low concentrations was provided by Burke (1986), who demonstrated protection by low gallamine concentrations of over 65% of the [^{3}H](−)3QNB sites in rat brain cortex and cerebellum from alkylation by PrBCM (Table 1). This notion is strengthened by the competitive inhibition of [^{3}H](−)3QNB binding by low gallamine concentrations in both brain (Burke, 1986) and heart (Dunlap and Brown, 1983) preparations (Table 1).

In summary, the emerging picture of the interactions of gallamine with muscarinic receptors is that, at low concentrations, gallamine competes for the muscarinic binding sites. At higher concentrations, it binds to an allosteric site (either on the muscarinic receptor itself or on an associated protein), through which it modulates both its own affinity and that of muscarinic ligands to the muscarinic binding sites. The extent of the modulation depends on the nature of the muscarinic ligand. Furthermore, gallamine appears to have a certain selectivity to a subpopulation of muscarinic binding sites, resulting in site heterogeneity of the muscarinic receptors toward gallamine binding, which is evident in the binding of [^{3}H](−)3QNB under low ionic strength conditions.

3.2. Bisquaternary Pyridinium Oximes as Allosteric Inhibitors of Rat Brain Muscarinic Receptors

Bisquaternary pyridinium oximes were originally designed as reactivators of acetylcholinesterase following organophosphate poisoning. They were shown to possess antimuscarinic activity (Kuhnen-Clausen, 1970; Amitai et al., 1980; Kuhnen-Clausen et al., 1983) and inhibit the binding of the muscarinic antagonist

[^{3}H]4NMPB. However, this inhibition did not fit a simple single-site competition model (Amitai et al., 1980), and the oxime-induced blockage of the acetylcholine-mediated response of smooth muscle had a noncompetitive character (Kuhnen-Clausen, 1970). These findings suggested the possibility of allosteric interactions of the oximes with muscarinic binding sites, and prompted studies on the mechanism of oxime-muscarinic receptor interactions. The results of these studies are summarized in Table 2.

3.2.1. Interactions with Antagonist-Occupied Muscarinic Receptors

The interaction of the bisquaternary pyridinium oximes 1-(2-hydroxy-iminoethylpyridinium)-1-(3-cyclohexylcarboxypyridinium) dimethyl ether (HGG-42) and 1-(2-hydroxyiminoethyl-pyridinium)-1-(3-phenylcarboxypyridinium) dimethyl ether (HGG-12) with muscarinic receptors in homogenates of rat cerebral cortex and brainstem was investigated (Kloog and Sokolovsky, 1985a). Below 10 μM oxime, the inhibition of [^{3}H]-4NMPB, [^{3}H](−)3QNB or [^{3}H](−)NMS binding by the oximes fitted simple competition on a homogeneous population of sites. At higher oxime concentrations, curvilinear Scatchard plots indicative of site heterogeneity or of allosteric interactions were obtained (Table 2). Moreover, the muscarinic antagonist binding capacity was reduced at high oxime concentrations (30 and 10% reduction in the brainstem and cortex, respectively) (Table 2). As discussed under section 1.2.2., a true reduction in B_{max} can occur only because of an irreversible inactivation of the receptors; alternatively, the reduction in B_{max} may be only an apparent phenomenon, resulting from the appearance of a subpopulation of sites with very low affinity such that the measurement of, for example, [^{3}H]4NMPB binding to them is not experimentally feasible. In this case, the first explanation appears to hold at least partially, since washing the samples free of the oximes did not elevate the B_{max} values back to the original level (Kloog and Sokolovsky, 1985a). The inactivation is most likely the result of reaction with the oxime moiety, which is capable of reacting with an ester bond(s), for example. Indeed, an analogous bispyridinium derivative lacking an oxime group had no irreversible effects on the muscarinic binding capacity (Kloog and Sokolovsky, 1985a). The reaction site probably differs from the muscarinic binding site itself, since atropine did not protect from the irreversible loss of muscarinic sites.

Table 2
Interactions of Bisquaternary Pyridinium Oximes with Muscarinic Receptors

Experiment	Preparation	Ligand	Allosteric interactions	Reference
Oxime blockade of ACh-mediated smooth muscle contraction	Guinea-pig ileum	ACh	+(noncompetitive inhibition)	Kuhnen-Clausen, 1970
Deviations from simple one-site competition	Mouse brain	[^{3}H]4NMPB	+	Amitai et al., 1980
	Rat cerebral cortex or brain stem	[^{3}H]4NMPB[a]	+(>10 μM oxime] −(<10 μM oxime)	Kloog and Sokolovsky, 1985a
Reduction in B_{max} of muscarinic antagonists	Rat brain stem or cerebral cortex	[^{3}H]4NMPB[a]	+(>10 μM oxime) −(<10 μM oxime)	Kloog and Sokolovsky, 1985a
Effect of oximes on antagonist dissociation rate	Rat brain stem or cerebral cortex	[^{3}H]4NMPB	+(>10 μM oxime) −(<10 μM oxime)	Kloog and Sokolovsky, 1985a
Irreversible effect of oximes on antagonist association rate	Rat brain stem or cerebral cortex	[^{3}H]4NMPB	+(200 μM oxime)	Kloog and Sokolovsky, 1985a

Irreversible inactivation by oximes of R_H-R_L interconversion	Rat brain stem or cerebral cortex	[3H]4NMPB inhibition by carbamylcholine	+(200 μM oxime)	Kloog and Sokolovsky, 1985b
Protection of R_H–R_L interconversion from oxime inactivation by occupation of R_H with carbamylcholine	Rat brain stem or cerebral cortex	[3H]4NMPB inhibition by carbamylcholine	+(200 μM oxime)	Kloog and Sokolovsky, 1985b

[a]In these experiments, similar results were obtained also with [3H](−)3QNB or [3H](−)NMS as ligands.

The existence of allosteric interactions between oxime and muscarinic binding sites is strongly supported by kinetic studies (Kloog and Sokolovsky, 1985a), which demonstrated a reduced [^{3}H]4NMPB dissociation rate in the presence of high oxime concentrations (Table 2). Further support for allosteric interactions is provided by the finding that high oxime concentrations induced a *persistent* reduction in [^{3}H]4NMPB association rate, which persisted even after washing the sample from the oxime; this reduction is therefore *not* the result of simple competitive inhibition (Kloog and Sokolovsky, 1985a).

3.2.2. Interactions with Agonist-Occupied Muscarinic Receptors

The irreversible effects of pretreatment with high HGG-12 concentrations on muscarinic agonist binding were examined in homogenates of rat brainstem or cerebral cortex (Kloog and Sokolovsky, 1985b). In both brain regions, the proportion of high affinity binding sites (%R_H) for carbamylcholine and oxotremorine was not affected, but the affinity of agonist binding was somewhat reduced. A more striking effect was observed for the ability of Gpp(NH)p to transfer R_H to R_L in the brainstem (where the %R_H is originally high), and for the ability of Co^{2+} to induce R_L to R_H transformation in the cortex (where R_L predominates). Both effects were completely abolished by the oxime treatment (Table 2). Moreover, low carbamylcholine concentrations that occupy almost exclusively R_H sites protected both the Gpp(NH)p and the Co^{2+}-mediated effects against inactivation by the oxime (Table 2). It was therefore proposed that agonist binding to R_H sites induces a conformational change that is, in turn, reflected in the site of oxime action; the latter may be on the muscarinic receptor itself, on the associated G-protein(s), or on another component associated with either one of these entities. The finding that the occupancy of R_H by carbamylcholine protects the Co^{2+}-mediated conversion of R_L to R_H in the cortex from inactivation by oxime is also suggestive of site–site interactions between R_H and R_L, either directly or through an intermediary component such as guanine nucleotide binding protein (Kloog and Sokolovsky, 1985b).

The studies summarized in Table 2 lead to a model analogous to that of gallamine-muscarinic receptor interaction. Thus, at low concentrations, the oximes appear to compete for the muscarinic binding sites. At higher concentrations, they bind to an allosteric site that interacts with the muscarinic binding sites and modulates their interactions with guanine nucleotide binding protein(s).

3.3. Allosteric Interactions of Other Drugs with Muscarinic Receptors

The general scheme of competition with muscarinic ligands along with binding to an allosteric site that modulates the properties of the muscarinic binding sites holds also for a variety of other drugs shown to interact with the muscarinic receptors. Studies leading to this conclusion are summarized in Table 3.

3.3.1. Verapamil

Verapamil is a potent antagonist of voltage-dependent calcium channels at submicromolar concentrations. At higher concentrations, it has a variety of additional inhibitory effects, among which is the inhibition of muscarinic receptors (Fairhurst et al., 1980; Karliner et al., 1982; El-Fakahany and Richelson, 1983). Initial studies conducted on rat atria or rat ventricle membranes at high ionic strength (Karliner et al., 1982) did not indicate allosteric effects, and the drug was classified as a competitive inhibitor, since B_{max} of [^{3}H](−)3QNB was not reduced. However, as discussed earlier (section 1.2.2.), this is not an indication for competitive or noncompetitive inhibition in binding studies. Indeed, later studies (Waelbroek et al., 1984b) demonstrated the existence of allosteric interactions between verapamil and muscarinic receptors (Table 3). In these studies, performed on rat heart homogenates at low ionic strength with [^{3}H](−)NMS and methyl[^{3}H]oxotremorine *M* ([^{3}H]OXO-*M*) as muscarinic ligands, a pure competitive inhibition was observed below 1 μM verapamil. However, high verapamil concentrations induced apparent site heterogeneity ($n_H < 1$) in the binding of the above muscarinic ligands, along with a significant reduction in their dissociation rates (Table 3). Similar effects on the dissociation rate of [^{3}H]-(−)NMS were observed in homogenates of rat cerebral cortex (but not in medulla-pons) (Baumgold, 1986). It therefore appears that an allosteric site for verapamil, to which it binds at concentrations above 1 μM and that interacts with the muscarinic binding site, exists in rat heart and cerebral cortex (Waelbroek et al., 1984b; Baumgold, 1986). This allosteric site is not specific for M_1 or M_2 muscarinic receptors, since the cerebral cortex is enriched

Table 3
Allosteric Interactions of Verapamil, Clomiphene, Local Anesthetics, and Antiarrhythmic Drugs with Muscarinic Receptors

Drug	Tissue and conditions	Ligand	Indications for allosteric interactions	Reference
Verapamil	Rat heart; low ionic strength	$[^3H](-)$NMS, $[^3H]$OXO-*M*	Deviations from simple one-site competition (>1 μM); reduced muscarinic ligand dissociation rate (at 10–100 μM drug)	Waelbroek et al., 1984b
	Rat cerebral cortex, low ionic strength	$[^3H](-)$NMS	Same as above	Baumgold, 1986
Clomiphene	Rat brain regions (median hypothalamus, medulla-pons, cortex); high ionic strength	$[^3H]$4NMPB	Deviations from simple one-site competition ($n_H > 1$); elevated $[^3H]$4NMPB dissociation rate (50 μM drug)	Ben-Baruch et al., 1982
H_2-HTX	N1E-115 neuroblastoma cells; high ionic strength	$[^3H]$SCOP	Deviations from simple one-site competition (>50 μM drug)	Burgermeister et al., 1978

Tetracaine	Same as above	[3H]SCOP	Same as above	Burgermeister et al., 1978
	Guinea-pig cerebral cortex synaptosomes; high ionic strength	[3H](−)3QNB	Same as above	Taylor et al., 1980
Quinidine	Rat heart; low ionic strength	[3H](−)NMS [3H]OXO-*M*	Deviations from simple one-sit competition[a] (>40 μM drug); reduction in muscarinic ligand dissociation rate	Waelbroek et al., 1984a
Quinidine, lidocaine	Rat atria, ventricle, brain stem, cerebral cortex; high ionic strength	[3H]4NMPB	Deviations from simple one-site competition (except in the cortex)[a]	Cohen-Armon et al., 1985a
Amiodarone	Same as above	[3H]4NMPB	Same as above[a]	Cohen-Armon et al., 1984

[a]In these experiments, much more pronounced allosteric interactions were observed with the agonist high-affinity binding sites.

with M_1, whereas only M_2 is detected in the heart; it could be located on the muscarinic receptor itself or on an associated protein (include a G-protein or a Ca^{2+}-channel). Interestingly, the plasmolytic agent secoverine, whose molecular structure partially resembles verapamil, shows very similar effects on muscarinic receptors in homogenates of rat heart or frontal cortex (Brunner et al., 1986). However, the effects of both verapamil and secoverine on the dissociation kinetics were demonstrated at about $10^{-4}M$ drug concentration (Table 3); thus, one cannot exclude the possibility of a nonspecific effect of the drugs on the general membrane properties. Such an effect would yield the same phenomena as allosteric interactions (*see* section 1.2.2.).

3.3.2. Clomiphene

The antiestrogen compound 1[-*p*-(β-diethylaminoethoxy)-phenyl]1,2-diphenyl-2-chloroethylene (clomiphene) is used as an ovulation inducer (Huppert, 1979). However, it has several other activities indicating effects on neuronal and/or smooth muscle function (Morris, 1985). This drug, which bears some structural homology to acetylcholine (Ben Baruch et al., 1982), was shown to inhibit the binding of [^{3}H]4NMPB to muscarinic receptors in homogenates of several rat brain regions (Ben Baruch et al., 1982). The inhibition of [^{3}H]4NMPB binding by clomiphene clearly involves allosteric interactions, as it yields n_H values well above unity and bell-shaped Scatchard plots in all the brain regions examined (medulla-pons, median hypothalamus, and cerebral cortex) (Table 3). This could be interpreted as positive cooperativity in clomiphene binding, where the drug first binds to an allosteric site, which increases its own affinity toward the muscarinic binding site. Alternatively, the binding of clomiphene to an allosteric site could *reduce* the affinity of the muscarinic ligand ([^{3}H]4NMPB) to the muscarinic site. The latter possibility is responsible for at least part of the observed allosteric effects, since clomiphene was found to increase the dissociation rate of [^{3}H]4NMPB (Table 3). Thus, the allosteric effects observed with clomiphene are the mirror image of those observed with, for example, gallamine. In the case of gallamine, the dissociation of muscarinic ligands was inhibited, although it was facilitated by clomiphene.

Clomiphene was also shown to inhibit noncompetitively the ACh-mediated contractions in guinea-pig ileum, in accord with the allosteric mechanism proposed in the brain (Morris, 1985).

However, it also inhibited contractions produced by noncholinergic spasmogens (histamine and bradykinin). It is therefore possible that clomiphene interacts allosterically not only with muscarinic receptors, but with other receptor systems as well. Alternatively, it could have nonspecific effects on the contraction of guinea-pig ileum, although the inhibition of this parameter was observed already at 1 μM clomiphene or less (Morris, 1985).

3.3.3. Local Anesthetics and Antiarrhythmic Drugs

High doses of local anesthetics were reported to inhibit noncompetitively the responses mediated by muscarinic receptors in a variety of tissues (Feinstein and Paimre, 1967; Fleisch and Titus, 1973; Richelson et al., 1978). These drugs, which also have antiarrhythmic activity, can act as blockers of voltage-sensitive sodium channels. Their effects on ligand binding to muscarinic receptors were tested in several preparations. In N1E-115 neuroblastoma cells, tetracaine and dihydroiso-histrionicotoxin (H_2-HTX, a sodium channel blocker known to interact with nicotinic ACh receptors that has effects similar to those of local anesthetics (*see* Kato and Changeux, 1976) inhibited the binding of [^{3}H](−)scopolamine ([^{3}H](−)SCOP) to its high affinity sites in a manner deviating from a simple one-site competition model (Burgermeister et al., 1978; Table 3). Similar deviations were reported (Taylor et al., 1980) for the effects of high tetracaine concentrations on [^{3}H](−)3QNB binding to a synaptosomal fraction from guinea-pig cerebral cortex (Table 3). The same study also detected a competitive inhibition by low tetracaine concentrations of carbamylcholine-induced contractions in guinea-pig ileum, and a noncompetitive inhibition at high concentrations. However, in both studies, the possibility of nonspecific effects on the membrane cannot be excluded in view of the high local anesthetic concentrations required to produce the allosteric behavior (Table 3). At these concentrations, both tetracaine and H_2-HTX inhibited also the nonspecific binding of the muscarinic ligands (Burgermeister et al., 1978, Taylor et al., 1980).

Further evidence for allosteric interactions of muscarinic sites with local anesthetics was supplied by Waelbroek et al. (1984b) in rat heart membranes at low ionic strength. This study investigated the effects of quinidine on the binding of [^{3}H](−)NMS and [^{3}H]OXO-*M*. The binding of the latter ligand (which binds to R_H sites under the assay conditions) was inhibited more strongly than [^{3}H](−)NMS binding, suggesting a preferential interaction with the high affinity agonist binding sites. The involvement of allosteric interactions in these effects was demonstrated by the marked reduction in the dissociation rates of both [^{3}H](−)NMS and [^{3}H]OXO-*M* at high quinidine concentrations (Table 3). Both the preferential interactions with R_H and the existence of an allosteric site to which the local anesthetics bind at high concentrations are corroborated by studies performed in homogenates of rat atria, ventricle, brain stem, and cerebral cortex, employing [^{3}H]4NMPB as a ligand (Cohen-Armon et al., 1985a). Quinidine and lidocaine inhibited [^{3}H]4NMPB binding less efficiently in the cortex, where the percentage of R_H is low, and competition experiments with carbamylcholine and oxotremorine demonstrated a more pronounced inhibition of agonist binding to R_H in the brain stem and heart. In the latter preparations (but not in the cortex), clear deviations from competitive inhibition on a homogenous site population were observed, including a reversible apparent reduction in B_{max} of [^{3}H]4NMPB (Table 3). Interestingly, very similar results were obtained on the same preparations with the antiarrhythmic drug 2-butyl-3-[3,5-diiodo-4-(β-diethylaminoethoxy)-benzoyl]benzofuran (amiodarone) (Cohen-Armon et al., 1984; Table 3). The emerging picture is that, at high concentrations, the above drugs bind to an allosteric site that modulates ligand binding to muscarinic binding sites; however, the allosteric interactions appear to affect more strongly the agonist high affinity binding sites (R_H). The location of this allosteric site(s) is unclear at present, and it may be on the muscarinic receptor itself or on associated membrane proteins. Since all of the drugs in Table 3 block voltage-sensitive sodium channels, it is possible that their allosteric effects are mediated through these channels, which are coupled to the muscarinic receptors (*see* section 5).

4. Interaction between Muscarinic Receptors and G-Proteins

Stimulation of receptors causes a variety of responses in intact tissues. Although many of the events that carry information into the cells from the various receptors are unknown, common biochemical consequences of receptor stimulation are inhibition of the accumulation of cyclic AMP, e.g., inhibition of adenylate cyclase or increased phosphodiesterase activity (*see*

Chapter 6); increase in phosphoinositide turnover reflecting an increase in phosphoinositidase C activity (*see* Chapter 7); an increase in cytosolic Ca^{2+} concentration (*see* Chapter 7); stimulation of Na^+/H^+ antiport leading to an increase in cytosolic pH; an increase in tyrosine-specific protein kinase activity. Some of these biochemical consequences of receptor stimulation by its agonist(s) have been identified and are probably important links in the chain of events of muscarinic response. In many cases this is achieved by receptor interactions with other membrane components, e.g., other receptors and ion channels and/or with regulatory proteins (Sokolovsky, 1984; Sokolovsky et al., 1986 and refs. therein). What emerged as a general pathway for the coupling of various membrane receptors to adenylate cyclase and other effectors (Rodbell, 1980) can also be applied to the muscarinic receptors, namely, that a guanine nucleotide binding protein (G-protein) is involved in the coupling process. Members of this family of closely related proteins regulate a diverse group of metabolic events (Gilman, 1984). Two of them, G_S and G_i, stimulate and inhibit adenylate cyclase activity, respectively (Rodbell, 1980; Gilman, 1984; Kahn and Gilman, 1984; Milligan and Klee, 1985). Other G-proteins are involved in diverse signal transduction systems (as reviewed by Stryer et al., 1981 and Bourne, 1986). Another member of this family, G_o, has no known function (Sternweis and Robishaw, 1984 and Florio and Sternweis, 1985), and activation of phospholipase C has been attributed to the action of an unidentified G-protein, G_P (Cockcroft and Gomperts, 1985; Straub and Gershengorn, 1986).

Studies with iterative computer methods for analyzing binding data derived from competition experiments employing ^{3}H-antagonist/agonist (Birdsall et al., 1978; Birdsall et al., 1980; Egozi et al., 1980; Wamsely et al., 1984) or direct binding measurements (Birdsall et al., 1980; Ehlert et al., 1980; Harden et al., 1983; Vickroy et al., 1984; Gurwitz et al., 1984a, 1985a) have shown that two agonist binding sites (high and low affinity) predominate in brain and heart, and that an additional site with higher affinity (super high) is also present in several preparations.

It is important to note that muscarinic agonists are capable of inducing different conformational changes in the receptor, a process that is dependent on the nature of the agonist under study, thereby leading to different modulation processes. The induction of different receptor conformations is in line with observations regarding:

1. the mode of binding of [^{3}H]acetylcholine ([^{3}H]ACh) (Gurwitz et al., 1985a)
2. the response of agonist binding to chemical modification of the receptor (Gurwitz and Sokolovsky, 1985b)
3. the binding of ACh, carbamylcholine, and oxotremorine to atrial muscarinic receptors is modulated differently by batrachotoxin (*see below*) (Cohen-Armon et al., 1985b, Cohen-Armon and Sokolovsky et al., 1986) and
4. the demonstration that oxotremorine differs from ACh and carbamylcholine in its effects on cyclic AMP formation and phosphatidylinositol hydrolysis (Fisher and Agranoff, 1981; Brown and Brown, 1984).

The notion that the receptor state depends on the structure of the agonist gains further support from a detailed analysis of the binding of various agonists to muscarinic receptors in heart preparations (Wong et al., 1986), which demonstrated that the distribution of the receptors between high and low affinity forms is determined at least in part by the nature of the agonists.

Guanine nucleotides—GTP, GDP and guanyl-5′-yl imidodiphosphate (Gpp(NH)p]—have an effect on agonist binding to the muscarinic receptors. In the presence of these nucleotides, agonist occupancy curves are shifted to the right, indicating the conversion of high affinity sites into low affinity sites (Berrie et al., 1979; Sokolovsky et al., 1980; Wei and Sulakhe, 1979, 1980; Ehlert et al., 1981; Waelbroeck et al., 1982; Burgisser et al., 1982; Harden et al., 1982). Direct binding experiments using [^{3}H]ACh (Gurwitz et al., 1984a, 1985a), methyl [^{3}H]oxotremorine *M* ([^{3}H]OXO-*M*) (Hulme et al., 1983), or [^{3}H]cis-methyldioxolane ([^{3}H]CP) (Ehlert et al., 1980), which measure the binding to the high affinity sites exclusively, indicate that the main effect of the guanine nucleotides is reduction in the binding of ^{3}H-agonist with minor changes in the corresponding K_D values. Thus, guanine nucleotides induce interconversion of agonist high affinity sites (R_H) to low affinity sites (R_L). This effect is variable with the region under investigation: in the heart, brainstem, and cerebellum (regions in which high affinity sites predominate), a substantial interconversion of R_H to R_L is observed. On the other hand, in the cortex, hippocampus, and caudate putamen (regions in which low affinity sites predominate), little or no interconversion is detected. It should be noted that, in most tissues, the conversion induced by guanine nucleotides is only partial, unlike in the heart receptors and unlike in other receptors where a complete conversion has been observed (Harden et al., 1982; Kurose et al., 1983). Thus, the agonist high affinity sites can be divided into guanine-nucleotide-sensitive and guanine-nucleotide-insensitive sites. The former are most probably those sites that are reversibly coupled to a G-protein, as discussed further below. We have recently shown that treatment with Cu^{2+} ions or with the reagent diamide (Gurwitz et al., 1984b), which oxidizes quite selectively cysteinyl residues to cystinyl residues, induces the formation of agonist high affinity sites, which are guanine nucleotide insensitive. Thus, the presence of such sites in various regions might be related to processes involving SH/S-S transformation, either intra- or intermolecular. Several reports by other workers (Aronstam and Eldefrawi, 1979; Uchida et al., 1982; Birdsall et al., 1983) have previously demonstrated dependence of agonist binding on the sulfhydryl redox state. The sulfhydryl residues may be on the receptor protein (R) or on a G-protein(s). The following scheme (Fig. 1) depicts three possible bridges that might result from intermolecular disulfide bond involved in the guanine-nucleotide-insensitive high affinity agonist binding state for muscarinic receptors.

At normal ionic strength, guanine nucleotides have no effect on the binding of antagonists to the muscarinic receptors. However, under nonphysiologic conditions (low ionic strength), a reciprocal effect on antagonist and agonist binding to cardiac muscarinic receptors was reported: increased affinity toward antagonists in the presence of guanine nucleotides, and in one case the appearance of positive cooperativity in antagonist binding (Burgisser et al., 1982; Martin et al., 1984; Mattera et al., 1985).

It should be noted that, in most of the studies carried out so far, guanine nucleotides induced interconversion of high affinity agonist binding sites to the low affinity form (R_H to R_L). However, as described above, we have recently observed a deviation from the typical behavior—namely, a guanine-nucleotide-induced conversion from R_L to R_H. Firstly, guanine nucleotides affect agonist binding to muscarinic receptors in the adenohypophysis of female rats at the proestrous stage by a conversion of R_L to R_H (Avissar and Sokolovsky, 1981), whereas at the pro and diestrous states, they induce the normal (R_H to R_L) conversion. Secondly, in young cultures of rat heart myocytes (cultured for 5 d), Gpp(NH)p induces a conversion of R_H to R_L, whereas a reverse effect, i.e., conversion of R_L to R_H, is obtained in homogenates of aged cultures (cultured for 14 d) (Moscona-

AGONIST HIGH AFFINITY STATE—
GTP INSENSITIVE

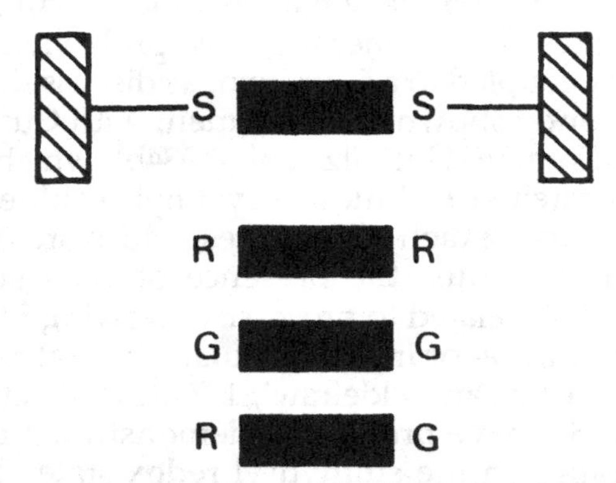

Fig. 1. A scheme depicting possible disulfide bridges involved in the formation of GTP-insensitive high affinity agonist binding state for the muscarinic receptor (R). G represents the putative guanine-nucleotide binding protein.

Amir et al., 1986). The effect of Gpp(NH)p appears to be strictly on the interconversion, since the dissociation constants (K_H and K_L) of the agonists were not significantly altered.

Divalent transition metal ions like Mn^{2+}, Co^{2+}, and Ni^{2+} can convert R_L to R_H with no detectable effect on antagonist binding parameters (Gurwitz and Sokolovsky, 1980; Nukada et al., 1983; Hulme et al., 1983; Gurwitz et al., 1984a; Gurwitz et al., 1985a). The effect of the metal ions is detected mainly in brain regions containing a larger population of low affinity binding sites, e.g., cortex and hippocampus. This effect can be reversed by removal of the ions or by guanine nucleotides. It would appear that the metal ions mimic the action of a possible endogenous modulator. Thus, there must be a physiological role for the different affinity states, which varies with the particular cell in which the muscarinic receptor resides, and the interconversion process is a tool for up- and downregulations of various responses.

The changes in binding brought about by guanine nucleotides reflect an interaction between the receptor and a G-protein. Sites of higher affinity for agonist appear to reflect a complex between the receptor and the G-protein(s), as described in Chapters 2 and 6. Three lines of evidence support this assumption. Firstly, in rat striatum, muscarinic inhibition of adenylate cyclase has been associated with the stimulation of high affinity GTPase (Onali et al., 1983). Secondly, islet-activating protein (pertussis toxin) which catalizes the ADP-ribosyation of G_i (Katada and Ui, 1982) and thereby prevents various receptors from inhibiting adenylate cyclase, has been shown to uncouple muscarinic receptors from adenylate cyclase (Kurose and Ui, 1983). Finally, reconstitution experiments employing purified G-protein(s) and muscarinic receptors yielded agonist sites of high affinity that reveal the sensitivity to guanine nucleotides characteristic of receptors in the native membrane (Florio and Sternweis, 1985; Haga et al., 1985; Haga et al., 1986). Importantly, Haga et al. (1986) demonstrated that the muscarinic receptors can interact with G_i and with G_o with the same site, which rules out the possibility that there are two kinds of muscarinic receptors, one interacting with G_i and the other with G_o. In spite of the wealth of information on the phenomenology of muscarinic receptor–agonist interaction and its modulation by G-protein(s) and guanine nucleotides, the molecular basis that gives rise to such complex interactions is not yet understood. It is worth noting that, because of the use of different experimental protocols, as well as because of differences in species and tissues used by the various laboratories (e.g., different Mg^{2+} ion concentrations, differences in the ionic composition of the buffer, in temperature, and so on), variations in the experimental results are often seen. Furthermore, we do not yet know whether the coupling mechanism is the same for the various G-proteins, or whether for each type of G-protein there is a specific mode of interaction.

The current popular concept for such an interaction, especially for cardiac muscarinic receptors that interact with G_i, is the ternary model consisting of agonist, receptor, and G-protein(s) (Ehlert, 1985, and references cited therein). The receptor is assumed to be in equilibrium between R and RG, where agonists bind with higher affinity to the RG complex. This model has been applied to numerous receptors that affect the adenylate cyclase system, the most notable case being that of the β-adrenergic receptors (De Lean et al., 1980; Lefkowitz et al., 1983). Recently, Mattera et al. (1985) very elegantly presented evidence that, in dog heart sarcolemmal membranes, there are at least two ternary complexes, one with and the other without cooperative interactions between the receptor sites. One of the features of the ternary complex model suggested along the lines of the model proposed for β-adrenergic receptors is that the binding of guanine nucleotides to the G-protein(s) is supposed to uncouple the receptor from the complex. The reverse interconversion induced by guanine nucleotides (R_L to R_H) in rat heart myocytes (Moscona-Amir et al., 1986) and in the adenohypophysis (Avissar and Sokolovsky, 1981) argues against the explanation given above, since it is unlikely that uncoupling would lead to opposite effects on the same receptor system in young and aged myocyte cultures or as a function of the estrous cycle in female rats. Moreover, again in comparison to the β-adrenergic model, one would expect that GDP and GTP will give rise to opposite effects. However, this is not the case, and it was shown that in membranes GDP decreases the affinity for muscarinic agonists in a manner similar to GTP (Gurwitz and Sokolovsky, 1980). The possible conversion of GDP to GTP in the membrane preparations was an obstacle in the interpretation of the above experiments. However, employing reconstituted systems, Haga et al. (1986) have demonstrated recently that indeed GDP like GTP decreases the affinity of muscarinic receptors reconstituted with G_i or G_o for the agonists. Moreover, the conversion of GDP to GTP during the incubation was less than 1%, indicating that the effect of GDP is not the result of its conversion to GTP. Thus, the binding of either GTP or GDP to the G-protein(s) has similar effects on the interaction of the latter with the muscarinic receptor. Finally, Lee et al. (1986) and Wong et al. (1986), in a very thorough and quantitative assessment of the ternary model, reached the conclusion that this model is compatible with the reported data on agonist binding to β-adrenergic receptors, but not to the muscarinic receptors. The foregoing considerations suggest that a simple ternary model (as proposed for the β-adrenergic receptors) is not consistent with the agonist binding parameters obtained for muscarinic interactions. A mechanism involving direct allosteric interactions between the muscarinic receptors and G-protein(s) is one reasonable possibility (Moscona-Amir et al., 1986; Lee et al., 1986).

5. Interaction between Muscarinic Receptors and Ion Channels

Ion channels can be coupled to receptors in two ways: (1) direct coupling as exemplified by the nicotinic receptor where the acetylcholine binding site and the ion channel are part of the same protein complex, and (2) indirect coupling where the physiological consequences of agonist binding are mediated by intracellular second messengers that ultimately convert the channel complex to its activated form. In the preceding section, we discussed the data related to the regulation of muscarinic receptors by G-protein(s). This section describes regulation of the muscarinic receptor by ion channels (K^+, Na^+, and Ca^{2+}) and the possible modulation of the latter interaction by G-protein(s).

A recent study (Cohen-Armon et al., 1985a) raised the possibility that antiarrhythmic and local anesthetic drugs affect muscarinic receptors through a site analogous to the voltage-sensitive Na^+-channel. In order to investigate this possibility, the effect of batrachotoxin (BTX) on the binding properties of muscarinic ligands was studied. BTX and several other alkaloids, e.g., veratridine and aconitine, interact with a site located on the voltage-sensitive sodium channel and induce a persistent activation (Catterall, 1977; McNeal et al., 1985 and references therein). In rat brainstem, cardiac ventricles, and atria (Cohen-Armon et al., 1985b; Cohen-Armon and Sokolovsky, 1986), BTX (1 μM), which induces an open state in sodium channels, enhanced the affinity of binding of several agonists to the muscarinic receptors. Analysis of the data indicated that the effect of BTX was to increase the affinity of the agonists toward the high affinity sites. Binding of antagonists was not affected by BTX. At higher concentrations of the toxin, the density of the high affinity muscarinic sites was also affected. The binding of agonists (but not of antagonists) to muscarinic receptors in turn enhanced the specific binding of [^{3}H]BTX to sodium channels. These effects on the muscarinic receptors and on the sodium channels were inhibited in the presence of Gpp(NH)p at concentrations lower than those bringing about conversion of binding sites from the high affinity to the low affinity conformation. On the basis of these findings, we proposed the existence of an interacting system containing three components: muscarinic receptors, G-protein(s), and Na^+-channel(s). The components of this complex interact with each other, and the interactions are modified allosterically upon ligand binding to the different components. The emerging picture of the mutual interactions within a three-component system is illustrated in Fig. 2. This scheme fits well with a previously proposed model describing cross-talk between receptors and other membrane components involved in signal transduction in the membrane (Sokolovsky, 1984; Sokolovsky et al., 1986).

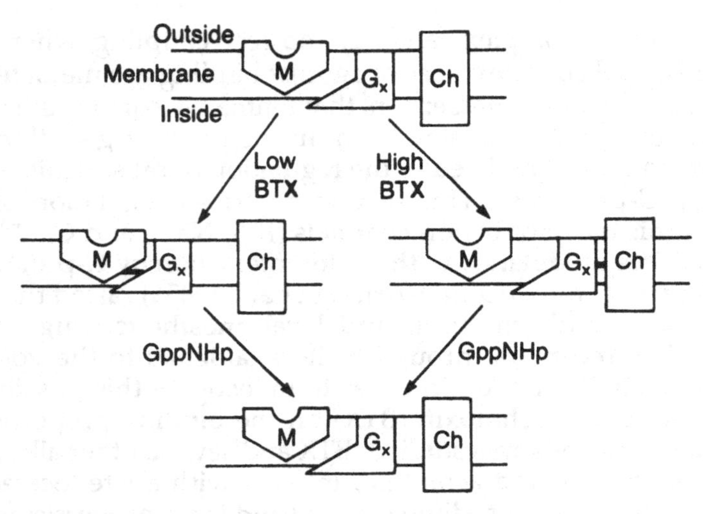

Fig. 2. Model depicting the putative interactions between the muscarinic receptors (M), G-protein(s) (G_x), and Na^+-channels (Ch).

Experimentation with brain stem synaptoneurosomes confirmed and extended these results. Cohen-Armon et al. (1988) showed that agonist binding to muscarinic receptors is voltage-dependent, i.e., that depolarization induces conversion of agonist binding sites from the high to the low affinity state, and that this process is mediated by G-protein(s). Moreover, the binding of, e.g., carbamylcholine to the muscarinic receptor induces $^{22}Na^+$ uptake via sodium channels. We thus have here a system with three interacting components: receptor, G-protein, and Na^+-channel. At resting potential, when the muscarinic receptors are in a state of high affinity for agonist binding, they activate the opening of the Na^+-channel. Opening of the channel leads to membrane depolarization, which in turn induces conversion of agonist binding sites to the low affinity state, thereby terminating the state of increased Na^+ permeability (i.e., closing the Na^+-channel) and inducing a return to resting potential.

Sorota et al. (1985) showed recently that pertussis toxin treatment blocks hyperpolarization by muscarinic agonists in chick atrium. This implicated a G-protein as an essential link that permits muscarinic receptors to regulate atrial potassium channels. Recently, two groups (Pfaffinger et al., 1985; Breitwieser and Szabo, 1985) have reported that the coupling of muscarinic receptors to specific K^+ channels is mediated by a G_i protein. Both groups used single atrial myocytes, dissociated enzymatically from embryonic chick hearts or frog heart. Using the whole cell patch-clamp technique, they showed that muscarinic activation (by ACh) of the K^+ conductance was inhibited when GTP was depleted from the cell. In addition, when the frog atrial cell, for example, was preloaded with Gpp(NH)p, application of ACh induced a persistent and irreversible activation of the K^+ conductance. Moreover, in the chick myocyte, ribosylation of the G-protein using the islet-activating protein prevented the activation of the K^+-channel. Thus, a G-protein is involved in coupling of the muscarinic receptors to the K^+-channel.

The identity of the G-protein that mediates the effect of ACh on K^+-channel function has been addressed by Yatani et al. (1987) and Logothetis et al. (1987). Both authors agree that a G_i-type protein (termed G_K by Yatani et al., 1987) plays a primary role in the process. Yatani et al. (1987) and Codina et al. (1987) suggest that it is the α-subunit of the G-protein that mediates the action of ACh. Logothetis et al. (1987) suggest that the β and γ subunits mediate the transducing action of ACh, whereas the α-subunit plays a regulatory role. The identity of the G-protein subunit(s) involved in the active mediation of signal transduction was debated recently by Birnbaumer (1987).

The third type of ion channel that might be coupled to muscarinic receptors is the Ca^{2+}-channel. Adenohypophyseal muscarinic receptors undergo changes with the progression of the estrous cycle identical with those occurring in vitro upon Ca^{2+} removal (Moscona-Amir et al., 1985). The main effect of Ca^{2+} on the muscarinic system in this tissue appears to be on the exposure or elimination (depending on the stage of the cycle) of a population of binding sites that is entirely (or mostly) high affinity with regard to agonist binding. The ability of the Ca^{2+} blocker D-600 to induce its maximal effect on the binding of muscarinic ligands also fluctuates during the estrous cycle. The cyclic variation in the effect of Ca^{2+} on D-600 sites, as observed in competition experiments between D-600 and [^{3}H]nitrendipine (Moscona-Amir and Sokolovsky, 1984), is consistent with the cyclic blocking effect of D-600 on Ca^{2+} channels, with a concomitant effect on the binding properties of adenohypophyseal muscarinic receptors. It was therefore suggested that the progression of the estrous cycle is accompanied by changes in the muscarinic receptors, which may in turn be coupled to Ca^{2+} channels.

References

Amitai, G., Kloog, Y., Balderman, D., and Sokolovsky, M. (1980) The interaction of bispyridinium oximes with mouse brain muscarinic receptors. *Biochem. Pharmacol.* **29**, 483–488.

Aronstam, R. S. and Eldefrawi, M. E. (1979) Reversible conversion between affinity states for agonists of the muscarinic acetylcholine receptor from rat brain. *Biochem. Pharmacol.* **28**, 701–703.

Avissar, S. and Sokolovsky, M. (1981) Guanine nucleotides preferentially inhibit binding of antagonist (male) agonists (female) to muscarinic receptors of rat adenohypophysis. *Biochem. Biophys. Res. Commun.* **102**, 753–760.

Avissar, S., Egozi, Y., and Sokolovsky, M. (1981) Biochemical characterization and sex dimorphism of muscarinic receptors in rat adenohypophysis. *Neuroendocrinology* **32**, 303–309.

Baumgold, J. (1986) Effects of verapamil on the binding characteristics of muscarinic receptor subtypes. *Eur. J. Pharmacol.* **126**, 151–154.

Ben-Baruch, G., Schreiber, G., and Sokolovsky, M. (1982) Cooperativity pattern in the interaction of the antiestrogen drug clomiphene with the muscarinic receptors. *Mol. Pharmacol.* **21**, 287–293.

Berrie, C. P., Birdsall, N. J. M., Burgen, A. S. V., and Hulme, E. C. (1979) Guanine nucleotides modulate muscarinic receptor binding in the heart. *Biochem. Biophys. Res. Commun.* **87**, 1000–1006.

Birdsall, N. J. M., Burgen, A. S. V., and Hulme, E. C. (1978) The binding of agonists to muscarinic receptors. *Mol. Pharmacol.* **14**, 723–726.

Birdsall, N. J. M., Burgen, A. S. V., Hulme, E. C., and Wong, E. H. F. (1983) The effects of *p*-chloromercuribenzoate on muscarinic receptors in the cerebral cortex. *Br. J. Pharmacol.* **80**, 187–196.

Birdsall, N. J. M., Hulme, E. C., and Burgen, A. S. V. (1980) The character of the muscarinic receptors in different regions of the rat brain. *Proc. R. Soc. Lond. B Biol. Sci.* **207**, 1–12.

Birnbaumer, L. (1987) Which G-protein subunits are the active mediators in signal transduction? *TIPS* **8**, 209–211.

Bourne, H. R. (1986) One molecular machine can transduce divers signals. *Nature* **321**, 814–817.

Boyer, J. L., Martinez-Carcamo, M., Monroy-Sanchez, J. A., Posadas, C., and Garcia-Sainz, J. A. (1986) Guanine nucleotide-induced positive cooperativity in muscarinic-cholinergic antagonist binding. *Biochem. Biophys. Res. Commun.* **134**, 172–177.

Breitwieser, G. and Szabo, G. (1985) Uncoupling of cardiac muscarinic and βadrenergic receptors from ion channels by guanine nucleotide analogue. *Nature* **317**, 538–540.

Brown, J .H. and Brown, S. L. (1984) Agonists differentiate muscarinic receptors that inhibit cyclic AMP formation from those that stimulate phosphoinositide metabolism. *J. Biol. Chem.* **259**, 3777–3181.

Brown, J. H. and Goldstein, D. (1986) Analysis of cardiac muscarinic receptors recognized selectively by nonquaternary but not by quaternary ligands. *J. Pharmacol. Exp. Ther.* **238**, 580–586.

Brunner, F., Waelbroek, M., and Christophe, J. (1986) Secovernine is a nonselective muscarinic antagonist on rat heart and brain receptors. *Eur. J. Pharmacol.* **127**, 17–25.

Burgermeister, W., Klein, W. L., Nirenberg, M., and Witkop, B. (1978) Comparative binding studies with cholinergic ligands and histrionicotoxin at muscarinic receptors of neural cell lines. *Mol. Pharmacol.* **14**, 751–767.

Burgisser, E., De Lean, A., and Lefkowitz, R. J. (1982) Reciprocal modulation of agonist and antagonist binding to muscarinic cholinergic receptor by guanine nucleotides. *Proc. Natl. Acad. Sci. USA* **79**, 1732–1737.

Burke, R. E. (1986) Gallamine binding to muscarinic M1 and M2 receptors, studied by inhibition of [^{3}H]pirenzepine and [^{3}H]quinuclidinyl benzilate binding to rat brain membranes. *Mol. Pharmacol.* **30**, 58–68.

Catterall, W. A. (1977) Activation of the action potential Na^+ ionophore by neurotoxins, an allosteric model. *J. Biol. Chem.* **252**, 8669–8676.

Clark, A. L. and Mitchelson, F. (1976) The inhibitory effect of gallamine on muscarinic receptors. *Br. J. Pharmacol.* **58**, 323–331.

Cockcroft, S. and Gomperts, B. D. (1985) Role of guanine nucleotide binding protein in the activation of polyphosphoinositide phosphodiesterase. *Nature* **314**, 534–536.

Codina, J., Yatani, A., Grenet, D., Brown, A. M., and Birnbaumer, L. (1987) The α subunit of the GTP binding protein G_k opens atrial potassium channels. *Science* **236**, 442–445.

Cohen-Armon, M. and Sokolovsky, M. (1986) Interactions between the muscarinic receptors, sodium channels, and guanine nucleotide binding proteins in rat atria. *J. Biol. Chem.* **261**, 12498–12505.

Cohen-Armon, M., Henis, Y. I., Kloog, Y., and Sokolovsky, M. (1985a) Interactions of quinidine and lidocaine with rat brain and heart receptors. *Biochem. Biophys. Res. Commun.* **127**, 326–332.

Cohen-Armon, M., Kloog, Y., Henis, Y. I., and Sokolovsky, M. (1985b) Batrachotoxin changes the properties of the muscarinic receptor in rat brain and heart: Possible interaction(s) between muscarinic receptors and sodium channels. *Proc. Natl. Acad. Sci. USA* **82**, 3524–3527.

Cohen-Armon, M., Schreiber, G., and Sokolovsky, M. (1986) Interaction of the antiarrhythmic drug amiodarone with the muscarinic receptor in rat heart and brain. *J. Cardiovasc. Pharmacol.* **6**, 1148–1155.

Cohen-Armon, M., Garty, H. and Sokolovsky, M. (1988) G-protein mediates voltage regulation of agonist binding to muscarinic receptors: Effect on receptor–Na^+ channel interaction. *Biochemistry* **27**, 368–374.

Dahlquist, F. W. (1978) Cooperativity in ligand binding to proteins showing association-dissociation phenomena: An application of the quantitative interpretation of the Hill coefficient. *Methods Enzymol.* **48**, 270–299.

De Lean, A., Stadel, J. M., and Lefkowitz, R. J. (1980) A ternary complex model explains the agonist-specific binding properties of adenylate cyclase-coupled β-adrenergic receptor. *J. Biol. Chem.* **255**, 7108–7117.

Dunlap, J. and Brown, J. H. (1983) Heterogeneity of binding sites on cardiac muscarinic receptors induced by the neuromuscular blocking agents gallamine and pancuronium. *Mol. Pharmacol.* **24**, 15–22.

Dunlap, J. and Brown, J. H. (1984) Divergences and similarities in muscarinic receptors of rat heart and retina: effects of agonists, guanine nucleotides, and *N*-ethylmaleimide. *J. Neurochem.* **43**, 214–220.

Egozi, Y., Kloog, Y., and Sokolovsky, M. (1980) Studies of postnatal changes of mucscarinic receptors in mouse brain, in *Neurotransmitters and Their Receptors* (Littauer, U. A., Dudai, Y., Silman, I., Teichberg, V. I., and Vogel, Z., eds), John Wiley, New York, pp. 201–215.

Ehlert, F. J. (1985) The relationship between muscarinic receptor occupancy and adenylate cyclase inhibition in the rabbit myocardium. *Mol. Pharmacol.* **28**, 410–421.

Ehlert, F. J., Dumont, Y., Roeske, W. R., and Yamamura, H. I. (1980) Muscarinic receptor binding in rat brain using the agonist [^{3}H]cis-methyldioxolane. *Life Sci.* **26**, 961–967.

Ehlert, F. J., Roeske, W. R., and Yamamura, H. I. (1981) Muscarinic receptor, regulation by guanine nucleotide, ions and *N*-ethylmaleimide. *Fed. Proc.* **40**, 153–161.

El-Fakahany, E. and Richelson, E. (1983) Effect of some calcium antagonists on muscarinic receptor-mediated cyclic GMP formation. *J. Neurochem.* **40**, 705–711.

Ellis, J. and Hoss, W. (1982) Competitive interaction of gallamine with multiple muscarinic receptors. *Biochem. Pharmacol.* **31**, 873–876.

Ellis, J. and Lenox, R. H. (1985a) Characterization of the interactions of gallamine with muscarinic receptors from brain. *Biochem. Pharmacol.* **34**, 2213–2217.

Ellis, J. and Lenox, R. H. (1985b) Quaternary forms of classical muscarinic antagonists distinguish subpopulations of muscarinic receptors: correlation with gallamine-defined subpopulations. *Biochem. Biophys. Res. Commun.* **126**, 1242–1250.

Fairhurst, A. S., Whittaker, M. L., and Ehlert, F. J. (1980) Interactions of D600 (methoxyverapamil) and local anesthetics with rat brain β-adrenergic and muscarinic receptors. *Biochem. Pharmacol.* **29**, 155–162.

Feinstein, M. B. and Paimre, M. (1967) Mode of anticholinergic action of local anesthetics. *Nature (Lond.)* **214**, 151–153.

Fisher, S. K. and Agranoff, B. W. (1981) Enhancement of the muscarinic synaptosomal phospholipid labeling effect by the ionophors A-23187. *J. Neurochem.* **37**, 968–977.

Fleisch, J. H. and Titus, E. (1973) Effect of local anaesthetics on pharmacologic receptor systems in smooth muscle. *J. Pharmacol. Exp. Ther.* **186**, 44–51.

Florio, V. A. and Sternweis, P. C. (1985) Reconstitution of resolved muscarinic cholinergic receptors with purified GTP-binding protein. *J. Biol. Chem.* **260**, 3477–3478.

Galper, J. B., Dziekan, L. C., O'Hara, D. S. and Smith, T. W. (1982) The biphasic response of muscarinic cholinergic receptors in cultured heart cells to agonists. *J. Biol. Chem.* **257**, 10344–10356.

Gilman, A. G. (1984) G-proteins and dual control of adeylate cyclase. *Cell* **36**, 577–579.

Gurwitz, D. and Sokolovsky, M. (1980) Agonist-specific reverse regulation of muscarinic receptors by transition metal ions and guanine nucleotides. *Biochem. Biophys. Res. Commun.* **96**, 1293–1301.

Gurwitz, D. and Sokolovsky, M. (1985b) Increased agonist affinity is induced in TNM-modified muscarinic receptors. *Biochemistry* **24**, 8086–8093.

Gurwitz, D., Baron, B., and Sokolovsky, M. (1984b) Copper ions and diamide induce a high affinity guanine nucleotides insensitive state for muscarinic agonists. *Biochem. Biophys. Res. Commun.* **120**, 271–277.

Gurwitz, D., Kloog, Y., and Sokolovsky, M. (1984a) Recognition of the muscarinic receptor by its endogenous neurotransmitter: binding of [^{3}H]-acetylcholine and its modulation by transition metal ions and guanine nucleotides. *Proc. Natl. Acad. Sci. USA.* **81**, 3650–3654.

Gurwitz, D., Kloog, Y., and Sokolovsky, M. (1985a) High affinity binding of [^{3}H]-acetylcholine to muscarinic receptors. Regional distribution and modulation by guanine nucleotides. *Mol. Pharmacol.* **28**, 297–295.

Haga, K., Haga, T., and Ichiyama, A. (1986) Reconstitution of muscarinic acetylcholine receptor. *J. Bio. Chem.* **261**, 10133–10140.

Haga, K., Haga, T., Ichiyama, A., Katada, T., Kurose, H., and Ui, M. (1985) Functional reconstitution of purified muscarinic receptors and inhibitory guanine nucleotide regulatory protein. *Nature* **316**, 731–733.

Hammer, R. and Giachetti, A. (1984) Selective muscarinic receptor antagonists. *Trends Pharmacol. Sci.* **5**, 18–20.

Hammer, R., Berrie, C. P., Birdsall, M. J. M., Burgen, A. S. V., and Hulme, E. C. (1980) Pirenzepine distinguishes between different subclassses of muscarinic receptors. *Nature (Lond.)* **283**, 90–92.

Harden, T. K., Meeker, R. B., and Martin, M. W. (1983) Interaction of radiolabeled agonists with cardiac muscarinic cholinergic receptors. *J. Pharmacol. Exp. Ther.* **227**, 570–577.

Harden, T. K., Scheer, A. G., and Smith, M. M. (1982) Differential modification of the interaction of cardiac muscarinic cholinergic and β-adrenergic receptors with a guanine nucleotide binding component(s). *Mol. Pharmacol.* **21**, 570–578.

Hedlund, B., Grynfarb, M., and Bartfai, T. (1982) Two ligands may bind simultaneously to the muscarinic receptors. *Naunyn-Schmiedeberg Arch. Pharmacol.* **320**, 3–13.

Henis, Y. I. and Levitzki, A. (1979) Ligand competition curves as a diagnostic tool for delineating the nature of site-site interactions: theory. *Eur. J. Biochem.* **102**, 449–465.

Henis, Y. I. and Sokolovsky, M. (1983) Muscarinic antagonists induce different receptor conformations. *Mol. Pharmacol.* **24**, 357–365.

Henis, Y. I., Galron, R., Avissar, S., and Sokolovsky, M. (1982) Interactions between antagonist-occupied muscarinic binding sites in rat adenohypophysis. *FEBS Lett.* **140**, 173–176.

Horsey, M. M. (1982) Regulation of antagonist binding to cardiac muscarinic receptors. *Biochem. Biophys. Res. Commun.* **107**, 314–321.

Hulme, E. C., Berrie, C. P., Birdsall, N. J. M., and Burgen, A. S. V. (1981) Two populations of binding sites for muscarinic antagonists in the rat heart. *Eur. J. Pharmacol.* **73**, 137–142.

Hulme, E. C., Berrie, C. P., Birdsall, N. J. M., Jameson, M., and Stockton, J. M. (1983) Regulation of muscarinic agonist binding by cations and guanine nucleotides. *Eur. J. Pharmacol.* **94**, 59–72.

Hulme, E. C., Birdsall, N. J. M., Burgen, A. S. V., and Mehta, P. (1978) The binding of antagonists to brain muscarinic receptors. *Mol. Pharmacol.* **14**, 737–750.

Huppert, L. C. (1979) Introduction of ovulation with clomiphene citrate. *Fertil. Steril.* **31**, 1–8.

Jarv, J., Hedlund, B., and Bartfai, T. (1980) Kinetic studies on muscarinic antagonist-agonist competition. *J. Biol. Chem. 255*, 2649–2651.

Kahn, R. A. and Gilman, A. G. (1984) ADP-ribosylation of Gs promotes the dissociation of its α and β subunits. *J. Biol. Chem.* **259**, 6235–6240.

Karliner, J. S., Motulsky, H. J., Dunlap, J., Brown, J. H., and Insel, P. A. (1982) Verapamil competitively inhibits α_1-adrenergic and muscarinic but not β-adrenergic receptors in rat myocardium. *J. Cardiovasc. Pharmacol.* **4**, 515–520.

Katada, T. and Ui, M. (1982) ADP ribosylation of the specific membrane protein of C6 cells by islet-activating protein associated with modification of adenylate cyclase activity. J. Biol. Chem. **257**, 7210–7216.

Kato, G. and Changeux, J. P. (1976) Studies on the effect of histrionicotoxin on the monocellular electroplax from *Electrophorus electricus* and on the Marmorata. *Mol. Pharmacol.* **12**, 92–100.

Kloog, Y. and Sokolovsky, M. (1978) Studies on muscarinic acetylcholine receptors from mouse brain: characterization of the interaction with antagonists. *Brain Res.* **144**, 31–48.

Kloog, Y. and Sokolovsky, M. (1985a) Bisquaternary pyridinium oximes as allosteric inhibitors of rat brain muscarinic receptors. *Mol. Pharmacol.* **27**, 418–428.

Kloog, Y. Sokolovsky, M. (1985b) Allosteric interactions between muscarinic agonist binding sites and effector sites demonstrated by the use of bisquaternary pyridinium oximes. *Life Sci.* **36**, 2127–2136.

Kloog, Y, Egozi, Y., and Sokolovsky, M. (1979) Characterization of muscarinic acetylcholine receptors from mouse brain: Evidence for regional heterogeneity and isomerization. *Mol. Pharmacol.* **15**, 545–559.

Kuhnen-Clausen, D. (1979) Investigation on the parasymaptholitic effect of toxogonin on the guinea pig isolated ileum. *Eur. J. Pharmacol.* **9**, 85–87.

Kuhnen-Clausen, D., Hagedron, I., Gross, G., Bayer, H., and Hucho, F. (1983) Interaction of bisquaternary pyridine salts (H-oximes) with cholinergic receptors. *Arch. Toxicol.* **54**, 171–179.

Kurose, H. and Ui, M. (1983) Functional uncoupling of muscarinic receptors from adenylate cyclase in rat cardiac membranes by the active component of islet-activating protein, pertussis toxin. *J. Cyclic Nucleotide Res.* **9**, 305–318.

Kurose, H., Katada, T., Amano, T., and Ui, M. (1983) Specific uncoupling by islet activating protein, pertussis toxin, of negative signal transduction via α-adrenergic, cholinergic and opiate receptors in neuroblastoma and glioma hybrid cells. *J. Biol. Chem.* **258**, 4870–4875.

Lee, J. H. and El-Fakahary, E. E. (1985) Heterogeneity of binding of muscarinic receptors antagonists in rat brain homogenates. *J. Pharmacol. Exp. Ther.* **239**, 707–714.

Lee, T. W. T., Sole, M. J., and Wells, J. W. (1986) Assessment of a ternary model for the binding of agonists to neurohumoral receptors. *Biochemistry* **25**, 7009–7020.

Lefkowitz, R. J., Stadel, J. M., and Caron, M. G. (1983) Adenylate cyclase-coupled Beta-adrenergic receptors: Stucture and mechanisms of activation and desensitization. *Annu. Rev. Biochem.* **52**, 159–186.

Levitzki, A. (1978) Quantitative aspects of allosteric mechanisms, in *Molecular Biology, Biochemistry and Biophysics* (Kleinzeller, A., Springer, G. F., and Wittmann, H. G., eds.) **Vol. 28**, Springer-Verlag, Berlin.

Levitzki, A. and Koshland, D. E., Jr. (1976) The role of negative cooperativity and half-of-the-sites reactivity in enzyme regulation, in *Current Topics in Cellular Regulation* (Horecker, B. L. and Stadtman, E. E., eds.) **Vol. x** Academic, pp. 1–40, New York.

Logothetis, D., Kurachi, Y., Galper, J., Neer, E. J., and Clapham, D. E. (1987) The βγ subunits of GTP-binding proteins activate the muscarinic K^+ channel in the heart. *Nature* **325**, 321–325.

Martin, M. W., Smith, M. M., and Harden, J. K. (1984) Modulation of muscarinic cholinergic receptor affinity for antagonists in rat heart. *J. Pharmacol. Exp. Ther.* **230**, 424–430.

Mattera, R., Pitts, B. J. R., Entman, M. L., and Birnbaumer, L. (1985) Guanine nucleotide regulation of a mammalian myocardial muscarinic receptors system. *J. Biol. Chem.* **260**, 7410–7421.

McNeal, E. T., Lewandowski, G. A., Daly, J. W., and Creveling, C. R. (1985) [^{3}H]-batrachotoxinin-A α-20-benzoate binding to voltage-sensitive sodium channels: A rapid and quantitative assay for local anesthetic activity in a variety of drugs. *J. Med. Chem.* **28**, 381–388.

Milligan, G. and Klee, W. A. (1985) The inhibitory guanine nucleotide binding protein (Ni) purified from bovine brain is a high affinity GTP-ase. *J. Cyclic Nucleotide Res.* **260**, 2057–2068.

Monod, J., Changeux, J.-P., and Jacob, F. (1963) Allosteric proteins and cellular control systems. *J. Mol. Biol.* **6**, 303–329.

Morris, I. D. (1985) Inhibition of intestinal smooth muscle function by tamoxifen and clomiphene. *Life Sci.* **37**, 273–278.

Moscona-Amir, E. M. and Sokolovsky, M. (1984) Interactions between Ca^{2+} antagonist binding sites in rat adenohypophysis: dependence on the estrous cycle. *Life Sci.* **35**, 1043–1050.

Moscona-Amir, E., Egozi, Y., Henis, Y. I., and Sokolovsky, M. (1985) Effect of Ca^{2+} on the binding characteristics of muscarinic receptors in rat adeohypophysis—variation during the estrous cycle. *Neuroendocrinology* **40**, 483–492.

Moscona-Amir, E., Henis, Y. I., Yechiel, E., Barenholz, Y., and Sokolovsky, M. (1986) Role of lipids in age-related changes in properties of muscarinic receptors in cultured rat heart myocytes. *Biochemistry* **25**, 8118–8125.

Mukherjee, A., Snyder, G., and McCann, S. M. (1980) Characterization of muscarinic receptors on intact rat anterior pituitary cells. *Life Sci.* **27**, 475–482.

Nedoma, J., Tucek, S., Danilov, A. F., and Shelkovnikov, S. A. (1986) Stabilization of antagonist binding to cardiac muscarinic receptors by gallamine and other neuromuscular blocking drugs. *J. Pharmacol. Exp. Ther.* **236**, 219–223.

Nukada, T., Haga, T., and Ichiyama, A. (1983) Muscarinic receptors in porcine caudate nucleus. 1. Enhancement by nickel and other cations of [^{3}H]cis-methyldioxolane binding to guanyl nucleotide-sensitive sites. *Mol. Pharmacol.* **24**, 366–373.

Onali, P., Olianas, M. C., Schwartz, J. P., and Costa, E. (1983) Involvement of a high affinity GTPase in the inhibitory coupling of striatal muscarinic receptors to adenylate cyclase. *Mol. Pharmacol.* **24**, 380–386.

Pfaffinger, P. J., Martin, J. M., Hunter, D. D., Nathanson, N. M., and Hille, B. (1985) GTP-binding proteins couple cardiac muscarinic receptors to a K channel.*Nature* **317**, 536–538.

Rathburn, F. J. and Hamiltion, J. T. (1970) Effect of gallamine on cholinergic receptors. *Can. Anaesth. Soc. J.* **17**, 574–590.

Richelson, E., Prendergast, F. G., and Divinetz-Romero, S. (1978) Muscarinic receptor-mediated cyclic GMP formation by cultured nerve cells—ionic dependence and effects of local anaesthetics. *Biochem. Pharmacol.* **27**, 2039–2048.

Riker, W. F. and Wescoe, W. C. (1951) The pharmacology of flaxedil, with observations on certain analogs. *Ann. N.Y. Acad. Sci.* **54**, 373–392.

Rodbell, M. (1980) The role of hormone receptors and GTP regulatory proteins in membrane signal transduction. *Nature* **284**, 17–22.

Snyder, S. H., Chang, K. J., Kuhar, M. J., and Yamamura, H. I. (1975) Biochemical identification of the mammalian muscarinic cholinergic receptor. *Fed. Proc.* **34**, 1915–1921.

Sokolovsky, M. (1984) Muscarinic receptors in the central nervous system. *Int. Rev. Neurobiol. 5*, 139–183.

Sokolovsky, M., Gurwitz, D., and Galron, R. (1980) Muscarinic receptors binding in mouse brain: regulation by guanine nucleotides. *Biochem. Biophys. Res. Commun.* **94**, 487–492.

Sokolovsky, M., Cohen-Armon, M., Egozy, Y., Gurwitz, D., Henis, Y. I., Kloog, Y., Moscona-Amir, E., and Schreiber, G. (1986) Modulation of muscarinic receptors and their interactions. *Trends. Pharmacol. Sci.* 7(suppl), 39–44.

Sorota, S., Tsuji, Y., Tajima, T., and Pappano, A. J. (1985) Pertussis toxin treatment blocks hyperpolarization by muscarinic agonists in chick atrium. *Circ. Res.* **57**, 748–758.

Sternweis, P. C. and Robishaw, J. D. (1984) Isolation of two proteins with high affinity for guanine nucleotides from membranes of bovine brain. *J. Biol. Chem.* **259**, 13806–13813.

Stockton, J. M., Birdsall, N. J. M., Burgen, A. S. V., and Hulme, E. C. (1983) Modification of the binding properties of muscarinic receptors by gallamine. *Mol. Pharmacol.* **23**, 551–557.

Straub, R. E. and Gershengorn, M. C. (1986) Thyrotropin-releasing hormone and GTP activate inositol trisphosphate formation in membranes isolated from rat pituitary cells. *J. Biol. Chem.* **261**, 2712–2717.

Stryer, L., Hurley, J. B., and Fung, B. K-K. (1981) Transducin: An amplifier protein in vision. *Trends. Biochem. Sci.* **6**, 245–247.

Taylor W. J., Wolf, A., and Young, J. M. (1980) The interaction of amine local anaesthetics with muscarinic receptors. *Br. J. Pharmacol.* **71**, 327–335.

Uchida, S., Matsumoto, K., Takeyasu, K., Higuchi, H., and Yoshida, H. (1982) Molecular mechanism of the effects of guanine nucleotides and sulfhydryl reagents on muscarinic receptors in smooth muscle studied by radiation inactivation. *Life Sci.* **1**, 201–209.

Vickroy, T. W., Roeske, W. R., and Yamamura, H. I. (1984) Pharmacological differences between the high affinity muscarinic agonist binding states of the rat heart and cerebral cortex labeled with (+)-[^{3}H]cismethyldioxolane. *J. Pharmacol. Exo. Ther.* **299**, 747–755.

Waelbroeck, M., De Neef, P., Robberecht, P., and Christophe, J. (1984a) Inhibitory effect of quinidine on rat heart muscarinic receptors. *Life Sci. 5*, 1069–1076.

Waelbroeck, M. Robberecht, P., Chatelain, P., and Christophe, J. (1982)

Rat cardiac muscarinic receptors. I. effects of guanine nucleotides on high- and low affinity binding sites. *Mol. Pharmacol.* **21**, 581–588.

Waelbroeck, M., Tobberecht, P., De Neef, P., and Christophe, J. (1984b) Effects of verapamil on the binding properties of rat heart muscarinic receptors: evidence for an allosteric site. *Biochem. Biophys. Res. Commun.* **121**, 340–345.

Wamsley, J. K., Zarbin, M., and Kuhar, M. J. (1984) Distribution of muscarinic cholinergic high and low affinity agonist binding sites: a light microscopic autoradiographic study. *Brain Res. Bull.* **1**, 233–243.

Wei, J.-W. and Sulkahe, P. V. (1979) Agonist-antagonist interactions with rat atrial muscarinic cholinergic receptor sites: differential regulation by guanine nucleotides. *Eur. J. Pharmacol.* **58**, 91–96.

Wei, J.-W. and Sulkahe, P. V. (1980) Cardiac muscarinic cholinergic receptor sites: opposing regulation by divalent cations and guanine nucleotides. *Eur. J. Pharmacol.* **62**, 345–349.

Wong, H-M. S., Sole, M. J., and Wells, J. W. (1986) Assessment of mechanistic proposals for the binding of agonists to cardiac muscarinic receptors. *Biochemistry* 5, 6995–7008.

Yatani, A., Codina, J., Brown, A. M., and Birnbaumer, L. (1987) Direct activation of mammalian atrial muscarinic potassium channels by GTP regulatory protein G_K. *Science* **235**, 207–211.

11

Regulation and Development of Muscarinic Receptor Number and Function

Neil M. Nathanson

1. Introduction

Knowledge of the factors involved in the regulation, development, expression, maintenance, localization, and function of neurotransmitter receptors is of obvious importance for understanding how signal processing in the nervous system may be controlled. The regulation of the number and action of muscarinic acetylcholine receptors (mAChR) present on cells and tissues is of interest both because of the fundamental neurobiological question of identifying the processes that determine the physiological sensitivity of a cell to cholinergic stimulation, as well as because of the possible therapeutic and pathological implications of changes in the mAChR. This chapter will summarize some of the studies carried out to identify the mechanisms that control the number and function of the mAChR.

2. Regulation by Exposure to Agonist

In many systems, exposure of cells, tissues, or whole animals to the continued presence of neurotransmitter or other receptor agonist can have both short-term and long-term effects on both the number and function of receptors present. The rapid loss of physiological responsiveness of the mAChR that occurs seconds or minutes after agonist exposure will be called receptor "desensitization," and the loss of cellular receptor binding sites that occurs minutes to hours after agonist exposure receptor will be called "downregulation."

2.1. Regulation of mAChR Number and Function in Cell Culture

As with many other hormone and neurotransmitter receptors, the number of mAChR on cells in culture is decreased after long-term (several hours) exposure to muscarinic agonists. Decreased mAChR number has been observed after prolonged agonist exposure of neuronal cell lines (Klein et al., 1979; Taylor et al., 1979; Shifrin and Klein, 1980), heart cells (Galper and Smith, 1980), neurons (Siman and Klein, 1979; Burgoyne and Pearce, 1981), smooth muscle (Takeyasu et al., 1981), thyroid cells (Champion and Mauchamp, 1982), and AtT-20 pituitary cells (Heisler et al., 1985). This agonist-induced decrease in receptor number ("downregulation") has been postulated to represent a mechanism for the regulation of receptor function in response to varying levels of physiological activity in vivo. The decrease in receptor number occurs over several hours and is thus distinct from short-term desensitization of receptor function, which occurs on a time scale of minutes without a decrease in total cellular receptor number (Taylor et al., 1979). The long-term decrease in mAChR number is accompanied by a decreased physiological sensitivity to cholinergic stimulation. For example, in neuronal cells, there are concomitant decreases in mAChR-mediated inhibition of adenylate cyclase (Nathanson et al., 1978), stimulation of guanylate cyclase (Taylor et al., 1979), and stimulation of phosphatidylinositol turnover (Siman and Klein, 1981). In cultured heart cells, there are concomitant decreases in mAChR-mediated negative chronotropic response, inhibiton of adenylate cyclase, and stimulation of ^{45}K and ^{86}Rb efflux (Galper et al., 1982a; Hunter and Nathanson, 1986).

Binding of an antagonist to the receptor does not decrease receptor number and can block the decrease caused by treatment with agonist. Thus, simple occupany of the ligand binding site without receptor activation does not promote loss of the mAChR. In neuronal cells, the apparent half-life of the receptor decreased from 11 h to 3 h after agonist addition (Klein et al., 1979). Following removal of agonist, a gradual recovery of mAChR number occurs, which can be prevented by inhibition of *de novo* protein synthesis (Klein et al., 1979; Taylor et al., 1979) or by inhibition of protein glycosylation (Liles and Nathanson, 1986). Galper et al., (1982b) examined the binding of the lipophilic ligand [^{3}H]QNB and the lipophobic ligand [^{3}H]NMS to intact cardiac cells after agonist exposure. Because of the different hydrophobicities of these two ligands, it is likely that the binding of [^{3}H]NMS to intact cells only labels receptors exposed on the cell surface, whereas [^{3}H]QNB labels the total receptor population. Agonists caused a rapid sequestration of the mAChR from the cell surface, so that the receptors were no longer available to bind [^{3}H]NMS, but could still bind [^{3}H]QNB; [^{3}H]QNB binding sites were lost from intact cells with a slower time course similar to the loss of binding sites seen in membrane homogenates. Maloteaux et al. (1983) and Feigenbaum and El-Fakahany (1985) demonstrated that the rapid loss of [^{3}H]NMS sites from intact neuronal cells was also rapidly reversible and was temperature sensitive. In 1321N1 astrocytoma cells, this conversion of the mAChR to a form that was not accessible to [^{3}H]NMS was accompanied by an alteration in the apparent subcelluar localization of the receptor. Instead of the receptor being associated with the plasma membrane after centrifugation, agonist exposure converted the receptor to a form that sedimented in sucrose gradients with light membrane vesicles (Harden et al., 1985). The conversion of the receptor to this light vesicle-associated form occurred with a half-time of approximately 10 min, and reversed following agonist removal with a half-time of 20 min. These results provide strong evidence for the hypothesis that agonist-induced decreases in mAChR number result from a rapid and initially reversible internalization of the receptor from the cell surface, followed by a slower degradation of the receptor.

In addition to the decreases in mAChR number and resulting decreased muscarinic responsiveness, long-term treatment of NG108-15 neuroblastoma–glioma hybrid cells with muscarinic agonists also increases adenylate cyclase activity. Both basal and PGE_1-stimulated cAMP formation in intact cells and adenylate cyclase activity in membrane homogenates are increased by up to two–threefold after 24–48 h (Nathanson et al., 1978). Similar increases have been reported following incubation with opiates and norepinephrine, which activate other receptors that inhibit adenylate cyclase activity in these cells (Sharma et al., 1975; Sabol and Nirenberg et al., 1979). It has been suggested that these increases in adenylate cyclase activity may represent a compensatory mechanism for regulation of cellular cyclic nucleotide levels or a biochemical model for narcotic addiction, although the actual physiological significance of this remains to be determined. Treatment of cells with pertussis toxin eliminates the increase in adenylate cyclase activity resulting from chronic carbachol treatment (Thomas and Hoffman, 1986). These results suggest that a functional G-protein (presumably G_i) is required for this effect of chronic agonist treatment and is consistent with the hypothesis that the increase in adenylate cyclase activity either results from and/or compensates for the initial inhibition of enzyme activity mediated by agonist exposure.

Treatment of cultured chick retinal neurons with cholinergic agonists appears to decrease selectively the number of muscarinic, but not nicotinic receptors. Long-term exposure of the neurons to carbachol decreased mAChR number, as measured by [^{3}H]QNB binding, by 70–84%. No effect of carbachol treatment on the number of putative nicotinic acetylcholine receptor binding sites, measured using either [^{125}I]α-bungarotoxin or [^{3}H]bromoacetylcholine, was observed in these cells (Betz, 1982; Siman and Klein, 1983).

2.2. Regulation of mAChR by In Vivo Agonist and Antagonist Administration

The number of mAChR can be altered in vivo by a variety of pharmacological treatments. Direct administration of muscarinic agonists decreases mAChR number in heart, brain, and spinal cord (Wise et al., 1980; Halvorsen and Nathanson, 1981; Marks et al., 1981; Meyer et al., 1982; Taylor et al., 1982). These decreases in mAChR number result in decreased physiological responsiveness. For example, in the heart, there is a decrease in the mAChR-mediated negative chronotropic response (Halvorsen and Nathanson, 1981: Nathanson et al., 1984) and in the mAChR-mediated inhibition of adenylate cyclase activity (Hunter and Nathanson, 1984). Similarly, the antinociceptive effect of intrathecal injection of muscarinic agonists is lost following agonist-induced downregulation of mAChR in the rat spinal cord (Taylor et al., 1982). Chronic administration of acetylcholinesterase inhibitors results in decreased numbers of mAChR in brain, ileum, and rat submandibular gland (Gazit et al., 1979; Dawson and Jarrott, 1981; Olianas et al., 1984; Costa and Murphy, 1985). These decreases in mAChR number are presumably the result of increased levels of endogenous ACh promoting agonist-induced downregulation of the receptor. These decreases in mAChR number also cause decreased physiological responsiveness. In the rat striatum, there is a reduction in mAChR-mediated inhibition of adenylate cyclase and stimulation of GTPase acitvity (Olianas et al., 1984). The decrease in mAChR number in the submandibular gland causes diminished mAChR stimulation of saliva secretion (Costa and Murphy, 1985). Rats that are selectively bred to exhibit increased sensitivity to the toxic effects of acetylcholinesterase inhibitors have increased numbers of mAChR in the striatum and hippocampus (Overstreet et al., 1984). Administration of either classical muscarinic antagonists or drugs such as tricyclic antidepressants that have mAChR antagonist activity increase the number of mAChR found in the brain and heart (Ben-Barak and Dudai, 1980; Wise et al., 1980; Rehavi et al., 1980; Westlind et al., 1981). This presumably is because of the antagonist preventing synaptically released ACh from interacting with the mAChR.

The regulation of mAChR number by direct administration of agonists in vivo exhibits similar characteristics as downregulation in cell culture. Kinetic studies and experiments with partial agonists on the loss of mAChR from the embryonic chick heart and brain indicate that the number of mAChR remaining after treatment with various concentrations of agonist is a reflection of the fraction of receptors occupied and activated by that particular concentration of agonist (Halvorsen and Nathanson, 1981; Meyer et al., 1982; Nathanson et al., 1984). Receptor number gradually returns to control values if further receptor–agonist interactions are blocked by administration of a muscarinic antagonist (Halvorsen and Nathanson, 1981; Meyer et al., 1982). This recovery of mAChR can be blocked by administration of protein synthesis inhibitors (Hunter and Nathanson, 1984), consistent with experiments in cell culture indicating that the receptors that reappear following agonist-induced decreases in receptor number represent newly synthesized receptors.

2.3. Rapid Agonist-Mediated Desensitization of Receptor Function

In contrast to the relatively long-term effects of agonist exposure described above, short-term treatment of some tissues and cells with muscarinic agonists can cause a rapid loss of mAChR-mediated physiological responses with no apparent change in receptor number. The negative chronotropic response mediated by the mAChR in the embryonic chick heart at early stages of development can completely desensitize within a minute, prior to the first detectable change in either total or cell surface receptors (Pappano and Skowronek, 1974; Halvorsen and Nathanson, 1981; Galper et al., 1982b). The desensitization of the mAChR-mediated activation of the inward rectifying potassium channel in rabbit cardiac Purkinje fibers has been studied using voltage-clamp recordings (Carmeliet and Mubagwa, 1986). Both the magnitude and rate of desensitization of this response elevated with increasing concentrations of agonist. At high concentration, the ACh-activated current decreased by 70–80% over 5 min. The rate of desensitization was relatively independent of membrane potential, whereas the extent of desensitization increased somewhat with membrane depolarization. The response resensitized with a half-time of about 2 min. Because desensitization was not seen for mAChR-mediated attenuation of the catecholamine-induced increase in calcium current, these authors also suggested that desensitization was not the result of agonist-mediated internalization of the receptor.

Agonist treatment of N1E-115 cells causes a loss of mAChR-mediated stimulation of cGMP synthesis with a half-time of 1–2 min (El-Fakahany and Richelson, 1980a). It has been suggested that this desensitization may be the result of either inactivation or calcium channels (El-Fakahany and Richelson, 1980b) or a rapid and reversible internalization of the receptor (Maloteaux et al., 1983; Feigenbaum and El-Fakahany, 1985), although the kinetic analyses of receptor internalization have not been performed at a sufficient level of resolution to unequivocally demonstrate that it could account for loss of receptor function. In astrocytoma cells, where agonist-mediated conversion of the mAChR to a form that associates with the light vesicle fraction has been demonstrated (*see above*), mAChR-mediated phosphatidylinositol breakdown is not diminished at a time when 50% of the receptor binding sites are apparently sequestered from the cell surface (Harden et al., 1985; Masters et al., 1985). The mAChR associated with the light vesicular fraction still exhibits guanine-nucleotide regulation of agonist binding, implying that the receptors remain associated with a G-protein. These results suggest that the mAChR in this cell type can still remain coupled to at least one physiological response following internalization.

Several studies have attempted to correlate changes in agonist binding properties with receptor desensitization, although no consistent results have appeared. Young (1974) demonstrated that incubation of smooth muscle strips with desensitizing concentrations of agonist resulted in a shift of carbachol–[^{3}H]QNB competition curves to the right with an increase in the Hill coefficient from 0.4 to 0.9, with no decrease in the total number of binding sites. These results could have been accounted for by a conversion of high affinity agonist sites to low affinity sites. Galper et al. (1982b) reported that short-term incubations with agonist caused a disappearance of high affinity agonist sites, which was suggested to result from the continued presence of bound agonist. Cioffi and El-Fakahany (1986) reported that short-term incubation of N1E-115 cells with agonist caused a preferential loss of low affinity agonist sites from the cell surface. The reason for these different observations and their relationship to agonist-mediated changes in receptor function require further investigation. It has also been reported that short-term desensitization of the mAChR in N1E-115 cells is accompanied by a selective loss of high affinity pirenzepine binding sites, but not low affinity pirenzepine sites (Cioffi and El-Fakahany, 1986). However, because the binding curve for pirenzepine in control cells, where there supposedly were two classes of binding sites, exhibited a Hill coefficient that was not significantly different from unity, the significance of this finding is not clear.

3. Physiological and Biochemical Properties of Newly Synthesized Receptors

Because long-term agonist exposure eliminates most of the mAChR in the treated cells and tissues, recovery from down-regulation following removal of agonist provides a system for the study of the properties of newly synthesized muscarinic receptors without the complicating presence of "older," preexisting mAChR. Taylor et al., (1979) compared the time course for the recovery of total mAChR binding sites (measured in membrane homogenates) with the time course for the recovery of mAChR-stimulated cGMP synthesis following agonist treatment in the neuroblastoma cell line N1E-115. They found that the reappearance of physiological responsiveness lagged behind the recovery of receptor number: the apparent $t_{1/2}$ for the recovery of receptor-mediated stimulation of cGMP synthesis was approximately 16 h. A similar approach was used to study the physiological responsiveness of newly synthesized mAChR in vivo in the embryonic chick heart. Embryos were first treated with carbachol, reducing receptor number by 85–90%, and then with atropine, allowing receptor number to return to control values by 14 h. The recovery of both the mAChR-mediated negative chronotropic response of isolated atria and mAChR-mediated inhibition of adenylate cyclase in membrane homogenates lagged behind the recovery of mAChR binding sites. At 20 h after atropine treatment, a greater than 10-fold higher concentration of carbachol was required to mediate both the negative chronotropic and cyclic nucleotide responses compared to controls, whereas less than threefold more carbachol was required by 29 h (Hunter and Nathanson, 1984).

In contrast to these results, it was reported that, in cardiac cell culture, mAChR number and function recovered concomitantly following downregulation (Galper et al., 1982a). However, these investigators only used a single saturating concentration of agonist, so that quantitative differences in dose–response curves observed in vivo could have been missed. A detailed study of newly synthesized mAChR appearing following recovery from downregulation in culture demonstrated that these mAChR do in fact exhibit diminished ability to mediate both receptor-mediated increases in potassium permeability and inhibition of adenylate cyclase (Hunter and Nathanson, 1986). Receptors that reappear following affinity alkylation of the mAChR with PrBCM also exhibit diminished physiological responsiveness, so that the diminished physiological responsiveness does not result from a nonspecific effect of agonist treatment on cell function. The changes in physiological sensitivity that occur after receptor number has returned to control values following either agonist or PrBCM treatment are not blocked by treatment with cyclohexemide and, thus, do not require *de novo* protein synthesis. The decreased physiological responsiveness does not appear to be the result of modification of the GTP-regulatory proteins. The "immature" and "mature" receptors do not differ significantly in molecular weight or isoelectric point, so that the diminished physiological sensitivity of newly synthesized mAChR is caused either by a minor modification of the receptor or a change in a component of the transduction system that has not yet been identified.

Biochemical studies of the mAChR have demonstrated that it is a glycoprotein (Peterson et al., 1986). Protein glycosylation can be important for the synthesis or stability of many membrane proteins. Treatment of neuroblastoma cells with tunicamycin, which inhibits *N*-glycosylation of nascent proteins, causes a selective depletion of cell surface mAChR and a corresponding increase in intracellular receptors without changing the degradation rate of the receptor. Protein glycosylation is thus required for the transport, insertion, and/or maintenance of the receptor in the cell membrane (Liles and Nathanson, 1986). Although these results suggest that glycosylation of the mAChR itself may be required, it cannot be excluded that glycosylation of another protein involved in the synthesis of the receptor is the critical step that is inhibited by tunicamycin.

A novel approach to the study of mAChR at different stages of their synthesis and degradation was provided by the demonstration that mAChR undergoing axonal transport would accumulate at the site of nerve ligature. Zarbin et al., (1982) used double-ligature experiments of the rat vagus nerve to identify and study the binding properties of both anterogradely and retrogradely transported receptors. They found that receptors undergoing anterograde transport (i.e., away from the cell body) exhibited high affinity and guanine-nucleotide-sensitive agonist binding, whereas receptors undergoing retrograde transport (i.e., from the nerve terminal region back to the soma) exhibited primarily low affinity, guanine nucleotide-insensitive, agonist binding. These results suggest that newly synthesized mAChR may be transported to the nerve terminal in association with the GTP-binding proteins required for their action, whereas recycled receptors are returned to the cell body in an uncoupled form without an associated G-protein. It is possible that this difference in apparent recycling, and therefore presumably also in rate of degradation between the mAChR and G-proteins, may be at least either responsible for or a reflection of the much higher levels of G-proteins compared to mAChR in the brain.

4. Regulation by Innervation, Hormones, and Factors

The changes in mAChR number observed following administration of cholinergic drugs suggests that innervation and denervation should also affect mAChR number. Lesions of the cholinergic input to various regions of the central nervous system have been reported to result in either no change (Kamiyaet al., 1981), increases (Westlind et al., 1981), or decreases (de Belleroche et al., 1985) in mAChR number. Both no change (Burt, 1978) and increases (Taniguchi et al., 1983) in mAChR number have been reported after denervation of sympathetic ganglia. These binding studies on tissue homogenates are difficult to interpret, however. As noted previously (Burt, 1978), the relative contributions of presynaptic, postsynaptic, and nonneuronal cells cannot be distinguished, and possible changes in the distribution of receptors are also ignored. Definitive resolution of these problems will probably require electron microscopic ultrastructural localization.

Many smooth muscles exhibit supersensitivity to ACh after denervation, but the contribution of changes in mAChR number to this altered responsiveness has not been unequivocaly established. An increase in mAChR that appeared to correlate with increased physiological sensitivity was observed after denervation in the guinea-pig vas deferens (Hata et al., 1980). However, parasympathetic denervation of the cat iris resulted in supersensitivity with no increase in receptor number, suggesting the involvement of a postreceptor process (Sachs et al., 1979). Electrophysiological studies indicate that changes in resting potential and electrical excitability of the membrane can be important in the development of supersensitivity after denervation of at least some smooth muscles. Denervation of the guinea-pig vas deferens results in decreased activity of the Na^+,K^+-ATPase. Because the ATPase is electrogenic, the membrane potential is depolarized by 5–10 mV. The resting membrane potential is thus closer to threshold, causing the tissue to become supersensitive to applied agonists (Gerthoffer et al., 1979). Denervation of the cholinergic sympathetic nerves innervating the sweat glands results in loss of physiological responsiveness to muscarinic agonists; it is not yet known if this is because of a decrease in mAChR number or a physiological uncoupling of the receptor (Kennedy et al., 1984).

Benson et al. (1989) have demonstrated, using voltage-clamp studies in hippocampal slices, that there was an increased sensitivity to muscarinic agonists following cholinergic denervation. Both the M_1-receptor mediated "leak current" and the M_2-receptor mediated calcium-dependent potassium channel exhibited supersensitivity. Because these experiments were carried out by intracellular recording from single cells under voltage-clamp conditions in the presence of tetrodotoxin using carbachol, the results cannot be the result of effects on presynaptic elements or acetylcholinesterase. Further work is required to demonstrate if these effects are the result of changes in the numbers of receptors, G-proteins, and/or channels.

Multiple effects of innervation of mAChR in the invertebrate central nervous system have been observed. Deafferentation of the cricket abdominal ganglion leads initially to a 30% increase in receptor number. Chronic denervation causes a 40% decrease in mAChR number. This loss of receptor sites appeared to be reversible because mAChR number increased towards control values upon partial reinnervation of the ganglion (Meyer et al., 1986). Because of the accessibility of invertebrates to experimental manipulation, this may provide an attractive system for the study of the regulatory mechanisms that control muscarinic receptors during development and synapse formation in the nervous system.

An unusual example of the regulation of mAChR by innervation is seen in the expansor secondarium, a muscle in the chick wing that receives only adrenergic innervation. Immediately after hatching, both muscarinic and adrenegic agonists cause muscle contraction. The physiological sensitivity to ACh, but not norepinephrine decreases over the next several weeks, because of a reduction in the number of mAChR present in the muscle (Kuromi and Hasegawa, 1975; Bennett et al., 1982; Rush et al., 1982). The developmental decreases in both cholinergic sensitivity and mAChR number could be prevented by denervation of the muscle (Kuromi and Hagihare, 1976; Rush et al., 1982). Organ culture experiments indicate that muscle contraction contributes to the effect of innervation on mAChR number (Gonoi, 1980), in a manner that may be analagous to the regulation of nicotinic AChR synthesis in skeletal muscle (Reiness and Hall, 1977).

Several studies have appeared on the possible role of trophic factors on the regulation of muscarinic acetylcholine receptor number. Treatment of the pheocromocytoma cell line PC12 with nerve growth factor (NGF) for several days leads to a four-fivefold increase in the specific activity of mucarinic receptor binding sites (Jumblatt and Tischler, 1982) and a 15-fold increase in the number of receptors per cell (Pozzan et al., 1986). There is also increased responsiveness following NGF treatment for both mAChR-mediated increases in cytosolic calcium levels and stimulation of phosphatidylinositol turnover (Pozzan et al., 1986; Vincentini et al., 1986). There is evidence that NGF also plays a role in the regulation of mAChR in sympathetic neurons both in vivo and in culture. Addition of NGF to sympathetic ganglion explants prevents the 50% decrease in mAChR number that normally occurs following dissection and culturing of the ganglia. Treatment of animals with 6-hydroxydopamine or axotomy decreases mAChR number in the ganglia by nearly 70%; these decreases are also partially prevented by coadministration of NGF (Dombrowski et al., 1983).

Treatment of the lateral septal region of the rat brain with thyrotropin-releasing hormone was reported to increase muscarinic receptor number by 44%, an effect that was postulated to be reponsible for the potentiation by TRH of the vassopressor effect of central administration of acetylcholine (Pirola et al., 1983). Because a 30-min treatment with TRH was sufficient to increase mAChR number, it seems unlikely that this increase could be the result of either an increased rate of synthesis or decreased rate of degradation of the receptor.

The level of thyroid hormones can regulate cardiac mAChR number and function. Hypothyroidism increases and hyperthyroidism decreases the number of cardiac muscarinic receptors (Sharma and Banjeree, 1977; Robberecht et al., 1982). The effects of varying thyroid hormone levels on mAChR number are in the opposite direction from that on the number of cardiac β-adrenergic receptors. These changes in mAChR number are accompanied by corresponding shifts in the mAChR-mediated negative inotropic response, with higher concentrations of agonist required in hyperthyroidism and lower concentrations required in hypothyroidism to evoke a physiological response (Ishac and Pennefather, 1983).

Treatment of ovariectomized rats with estrogen resulted in a 26% increase in mAChR number in the medial basal hypothalamus, and a 40% decrease in the medial preoptic area. No change in receptor number was seen in these areas in male rats, suggesting a sexual dimorphism in the regulation of muscarinic receptors by estrogen (Dohanich et al., 1982). Direct effects of steroid hormones on the binding properties of the mAChR have also been reported (Sokolovsky et al., 1981).

Chronic treatment with lithium causes increased numbers

of mAChR both in vivo and in cell culture. Administration of lithium to rats increased pilocarpine-induced catalepsy and pilocarpine-induced hypothermia. These increased responses to mAChR agonists were associated with increased numbers of mAChR binding sites in the striatum (Lerer and Stanley, 1985). Chronic treatment of N1E-115 neuroblastoma cells with lithium also increased mAChR number. Lithium treatment also attenuated agonists-mediated but not phorbol-ester-mediated decreases in receptor number (Liles and Nathanson, 1988). Although the relationship of these effects to the therapeutic actions of lithium is unclear, the alteration of metabolism of the phosphatidyl inositol pathway may result in the effects of lithium on the number and regulation of the mAChR.

5. Regulation by Electrical Activity

Electrical activity and membrane depolarization of excitable cells can regulate the expression of proteins involved in neurotransmission. For example, increased electrical activity decreases the synthesis of nicotinic acetylcholine receptors in skeletal muscle (Reiness and Hall, 1977). Luqmani et al. (1979) demonstrated that depolarization of brain synaptosomes by either electrical stimulation or with veratrine decrease mAChR number by up to 50%. This decrease was blocked by the sodium channel blocker tetrodotroxin. Depolarization with high potassium medium did not decrease mAChR number, suggesting that sodium entry might be required. Milligan and Strange (1984) studied the effects of sodium channel activators on mAChR in N1E-115 cells. They reported that incubation of the cells with these drugs decreased the amount of [^{3}H]NMS bound to the cells not by decreasing the actual number of mAChR, but merely by competing with the radioligand for binding to the receptor. No effect on receptor number was found after incubations with these drugs for up to 6 h, so that it was concluded that there was no effect of electrical activity on mAChR number in these cells. Further work has shown that depolarization of N1E-115 cells with either veratridine or high potassium medium caused a 50–200% increase in both total and cell surface mAChR number over a 12-h period, with a 6–8 h lag before receptor number began to increase. Inhibitors of either protein synthesis or protein glycosylation did not block the increase in mAChR number, indicating that the increase in mAChR number did not require *de novo* receptor synthesis and thus may be due not to an increase in the rate of synthesis, but rather a decrease in the rate of degradation of the receptor. Determination of the rate of dissappearance of the receptor following treatment of cells with cyclohexemide indicated that depolarization increases mAChR number by decreasing the rate of degradation of the receptor. The effects of membrane depolarization can be mimicked by treatment of cells with either organic or inorganic calcium channel blockers, and can be prevented by treatment with the calcium ionophore A23187. It has been suggested that depolarization increases mAChR number by inactivating voltage-sensitive calcium channels, decreasing calcium flux, and thereby decreasing the rate of degradation of the receptor (Liles and Nathanson, 1987).

Decreased numbers of mAChR in the central nervous system have been reported following the induction of kindling, a model for epilepsy in which the repeated administration of initially subconvulsive electrical stimuli eventually induces seizure activity. Decreased numbers of mAChR in the hippocampus were found following seizures generated by amygdala kindling, electroconvulsive shock, entorhinal kindling, and entorhinal lesion (Dashieff et al., 1982). The loss of mAChR has been localized by autoradiography to occur on granule cells in the dendate gyrus (Savage et al., 1983). Because this decrease in mAChR number is not blocked by lesion of the cholinergic inputs to the hippocampus but is blocked by lesion of noncholinergic afferent fibers, it has been suggested that this is an agonist-independent phenomenon secondary to the increased electrical activity that occurs during seizures (Savage et al., 1985). Decreased numbers of mAChR in the central nervous system in the mouse mutant *totterer*, an animal model of epilepsy, have been observed with a developmental time course suggesting that the changes in receptor number are also secondary to increased neuronal activity. (Liles et al., 1986b). Abdul-Ghani et al. (1981) demonstrated that central administration of the sodium channel activator tityus toxin caused a tetrodotoxin-sensitive decrease in receptor number in the rat sensory-motor cortex. Electrical stimulation also decreased mAChR number in this brain region. Thus, direct electrical or pharmacological depolarization can decrease mAChR number in the central nervous system.

6. Possible Roles of Phosphorylation in Receptor Regulation

There have been several reports suggesting that protein phosphorylation may regulate mAChR number and function. Burgoyne (1980, 1981) reported that "phosphorylating conditions" (i.e., the presence of ATP) decreased [^{3}H]QNB binding to rat cortex membranes by 20–40%. The mAChR binding sites that remained had altered agonist binding and guanine nucleotide sensitivity (Burgoyne, 1983). Because the addition of both cAMP and calmodulin was required for this loss of binding sites, it was suggested that both cAMP-dependent and calmodulin-dependent protein kinases were involved in receptor down-regulation. The relationship of the loss of binding sites observed in these studies to the agonist-induced decreases in receptor number described above is not clear. Burgoyne and Pearce (1981) reported that a polypeptide with a similar electrophoretic mobility as the mAChR was phosphorylated in crude membrane fractions by incubation of neurons with muscarinic agonists. Because of the very low density of the mAChR in these cells, however, it seems unlikely that phosphorylation of the receptor could be detected by SDS gel electrophoresis of crude membrane fractions.

The calcium- and phospholipid-dependent enzyme protein kinase C may regulate mAChR number and function. Tumor-promoting phorbol esters, which can activate protein kinase C in intact cells, act in a synergistic fashion with the calcium ionophore A23187 to decrease mAChR number in N1E-115 cells. The ionophore and phorbol ester cause a rapid initial internalization and subsequent degradation of the mAChR with the same time course as agonist treatment (Liles et al., 1986a). Addition of muscarinic agonists to the cells rapidly activates protein kinase C before agonist-mediated receptor internalization occurs; the kinase remains activated during the subsequent degradation of the receptor. Because of analogous results seen in other systems (Leeb-Lundberg et al., 1985), these results suggest that protein kinase C-mediated phosphorylation of the mAChR may be involved in agonist-induced receptor loss in these cells. Confirmation of this hypothesis would require the demonstration that addition of phorbol esters and agonists cause the phosphorylation of the mAChR with similar time course, concentration dependence, and pharmacological specificity for different esters and agonists as the internalization and degradation of the receptor mediated by these drugs in N1E-115 cells.

The β-adrenergic receptor is specifically phosphorylated by a novel protein kinase only when the receptor is occupied by agonist. It is believed that this may represent a mechanism for the regulation of β-adrenergic receptor function analogous to the regulation of rhodopsin function by rhodopsin kinase, thus playing a role in desensitization and light adaptation, respectively (Benovic et al., 1986a,b). Because of the many similarities between signal transduction in the β-adrenergic receptor, transducin, and muscarinic receptor systems, it is possible that a similar receptor kinase may regulate the function of the muscarinic receptor.

Kwatra and Hosey (1986) have obtained direct evidence for phosphorylation of the muscarinic receptor *in situ*. They found that, after incubation of chick hearts in medium containing ^{32}P followed by affinity purification of the receptor, there was 0.2–0.4 mol of phosphorus/mol of mAChR. After incubation of the hearts with agonist for 30 min, there was 2.5–5.0 mol of phosphorus/mol of receptor. Kwatra et al. (1987) subsequently showed that phosphorylation was detectable within 1 min of carbachol treatment and reached a maximum by 4–5 min. Carbachol treatment caused a decrease in high affinity agonist binding to the mAChR with a similar time course and concentration dependence as the phosphorylation of the receptor. These results demonstrate that phosphorylation of the muscarinic receptor can be stimulated in vivo by the presence of agonist over a time scale consistent with its possible role in receptor desensitization downregulation. Agonist pretreatment also caused a threefold increase in the concentration of agonist required to elicit a negative inotropic effect. Receptor phosphorylation was not affected by activation of protein kinase C, by elevation of cyclic nucleotides, or by calmodulin antagonists. These results support the possibility that a specific receptor kinase may be involved in mAChR phosphorylation and desensitization.

In other cell types, activation of protein kinase C may regulate receptor function without directly acting on the mAChR itself. The tumor-promoting phorbol esters can block mAChR-stimulation of phosphoinositide turnover and calcium mobilization in hippocampal slices (Labarca et al., 1984), pheochromocytoma cells (Vincentini et al., 1985), adrenal medullary cells (Misbahuddin et al., 1985), and astrocytoma cells (Orellana et al., 1985).

In contrast to the loss of mAChR seen after phorbol ester treatment of N1E-115 cells, the blockade of mAChR function in these studies was not accompanied by significant changes in either cell surface or total mAChR number. Treatment of lymphoma cells and platelets with phorbol esters appears to causes the phosphorylation by protein kinase C of the α subunit of G_i, thereby inactivating the G-protein (Katada et al., 1985). The inhibition of mAChR function in these cells and tissues may be the results of a similar phosphorylation and inactivation of the coupling proteins involved in these responses.

Limas and Limas (1985) reported that incubation of cardiac myocytes with carbachol leads to a rapid (within 20 min) decrease in the number of β-adrenergic receptor binding sites on the cell surface, with a concomitant redistribution of β-adrenergic receptors from the plasma membrane to a form that sediments with a light vesicle fraction. It was suggested that the pharmacological specificity of the receptor mediating this effect was the M_1 subtype. Because M_1 receptors are frequently (although not always) linked to phosphatidylinositol hydrolysis, it was suggested that this apparent heterologous desensitization was the result of activation of protein kinase C by the mAChR, and subsequent phosphorylation and desensitization of the β-adrenergic receptor.

Molecular cloning experiments have demonstrated that there are multiple forms of protein kinase C arising from the expression of multiple genes in a tissue-specific manner (Coussens et al., 1986). It is possible that these forms differ in their subcellular distribution, substrate specificity, ability to be activated by different receptor systems or phorbol esters, and so on. This necessarily complicates the interpretation of the effects of phorbol ester treatment on mAChR number and function, and provides a potential explanation of the different effects of phorbol ester treatment observed in different tissues and cells. It may also explain why the effects of phorbol ester treatment would not mimic the physiological effects owing to receptor-mediated generation of diacylglycerol, which could selectively activate only a subset of the protein kinase C subtypes in a cell. Furthermore, studies using expression of cloned mAChR genes indicate that activation of protein kinase C has selective effects on specific mAChR subtypes: although both the M_1 and M_2 receptors undergo internalization following agonist treatment, the M_1, but not the M_2, receptor is internalized following phorbol ester treatment (Scherer et al., 1988).

7. Developmental Changes In mAChR Numbers and Function

The study of the changes that occur in the number, function, and structure of a neurotransmitter receptor during development can provide interesting information both on the factors involved in the regulation of the macromolecules necessary for synaptic activity as well as the role of various components of the receptor system in the mediation of physiological responses. Several examples of developmental regulation of mAChR have been reported.

7.1. Developmental Changes in the mAChR in the Central Nervous System

Several studies have examined the appearance of mAChR in the central nervous system during development. Muscarinic receptors in the mouse cerebellum reach their maximum density approximately 3 wk after birth (East and Dutton, 1980). Yavin and Harel (1979) reported that the density of mAChR binding sites in a number of discrete areas of the rabbit brain was elevated in newborn compared to prenatal tissues, and was further elevated in adult tissues, except for cerebellum, where mAChR density declined after birth. Kuhar et al. (1980) used the binding of [^{3}H]NMS and [^{3}H]PrBCM to measure total mAChR number in the rat cerebral cortex, diencephalon, and pons-medulla, and the binding of [^{3}H]oxotremorine-*M* to quantitate high affinity agonist sites. The number of total mAChR binding sites in the medulla approximately doubled between birth and 10 d of age, when the adult level was attained, whereas receptor number in the cortex and diencephalon at birth was considerably less than the adult level and had not reached the adult number by day 20. A subsequent autoradiographic study by Hohmann et al. (1985) using [^{3}H]PrBCM demonstrated that the adult density and distribution of mAChR was reached by one month after birth. High affinity agonist sites did not appear until 6–7 d after antagonist binding sites (Kuhar et al., 1980), raising the possibility that a coupling protein may appear after the mAChR binding site. It has not yet been determined if there is in fact a lag in the onset of physiological responsiveness mediated by the muscarinic receptor during development. Evans et al. (1985)

reported that the total number of mAChR binding sites in the cortex increased during development in parallel with the number of M_1-subtype sites. Because molecular biological studies indicate that the M_1 and M_2 muscarinic receptors are different proteins encoded by different genes (Kubo et al., 1986a,b), it will be interesting to determine if the different receptor subtypes are differentially regulated in tissues during development.

Changes in the density, distribution, and structure of muscarinic receptors have been reported to occur during development of the chick embryonic retina. The density of mAChR binding sites goes up 30-fold between embryonic day 6 and 14, with a relatively slight decrease occurring after hatching. Autoradiography using [^{3}H]QNB demonstrated two bands of mAChR binding sites in the inner plexiform layer in embryonic retina and three bands in the inner plexiform layer in adult retina (Sugiyama et al., 1977). A change in the apparent isoelectric point of the solubilized receptor was reported to occur during development. Muscarinic receptors from retina of 8–13 d embryos exhibited a pI that was 0.1–0.25 pH U more acidic than that of post-hatched retinal mAChR, with two forms present at intermediate ages. Muscarinic receptors from both embryonic and hatched chick retina exhibited equivalent responsiveness in mediating stimulation of phosphatidylinositol turnover (Large et al., 1985b). The apparent molecular weight of the receptor also changed during development (Large et al., 1985b). Embryonic receptors labeled with [^{3}H]PrBCM migrated on SDS gels mainly with a mol w of 86,000, whereas hatched receptors migrated predominantly with a mol w of 72,000. The two forms exhibited different stabilities on cultured retinal cells, with $t_{1/2}$ of 4.6 and 18.6 h for the larger and smaller forms, respectively. Because synapse formation in the retina occurs during this period of development, it was suggested by Large et al. (1985a,b) that synaptogenesis may regulate the structure of the mAChR. It is not known whether the two forms of receptor are related by proteolysis or some other covalent modification, or represent different primary translation products. Because of the extensive posttranslational processing (i.e., glycosylation) that the mAChR undergoes and because the primary translation products of both the M_1 and M_2 receptors are considerably smaller than the mature receptors (Peterson et al., Kubo et al., 1986a,b), electrophoretic differences may be the result of differences in posttranslational processing.

Large et al. (1986) have also reported changes in the number and isoelectric point of mAChR in the rat olfactory bulb with development. The number of binding sites increased approximately eightfold from birth to adulthood. While no change in the apparent mol w of [^{3}H]PrBCM-labeled receptors was observed, the isoelectric point of the receptors decreased during development. Because the level of mAChR and choline acetyltransferase increased in parallel during development, it was suggested that these changes are also a consequence of synapse formation.

7.2. Developmental Changes in and Regulation of Cardiac mAChR

Muscarinic receptors in the chick heart undergo several physiological changes during the course of embryonic development. Atria from 3–4 d embryos exhibit a smaller negative chronotropic response to muscarinic agonists than atria from older embryos, even though the density of mAChR binding sites is similar in both younger and more responsive older hearts (Pappano and Skowronek, 1974; Galper et al., 1977; Renaud et al., 1980). Muscarinic receptors at both 3–4 d and older ages exhibit similar sensitivities for inhibition of adenylate cyclase and stimulation of phosphoinositide turnover (Halvorsen and Nathanson, 1984; Orellana and Brown 1985), suggesting that the mAChR itself is not altered at 3–4 d. Detailed examination of the function of the G-proteins in atrial membranes of different ages suggested that the interaction of the mAChR with G_i/G_o was impaired at 4 d of development relative to older ages. For example, at 4 d of development, much higher concentrations of guanine nucleotides were required in order to demonstrate guanine nucleotide regulation of agonist binding to the mAChR, and the magnitude of the effect of nucleotide on binding was also greatly reduced. In addition, higher concentrations of guanine nucleotides were required to elicit receptor-independent inhibition of adenylate cyclase activity in younger atria. One- and two-dimensional gel electrophoretic analysis of the polypeptides labeled by IAP in atrial membranes of different ages demonstrated that the electrophoretic mobility of the GTP-regulatory proteins associated with the receptor also changes during this time (Halvorsen and Nathanson, 1984). These results indicate that there are both functional and physical changes in these GTP-binding proteins that occur with the same time course during embryonic development as the onset of the mAChR-mediated negative chronotropic response, and suggest that these changes may contribute to the onset of physiological responsiveness. Because changes in cyclic nucleotides are not involved in coupling mAChR to potassium channels in the heart, these results also suggested that the GTP-binding proteins may directly link muscarinic receptors to potassium channels in the heart; subsequent biochemical and electrophysiological experiments have confirmed this hypothesis (Pfaffinger et al., 1985; Breitweiser and Szabo, 1985; Martin et al., 1985; Endoh et al., 1985; Sorota et al., 1985; Kurachi et al., 1986).

Several other properties of the mAChR change during development of the embryonic chick heart. The negative chronotropic response to muscarinic agonists rapidly desensitizes in atria younger than embryonic day 12, but does not desensitize at older ages (Pappano and Skowronek, 1974). In the atrium, muscarinic receptors decrease contractility and cAMP levels in the absence of β-adrenergic stimulation both before and after hatching. Muscarinic receptors in the ventricle inhibit β-adrenergic stimulation of cardiac contractility both before and after hatching, but exert direct effects on contractility in the absence of β-adrenergic stimulation only after hatching (Biegon and Pappano, 1980). It was suggested that differences in the G-proteins between atria and ventricles could be responsible for these developmentally regulated differences in physiological coupling. However, detailed examination of the functional and biochemical properties of the G-proteins in 8-d embryonic atria and ventricles indicate that differences in the physiological actions of muscarinic receptors are not the result of differences in G_i/G_o between the two tissues (Martin et al., 1987). The properties of the mAChR in the (hatched) chick atrium and ventricle are also similar (Sorota et al., 1986), indicating that differences in the physiological responses in the two tissues are also not the result of differences in the receptor itself.

Kirby and Aronstam (1983) reported that there is a significant increase in mAChR number in chick atria, but not ventricle between embryonic days 11–14, which is at the time of the onset of functional cholinergic innervation of the chick heart. This atrial-specific developmental increase in mAChR number could be blocked by atropine treatment proir to the onset of innervation, but atropine had no effect if administered after the onset of innervation. It was suggested that there may be an important developmental effect of innervation to regulate mAChR number that is blocked by atropine treatment. Using antibodies specific for $G_{o\alpha}$ and G_{β}, Luetje et al. (1987) demonstrated that the amount of these subunits of the GTP-regulatory proteins also increase in atria but not in ventricles during this period of embryonic development, raising the possibility that both the mAChR and its coupling proteins may be regulated in a coordinate fashion during development and innervation of the heart.

A novel type of experiment suggesting a role for cholinergic innervation in the regulation of cardiac mAChR number was provided by the work of Kirby et al. (1985). They used microsurgery to ablate part of the chick embryonic neural crest early in development, resulting in a disruption of parasympathetic innervation, as evidenced by a decrease in both the amount of ACh and the number of ganglion cells present in the heart. This treatment also decreased the density of muscarinic receptors present at 15 d of development.

The number of mAChR in the embryonic chick ventricle has been reported to increase around the time of hatching (Hosey et al., 1985; Sullivan et al., 1985). Liang et al. (1986) measured the ability of chick cardiac mAChR to inhibit adenylate cyclase in membranes prepared from whole heart at different ages of development, and attempted to correlate this with changes in the levels of G-proteins and muscarinic receptors during development. These authors concluded that increases in $G_{o\alpha}$ were responsible for increases in biochemical responsiveness early in development and increases in mAChR number were responsible for increases in responsiveness later in development. Because differential regulation of both receptors and G-proteins in atria compared to ventricles occurs during development, however, important information on both regulatory and coupling mechanisms could have been missed in this study using whole undissected hearts.

Siegel and Fischbach (1984) compared the biochemical and physiological properties of mAChR on isolated cultured embryonic chick atrial and ventricular cells. Binding and autoradiographic studies with [^{3}H]NMS and [^{3}H]QNB showed that there were equivalent numbers of receptor binding sites on atrial and ventricular cells. Autoradiography combined with anti-myosin immunocytochemistry demonstrated that the density of muscarinic receptors on myocytes was 10-fold higher than on cardiac fibroblasts. As in the intact tissue, muscarinic agonist caused hyperpolarization of atrial cells but not ventricular cells, indicating that the mAChR was not coupled to potassium channels in the ventricle. Most interestingly, the addition of a saline extract from chick brain to the ventricular cells caused a dramatic increase in mAChR-mediated hyperpolarization without affecting the number of mAChR binding sites, suggesting the presence of one or more components in the extract, which can regulate muscarinic responsiveness at a step distal to the receptor itself.

Several studies have shown that the physiological responsiveness of mAChR in cultured cardiac cells can be regulated by factors in the serum used in the growth medium. Galper et al. (1984) found that cardiac cultures from 3½-d chick embryos possessed mAChR that were relatively insensitive in stimulating increases in potassium permeability, but that growth in certain lots of horse serum induced a large increase in this response. Similarily, growth of cells in different lots of fetal calf serum can cause the concentration of agonist required for both stimulation of $^{86}Rb^{+}$ efflux and inhibition of adenylate cyclase to vary over 2–3 orders of magnitude (Hunter and Nathanson, 1986). Renaud et al. (1982) reported that chick embryonic ventricular cells grown in untreated fetal calf serum did not exhibit a mAChR-mediated negative chronotropic response, whereas cells grown in lipoprotein-depleted serum did respond. Cells grown in lipoprotein-depleted serum exhibited a decreased number of receptors even though the rate of degradation appeared to be greatly reduced, suggesting a large decrease in the rate of synthesis of the receptor compared to cells grown in normal serum. The authors have suggested that these effects of lipoprotein-depleted serum resulted from an increased cholesterol content of the plasma membrane.

A regional-specific effect of drug-induced diabetes has been reported on cardiac muscarinic receptors. Treatment of rats with streptozotocin decreased mAChR number in atria by 30%, but had no effect on receptor number in ventricle. Interestingly, there was an *increased* sensitivity to the mAChR-mediated negative chronotropic response of isolated atria, suggesting that the remaining receptors were much more efficiently coupled to the transduction system mediating this effect (Carrier et al., 1984).

Attempts to determine the effect of sympathectomy on mAChR number in the heart have not yielded consistent findings. Sharma and Banerjee (1978) reported that treatment of rats with 6-hydroxydopamine decreased cardiac mAChR number by

30%. Because treatment with reserpine did not decrease receptor number, they concluded that the decrease in mAChR number represented the loss of presynaptic receptors on sympathetic nerve terminals. Nomura et al. (1979) found a decrease both in the number of mAChR binding sites and the mAChR-mediated negative inotropic response following 6-hydroxydopamine treatment. These authors concluded that a significant proportion of the receptor binding sites that were lost were located on the cardiac membrane itself and not on nerve terminals. In contrast, increases in mAChR number (Yamada et al., 1980) and physiological reponsiveness (Ishii et al., 1985) have also been reported following chemical or immunosympathectomy. Waelbroeck et al. (1983) have reported that treatment of spontaneously hypertensive (but not normotensive) rats with 6-hydroxydopamine selectively decreased the number of high affinity agonist sites without changing the total number of cardiac mAChR binding sites. The reasons for the discrepancies in the findings reported by different laboratories are not apparent.

8. Conclusions

Although numerous studies on the regulation of the number and function of muscarinic receptors have appeared, there has been relatively little detailed information identifying the underlying molecular basis for many of the observed phenomena. There has been an explosive growth in the last several years in the knowledge of and tools available for continued study of the muscarinic receptor. Imaging techniques that allow the visualization of the number and distribution of receptors in living animals (Eckelman et al., 1984; Syrota et al., 1985) will permit studies on the regulation of mAChR in a variety of normal and pathological conditions. The ability to purify the receptor from relatively small amount of tissues and to reconstitute the mAChR with effector proteins (Peterson et al., 1984; Haga et al., 1985; Kwatra and Hosey, 1986) will allow the testing of molecular models for the basis of functional desensitization of the receptor as well as the mechanism of receptor action. The development of monoclonal antibodies (Lieber et al., 1984; Luetje et al., 1987a), that recognize the muscarinic receptor makes feasible detailed cell biological studies on the synthesis, and processing of the receptor, as well as localization of the receptor at the electron microscopic level using immunocytochemical techniques. Finally, the cloning and sequencing of cDNAs encoding the receptors (Kubo et al., 1986a,b) makes available the powerful tools of recombinant DNA technology to the field of muscarinic receptor research.

Acknowledgments

Research in the author's laboratory was supported by a Grant-in-Aid from the American Heart Association, and by Grant HL30639 from the National Institutes of Health, and by the United States Army Research Office. The author is an Established Investigator of the American Heart Association.

References

Abdul-Ghani, A. S., Boyar, M. M., Coutinho-Netto, J., Bradford, H. F., Bernie, C. P., Hulme, E. D., and Birdsall, N. J. M. (1981) Effect of Tityus toxin and sensory stimulation on muscarinic cholinergic receptors *in vivo*. *Biochem. Pharmacol.* **30**, 2713–2714.

Ben-Barak, J. and Dudai, Y. (1980) Scopolamine induces an increase in muscarinic receptor level in rat hippocampus. *Brain Res.* **193**, 309–313.

Bennett, T., Lot, T. Y., and Strange, P. G. (1982) The effects of noradrenergic denervation on muscarinic receptors of smooth muscle. *Brit. J. Pharmacol.* **76**, 177–183.

Benovic, J. L., Mayor, F. Jr., Somers, R. L., Caron, M. G., and Lefowitz, R. J. (1986a) Light-dependent phosphorylation of rhodopsin by β-adrenergic receptor kinase. *Nature* **321**, 869–875.

Benovic, J. L., Strasser, R. H., Caron, M. G., and Lefkowitz, R. J. (1986b) β-adrenergic receptor kinase: Identification of a novel protein kinase that phosphorylates the agonist-occupied form of the receptor. *Proc. Natl. Acad. Sci. USA* **83**, 2797–2801.

Benson, D. M., Blitzer, R. D., Haroutunian, V., and Landau, E. M. (1989) Functional muscarinic supersensitivity in denervated rat hippocampus. *Brain Res.* **478**, 399–402.

Betz, H. (1982) Differential down-regulation by carbamylcholine of putative nicotinic and muscarinic acetylcholine receptor sites in chick retina cultures. *Neurosci. Lett.* **28**, 265–268.

Biegon, R. L. and Pappani, A. J. (1980) Dual mechanism for inhibition of calcium-dependent action potentials by acetylcholine in avian ventricular muscle: Relationship to cyclic AMP. *Cir. Res.* **46**, 353–362.

Breitwieser, G. E. and Szabo, G. (1985) Uncoupling of cardiac muscarinic and β-adrenergic receptors from ion channels by a guanine nucleotide analogue. *Nature* **317**, 538–540.

Burgoyne, R. D. (1980) A possible role of synaptic-membrane protein phosphorylation in the regulation of muscarinic acetylcholine receptors. *FEBS Lett.* **122**, 288–292.

Burgoyne, R. D. (1981) The loss of muscarinic acetylcholine receptors in synaptic membranes under phosphorylating conditions is dependent on calmodulin. *FEBS Lett.* **127**, 144–148.

Burgoyne, R. D. (1983) Regulation of the muscarinic acetylcholine receptor: Effects of phosphorylating conditions on agonist and antagonist binding. *J. Neurochem.* **40**, 324–331.

Burgoyne, R. D. and Pearce, B. (1981) Muscarinic acetylcholine receptor regulation and protein phosphorylation in primary cultures of rat cerebellum. *Dev. Brain Res.* **2**, 55–63.

Burt, D. (1978) Muscarinic receptor binding in rat sympathetic ganglia is unaffected by denervation. *Brain Res.* **143**, 573–579.

Carmeliet, E. and Mubagwa, K. (1986) Desensitization of the acetylcholine-induced increase of potassium conductance in rabbit cardiac Purkinje fibres. *J. Physiol.* **371**, 239–255.

Carrier, G. O., Edwards, A. D., and Aronstam, R. S. (1984) Cholinergic supersensitivity and decreased number of muscarinic receptors in atria from short-term diabetic rats. *J. Mol. Cell Cardiol.* **16**, 963–965.

Champion, S. and Mauchamp, J. (1982) Muscarinic cholinergic receptors on cultured thyroid cells. II. Carbachol-induced desensitization. *Mol. Pharmacol.* **21**, 66–72.

Cioffi, C. L. and El-Fakahany, E. E. (1986) Short-term desensitization of muscarinic cholinergic receptors in mouse neuroblastoma cells: Selective loss of agonist low-affinity and pirenzepine high-affinity sites. *J. Pharmacol. Exp. Therapeu.* **238**, 916–923.

Costa, L. G. and Murphy, S. D. (1985) Characterization of muscarinic cholinergic receptors in the submandibular gland of the rat. *J. Autonomic Nervous Syst.* **13**, 287–301.

Coussens, L., Parker, P. J., Rhee, L., Yang-Feng, Y., Chen, E., Waterfield, M. D., Francke, U., and Ullrich, A. (1986) Multiple, distinct forms of bovine and human protein kinase C suggest diversity in cellular signaling pathways. *Science* **233**, 859–866.

Dasheiff, R. M., Savage, D. D., and McNamara, J. O. (1982) Seizures down-regulate muscarinic cholinergic receptors in hippocampal formation. *Brain Res.* **235**, 327–334.

Dawson, R. M. and Jarrott, B. (1981) Response of muscarinic cholinoceptors of guinea pig brain and ileum to chronic administration of carbamate or organophosphate cholinesterase inhibitors. *Biochem. Pharmacol.* **30**, 2365–2368.

de Belleroche, J., Gardnier, I. M., Hamilton, M. H., and Birdsall, N. J. M. (1985) Analysis of muscarinic receptor concentration and subtypes following lesion of rat substantia innominata. *Brain Res.* **340**, 201–209.

Dohanich, G. P., Witcher, J. A., Weaver, D. R., and Clemens, L. G. (1982) Alteration of muscarinic binding in specific brain areas following estrogen treatment. *Brain Res.* **241**, 347–350.

Dombrowski, A. M., Jerkins, A. A., and Kauffman, F. C. (1983) Muscarinic receptor binding and oxidative enzyme activities in the adult rat superior cervical ganglion: Effects of 6-hydroxydopamine and nerve growth factor. *J. Neurosci.* **3**, 1963–1970.

East, J. M. and Dutton, G. R. (1980) Muscarinic binding sites in developing normal and mutant mouse cerebellum. *J. Neurochem.* **34**, 657–661.

Eckelman, W. C., Reba, R. C., Rzeszotarski, W. J., Gibson, R. E., Hill, T., Holman, B. L., Budinger, T., Conklin, J. J., Eng, R., and Grissom, M. P. (1984) External imaging of cerebral muscarinic acetylcholine receptors. *Science* **223**, 291–293.

El-Fakahany, E. and Richelson, E. (1980a) Temperature dependence of muscarinic acetylcholine receptor activation, desensitization, and resensitization. *J. Neurochem.* **34**, 1288–1295.

El-Fakahany, E. and Richelson, E. (1980b) Involvement of calcium channels in short-term desensitization of muscarinic receptor-mediated cyclic GMP formation in mouse neuroblastoma cells. *Proc. Natl. Acad. Sci. USA* **77**, 6897–6901.

Endoh, M., Maruyama, M., and Iijima, T. (1985) Attenuation of muscarinic cholinergic inhibition by islet-activating protein in the heart. *Am. J. Physiol. (Heart Circ. Physiol.* **18)** **249**, H309–H320.

Evans, J. B., Watson, M., Yamamura, H. I., and Roeske, W. R. (1985) *J. Pharmacol. Exp. Ther.* **235**, 612–618.

Feigenbaum, P. and El-Fakahany, E. E. (1985) Regulation of muscarinic cholinergic receptor density in neuroblastoma cells by brief exposure to agonist: Possible involvement in desensitization of receptor function. *J. Pharmacol. Exp. Ther.* **233**, 134–140.

Galper, J. B. and Smith, T. W. (1980) Agonist and guanine nucleotide modulation of muscarinic cholinergic receptors in cultured heart cells. *J. Biol. Chem.* **255**, 9571–9579.

Galper, J. B., Dziekan, L. C., Miura, D. S., and Smith, T. W. (1982a) Agonist-induced changes in the modulation of K^+ permeability and beating rate by muscarinic agonists in cultured heart cells. *J. Gen. Physiol.* **80**, 231–256.

Galper, J. B., Dziekan, L. C., O'Hara, D. S., and Smith, T. W. (1982b) The biphasic response of muscarinic cholinergic receptors in cultured heart cells to agonists. *J. Biol. Chem.* **257**, 10344–10356.

Galper, J. B., Dziekan, L. C., and Smith, T. W. (1984) The development of physiological responsiveness to muscarinic agonists in chick heart cell cultures. Role of high affinity receptors and sensitivity to guanine nucleotides. *J. Biol. Chem.* **259**, 7382–7390.

Galper, J. B., Klein, W., and Catterall, W. A. (1977) Muscarinic acetylcholine receptors in developing chick heart. *J. Biol. Chem.* **252**, 8692–8699.

Gazit, H., Silman, I., and Dudai, Y. (1979) Administration of an organophosphate causes a decrease in muscarinic receptor levels in rat brain. *Brain Res.* **174**, 351–356.

Gerthoffer, W. T., Fedan, J. S., Westfall, D. P., Goto, K., and Fleming, W. W. (1979) Involvement of the sodium-potassium pump in the mechanism of postjuctional supersensitivity of the vas deferens of the guinea pig. *J. Pharmacol. Exper. Ther.* **210**, 27–36.

Gonoi, T. (1980) Changes in cholinergic and adrenergic responses of organ-cultured chick smooth muscle. *Eur. J. Pharmacol.* **68**, 287–293.

Haga, K., Haga, T., Ichiyama, A., Katada, T., Kurose, H., and Ui, M. (1985) Functional reconstitution of purified muscarinic receptors and inhibitory guanine nucleotide regulatory protein. *Nature* **316**, 731–733.

Halvorsen, S. E. and Nathanson, N. M. (1981) *In vivo* regulation of muscarinic acetylcholine receptor number and function in embryonic chick heart. *J. Biol. Chem.* **256**, 7941–7948.

Halvorsen, S. W. and Nathanson, N. M. (1984) Ontogenisis of physiological responsiveness and guanyl nucleotide sensitivity of cardiac muscarinic acetylcholine receptors during chick embryonic development. *Biochemistry* **23**, 5813–5821.

Harden, T. K., Petch, L. A., Traynelis, S. F., and Waldo, G. L. (1985) Agonist-induced alteration in the membrane form of muscarinic cholinergic receptors. *J. Biol. Chem.* **260**, 13060–13066.

Hata, F., Takeyasu, K., Morikawa, Y., Lai, R. T., Ishida, H., and Yoshida, H. (1980) Specific changes in the cholinergic system in guinea-pig vas deferens after denervation. *J. Pharmacol. Exp. Ther.* **215**, 716–722.

Heisler, S., Desjardins, D., and Nguyen, M. H. (1985) Muscarinic receptors in mouse pituitary tumor cells: Prolonged agonist pretreatment decreases receptor content and increases forskolin- and hormone-stimulated cyclic AMP synthesis and adrenocorticotropin secretion. *J. Pharmacol. Exp. Ther.* **232**, 232–238.

Hohmann, C. F., Pert, C., C., and Ebner, F. F. (1985) Developement of cholinergic markers in mouse forebrain. II. Muscarinic receptor binding in cortex. *Develop. Brain Res.* **23**, 243–253.

Hosey, M. M., McMahon, K. K., Danckers, A. M., O'Callahan, C. M., Wong, J., and Green, R. D. (1985) Differences in the properties of muscarinic cholinergic receptors in the developing chick myocardium. *J. Pharmacol. Exp. Ther.* **232**, 795–801.

Hunter, D. D. and Nathanson, N. M. (1984) Decreased physiological sensitivity mediated by newly synthesized muscarinic acetylcholine receptors in the embryonic chick heart. *Proc. Natl. Acad. Sci. USA* **81**, 3582–3586.

Hunter, D. D. and Nathanson, N. M. (1986) Biochemical and physical analyses of newly synthesized muscarinic acetylcholine receptors in cultured embryonic chicken cardiac cells. *J. Neurosci.* **6**, 3737–3748.

Ishac, E. J. N. and Pennefather, J. N. (1983) The effects of altered thyroid state upon responses mediated by atrial muscarinic receptors in the rat. *Br. J. Pharmacol.* **79**, 451–459.

Ishii, K., Ishii, N., Shigenobu, K., and Kasuya, Y. (1985) Acetylcholine supersensitivity in the rat heart produced by neonatal sympathectomy. *Can. J. Physiol. Pharmacol.* **63**, 898–899.

Jumblatt, J. E. and Tischler, A. S. (1982) Regulation of muscarinic ligand binding sites by nerve growth factor in PC12 phaeochromocytoma cells. *Nature* **297**, 151–154.

Kamiya, H., Rotter, A., and Jacobowitz, D. M. (1981) Muscarinic receptor binding following cholinergic nerve lesions of the cingulate cortex and hippocampus of the rat. *Brain Res.* **209**, 432–439.

Katada, T., Gilman, A. G., Watanabe, Y., Bauer, S., and Jakobs, K. H. (1985) Protein kinase C phosphorylates the inhibitory guanine-nucleotide-binding regulatory component and apparently suppresses its function in hormonal inhibition of adenylate cyclase. *Eur. J. Biochem.* **151**, 431–439.

Kennedy, W. R., Sakuta, M., and Quick, D. C. (1984) Rodent eccrine sweat glands: A case of multiple efferent innervation. *Neurosci.* **11**, 741–749.

Kirby, M. L. and Aronstam, R. S. (1983) Atropine-induced alterations of normal development of muscarinic receptors in the embryonic chick heart. *J. Mol. Cell. Cardiol.* **15**, 685–696.

Kirby, M. L., Aronstam, R. S., and Buccafusco, J. J. (1985) Changes in cholinergic parameters associated with failure of conotruncal septation in embryonic chick hearts after neural crest ablation. *Circul. Res.* **56**, 392–401.

Klein, W. L., Nathanson, N. M., and Nirenberg, M. (1979) Muscarinic acetylcholine receptor regulation by accelerated rate of receptor loss. *Biochem. Biophys. Commun.* **90**, 506–512.

Kubo, T., Fukuda, K., Mikami, A., Maeda, A., Takahashi, H., Mishina, M., Haga, T., Haga, K., Ichiyama, A., Kaangawa, K., Kojima, M., Matsuo, H., Hirose, T., and Numa, S. (1986a) Cloning, sequencing, and expression of complementary DNA encoding the muscarinic acetylcholine receptor. *Nature* **323**, 411–416.

Kubo, T., Maeda, A., Sugimoto, K., Akiba, I., Mikami, A., Takahashi, H., Haga, T., Haga, K., Ichiyama, A., Kangawa, K., Matsuo, H., Hirose, T., and Numa, S. (1986b) Primary sequence of porcine cardiac muscarinic acetylcholine receptor deduced from the cDNA sequence. *FEBS Lett.* **209**, 367–372.

Kubar, M. J., Birdsall, N. J. M., Burgen, A. S. V., and Hulme, E. C. (1980) Ontogeny of muscarinic receptors in rat brain. *Brain Res.* **184**, 375–383.

Kurachi, Y., Nakajima, T., and Sugimoto, T. (1986) On the mechanism of activation of muscarinic K+ channels by adenosine in isolated atrial cells: involvement of GTP-binding proteins. *Pflugers Arch.* **406**, 264–274.

Kuromi, H. and Hasegawa, S. (1975) Changes in acetylcholine and noradrenaline sensitivity of chick smooth muscle wholly innervated by sympathetic nerve during development. *Eur. J. Pharmacol.* **33**, 41–48.

Kuromi, H. and Hagihara, Y. (1976) Influence of sympathetic nerves on development of responsiveness of the chick smooth muscle to drugs. *Eur. J. Pharmacol.* **36**, 55–59.

Kwatra, M. M. and Hosey, M. M. (1986) Phosphorylation of the cardiac muscarinic receptor in intact chick heart and its regulation by a muscarinic receptor. *J. Biol. Chem.* **261**, 12429–12432.

Kwatra, M. M. Leung, E., Maan, A. C., McMahon, M. K., and Ptasienski, J., Green, R. D., and Hosey, M. M. (1987) Correlation of agonist-induced phosphorylation of chick heart muscarinic receptors with receptor desensitization. *J. Biol. Chem.* **262, 163**, 14–16321.

Labarca, R., Janowsky, A., Patel, J., and Paul, S. M. (1984) Phorbol esters inhibit agonist-induced [3H]inositol-1-phosphate accumulation in rat hippocampal slices. *Biochem. Biophys. Res. Commun.* **123**, 703–709.

Large, T. H., Cho, N. J., De Mello, F. G., and Klein, W. L. (1985a) Molecular alteration of a muscarinic acetylcholine receptor system during shnaptogenesis. *J. Biol. Chem.* **260**, 8873–8881.

Large, T. H., Cho, N. J., De Mello, F. G., and Klein, W. L. (1985b) Two molecular weight forms of muscarinic acetylcholine receptors in the avian central nervous system: switch in predominant form during differentiation of synapses. *Proc. Natl. Acad. Sci. USA* **82**, 8785–8789.

Large, T. H., Lambert, M. P., Gremillion, M. A., and Klein, W. L. (1986) Parallel postnatal development of choline acetyltransferase activity and muscarinic acetylcholine receptors in the rat olfactory bulb. *J. Neurochem.* **46**, 671–680.

Leeb-Lundberg, L. M. F., Cotecchia, S., Lomasney, J. W., DeBernadis, J. F., Lefkowitz, R. J., and Caron, M. C. (1985) Phorbol esters promote α_1-adrenergic receptor phosphorylation and uncoupling from inositol phospholipid metabolism. *Proc. Natl. Acad. Sci. USA* **81**, 4331–4334.

Lerer, B. and Stanley, M. (1985) Effect of chronic lithium or cholinergically mediated responses and [3H]QNB binding in rat brain. *Brain Res.* **344**, 211–219.

Liang, B. T., Hellmich, M. R., Neer, E. J., and Galper, J. B. (1986) Development of muscarinic cholinergic inhibition of adenylate cyclase in embryonic chick heart. Its relationship to changes in the inhibitory guanine nucleotide regulatory protein. *J. Biol. Chem.* **261**, 9011–9021.

Lieber, D., Harbon, S., Guillet, J. G., Andre, C., and Strosberg, A. D. (1984) Monoclonal antibodies to purified muscarinic receptor display agonist-like activity. *Proc. Natl. Acad. Sci. USA* **81**, 4331–4334.

Liles, W. C. and Nathanson, N. M. (1986) Regulation of neuronal muscarinic receptor number by protein glycosylation. *J. Neurochem.* **46**, 86–95.

Liles, W. C. and Nathanson, N. M. (1987) Regulation of muscarinic receptors in cultured neuronal cells by chronic membrane depolarization. *J. Neurosci.* **7**, 2556–2563.

Liles, W. C., Hunter, D. D., Neier, K. E., and Nathanson, N. M. (1986a) Activation of protein kinase C induces rapid internalization and subsequent degradation of muscarinic acetylcholine receptors in neuroblastoma cells. *J. Biol. Chem.* **261**, 5307–5313.

Liles, W. C. and Nathanson, N. M. (1988) Altered regulation of neuronal muscarinic receptor induced by chronic lithium treatment. *Brain Res.* **439**, 88–94.

Liles, W. C., Taylor, S., Finnell, R., Lai, H., and Nathanson, N. M. (1986b) Decreased muscarinic acetylcholine receptor number in the central nervous system of the tottering (*tg/tg*) mouse. *J. Neurochem.* **46**, 977–982.

Limas, C. J. and Limas, C. (1985) Carbachol induces desensitization of cardiac β-adrenergic receptors through muscarinic M_1 receptors. *Biochem. Biophys. Res. Commun.* **128**, 699–704.

Luetje, C. W., Brumwell, C., Norman, M. G., Peterson, G. L., Schimerlik, M. I., and Nathanson, N. M. (1987a) Isolation and characterization of monoclonal antibodies specific for the cardiac muscarinic receptor. *Biochemistry* **26**, 6892–6896.

Luetje, C. W., Gierschik, P., Milligan, G., Unson, C., Speigel, A., and Nathanson, N. M. (1987b) Tissue specific regulation of GTP-binding protein and muscarinic acetylcholine receptor levels during cardiac development. *Biochemistry* **26**, 4876–4884.

Luqmani, Y. A., Bradford, H. F., Birdsall, N. J. M., and Hulme, E. C. (1979) Depolarization-induced changes in muscarinic cholinergic receptors in synaptosomes. *Nature* **277**, 481–483.

Maloteaux, J. M., Gossuin, A., Pauwels, P. J., and Laduron, P. M. (1983) Short-term disappearance of muscarinic cell surface receptors in carbachol-induced desensitization. *FEBS Lett.* **156**, 103–107.

Marks, M. J., Artman, L. D., Patinkin, D. M., and Collins, A. C. (1981) Cholinergic adaptations to chronic oxotremorine infusion. *J. Pharmacol. Exp. Ther.* **218**, 337–343.

Martin, J. M.,, Hunter, D. D., and Nathanson, N. M. (1985) Islet activating protein inhibits physiological responses evoked by cardiac muscarinic acetylcholine receptors. Role of GTP-binding proteins in regulation of potassium permeability. *Biochem.* **23**, 5813–5821.

Martin, J. M., Subers, E. M., Halvorsen, S. H., Nathanson, N. M. (1987) Functional and physical properties of atrial and ventricular GTP-binding proteins: relationship to muscarinic acetylcholine receptor mediated responses. *J. Pharmacol. Exp. Ther.* **240**, 683–688.

Masters, S. B., Quinn, M. T., and Brown, J. H. (1985) Agonist-mediated desensitization of muscarinic receptor-mediated calcium efflux without concomitant desensitization of phosphoinositide hydrolysis. *Mol. Pharmacol.* **27**, 325–332.

Meyer, M. R., Gainer, M. W., and Nathanson, N. M. (1982) *In vivo* regulation of muscarinic cholinergic receptors in embryonic chick brain. *Mol. Pharmacol.* **21**, 280–286.

Meyer, M. R., Reddy, G. R., and Edwards, J. S. (1986) Metabolic changes in deafferented central neurons of an insect, *Acheta domesticus*. II. Effects on cholinergic binding sites and acetylcholinesterase. *J. Neurosci.* **6**, 1676–1684.

Milligan, G. and Strange, P. G. (1984) Muscarinic acetylcholine receptors in neuroblastoma cells: Lack of effect of *Veratrum* alkaloids on receptor number. *J. Neurochem.* **43**, 33–41.

Misbahuddin, M., Isosaki, M., Houchi, H., and Oka, M. (1985) Muscarinic receptor-mediated increase in cytoplasmic free C^{2+} in isolated bovine adrenal medullary cells. *FEBS Lett.* **190**, 25–28.

Nathanson, N. M., Holttum, J., Hunter, D. D., and Halvorsen, S. W. (1984) Partial agonist activity of oxotremorine at muscarinic acetylcholine receptors in the embryonic chick heart. *J. Pharmacol. Exp. Ther.* **229**, 455–458.

Nathanson, N. M., Klein, W. L., and Nirenberg, M. (1978) Regulation of adenylate cyclase activity mediated by muscarinic acetylcholine receptors. *Proc. Natl. Acad. Sci. USA* **75**, 1788–1791.

Nomura, Y., Kajiyama, H., and Segawa, T. (1979) Decrease in muscarinic cholinergic response of the rat heart following treatment with 6-hydroxydopamine. *Eur. J. Pharmacol.* **60**, 323–327.

Olianas, M. C., Onali, P., Schwartz, J. P., Neff, N. H., and Costa, E. (1984) The muscarinic receptor adenylate cyclase complex of rat striatum: Desensitization following chronic inhibition of acetylcholinesterase activity. *J. Neurochem.* **42**, 1439–1443.

Orellana, S. A. and Brown, J. H. (1985) Stimulation of phosphoinositide hydrolysis and inhibition of cyclic AMP formation by muscarinic agonists in developing chick heart. *Biochem. Pharmacol.* **34**, 1321–1324.

Orellana, S. A., Solski, P. A., and Brown, J. H. (1985) Phorbol ester inhibits phosphoinositide hydrolysis and calcium mobilization in cultured astrocytoma cells. *J. Biol. Chem.* **260**, 5236–5239.

Overstreet, D. H., Russell, R. W., Crocker, A. D., and Schiller, G. D. (1984) Selective breeding for differences in cholinergic function: Pre- and postsynaptic mechanisms involved in sensitivity to the anticholinesterase, DFP. *Brain Res.* **294**, 327–332.

Pappano, A. J. and Skowronek, C. A. (1974) Reactivity of chick embryo heart to cholinergic agonists during ontogenesis: Decline in desensitization at the onset of cholinergic transmission. *J. Pharmacol. Exp. Ther.* **191**, 109–118.

Peterson, G. L., Herron, G. S., Yamaki, M., Fullerton, D. S., and Schimerlik, M. I. (1984) Purification of the muscarinic acetylcholine receptor from porcine atria. *Proc. Natl. Acad. Sci. USA* **81**, 4993–4997.

Peterson, G. L., Rosenbaum, L. C., Broderick, D. J., and Schimerlik, M. I. (1986) Physical properties of the purified cardiac muscarinic acetylcholine receptor. *Biochem.* **25**, 3189–3202.

Pfaffinger, P. J., Martin, J. M., Hunter, D. D., Nathanson, N. M., and Hille, B. (1985) GTP-binding proteins couple cardiac muscarinic receptors to a K channel. *Nature* **317**, 536–538.

Pirola, C. J., Balda, M. S., Finkelman, S., and Nahmod, V. E. (1983) Thyrotropin-releasing hormone increases the number of muscarinic receptors in the lateral septal area of the rat brain. *Brain Res.* **273**, 387–391.

Pozzan, T., Di Virgilia, F., Vincentini, L. M., and Meldoleshi, J. (1986) Activation of muscarinic receptors in PC12 cells. Stimulation of Ca^{2+} influx and redistribution. *Biochem. J.* **234**, 547–553.

Rehavi, M., Ramot, O., Yavetz, B., and Sokolovsky, M. (1980) Amitripyline: Long-term treatment elevates α-adrenergic and muscarinic receptor binding in mouse brain. *Brain Res.* **194**, 443–453.

Reiness, C. G. and Hall, Z. W. (1977) Electrical stimulation of denervated muscles reduces incorporation of methionine into ACh receptor. *Nature* **231**, 296–301.

Renaud, J. F., Barhanin, J., Cavey, D., Fosset, M., and Lazdunski, M. (1980) Comparative properties of the *in ovo* and *in vitro* differentiation of the muscarinic cholinergic receptor in embryonic heart cells. *Dev. Biol.* **78**, 184–200.

Renaud, J. F., Scanu, A. M., Kazazoglou, A., Romey, G., and Lazdunski, M. (1982) Normal serum and lipoprotein-deficient serum give different expressions of excitability, corresponding to different stages of differentiation, in chicken cardiac cells in culture. *Proc. Natl. Acad. Sci. USA* **79**, 7768–7772.

Robberecht, P., Waelbroeck, M., Claeys, M., Huu, A. N., Chatelain, P., and Christophe, J. (1982) Rat cardiac muscarinic receptors. II. Influence of thyroid status and cardiac hypertrophy. *Mol. Pharmacol.* **21**, 589–593.

Rush, R. A., Crouch, M. F., Morris, C. P., and Gannon, B. J. (1982) Neural regulation of muscarinic receptors in chick expansor secondarium muscle. *Nature* **296**, 569–570.

Sabol, S. L. and Nirenberg, M. (1979) Regulation of adenylate cyclase activity of neuroblastoma × glioma hybrid cells by α-adrenergic receptors. II. Long-lived increase of adenylate cyclase activity mediated by α receptors. *J. Biol. Chem.* **254**, 1921–1926.

Sachs, D. I., Kloog, Y., Korczyn, A. D., Heron, D. S., and Sokolovsky, M. (1979) Denervation, supersensitivity, and muscarinic receptors in the cat iris. *Biochem. Pharmacol.* **28**, 1513–1518.

Savage, D. D., Dasheiff, R. M., and McNamara, J. O. (1983) Kindled seizure-induced reduction of muscarinic cholinergic receptors in rat hippocampal formation: evidence for localization to dentate granule cells. *J. Comp. Neurol.* **221**, 106–112.

Savage, D. D., Rigsbee, L. C., and McNamara, J. O. (1985) Knife cuts of entorhinal cortex: Effects on development of amygdaloid kindling and seizure-induced decreased of muscarinic cholinergic receptors. *J. Neurosci.* **5**, 408–413.

Scherer, N. M., Shapiro, R. A., and Nathanson, N. M. (1988) Modulation of M_1 and M_2 mAChRs in wild type and Kin8 adrenal carcinoma Y1 cells. *Soc. for Neurosci. Abstr.* **14**, 269.

Sharma, V. K. and Banerjee, S. P. (1977) Muscarinic cholinergic receptors in rat heart. Effects of thyroidectomy. *J. Biol. Chem.* **252**, 7444–7446.

Sharma, V. K. and Banerjee, S. P. (1978) Presynaptic muscarinic cholinergic receptors. *Nature* **272**, 276–278.

Sharma, S. K., Klee, W. A., and Nirenberg, M. (1975) Dual regulation of adenylate cyclase accounts for narcotic dependence and tolerance. *Proc. Natl. Acad. Sci. USA* **72**, 3092–3096.

Shifrin, G. S. and Klein, W. L. (1980) Regulation of muscarinic acetylcholine receptor concentration in cloned neuroblastoma cells. *J. Neurochem.* **34**, 993–999.

Siegel, R. E. and Fischbach, G. D. (1984) Muscarinic receptors and responses in intact embryonic chick atrial and ventricular heart cells. *Dev. Biol.* **101**, 346–356.

Siman, R. G. and Klein, W. L. (1979) Cholinergic activity regulates muscarinic receptors in central nervous system cultures. *Proc. Natl. Acad. Sci. USA* **76**, 4141–4145.

Siman, R. G. and Klein, W. L. (1981) Specificity of muscarinic acetylcholine receptor regulation by receptor activity. *J. Neurochem.* **37**, 1099–1108.

Siman, R. G. and Klein, W. L. (1983) Differential regulation of muscarinic and nicotinic receptors by cholinergic stimulation in cultured avian retina cells. *Brain Res.* **262**, 99–108.

Sokolovsky, M., Egozi, Y., and Avissar, S. (1981) Molecular regulation of receptors: Interaction of β-estradiol and progesterone with the muscarinic system. *Proc. Natl. Acad. Sci. USA* **78**, 5554–5558.

Sorota, S., Adam, L. P., and Pappano, A. J. (1986) Comparison of muscarinic receptor properties in hatched chick atrium and ventricle. *J. Pharmacol. Exp. Ther.* **236**, 602–609.

Sorota, S., Tsuji, Y., Tajima, T., and Pappano, A. J. (1985) Pertussis toxin treatment blocks hyperpolarization by muscarinic agonists in chick atrium. *Circul. Res.* **57**, 748–758.

Sugiyama, H., Daniels, M. P., and Nirenberg, M. (1977) Muscarinic acetylcholine receptors of the developing retina. *Proc. Natl. Acad. Sci. USA* **74**, 5524–5528.

Sullivan, J. K., Sorota, S., Zotter, C., and Pappano, A. J. (1985) Is muscarinic receptor number in avian ventricle regulated by vagal innervation, in *Cardiac Morphogenesis* (Ferrans, V. J., Rosenquist, G. L., and Weinstein, C., eds.) Elsevier, New York, pp. 202–207.

Syrota, A., Comar, D., Paillotin, G., Dary, J. M., Aumont, M. C., Stulzaft, O., and Maziere, B. (1985) Muscarinic cholinergic receptor in the human heart evidenced under physiological conditions by position emission tomography. *Proc. Natl. Acad. Sci. USA* **82**, 584–588.

Takeyasu, K., Uchida, S., Lai, R. T., Higuchi, H., Noguchi, Y., and Yoshida, H. (1981) Regulation of muscarinic acetylcholine receptors and contractility of guinea pig vas deferens. *Life Sci.* **28**, 527–540.

Taniguchi, T., Kurahashi, K., and Fujiwara, M. (1983) Alterations in muscarinic cholinergic receptors after preganglionic denervation of the superior cervical ganglion in cats. *J. Pharmacol. Exp. Ther.* **224**, 674–678.

Taylor, J. E., El-Fakahany, E., and Richelson, E. (1979) Long-term regulation of muscarinic acetylcholine receptors on cultured nerve cells. *Life Sci.* **25**, 2181–2187.

Taylor, J. E., Taksh, T. L., and Richelson, E. (1982) Agonist regulation of muscarinic acetylcholine receptors in rat spinal cord. *J. Neurochem.* **39**, 521–524.

Thomas, J. M. and Hoffman, B. B. (1986) Muscarinic cholinergic receptor-induced enhancement of PGE_1-stimulated cAMP accumulation in neuroblastoma × glioma cells: Prevention by pertussis toxin. *J. Cyclic Nucl. Protein Phosphor. Res.* **11**, 317–325.

Vincentini, L. M., Ambrosini, A., Di Virgilio, F., Meldolesi, J., and Pozzan, T. (1986) Activation of muscarinic receptors in PC12 cells. Correlation between cytosolic Ca^{2+} rise and phosphoinositide hydrolysis. *Biochem. J.* **234**, 555–562.

Vincentini, L. M., Di Virgilio, F., Ambrosini, A., Pozzan, T., and Meldolesi, J. (1985) Tumor promoter phorbol 12-myristate, 13-acetate inhibits phosphoinositide hydrolysis and cytosolic Ca^{2+} rise induced by the activation of muscarinic receptors in PC12 cells. *Biochem. Biophys. Res. Commun.* **127**, 310–317.

Waelbroeck, M., Robberecht, P., Christophe, J., Chatelain, P., Huu, A. N., and Roba, J. (1983) Effects of a chemical sympathectomy on cardiac muscarinic receptors in normotensive (WKY) and spontaneously hypertensive (SHR) rats. *Biochem. Pharmacol.* **32**, 1805–1807.

Westlind, A., Grynfarh, M., Hedlund, B., Bartfai, T., and Fuxe, K. (1981) Muscarinic supersensitivity induced by septal lesions or chronic atropine treatment. *Brain Res.* **225**, 131–141.

Wise, B. C., Shoji, M., and Kuo, J. F. (1980) Decrease or increase in cardiac muscarinic cholinergic receptor number in rats treated with methacholine or atropine. *Biochem. Biophys. Res. Commun.* **92**, 1136–1142.

Yamada, S., Yamamura, H. I., and Roeske, W. R. (1980) Alterations in cardiac autonomic receptors following 6-hydroxydopamine treatment in rats. *Mol. Pharmacol.* **18**, 185–192.

Yavin, E. and Harel, S. (1979) Muscarinic binding sites in the developing rabbit brain. Regional distribution and ontogenesis in the prenatal and early neonatal cerebellum. *FEBS Lett.* **97**, 151–154.

Young, J. M. (1974) Desensitization and agonist binding to cholinergic receptors in intestinal smooth muscle. *FEBS. Lett.* **46**, 354–356.

Zarbin, M. A., Wamsley, J. K., and Kuhar, M. J. (1982) Axonal transport of muscarinic cholinergic receptors in rat vagus nerve: High and low affinity agonist receptors move in opposite directions and differ in nucleotide sensitivity. *J. Neurosci.* **2**, 934–941.

SECTION 5

FUTURE DIRECTIONS

12

Future Directions

N. J. M. Birdsall and E. C. Hulme

The purification of muscarinic receptor species from different tissues (Peterson et al., 1984; Haga and Haga, 1985; Berrie et al., 1985; Wheatley et al., 1987; Baumgold et al., 1987) and the cloning of different muscarinic receptor proteins from pigs (Kubo et al., 1986a,b; Peralta et al., 1987b), humans (Bonner et al., 1987; Peralta et al., 1987a, 1988), and rats (Bonner et al., 1987) have opened the way for the investigation and clarification of a range of problems that could not previously be tackled. Within a short time, we will know the number of different muscarinic receptor proteins that are present in the genome of several species. To date, four muscarinic receptor proteins have been cloned and expressed (Kubo et al., 1986a,b; Fukuda et al., 1987; Bonner et al., 1987; Peralta et al., 1987a, 1988). From genomic blots of rat and human DNA, there may be 2–5 additional members of the muscarinic receptor gene family. These will need to be cloned, sequenced, and possibly expressed in order to determine whether some are pseudogenes. Between species, there is a very high homology between any given molecular subtype. For example, one rat amino acid sequence has a 98% identity to that cloned by Kubo et al. (1986a) from porcine brain. Between molecular subtypes, there is only 35–45% identity. Therefore, it is to be expected that a given molecular subtype will exhibit very similar binding and functional properties, irrespective of the mammalian species from which it is obtained. The homology between say, insect, avian, Torpedo, and mammalian muscarinic receptor sequences remains to be determined, although the binding properties of the receptors seem to be rather similar.

There is already duplication regarding the nomenclature of the molecular species. Kubo et al. (1986a,b) name the muscarinic receptors cloned from porcine brain and heart as mAChR I and mAChR II, respectively. Bonner et al. (1987) refer to the equivalent species from rat, pig, and human as m1 and m2; the two additional species being named as m3 and m4. Peralta et al. (1987b, 1988) refer to the porcine and human sequences as PMn and HMn (n = 1 – 4), respectively. Their nomenclature for HM3 and HM4 corresponds to that denoted by Bonner et al. as m4 and m3, respectively. In other words, the receptor subtype that is ~120 amino acids longer than the other subtypes is m3 in one terminology and HM4 in another. In view of the cross-species similarity between equivalent molecular subtypes, we will use the m1–m4 nomenclature in this chapter.

The knowledge of DNA sequences of the subtypes means that one can design appropriate probes to be used for *in situ* hybridization studies for the localization of the receptor mRNA within cell bodies in the CNS or PNS or within cells in culture. Semiquantitative estimates of mRNA levels and change in these levels consequent upon maturation, synapse formation, and exposure to other factors that may regulate muscarinic gene transcription are now possible. In other words, the whole range of modern molecular biological techniques are now available for the study of muscarinic receptors.

The amino acid sequences of the receptor subtypes allow the generation of specific antisequence antibodies using oligopeptides coupled to an appropriate carrier. These antipeptide antibodies will complement antireceptor antibodies that have been raised (with considerable difficulty) against the whole receptor molecule (Andre et al., 1984; Venter et al., 1984; Luetje et al., 1987; Subers et al., 1988). The antibodies may be used to immunoprecipitate (and hence purify) receptor subtypes, recognize the subtypes (or proteolytic fragments) on Western blots, and to localize receptor molecules at the light microscope and electron microscope level in different tissues. The presynaptic/postsynaptic location of the receptor subtypes will be investigated. It may also be possible, using antisequence antibodies and immunogold staining to determine whether specific sequences of the receptors are extracellularly or cytoplasmically located.

This leads naturally to the study of muscarinic receptor structure. It appears that muscarinic receptors are members of a superfamily of receptors that interact with and activate one or more members of the superfamily of G-proteins. With the exception of the opsin family, the receptors cloned to date do not have introns in their coding regions. The reason for this unusual motif of intronless genes is unknown and of obvious interest, but makes the isolation and sequencing of clones from genomic libraries relatively easy. There are considerable homologies between the four muscarinic receptor sequences and those of the avian β- (Yarden et al., 1986), and mammalian β_1 - (Frielle et al., 1987), β_2- (Dixon et al., 1986; Kobilka et al., 1987a, Chung et al., 1987), α_2-adrenergic receptors (Kobilka et al., 1987c), the substance K receptor (Masu et al., 1987), an unassigned receptor protein, G21 (Kobilka et al., 1987b), and members of the opsin family (Nathans and Hogness, 1983; Nathans et al., 1986). The similarities in the hydropathicity profiles suggest the existence of seven transmembrane α-helical segments in all these receptor proteins, as has been shown for bacteriorhodopsin (Henderson and Unwin, 1975), which however is a proton translocator and not a G-protein-coupled receptor. All the members of the receptor family lack a signal peptide, but appear to be heavily glycosylated at site(s) close to the NH_2-terminus. It is possible to superimpose the models of these receptors directly into the model of rhodopsin developed by Findlay and his colleagues (Findlay and Pappin, 1986). This is illustrated for the m1 sequence reported by Kubo et al. 1986a (Fig. 1).

We have discussed the marked sequence differences and similarities between the muscarinic receptor subtypes (Hulme and Birdsall, 1986; Hulme et al., 1987; Wheatley et al., 1988). At least 44 residues are identical in the sequences or are closely homologous. These residues are located in or near the putative transmembrane segments or within the short loops, and include many of the proline and glycine residues as well as tryptophan and charged residues at the putative cytoplasmic boundaries of the transmembrane domains. Two cysteine residues are conserved within the 2-3 and 4-5 loops, and these may form a disulphide bond. There is an interesting and undoubtedly important conservation of acidic residues in transmembrane segments 2 and 3 (shown in Fig. 1).

The receptor sequences differ most at the NH_2- and C-terminals and the 4-5 and 6-7 loops. However, the most notable differences are found in the 5-6 loop, where there is an extraordinary variability in the length of this loop and essentially no homology except at the regions thought to be close to the trans-

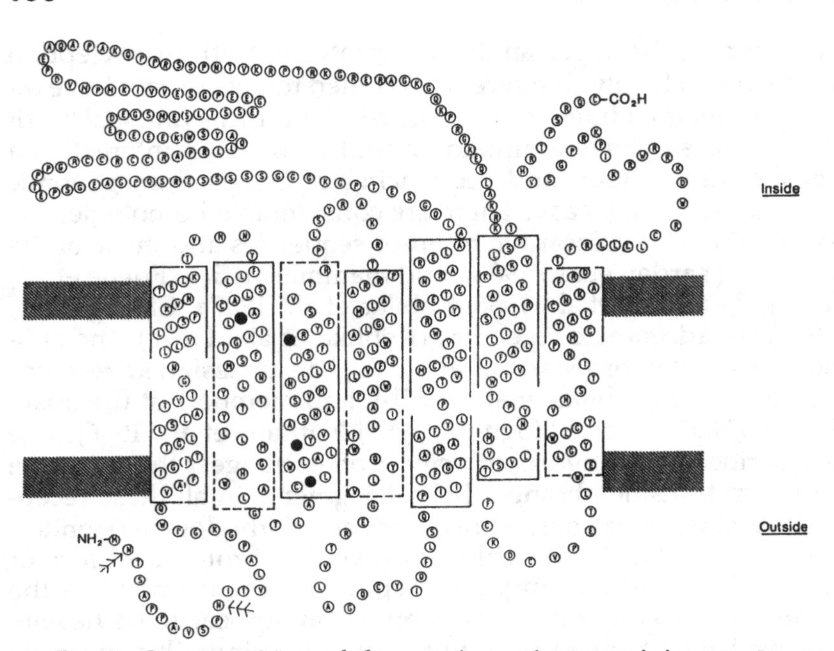

Fig. 1. Superposition of the porcine m1 muscarinic receptor sequence into the model of opsin proposed by Findlay and Pappin (1986). The conserved aspartate residues in the m1–m4 sequences are shown as filled circles (from Wheatley, et al. 1988).

membrane segments. This loop may have functions quite different from those of G-protein recognition or activation: we have speculated that it may contain trafficking signals, regions that could interact with cytoskeletal elements or with other unknown proteins that may be involved in the coupling process, or with desensitization (Hulme and Birdsall, 1986).

The structure of these proteins is being investigated by molecular genetic, peptide mapping and protein modification studies. Essentially, all of the molecular genetic studies are being carried out on β-adrenergic receptors at present. Much of the longer extra helical segments of the 5-6 loop and the NH_2- and C-terminals may be deleted without affecting ligand binding (Dixon et al., 1987b). However, several deletions and single amino acid substitutions within the transmembrane segments alter or abolish ligand binding (Dixon et al., 1987a,b; Strader et al., 1987b) or drastically affect the processing and membrane localization of the receptor. The C-terminus of the β-receptor, by analogy with rhodopsin, has been postulated to be the region phosphorylated by a receptor-specific kinase (Benovic et al., 1986). The phosphorylation was thought to result in desensitization and/or internalization of the receptor. Analogous phosphorylation events are thought to occur with muscarinic receptors (Liles et al., 1986; Kwatra and Hosey, 1986). However, deletion of the C-terminus of the hamster β-adrenergic receptor did not block internalization (Strader et al., 1987a), suggesting that phosphorylation of the 5-6 loop rather than the C-terminus might be important. In most of these mutagenesis studies, only the binding of one antagonist and one agonist has been determined, and functional studies have not been carried out. Nevertheless, one is beginning to have some idea of the residues and regions of the β-adrenergic receptor, which may be important for binding and function. Current dogma suggests that the general conclusions from these molecular biological studies on β_2 receptors will be applicable to other G-protein-coupled receptors including the muscarinic receptor. However, there is tremendous scope for molecular biological manipulations of the muscarinic receptor subtypes.

A complementary approach is to use peptide mapping and protein chemistry techniques to localize residues on the receptor that are important for ligand binding and function, and that may be labeled. We have outlined a peptide mapping approach to obtain this information (Hulme et al., 1987; Wheatley et al., 1988; Curtis et al., 1989). The procedure involves the use of proteases that are active in SDS to generate from the receptor a series of peptides containing different numbers of transmembrane segments. The glycosylated peptides are separated on a wheat germ agglutinin column, and they form a series of N-terminal peptides of increasing molecular weight, which can be separated by SDS-PAGE to generate a "ladder" of peptides. By this means, a labeled residue may be located to a specific region of the receptor and, if necessary, microsequencing may be used (after deglycosylation, chemical cleavage and/or further protease treatment) to pinpoint the precise residue that has been modified. This protocol is a way of overcoming handling problems of very hydrophobic peptides. It is perhaps not surprising that most, if not all, of the peptides that have been sequenced to date by proteolytic enzyme or CNBr treatment of G-protein-coupled receptors have come from the 5-6 loop. This is because only water-soluble peptides can be separated by reversed-phase HPLC.

Our results suggest that the predominant site of alkylation of [^{3}H]propylbenzilylcholine mustard on muscarinic receptors is a conserved, buried aspartate residue probably in the third transmembrane segment (Curtis et al., 1989) (Fig. 1). This result together with the molecular biology studies indicates that the binding site for muscarinic ligands, including acetylcholine, is buried within the transmembrane segments. Hence, the residues that are responsible for the selective binding of drugs such as pirenzepine presumably lie in this region. With increasing information regarding the residues that are important for binding and selectivity, it may be possible to model the binding site and deduce the mechanism by which the binding of the agonist leads to the activation of the G-protein. It may be that this will have to await the breakthrough of obtaining high-resolution, three-dimensional structural information (e.g., X-ray diffraction data) from such receptors.

A more immediate problem is the assignment of a pharmacological and functional profile to each molecular species of muscarinic receptor. At the time of writing this chapter, we know of four molecular species, maybe four pharmacological subtypes, and the fact that muscarinic receptors, via their interaction with several G-proteins [at least 4: G_i, G_o, G_n (Haga et al., 1988), and a nonpertussis toxin sensitive G-protein (Hughes et al., 1984)], can modulate adenylate cyclase activity, open K^+ channels, and stimulate the production of inositol trisphosphate. These complexities may be resolved by examining the binding and functional properties of muscarinic receptors in clonal cell lines derived from neural tissue. Alternatively, the receptors may be expressed by injection of cRNA into oocytes or by transfection of DNA into fibroblast or other cell lines that may or may not express their "own" muscarinic receptors. The final approach is to reconstitute purified muscarinic receptor proteins with different G-proteins.

In theory, these different approaches should give similar or identical results. In practice, the picture is most confused both with regard to the pharmacological profile of the different molecular species and the selectivity of their interaction with different G-proteins. The basic problems are outlined below.

On one hand, the purified M_1 and M_2 receptors obtained from porcine brain and heart exhibit identical binding properties, as far as QNB, pirenzepine, and acetylcholine are concerned, when the binding properties are measured in the soluble state or when reconstituted into membranes (Haga et al., 1988). On the other hand, when the cRNA from the cloned porcine m1 and m2 sequences are expressed in oocytes, the binding properties of two muscarinic receptors are different, and are precisely those found in the porcine cerebrum and heart (Fukuda et al., 1987). Similar discrepancies arise when one examines the binding properties of the m1–m4 sequences expressed in COS-7 cells (Bonner et al., 1987), CHO cells (Peralta et al., 1987a, 1988), or A9-L cells (Brann et al., 1987).

Bonner and his coworkers found that three of the subtypes, m1, m3, and m4, had a pirenzepine dissociation constant 1-2 × $10^{-8}M$. These fall into the range of affinity found for the M_1 pharmacological subtype. Interestingly, the [^{3}H]QNB affinity differed by a factor of 6 between the different subtypes. Does this mean that there are three subtypes of M_1 receptors? The data of Peralta et al. (1987a, 1988) also showed that the pirenzepine affinity for the m1, m3, and m4 subtypes was very similar, and the same interpretation was made—three M_1 subtypes. However, the reported dissociation constants for pirenzepine were $\sim 10^{-6}M$, about a factor of 100 higher than those found either by Bonner et al. (1987) or in functional studies in mammalian tissue. The m2 receptor exhibited a "cardiac" M_2 receptor profile, in the sense that AF-DX 116 had a higher affinity than pirenzepine (Peralta et al., 1988). However, the AF-DX 116 affinity was somewhat higher than that found in whole tissue and the pirenzepine affinity 3–6-fold lower.

Relevent information may also be obtained from the properties of muscarinic receptors in clonal cell lines. The neuroblastoma-glioma cell line NG 108-15 has been shown by Northern blots to contain predominantly, if not exclusively, the mRNA for the m4 (HM3) subtype (Peralta et al., 1988). However, this muscarinic receptor exhibits an "M_2" profile on the basis of the affinity of pirenzepine in functional assay (Evans et al., 1984). In a recent study, Liang et al., (1987) have shown that the NG 108-15 muscarinic receptor has a relatively low apparent size on SDS-PAGE, 66 kdaltons and 45 kdaltons before and after endoglycosidase F treatment, respectively. These data are compatible with the mol wt predicted for a m4 receptor. In contrast, the 1321N1 cell line contains an apparantly larger species (92 kdaltons and 77 kdaltons before and after endo F treatment). It seems that the 1321N1 muscarinic receptor represents a higher molecular weight muscarinic receptor, such as m3. This may be the "glandular" receptor, in view of the fact that a higher molecular weight receptor ($M_r > 80$ kdaltons) than "usual" (Birdsall et al., 1979; Venter, 1983; Dadi and Morris, 1984) has been found in exocrine glands (Hootman et al., 1985), and both the "glandular" receptor and the 1321N1 receptor appear to couple very efficiently to triphosphoinositide phospholipase C. Further, m3 (but not m1, m2, or m4) mRNA is expressed in the pancreas (Peralta et al., 1987a, 1988). However, none of the cloned and expressed receptors to date appear to have the pharmacology expected of the glandular receptor subtype, at least in the cell lines that have been transfected.

Similar complications arise when muscarinic receptor-G-protein selectivity is investigated. No selectivity has been detected in reconstitution studies (Haga et al., 1986, 1988); "absolute" selectivity is claimed for the muscarinic receptors in NG 108-15 cells and 1321N1 cells to inhibit adenylate cyclase (via G_i) and increase phosphoinositide hydrolysis (via the putative G_p), respectively, despite G_i and "G_p" being present in both cell lines (Hepler et al., 1987). The m2 receptor, when expressed in CHO cells, will couple to both adenylate cyclase and phosphoinositide turnover, but with higher efficacy to the former response (Ashkenazi et al., 1987). The two responses are claimed to result from the coupling of one receptor type to two different pertussis toxin-sensitive G-proteins. Finally, less directly coupled responses measured in electrophysiological studies of m1 and m2 receptors expressed in Xenopus oocytes nevertheless suggest that these receptors selectively activate different effector mechanisms and open different ion channels (Fukuda et al., 1987). There seem, therefore, to be several as yet undetermined factors that are responsible for determining the drug and G-protein binding selectivity of the different molecular subtypes.

It may seem a pity to end on such a note of confusion. However, this is only the current state of knowledge (ignorance). We feel very optimistic about the future of this area of research. The most fundamental questions are now being addressed: What is the structure of muscarinic receptors? How do drugs bind to the receptors? What is the basis of the interaction of selective drugs with a given receptor subtype? Which G-proteins interact with a given receptor subtype, and what is the molecular basis for this selectivity? What is the nature of the conformational change in the receptor induced by the binding of an agonist, and how does this lead to the activation of the G-protein? Where precisely are the receptors located? How and at what levels are the expression of the receptors regulated? The plurality of muscarinic receptor subtypes and our rapidly increasing knowledge means that in addition, there is a tremendous potential for the development of novel selective muscarinic drugs.

Update (October 1988)

The human and rat genes for a fifth muscarinic receptor, m5, have been cloned and expressed (Bonner et al., 1988). This receptor subtype is most closely related to the m3 receptor. The porcine m3 sequence (Akiba et al., 1988) and mouse m1 sequence (Lai et al., 1988) have also been described, as well as confirmatory reports of the sequences of rat m2 (Gocayne et al., 1987), m4 (Braun et al., 1987), m1, m3 (Stein et al., 1988), and human m1 (Allard et al., 1987) receptors.

Selectivity in the responses coupled to the different receptor subtypes has been demonstrated by biochemical and electrophysiological experiments. Most experiments point to m1, m3, and m5 receptors being efficiently coupled to phosphoinositide hydrolysis (Peralta et al., 1988; Bonner et al., 1988; Lai et al., 1988; Fukda et al., 1988) with no detectable ability to inhibit adenylate cyclase. In fact, in transfected human embryonic kidney cells, the m1 and m3 receptors will *stimulate* adenylate cyclase activity in the presence of a phosphodiesterase inhibitor, but this is probably a consequence of the production of inositol trisphosphate and/or diacylglycerol. The m2 and m4 receptors readily inhibit adenylate cyclase and couple only weakly to phosphoinositide hydrolysis (Peralta et al., 1988; Fukuda et al., 1988). In NG108-15 cells, transfected with m1–m4 cDNA or genomic DNA, m1 and m3 receptor activate a Ca^{2+}-dependent K^+ current and inhibit the M-current (Fukuda et al., 1988). Both these responses are thought to be mediated via phosphoinositide hydrolysis. This coupling selectivity pattern follows sequence homologies among m1, m3, and m5 in the long third intracellular loop, and may indicate that structural features in this region of the receptor are responsible for the G-protein selectivity. In contrast to this picture, Stein et al. (1988) have reported that the rat m1 receptor expressed in RAT-1 cells couples efficiently to adenylate cyclase and weakly to phosphoinositide hydrolysis! Only the former response was blocked by pertussis toxin, suggesting that the two responses were mediated by different G-proteins. The only conclusion to be drawn at present from these conflicting results is that there may be unknown cellular factors that are responsible for the determination of receptor-G-protein selectivity.

Finally, more recent studies on the pharmacology of the m1–m5 receptors and the tissue localization of the mRNA for these species, point to m3 as being the "glandular" receptor (Akiba et al., 1988; Bonner et al., 1988; Buckley et al., personal communication).

References

Akiba, I., Kubo, T., Maeda, A., Bujo, H., Nakai, J., Mishina, M., and Numa, S. (1988) Primary structure of porcine muscarinic acetylcholine receptor III and antagonist binding studies. *FEBS Lett.* **235**, 257–261.

Allard, W. J., Sigal, I. S., and Dixon, R. A. (1987) Sequence of the gene encoding the human M1 muscarinic acetylcholine receptor. *Nucleic Acids Res.* **15**, 10604.

Andre, C., Guillet, J. G., De Backer, J.-P., Vanderheyden, P., Hobecke, J., and Strosberg, A. D. (1984) Monoclonal antibodies against the native or denatured forms of muscarinic acetylcholine receptors. *EMBO J.* **3**, 17–21.

Ashkenazi, A., Winslow, J. W., Peralta, E. G., Peterson, G. L., Schimerlik, M. I., Capon, D. J., and Ramachandran, J. (1987) An M2 muscarinic receptor subtype coupled to both adenylyl cyclase and phosphoinositide turnover. *Science* **238**, 672–675.

Baumgold, J., Merril, C., and Gershon, E. S. (1987) Loss of pirenzepine regional selectivity following solubilization and partial purification of the putative M1 and M2 muscarinic receptor subtypes. *Molec. Brain Res.* **2**, 7–14.

Benovic, J. L., Strasser, R. H., Caron, M. G., and Lefkowitz, R. J. (1986) Beta-adrenergic receptor kinase: identification of a novel protein kinase that phosphorylates the agonist-occupied form of the receptor. *Proc. Natl. Acad. Sci. USA* **83**, 2797–2801.

Berrie, C. P., Birdsall, N. J. M., Dadi, H. K., Hulme, E. C., Morris, R. J., Stockton, J. M., and Wheatley, M. (1985) Purification of the muscarinic acetylcholine receptor from rat forebrain. *Trans. Biochem. Soc.* **13**, 1101–1103.

Birdsall, N. J. M., Burgen, A. S. V., and Hulme, E. C. (1979) A study of the muscarinic receptor by gel electrophoresis. *Brit. J. Pharmacol.* **66**, 337–342.

Bonner, T. I., Young, A. C., Brann, M. R., and Buckley, N. J. (1988) Cloning and expression of the human and rat m5 muscarinic acetylcholine receptor genes. *Neuron* **1**, 403–410.

Bonner, T. I., Buckley, N. J., Young, A. C., and Brann, M. R. (1987) Identification of a family of muscarinic acetylcholine receptor genes. *Science* **237**, 527–532.

Brann, M. R., Buckley, N. J., Jones, S. V. P., and Bonner, T. I., (1987) Expression of a cloned muscarinic receptor in A9 cells. *Mol. Pharmacol.* **32**, 450–455.

Braun, T., Schofield, P. R., Shivers, B. D., Pritchett, D. B., and Seeburg, P. H. (1987) A novel subtype of muscarinic receptor identified by homology screening. *Biochem. Biophys. Res. Commun.* **49**, 125–132.

Chung, F.-Z., Lentes, K.-U., Gocayne, J., Fitzgerald, M., Robinson, D., Kerlavage, A. R., Fraser, C. M., and Venter, J. C. (1987) Cloning and sequence analysis of the human brain β-adrenergic receptor. *FEBS Letters* **211**, 200–206.

Curtis, C. A. M., Wheatley, M., Bansal, S., Birdsall, N. J. M., Pedder, E. K., Poyner, D., and Hulme, E. C. (1989) Propylbenzilylcholine mustard labels an acidic residue in transmembrane helix three of the muscarinic receptor. *J. Biol. Chem.*, **264**, 489–495.

Dadi, H. K. and Morris, R. J. (1984) Muscarinic cholinergic receptor of rat brain. *Eur. J. Biochem.* **144**, 617–628.

Dixon, R. A. F., Kobilka, B. K., Strader, D. J., Benovic, J. L., Dohlman, H. G., Frielle, T., Bolanowski, M. A., Bennett, C. D., Rands, E., Diehl, R. E., Mumford, R. A., Slater, E. E., Sigal, J. S., Caron, MG., Lefkowitz, R. J., and Strader, C. D. (1986) Cloning of the gene and cDNA for mammalian β-adrenergic receptor and homology with rhodopsin. *Nature* **321**, 75–79.

Dixon, R. A. F., Sigal, I. S., Candelore, M. R., Register, B. R., Scattergood, W., Rands, E., and Strader, C. D. (1987a) Structural features required for ligand binding to the β-adrenergic receptor. *EMBO J.* **6**, 3269–3275.

Dixon, R. A. F., Sigal, I. S., Rands, E., Register, R. B., Candelore, M. R., Blake, A. D., and Strader, C. D. (1987b) Ligand binding to the β-adrenergic receptor involves its rhodopsin-like core. *Nature* **326**, 73–77.

Evans, T., Smith, M. M., Tanner, L. I., and Harden, T. K. (1984) Muscarinic cholinergic receptors of two cell lines that regulate cyclic AMP metabolism by different molecular mechanisms. *Mol. Pharmacol.* **26**, 395–404.

Findlay, J. B. C. and Pappin, D. J. C. (1986) The opsin family of proteins. *Biochem. J.* **238**, 625–642.

Frielle, T., Collins, S., Daniel, K. W., Caron, M. G., Lefkowitz, R. J., and Kobilka, B. K. (1987) Cloning of the cDNA for the human β_1-adrenergic receptor. *Proc. Natl. Acad. Sci., USA* **84**, 7920–7924.

Fukuda, K., Kubo, T., Akiba, I., Maeda, A., Mishina, M., and Numa, S. (1987) Molecular distinction between muscarinic acetylcholine receptor subtypes. *Nature* **327**, 623–625.

Fukada, K., Higashida, H., Kubo, T., Maeda, A., Akiba, I., Bujo, H., Mishina, M., and Numa, S. (1988) Selective coupling with K^+ currents of muscarinic receptor subtypes in NG108-15 cells. *Nature* **335**, 355–358.

Gocayne, J., Robinson, D. A., FitzGerald, M. G., Chung, F. Z., Kerlavage, A. R., Lentes, K. V., Lai, J., Wang, C. D., Fraser, C. M., and Vente, J. C. (1987) Primary structure of rat cardiac beta-adrenergic and muscarinic cholinergic receptors obtained by automated DNA sequence analysis: further evidence of a multigene family. *Proc. Natl. Acad. Sci. USA* **84**, 8296–8300.

Haga, K. and Haga, T. (1985) Purification of the muscarinic acetylcholine receptor from porcine brain. *J. Biol. Chem.* **260**, 7927–7935.

Haga, T., Haga, K., Berstein, G., Nishiyama, T., Uchiyama, H., and Ichiyama, A. (1988) Molecular properties of muscarinic receptors. *Trends. Pharmacol. Sci.* **Suppl. III**,12–18.

Henderson, R. and Unwin, P. N. T. (1975) Three-dimensional model of purple membrane obtained by electron microscopy. *Nature* **257**, 28–32.

Hepler, J. R., Hughes, A. R., and Harden, T. K. (1987) Evidence that muscarinic cholinergic receptors selectively interact with either the cyclic AMP or the inositol phosphate second-messenger response systems. *Biochem. J.* **247**, 793–796.

Hootman, S. R., Picado-Leonard, T. M., and Burnham, D. B. (1985) Muscarinic acetylcholine receptor structure in acinar cells of mammalian exocrine glands. *J. Biol. Chem.* **260**, 4186–4194.

Hughes, A. R., Martin, M. W., and Harden, T. K. (1984) Pertussis toxin differentiates between two mechanisms of attenuation of cyclic AMP accumulation by muscarinic cholinergic receptors. *Proc. Natl. Acad. Sci. USA* **81**, 5680–5684.

Hulme, E. C. and Birdsall, N. J. M. (1986) Distinctions in acetylcholine receptor activity. *Nature* **323**, 396–397.

Hulme, E. C., Wheatley, M., Curtis, C., and Birdsall, N. J. M. (1987) The muscarinic acetylcholine receptors: structure, function and location of the ligand binding site, in: *International Symposium on Muscarinic Cholinergic Mechanisms* (Cohen, S. and Sokolovsky, M., eds.), Freund, London, pp 192–211.

Kobilka, B. K., Dixon, R. A. F., Frielle, T., Dohlman, H. G., Bolarowski, M. A., Sigal, I. S., Yang-Feng, T. L., Francke, U., Caron, M. G., and Lefkowitz, R. J. (1987a) cDNA for the human β_2-adrenergic receptor: a protein with multiple membrane-spanning domains and encoded by a gene whose chromosomal location is shared with that of the receptor for platelet-derived growth factor. *Proc. Natl. Acad. Sci., USA* **84**, 46–50.

Kobilka, B. K., Frielle, T., Collins, S., Yang-Feng, T., Kobilka, T. S., Franck, U., Lefkowitz, R. J., and Caron, M. G. (1987b) An intronless gene encoding a potential member of the family of receptors coupled to guanine nucleotide regulatory proteins. *Nature* **329**, 75–79.

Kobilka, B. K., Matsui, H., Kobilka, T. S., Yang-Feng, T. L., Francke, U., Caron, M. G., Lefkowitz, R. J., and Regan, J. W. (1987c) Cloning, sequencing and expression of the gene coding for the human platelet α_2-adrenergic receptor. *Science* **238**, 650–656.

Kubo, T., Fukuda, K., Mikami, A., Maeda, A., Takahashi, H., Mishina, M., Haga, T., Haga, K., Ichiyama, A., Kangawa, K., Kojima, M., Matsuo, H., and Numa, S. (1986a) Cloning, sequencing and expression of the complementary DNA encoding the muscarinic acetylcholine receptors. *Nature* **323**, 411–416.

Kubo, T., Maeda, A., Sugimoto, K., Akiba, I., Mikami, A., Takahashi, H., Haga, T., Haga, K., Ichiyama, A., Kangawa, K., Matsuo, H., Hirose, T., and Numa, S. (1986b) Primary structure of porcine cardiac muscarinic acetylcholine receptor deduced from the cDNA sequence. *FEBS Letters* **209**, 367–372.

Kwatra, M. M. and Hosey, M. M. (1986) Phosphorylation of the cardiac muscarinic receptor in intact chick heart and its regulation by a muscarinic agonist. *J. Biol. Chem.* **261**, 12429–12432.

Lai, J., Mei, L., Roeske, W. R., Chung, F. Z., Yamamura, H. I., and Venter, J. C. (1988) The cloned murine M1 muscarinic receptor is associated with the hydrolysis of phosphatidylinositols in transfected murine B82 cells. *Life Sci.* **42**, 2489–2502.

Liang, M., Martin, M. W., and Harden, T. K. (1987) [^{3}H]Propylbenzilylcholine mustard-labelling of muscarinic cholinergic receptors that selectively couple to phospholipase C or adenylate cyclase in two cultured cell lines. *Mol. Pharmacol.* **32**, 433–449.

Liles, W. C., Hunter, D. D., Meier, K. E., and Nathanson, N. M. (1986) Activation of protein kinase C induces rapid internalization and subsequent degradation of muscarinic acetylcholine receptors in neuroblastoma cells. *J. Biol. Chem.* **261**, 5307–5313.

Luetje, C. W., Brumwell, C., Norman, M. G., Peterson, G. L., Schimerlik, M. I., and Nathanson, N. M. (1987) Isolation and characterization of monoclonal antibodies specific for the cardiac muscarinic acetylcholine receptors. *Biochemistry* **26**, 6892–6896.

Masu, Y., Nakayama, K., Tamaki, H., Harada, Y., Kuno, M., and Nakanishi, S. (1987) cDNA cloning of bovine substance-K receptor through oocyte expression system. *Nature* **329**, 836–838.

Nathans, J. and Hogness, D. G. (1983) Isolation, sequence analysis and intron-exon arrangment of the gene encoding bovine rhodopsin. *Cell* **34**, 807–814.

Nathans, J., Thomas, D., and Hogness, D. S. (1986) Molecular genetics of human color vision: the genes encoding blue, green and red pigments. *Science* **232**, 193–202.

Peralta, E. G., Ashkenazi, A., Winslow, J. W., Ramachandran, J., and Capon, D. J. (1988) Differential regulation of PI hydrolysis and adenylate cyclase by muscarinic receptor subtypes. *Nature* **343**, 434–437.

Peralta, E. G., Ashkenazi, A., Winslow, J. W., Smith, D. H., Ramachandran, J., and Capon, D. (1987a) Distinct primary structures, ligand-binding properties and tissue-specific expression of four human muscarinic acetylcholine receptors. *EMBO J.* **13**, 3923–3929.

Peralta, E. G. Winslow, J. W. Peterson, G. L., Smith, D. H., Ashkenazi, A., Ramachandran, J., Schimerlik, M. I., and Capon, D. J. (1987b) Primary structure and biochemical properties of an M_2 muscarinic receptor. *Science* **236**, 600–605.

Peralta, E. G., Winslow, J. W., Ashkenazi, A., Smith, D. H., Ramachandran, J., and Capon, D. J. (1988) Structural basis of muscarinic acetylcholine receptor subtype diversity. *Trends. Pharmacol. Sci.* **Suppl. III**, 6–11.

Peterson, G. L., Herron, G. S., Yamaki, M., Fullerton, D. S., and Schimerlik, M. I. (1984) Purification of the muscarinic acetylcholine receptor from porcine atria. *Proc. Natl. Acad. Sci., USA* **81**, 4993–4997.

Stein, R., Pinkas-Kramarski, R., and Sokolovsky, M. (1988) Cloned M1 muscarinic receptors mediate both adenylate cyclase inhibition and phosphoinositide turnover. *EMBO J.* **7**, 3031–3035.

Strader, C. D., Sigal, I. S., Blake, A. D., Cheung, A. H., Register, R. B., Rands, E., Zemcik, B. A., Candelore, M. R., and Dixon, R. A. F. (1987a) The carboxyl terminus of the hamster β-adrenergic receptor expressed in mouse L cells is not required for receptor sequestration. *Cell* **49**, 855–863.

Strader, C. D., Sigal, I. S., Register, R. B., Candelore, M. R., Rands, E., and Dixon, R. A. F. (1987b) Identification of residues required for ligand binding to the β-adrenergic receptor. *Proc. Natl. Acad. Sci., USA* **84**, 4384–4388.

Subers, E. M., Liles, W. C., Luetje, C. W., and Nathanson, N. M. (1988) Biochemical and immunological studies on the regulation of cardiac and neuronal muscarinic acetylcholine receptor number and function. *Trends. Pharmacol. Sci.* **Suppl. III**, 25–34.

Venter, J. C. (1983) Muscarinic cholinergic receptor structure. Receptor size, membrane orientation and absence of major phylogenetic structural diversity. *J. Biol. Chem.* **258**, 4842–4848.

Venter, J. C., Eddy, B., Hall, L. M., and Fraser, C. M. (1984) Monoclonal antibodies detect the conservation of muscarinic cholinergic receptor structure from *Drosophila* to human brain and detect possible structural homology with α_1-adrenergic receptors. *Proc. Natl. Acad. Sci., USA* **81**, 272–276.

Wheatley, M., Birdsall, N. J. M., Curtis, C., Eveleigh, P., Pedder, E. K., Poyner, D., Stockton, J. M., and Hulme, E. C. (1987) The structure and properties of the purified muscarinic acetylcholine receptor from rat forebrain. *Biochem. Soc. Trans.* **15**, 113–116.

Wheatley, M., Hulme, E. C., Birdsall, N. J. M., Curtis, C. A. M., Eveleigh, P., Pedder, E. K., and Poyner, D. (1988) Peptide mapping studies on muscarinic receptors: receptor structure and location of the ligand binding site. *Trends. Pharmacol. Sci.* **Suppl. III**, 19–24.

Yarden, Y., Rodriguez, H., Wong, S. K.-F., Brandt, D. R., May, D. C., Burnier, J., Harkins, R. N., Chen, E. Y., Ramachandran, J., Ullrich, A., and Ross, E. M. (1986) The avian β-adrenergic receptor: primary structure and membrane topology. *Proc. Natl. Acad. Sci., USA* **83**, 6795–6799.

Index

A

T

V